DYNAMICS OF STRUCTURES

**McGRAW-HILL
BOOK COMPANY**

New York
St. Louis
San Francisco
Auckland
Düsseldorf
Johannesburg
Kuala Lumpur
London
Mexico
Montreal
New Delhi
Panama
Paris
São Paulo
Singapore
Sydney
Tokyo
Toronto

Dynamics
of Structures

RAY W. CLOUGH

*Professor of Civil Engineering
University of California, Berkeley*

JOSEPH PENZIEN

*Professor of Structural Engineering
University of California, Berkeley*

This book was set in Times New Roman.
The editors were B. J. Clark and Michael Gardner;
the cover was designed by Pencils Portfolio, Inc.;
the production supervisor was Charles Hess.
The drawings were done by ANCO Technical Services.
Kingsport Press, Inc., was printer and binder.

Library of Congress Cataloging in Publication Data

Clough, Ray W   date
    Dynamics of structures.

    Includes index.
    1. Structural dynamics.   I.   Penzien, J., joint
author. II.   Title.
TA654.C6   624'.171   74-26963
ISBN 0-07-011392-0

**DYNAMICS
OF STRUCTURES**

890KPKP84321

CONTENTS

Preface   xv

List of Symbols   xix

1   Overview of Structural Dynamics   1

1-1   Fundamental Objective of Structural-Dynamics Analysis   1
1-2   Types of Prescribed Loadings   2
1-3   Essential Characteristics of a Dynamic Problem   4
1-4   Methods of Discretization   5
        Lumped-Mass Procedure   5
        Generalized Displacements   6
        The Finite-Element Concept   7
1-5   Formulation of the Equations of Motion   9
        Direct Equilibration Using d'Alembert's Principle   9
        Principle of Virtual Displacements   10
        Hamilton's Principle   11
        Summary   11
1-6   Organization of the Text   12

Part One   SINGLE-DEGREE-OF-FREEDOM SYSTEMS

2   Formulation of the Equation of Motion   **17**

2-1   Components of the Basic Dynamic System   17
2-2   Methods of Formulation   18
      Direct Equilibration   18
      Virtual-Work Analysis   19
      Application of Hamilton's Principle   19
2-3   Influence of Gravitational Forces   20
2-4   Influence of Support Excitation   21
2-5   Generalized SDOF Systems: Rigid-Body Assemblage   23
2-6   Generalized SDOF Systems: Distributed Flexibility   29
2-7   Expressions for Generalized System Properties   34

3   Free-Vibration Response   **41**

3-1   Solution of the Equation of Motion   41
3-2   Undamped Free Vibrations   42
3-3   Damped Free Vibrations   44
      Critical Damping   44
      Underdamped Systems   45
      Overdamped Systems   48

4   Response to Harmonic Loading   **52**

4-1   Undamped System   52
      Complementary Solution   52
      Particular Solution   53
      General Solution   53
      Response Ratio   54
4-2   Damped System   54
4-3   Resonant Response   60
4-4   Accelerometers and Displacement Meters   62
4-5   Vibration Isolation   64
4-6   Evaluation of Damping in SDOF Systems   69
      Free-Vibration Decay   70
      Resonant Amplification   70
      Half-Power (Bandwidth) Method   72
      Energy Loss per Cycle (Resonance Testing)   73
      Hysteretic Damping   76

5 Response to Periodic Loadings   **80**

5-1 Fourier Series Expression of the Loading   80
5-2 Response to the Fourier Series Loading   81
5-3 Exponential Form of Fourier Series Solution   83

6 Response to Impulsive Loads   **87**

6-1 General Nature of Impulsive Loads   87
6-2 Sine-Wave Impulse   88
6-3 Rectangular Impulse   91
6-4 Triangular Impulse   92
6-5 Shock or Response Spectra   94
6-6 Approximate Analysis of Impulsive-Load Response   96

7 Response to General Dynamic Loading   **100**

7-1 Duhamel Integral for an Undamped System   100
7-2 Numerical Evaluation of the Duhamel Integral for an Undamped System   102
7-3 Response of Damped Systems   105
7-4 Response Analysis through the Frequency Domain   108
7-5 Numerical Analysis in the Frequency Domain   113
     Discrete Fourier Transforms   114
     Fast Fourier Transform Analysis   114

8 Analysis of Nonlinear Structural Response   **118**

8-1 Analysis Procedure   118
8-2 Incremental Equation of Equilibrium   119
8-3 Step-by-Step Integration   121
8-4 Summary of Procedure   123

9 Vibration Analysis by Rayleigh's Method   **129**

9-1 Basis of the Method   129
9-2 Approximate Analysis of a General System   131
9-3 Selection of the Vibration Shape   133
9-4 Improved Rayleigh Method   137

Part Two MULTI-DEGREE-OF-FREEDOM SYSTEMS

10 Formulation of the MDOF Equations of Motion   **145**

10-1 Selection of the Degrees of Freedom   145
10-2 Dynamic-Equilibrium Condition   146
10-3 Axial-Force Effects   149

11   Evaluation of Structural-Property Matrices   **151**

11-1   Elastic Properties   151
    Flexibility   151
    Stiffness   152
    Basic Structural Concepts   153
    Finite-Element Stiffness   155
11-2   Mass Properties   160
    Lumped-Mass Matrix   160
    Consistent-Mass Matrix   161
11-3   Damping Properties   165
11-4   External Loading   165
    Static Resultants   166
    Consistent Nodal Loads   166
11-5   Geometric Stiffness   167
    Linear Approximation   167
    Consistent Geometric Stiffness   169
11-6   Choice of Property Formulation   172

12   Undamped Free Vibrations   **176**

12-1   Analysis of Vibration Frequencies   176
12-2   Analysis of Vibration Mode Shapes   179
12-3   Flexibility Formulation of Vibration Analysis   182
12-4   Influence of Axial Forces   182
    Free Vibrations   182
    Buckling Load   183
    Buckling with Harmonic Excitation   184
12-5   Orthogonality Conditions   185
    Basic Conditions   185
    Additional Relationships   186
    Normalizing   187

13   Analysis of Dynamic Response   **191**

13-1   Normal Coordinates   191
13-2   Uncoupled Equations of Motion: Undamped   193
13-3   Uncoupled Equations of Motion: Damped   194
    Derivation of the Equations   194
    Conditions for Damping Orthogonality   195
    Damping Coupling   198
13-4   Summary of the Mode-Superposition Procedure   199

14 Practical Vibration Analysis — 208

14-1	Preliminary Comments	208
14-2	Stodola Method	209
	Fundamental-Mode Analysis	209
	Proof of Convergence	213
	Analysis of Second Mode	215
	Analysis of Third and Higher Modes	219
	Analysis of Highest Mode	220
14-3	Buckling Analysis by Matrix Iteration	222
14-4	Holzer Method	226
	Basic Procedure	226
	Transfer-Matrix Procedure	230
	Holzer-Myklestad Method	232
14-5	Reduction of Degrees of Freedom	235
	Background	235
	Discrete-Mass Allocations	236
	Rayleigh Method in Discrete-Coordinate Systems	237
	Rayleigh-Ritz Method	239
14-6	Basic Concepts in Matrix Iteration	243
	Eigenproperty Expansion of the Dynamic Matrix	243
	Iterative Solution of the Eigenproblem	245
	Iteration with Shifts	247
	Subspace Iteration	250
14-7	Symmetric Form of Dynamic Matrix	252
	Diagonal Mass Matrix	253
	Consistent-Mass Matrix	254
14-8	Analysis of Unconstrained Structures	255

15 Analysis of Nonlinear Systems — 260

15-1	Introduction	260
15-2	Incremental Equilibrium Equations	262
15-3	Step-by-Step Integration: Linear-Acceleration Method	263
15-4	Unconditionally Stable Linear-Acceleration Method	265
15-5	Performance of the Wilson θ Method	268

16 Variational Formulation of the Equations of Motion — 271

16-1	Generalized Coordinates	271
16-2	Lagrange's Equations of Motion	273
16-3	Derivation of the General Equations of Motion	279
16-4	Constraints and Lagrange Multipliers	284

Part Three   DISTRIBUTED-PARAMETER SYSTEMS

17   Partial Differential Equations of Motion   293

17-1   Introduction   293
17-2   Beam Flexure: Elementary Case   294
17-3   Beam Flexure: Including Axial-Force Effects   296
17-4   Beam Flexure: Including Shear Deformation and Rotatory
         Inertia   298
17-5   Beam Flexure: Including Viscous Damping   301
17-6   Beam Flexure: Generalization Support Excitations   302
17-7   Axial Deformations   305

18   Analysis of Undamped Free Vibrations   308

18-1   Beam Flexure: Elementary Case   308
18-2   Beam Flexure: Including Axial-Force Effects   317
18-3   Beam Flexure: Including Shear Deformation and Rotatory
         Inertia   318
18-4   Beam Flexure: Orthogonality of Vibration Mode Shapes   321
18-5   Free Vibrations in Axial Deformation   323
18-6   Orthogonality of Axial Vibration Modes   325

19   Analysis of Dynamic Response   328

19-1   Normal Coordinates   328
19-2   Uncoupled Flexural Equations of Motion: Undamped   331
19-3   Uncoupled Flexural Equations of Motion: Damped   336
19-4   Uncoupled Axial Equations of Motion: Undamped   338

20   The Dynamic Direct-Stiffness Method   345

20-1   Introduction   345
20-2   Dynamic Flexural-Stiffness Matrix   346
20-3   Dynamic Stiffness for Flexure and Rigid Axial Displacements   353
20-4   Dynamic Axial-Deformation Stiffness Matrix   357
20-5   Combined Flexural- and Axial-Deformation Stiffness   358
20-6   Axial-Force Effects on Transverse-Bending Stiffness   360

21   Wave-Propagation Analysis   364

21-1   Basic Axial-Wave-Propagation Equation   364
21-2   Consideration of Boundary Conditions   369
21-3   Discontinuity in Bar Properties   371
21-4   Stress Waves Developed During Pile Driving   377
21-5   Shear-Wave Propagation in Buildings   382

Part Four RANDOM VIBRATIONS

22 Probability Theory **389**

22-1 Single Random Variable 389
22-2 Important Averages of a Single Random Variable 394
22-3 One-dimensional Random Walk 396
22-4 Two Random Variables 403
22-5 Important Averages of Two Random Variables 412
22-6 Scatter Diagram and Correlation of Two Random Variables 416
22-7 Two-dimensional Random Walk 419
22-8 *m* Random Variables 430
22-9 Linear Transformations of Normally Distributed Random Variables 432

23 Random Processes **436**

23-1 Definition 436
23-2 Stationary and Ergodic Processes 438
23-3 Autocorrelation Function for Stationary Processes 444
23-4 Power Spectral Density Function for Stationary Processes 449
23-5 Relationship between Power Spectral Density and Autocorrelation Functions 451
23-6 Power Spectral Density and Autocorrelation Functions for Derivatives of Processes 454
23-7 Superposition of Stationary Processes 455
23-8 Stationary Gaussian Processes: One Independent Variable 457
23-9 Stationary Gaussian White Noise 464
23-10 Probability Distribution for Maxima 468
23-11 Probability Distribution for Extreme-Values 472
23-12 Nonstationary Gaussian Processes 476
23-13 Stationary Gaussian Process: Two or More Independent Variables 476

24 Stochastic Response of Linear SDOF Systems **482**

24-1 Transfer Functions 482
24-2 Relationship between Unit-Impulse- and Complex-Frequency-Response Functions 485
24-3 Relationship between Input and Output Autocorrelation Functions 489
24-4 Relationship between Input and Output Power Spectral Density Functions 492

24-5   Response Characteristics for Narrowband Systems   495
24-6   Nonstationary Mean Square Response Resulting from Zero
       Initial Conditions   498
24-7   Fatigue Predictions for Narrowband Systems   502

25   Stochastic Response of Linear MDOF Systems   **508**

25-1   Time-Domain Response for Linear Systems   508
25-2   Frequency-Domain Response for Linear Systems   510
25-3   Response to Discrete Loadings   511
25-4   Response to Distributed Loadings   515

**Part Five   ANALYSIS OF STRUCTURAL RESPONSE TO
              EARTHQUAKES**

26   Seismological Background   **521**

26-1   Introductory Note   521
26-2   Seismicity   522
26-3   Elastic-Rebound Theory of Earthquakes   525
26-4   Earthquake Waves   528
26-5   Measures of Ground-Motion Characteristics   531
26-6   Selection of Design Earthquakes   539

27   Deterministic Analysis of Earthquake Response   **546**

27-1   Earthquake Input Mechanisms   544
27-2   Excitation by Rigid-Base Translation   546
       Lumped SDOF Systems   546
       Generalized SDOF Systems   549
       Lumped MDOF Systems   555
       Comparison with Uniform Building Code Requirements   570
       Distributed-Parameter Systems   572
27-3   Excitation by Rigid-Base Rotation   574
27-4   Multiple-Support Excitation   575
27-5   Influence of Foundation Medium in Earthquake Response   578
       Modeling of the Foundation Medium   578
       Soil Modification of Earthquake Motions   581
       Soil-Structure Interaction: Equations of Motion   584
       Soil-Structure Interaction: Response Analysis   591
27-6   Nonlinear Response to Earthquakes   594
       Need for Nonlinear Analysis   594
       Method of Nonlinear Analysis   597

Typical Inelastic-Response Behavior 598
Influence of Strength Variations 599
Ductility-Factor Method 602

28 Nondeterministic Analysis of Earthquake Response 611

28-1 Stochastic Modeling of Strong Ground Motions 611
Stationary White Noise 611
Stationary Filtered White Noise 613
Nonstationary Filtered White Noise 615
28-2 Analysis of Linear Systems 616
SDOF Systems 616
MDOF Systems 617
28-3 Analysis of Nonlinear Systems 617
28-4 Extreme-Value Response of SDOF Systems 618
Stationary White Noise Excitation 618
Stationary Filtered White Noise Excitation 619
28-5 Extreme-Value Response of MDOF Systems 626

Index **629**

PREFACE

This book is the result of more than 25 years spent in developing the program of instruction in structural dynamics at the University of California, Berkeley. Understandably, the material has undergone considerable revision during that time. Three different sets of notes have been prepared and distributed to the classes at widely separated intervals, and local versions of these have been used as class notes in the curricula presented at such diverse locations as Santiago, Chile, Trondheim, Norway, and Tokyo, Japan.

In the initial organization of this material, Professor Clough was strongly influenced by the lectures presented by Professor R. L. Bisplinghoff at the Massachusetts Institute of Technology, and he acknowledges his indebtedness to that superbly prepared course on the dynamics of airplane structures. Subsequent orientation of the text toward civil engineering problems in structural dynamics reflected the work of Hohenemser and Prager in their pioneering treatise "Dynamik der Stabwerke."[1] Similarly, Professor Penzien acknowledges the great benefit he received from the lectures on random vibrations given by Professor S. H. Crandall at the Massachusetts Institute of Technology. The continuing development of this

---

[1] K. Hohenemser and W. Prager, "Dynamic der Stabwerke," Julius Springer, Berlin, Germany, 1933.

subject matter, however, has been essentially the work of both Clough and Penzien. Contributions to the literature by numerous writers have been incorporated appropriately into the sequence of lectures; most such contributions are so well established in the field of structural dynamics that it is difficult to assign credit for them. Consequently, few credit references are given, and the authors offer their apologies to those who may feel slighted.

Although this text material has undergone continuous revision during its development, the general organization has remained unchanged. The logical transition from structures with a single degree of freedom, to generalized single-degree systems, to the mode-superposition analysis of multidegree discrete-coordinate structures has provided a simple route for the "statics" trained structural engineer to follow in being introduced to the special problems arising from dynamic loadings. Moreover, it has always been considered essential to emphasize transient dynamic response analysis rather than to be concerned only with vibration analysis. As a prerequisite to the study of structural dynamics, it has proven effective to require a solid background of static structures theory, including matrix methods, and it is assumed that the readers of this text have had such preparation.

Probably the most obvious and far-reaching change that has occurred during the development of this text material has been the acceptance of high-speed digital computers as a standard tool for structural analysis. Before computers were widely used in structural offices, major emphasis in structural dynamics was placed on efficient methods for slide-rule and desk-calculator analyses. Such methods still have a prominent place in this text because the authors are convinced of their value in the study of the subject. If the details of a hand-solution procedure are thoroughly understood, it is not difficult to write or use a corresponding computer code, but it may be impossible to use a "black box" computer program effectively without knowing the computational details. Nevertheless, it is recognized that any significant, practical dynamic response analysis requires so much numerical effort as to be economically feasible only by computer. Therefore, the solution techniques emphasized herein are generally those that may be employed effectively with a high-speed computer, as well as by hand. The purpose of the presentation is to explain the basis of the methods; coding techniques and efficient computer usage are not discussed in detail.

The subject matter of this text is the basis of a sequence of graduate-level quarter-length courses given at the University of California; however, most of it could be covered just as well at the advanced undergraduate (fourth-year) level. The basic course on dynamics of structures treats the material in Part 1 and in Part 2 through the first sections of Chap. 14. It is expected that nearly all Masters-Degree students in structural engineering should have at least this much contact with the dynamics field. The subsequent "advanced" structural dynamics course treats the remainder of Part 2 and all of Part 3. In addition, much of the earthquake engineering material in Chaps. 26 and 27 is presented in these first two courses to provide practical applications of the theory. The basic material of Part 4 is covered in a single-quarter course on random vibrations, together with some of the earthquake-engineering applications from Chap. 28. Finally, the entire book serves as background and reference material for a course on structural design for dynamic loads; the prerequisite for this design course is the basic course on dynamics of structures. Although the frame of reference for most of this material is the field of civil-engineering applications, the same basic techniques of structural dynamics are applicable in aerospace engineering, naval architecture, automotive engineering, and any field in which structural systems are subjected to dynamic loads.

A large number of examples have been incorporated into the text because the authors have found that much of this material may be taught most effectively by that means. Moreover, many homework problems have been provided with most of the chapters because it is essential for the student to actually make use of these analytical techniques to master them fully. However, such problems must be assigned sparingly because dynamic-response analyses are notoriously time consuming. The authors have found that from one to four problems may constitute an adequate weekly assignment, depending on the subject matter and type of solution procedure required. Consequently, the book includes many more problems than can be assigned during a one-year sequence of courses on structural dynamics.

RAY W. CLOUGH

JOSEPH PENZIEN

LIST OF SYMBOLS

a	distance
a_o, a_n	Fourier coefficients, constants
A	area, constant
A_1, A_2	constants
b	distance, integer
b_o, b_n	Fourier coefficients, constants
B	constant
c	damping coefficient
c^*	generalized damping coefficient
c_c	critical damping coefficient
c_n	Fourier coefficients
c_{ij}	damping influence coefficients
c_n	normal mode generalized damping coefficients
D	dynamic magnification factor, plate stiffness
$\mathbf{D}$	dynamic matrix $= \mathbf{k}^{-1}\mathbf{m}$
D_1, D_2	constants
e	axial displacement
E	Young's modulus
$\mathbf{E}$	dynamic matrix $= \mathbf{D}^{-1}$

$E[\ ]$	expected value, ensemble average
EI	flexural stiffness
f	natural cyclic frequency
$\tilde{f}_{ij}$	flexibility influence coefficients
f_I, f_D, f_S	inertia, damping, and spring forces, respectively
g	acceleration of gravity
g_i	general displacement coordinates, stress wave functions
G	shear modulus
G, G_1, G_2	constants
h	plate thickness, story height
$h(t)$	unit impulse response function
$H(\omega), H(i\omega)$	complex frequency response function
Hz	Hertz (measure of frequency, cycles per second)
i	integer
I	impulse, moment of inertia
$\mathbf{I}$	identity matrix
j	integer
k, k_i	spring constants
$k^*, \bar{k}^*$	generalized spring constants
$\tilde{k}(t)$	effective stiffness
k_G	geometric stiffness
k_{ij}	stiffness influence coefficients
$\bar{k}_{ij}$	combined stiffness influence coefficients
$k_{G_{ij}}$	geometric stiffness influence coefficients
K_n	generalized stiffness of nth normal mode
L	length
$\mathcal{L}$	earthquake excitation factor
m	mass, integer
m_i	mass
m_{ij}	mass influence coefficients
$\bar{m}$	uniform mass/unit length
$\bar{m}_I$	rotatory mass moment of inertia
M_n	generalized mass of nth normal mode
m^*	generalized mass
$\mathfrak{M}, \mathfrak{M}_i$	internal moment at a section
n	integer, constant
N	axial load, number of time increments, number of degrees of freedom
N_{cr}	critical axial load
$\mathcal{N}$	time varying axial force

p, p_o	load
$\bar{p}$	uniform loading/unit length
p^*	generalized loading
p_{eff}	effective loading
$p(x)$	probability density function
$p(x, y)$	joint probability density function
$p(x \mid y)$	conditional probability density function
$p(x_1, x_2, \ldots, x_m)$	multivariate probability density function
$P_n(t)$	forcing function of nth normal mode
$P(X), P(X, Y)$	probability distribution functions
Pr	probability
q_i	ith generalized coordinate
Q_i	ith generalized forcing function
r	radius of gyration
$R(t)$	response ratio
$R_x(\tau)$	autocorrelation function
$R_{xy}(\tau)$	cross-correlation function
s	constant
$S_x(\bar{\omega})$	power-spectral density function
$S_{xy}(\bar{\omega})$	cross-spectral density function
S_a	spectral acceleration response
S_d	spectral displacement response
S_v	spectral pseudovelocity response
SI	response-spectrum intensity
t, t_i	time
t_1	impulse duration
t_{ij}	transfer influence coefficients
T	period of vibration, kinetic energy
T_n	period of nth normal mode
T_p	period of loading
TR	transmissibility
u	displacement in x direction
U	strain energy
v	displacement in y direction
v^t	total displacement
v_g, v_{go}	ground displacement
v_{st}	static displacement
V, V_f, V_n	potential energy
$\mathcal{V}$	internal shear force at a section
w	displacement in z direction

W	work, weight
W_{nc}	work by nonconservative forces
W_N	work by axial load N
x	space coordinate
$\bar{x}$	mean value of x
$\overline{x^2}$	mean square value of x
$x(t)$	random process
y	space coordinate
$y(t)$	random process
Y_n	generalized displacement of nth normal mode
z	space coordinate
Z, Z_n, Z_o	generalized coordinates
β	frequency ratio
γ	weight/unit area
δ	log decrement, variation, residual
Δ	increment
Δ_{st}	static displacement
ε	strain
ζ	time function, hysteretic damping coefficient
λ_G	axial-load factor
λ_i	Lagrange multiplier
λ_n	nth eigenvalue
θ	phase angle, slope, rotation
μ	ductility factor
μ_{xy}	covariance
v	Poisson's ratio
ξ, ξ_n	damping ratio
ρ	vector amplitude, mass/unit volume
ρ_{xy}	correlation coefficient
σ	stress
σ_x	standard deviation
$\sigma_x{}^2$	variance
τ	time
ϕ_{ij}	modal displacement
ϕ_n	nth mode shape
Φ	mode shape matrix
ψ, ψ_n	generalized displacement function
$\boldsymbol{\psi_n}$	generalized displacement vector
ω, ω_n	undamped natural circular frequency
ω_D, ω_{Dn}	damped natural circular frequency
$\bar{\omega}$	circular frequency of harmonic forcing function

OVERVIEW OF STRUCTURAL DYNAMICS

1-1 FUNDAMENTAL OBJECTIVE OF STRUCTURAL-DYNAMICS ANALYSIS

The primary purpose of this book is to present methods for analyzing the stresses and deflections developed in any given type of structure when it is subjected to an arbitrary dynamic loading. In one sense, the objective may be considered to be the extension of standard methods of structural analysis, which generally are concerned only with static loading, to permit consideration of dynamic loads as well. In this context, the static-loading condition may be looked upon merely as a special form of dynamic loading. However, in the analysis of linear structures it is convenient to distinguish between the static and the dynamic components of the applied loading, to evaluate the response to each type of loading separately, and then to superpose the two response components to obtain the total effect. When treated thus, the static and dynamic methods of analysis are fundamentally different in character.

For the purposes of this presentation the term *dynamic* may be defined simply as time-varying; thus a dynamic load is any load of which the magnitude, direction, or position varies with time. Similarly, the structural response to a dynamic load, i.e., the resulting deflections and stresses, is also time-varying, or dynamic.

Two basically different approaches are available for evaluating structural response to dynamic loads: deterministic and nondeterministic. The choice of method to be used in any given case depends upon how the loading is defined. If the time variation of loading is fully known, even though it may be highly oscillatory or irregular in character, it will be referred to herein as a *prescribed dynamic loading*; and the analysis of the response of any specified structural system to a prescribed dynamic loading is defined as a deterministic analysis. On the other hand, if the time variation is not completely known but can be defined in a statistical sense, the loading is termed a *random dynamic loading*; a nondeterministic analysis correspondingly is the analysis of response to a random dynamic loading. The principal emphasis in this text is placed on development of methods of deterministic dynamic analysis. However, Part Four is devoted to presenting an introduction to nondeterministic methods of analysis. In addition, a chapter on nondeterministic earthquake-response analysis is included in Part Five, which deals with applications of methods of structural dynamics in the field of earthquake engineering.

In general, the structural response to any dynamic loading is expressed basically in terms of the displacements of the structure. Thus a deterministic analysis leads to a displacement-time history corresponding to the prescribed loading history; other aspects of the deterministic structural response, such as stresses, strains, internal forces, etc., are usually obtained as a secondary phase of the analysis, from the previously established displacement patterns. On the other hand, a nondeterministic analysis provides statistical information about the displacements which result from a statistically defined loading. In this case, the time variation of the displacements is not determined, and other aspects of the response, such as stresses, internal forces, etc., must be evaluated directly by independent nondeterministic analysis rather than from the displacement results.

1-2 TYPES OF PRESCRIBED LOADINGS

Almost any type of structural system may be subjected to one form or another of dynamic loading during its lifetime. From an analytical standpoint, it is convenient to divide prescribed or deterministic loadings into two basic categories, periodic and nonperiodic. Some typical forms of prescribed loadings and examples of situations in which such loadings might be developed are shown in Fig. 1-1.

As is indicated in Fig. 1-1a and b, periodic loadings are repetitive loads which exhibit the same time variation successively for a large number of cycles. The simplest periodic loading is the sinusoidal variation shown in Fig. 1-1a, which is termed *simple harmonic*; such loadings are characteristic of unbalanced-mass effects in rotating

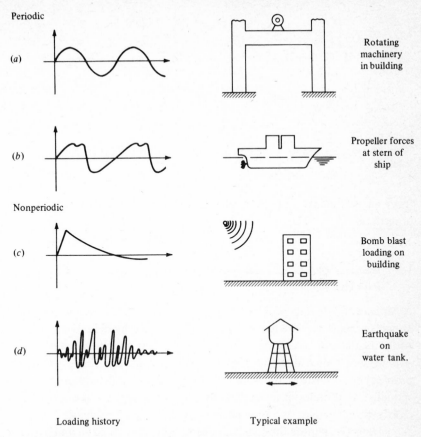

Periodic

(a)

(b)

Nonperiodic

(c)

(d)

Loading history

Typical example

Rotating machinery in building

Propeller forces at stern of ship

Bomb blast loading on building

Earthquake on water tank.

FIGURE 1-1
Characteristics and sources of typical dynamic loadings: (a) simple harmonic; (b) complex; (c) impulsive; (d) long-duration.

machinery. Other forms of periodic loading, e.g., those caused by hydrodynamic pressures generated by a propeller at the stern of a ship or by inertial effects in reciprocating machinery, frequently are more complex. However, by means of a Fourier analysis any periodic loading can be represented as the sum of a series of simple harmonic components; thus, in principle, the analysis of response to any periodic loading follows the same general procedure.

Nonperiodic loadings may be either short-duration *impulsive* loadings or long-duration general forms of loads. A blast or explosion is a typical source of impulsive load; for such short-duration loads, special simplified forms of analysis may be employed. On the other hand, a general, long-duration loading such as might result

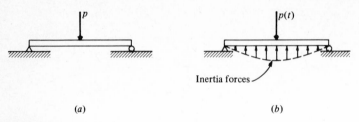

(a)   (b)

FIGURE 1-2
Basic difference between static and dynamic loads: (*a*) static loading; (*b*) dynamic loading.

from an earthquake can be treated only by completely general dynamic-analysis procedures.

1-3 ESSENTIAL CHARACTERISTICS OF A DYNAMIC PROBLEM

A structural-dynamic problem differs from its static-loading counterpart in two important respects. The first difference to be noted, by definition, is the time-varying nature of the dynamic problem. Because the load and the response vary with time, it is evident that a dynamic problem does not have a single solution, as a static problem does; instead the analyst must establish a succession of solutions corresponding to all times of interest in the response history. Thus a dynamic analysis is clearly more complex and time-consuming than a static analysis.

However, a more fundamental distinction between static and dynamic problems is illustrated in Fig. 1-2. If a simple beam is subjected to a static load *p*, as shown in Fig. 1-2*a*, its internal moments and shears and deflected shape depend directly upon the given load and can be computed from *p* by established principles of force equilibrium. On the other hand, if the load *p*(*t*) is applied dynamically, as shown in Fig. 1-2*b*, the resulting displacements of the beam are associated with accelerations which produce inertia forces resisting the accelerations. Thus the internal moments and shears in the beam in Fig. 1-2*b* must equilibrate not only the externally applied force but also the inertia forces resulting from the accelerations of the beam.

Inertia forces which resist accelerations of the structure in this way are the most important distinguishing characteristic of a structural-dynamics problem. In general, if the inertia forces represent a significant portion of the total load equilibrated by the internal elastic forces of the structure, then the dynamic character of the problem must be accounted for in its solution. On the other hand, if the motions are so slow that the inertia forces are negligibly small, the analysis for any desired instant of time may be made by static structural-analysis procedures even though the load and response may be time-varying.

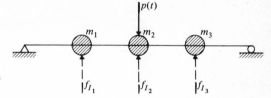

FIGURE 1-3
Lumped-mass idealization of a simple
beam.

1-4 METHODS OF DISCRETIZATION

Lumped-Mass Procedure

In the dynamic system of Fig. 1-2b, the analysis obviously is greatly complicated by the fact that the inertia forces result from structural displacements which in turn are influenced by the magnitudes of inertia forces. This closed cycle of cause and effect can be attacked directly only by formulating the problem in terms of differential equations. Furthermore, because the mass of the beam is distributed continuously along its length, the displacements and accelerations must be defined for each point along the axis if the inertia forces are to be completely defined. In this case, the analysis must be formulated in terms of partial differential equations because the position along the span as well as the time must be taken as independent variables.

On the other hand, if the mass of the beam were concentrated in a series of discrete points or lumps, as shown in Fig. 1-3, the analytical problem would be greatly simplified because inertia forces could be developed only at these mass points. In this case it is necessary to define the displacements and accelerations only at these discrete points.

The number of displacement components which must be considered in order to represent the effects of all significant inertia forces of a structure may be termed the *number of dynamic degrees of freedom* of the structure. For example, if the system of Fig. 1-3 were constrained so that the three mass points could move only in a vertical direction, this would be called a three-degree-of-freedom (3 DOF) system. On the other hand, if these masses were not concentrated in points but had finite rotational inertia, the rotational displacements of the three points would also have to be considered and the system would have 6 DOF. If axial distortions of the beam also were significant, displacements parallel with the beam axis would result and the system would have 9 DOF. More generally, if the structure could deform in three-dimensional space, each mass would have 6 DOF and the system would have 18 DOF. On the other hand, if the masses were concentrated in points so that the rotational inertia might be ignored, the three-dimensional system would then have 9 DOF. On the basis of these considerations, it is clear that a system with continuously distributed mass, as in Fig. 1-2b, has an infinite number of degrees of freedom.

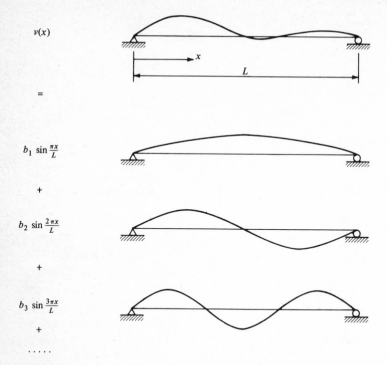

$v(x)$

$=$

$b_1 \sin \frac{\pi x}{L}$

$+$

$b_2 \sin \frac{2\pi x}{L}$

$+$

$b_3 \sin \frac{3\pi x}{L}$

$+$

. . . . .

FIGURE 1-4
Sine-series representation of simple beam deflection.

Generalized Displacements

The lumped-mass idealization described above provides a simple means of limiting the number of degrees of freedom that must be considered in the analysis of arbitrary problems in structural dynamics. The lumping procedure is most effective in treating systems in which a large proportion of the total mass actually is concentrated in a few discrete points. Then it may be assumed that the mass of the structure which supports these point concentrations can also be included in the lumps and the structure itself considered to be weightless.

In cases where the mass of the system is quite uniformly distributed throughout, however, an alternative approach to limiting the degrees of freedom may be preferable. This procedure is based on the assumption that the deflected shape of the structure can be expressed as the sum of a series of specified displacement patterns; these patterns then become the displacement coordinates of the structure. A simple example of this approach to expressing deflections in structures is the trigonometric-series representation of the deflection of a simple beam. In this case, the deflection may be expressed as the sum of independent sine-wave contributions, as shown in Fig. 1-4, or, in mathematical form,

$$v(x) = \sum_{n=1}^{\infty} b_n \sin \frac{n\pi x}{L} \qquad (1\text{-}1)$$

In general, any arbitrary shape compatible with the prescribed support conditions can be represented by an infinite series of such sine-wave components. The amplitudes of the sine-wave shapes may be considered to be the coordinates of the system, and the infinite number of degrees of freedom of the actual beam are represented by the infinite number of terms included in the series. The advantage of this approach is that a good approximation to the actual beam shape can be achieved by a truncated series of sine-wave components; thus a 3 DOF approximation would contain only three terms in the series, etc.

This concept can be further generalized by recognizing that the sine-wave shapes used as the assumed displacement patterns were an arbitrary choice in this example. In general, any shapes $\psi_n(x)$ which are compatible with the prescribed geometric-support conditions and which maintain the necessary continuity of internal displacements may be assumed. Thus a generalized expression for the displacements of any one-dimensional structure might be written

$$v(x) = \sum_n Z_n \psi_n(x) \qquad (1\text{-}2)$$

For any assumed set of displacement functions $\psi_n(x)$, the resulting shape of the structure depends upon the amplitude terms Z_n, which will be referred to as *generalized coordinates*. The number of assumed shape patterns represents the number of degrees of freedom considered in this form of idealization. In general, better accuracy can be achieved in a dynamic analysis for a given number of degrees of freedom by using the shape-function method of idealization than by the lumped-mass approach. However, it also should be recognized that greater computational effort is required for each degree of freedom when such generalized coordinates are employed.

The Finite-Element Concept

A third method of expressing the displacements of any given structure in terms of a finite number of discrete displacement coordinates, which combines certain features of both the lumped-mass and the generalized-coordinate procedures, has now become popular. This approach, which is the basis of the finite-element method of analysis of structural continua, provides a convenient and reliable idealization of the system and is particularly effective in digital-computer analyses.

The finite-element type of idealization is applicable to structures of all types: framed structures, which comprise assemblages of one-dimensional members (beams, columns, etc.); plane-stress or plate- or shell-type structures, which are made up of two-dimensional components; and general three-dimensional solids. For simplicity, only the one-dimensional type of structural components will be considered in the present discussion, but the extension of the concept to two- and three-dimensional structural elements is straightforward.

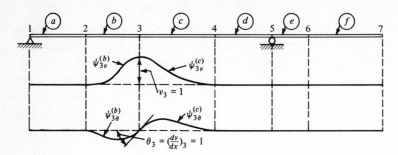

FIGURE 1-5
Typical finite-element beam coordinates.

The first step in the finite-element idealization of any structure, e.g., the beam shown in Fig. 1-5, involves dividing it into an appropriate number of segments, or elements, as shown. Their size is arbitrary; they may be all of the same size or all different. The ends of the segments, at which they are interconnected, are called *nodal points*. The displacements of these nodal points then become the generalized coordinates of the structure.

The deflection of the complete structure can now be expressed in terms of these generalized coordinates by means of an appropriate set of assumed displacement functions, using an expression similar to Eq. (1-2). In this case, however, the displacement functions are called *interpolation functions* because they define the shape between the specified nodal displacements. For example, in Fig. 1-5 are shown the interpolation functions associated with the two degrees of freedom of point 3, which produce transverse displacements in the plane of the figure. In principle, these interpolation functions could be any curve which is internally continuous and which satisfies the geometric displacement conditions imposed by the nodal displacements. For one-dimensional elements it is convenient to use the shapes which would be produced by these nodal displacements in a uniform beam (these are cubic hermitian polynomials and are sketched in Fig. 1-5).

Because the displacement functions used in this procedure satisfy the requirements stated in the preceding section, it should be apparent that coordinates used in the finite-element method are just a special form of generalized coordinates. The advantages of this special procedure are as follows:

1 Any desired number of generalized coordinates can be introduced merely by dividing the structure into an appropriate number of segments.

2 Since the displacement functions chosen for each segment may be identical, computations are simplified.

3 The equations which are developed by this approach are largely uncoupled because each nodal displacement affects only the neighboring elements; thus the solution process is greatly simplified.

In general, the finite-element approach provides the most efficient procedure for expressing the displacements of arbitrary structural configurations by means of a discrete set of coordinates.

1-5 FORMULATION OF THE EQUATIONS OF MOTION

As mentioned earlier, the primary objective of a deterministic structural-dynamic analysis is the evaluation if the displacement-time history of a given structure subjected to a given time-varying load. In most cases, an approximate analysis involving only a limited number of degrees of freedom will provide sufficient accuracy, and the problem thus can be reduced to the determination of the time history of these selected displacement components. The mathematical expressions defining the dynamic displacements are called the *equations of motion* of the structure, and the solution of these equations of motion provides the required displacement histories.

The formulation of the equations of motion of a dynamic system is possibly the most important (and sometimes the most difficult) phase of the entire analysis procedure. In this text, three different methods will be employed for the formulation of these equations, each having advantages in the study of special classes of problems. The fundamental concepts associated with each of these methods are described in the following paragraphs.

Direct Equilibration Using d'Alembert's Principle

The equations of motion of any dynamic system represent expressions of Newton's second law of motion, which states that the rate of change of momentum of any mass m is equal to the force acting on it. This relationship can be expressed mathematically as the differential equation

$$\mathbf{p}(t) = \frac{d}{dt}\left(m\,\frac{d\mathbf{v}}{dt}\right) \qquad (1\text{-}3)$$

where $\mathbf{p}(t)$ is the applied force vector and $\mathbf{v}(t)$ is the position vector of the mass m. For most problems in structural dynamics it may be assumed that the mass does not vary with time, in which case Eq. (1-3) may be written

$$\mathbf{p}(t) = m\,\frac{d^2\mathbf{v}}{dt^2} \equiv m\ddot{\mathbf{v}}(t) \qquad (1\text{-}3a)$$

where the dots represent differentiation with respect to time. Equation (1-3a), the familiar expression that force is equal to the product of mass and acceleration, may also be written

$$\mathbf{p}(t) - m\ddot{\mathbf{v}}(t) = \mathbf{0} \qquad (1\text{-}3b)$$

in which case the second term $m\ddot{\mathbf{v}}(t)$ is called the *inertia force* resisting the acceleration of the mass.

The concept that a mass develops an inertia force proportional to its acceleration and opposing it is known as *d'Alembert's principle*. It is a very convenient device in problems of structural dynamics because it permits the equations of motion to be expressed as equations of dynamic equilibrium. The force $\mathbf{p}(t)$ may be considered to include many types of force acting on the mass: elastic constraints which oppose displacements, viscous forces which resist velocities, and independently defined external loads. Thus if an inertia force which resists acceleration is introduced, the expression of the equation of motion is merely an expression of the equilibration of all of the forces acting on the mass. In many simple problems the most direct and convenient way of formulating the equations of motion is by means of such direct equilibrations.

Principle of Virtual Displacements

If the structural system is reasonably complex, however, and involves a number of interconnected mass points or bodies of finite size, the direct equilibration of all the forces acting in the system may be difficult. Frequently, the various forces involved may readily be expressed in terms of the displacement degrees of freedom, but their equilibrium relationships may be obscure. In this case, the principle of virtual displacements can be used to formulate the equations of motion as a substitute for the equilibrium relationships.

The principle of virtual displacements may be expressed as follows. If a system which is in equilibrium under the action of a set of forces is subjected to a virtual displacement, i.e., any displacement compatible with the system constraints, the total work done by the forces will be zero. With this principle, it is clear that the vanishing of the work done during a virtual displacement is equivalent to a statement of equilibrium. Thus, the response equations of a dynamic system can be established by first identifying all the forces acting on the masses of the system, including inertia forces defined in accordance with d'Alembert's principle. Then the equations of motion are obtained by introducing virtual displacements corresponding to each degree of freedom and equating the work done to zero. A major advantage of this approach is that the virtual-work contributions are scalar quantities and can be added algebraically, whereas the forces acting on the structure are vectorial and can only be superposed vectorially.

Hamilton's Principle

Another means of avoiding the problems of establishing the vectorial equations of equilibrium is to make use of scalar energy quantities in a variational form. The most generally applicable variational concept is Hamilton's principle, which may be expressed as

$$\int_{t_1}^{t_2} \delta(T - V) \, dt + \int_{t_1}^{t_2} \delta W_{nc} \, dt = 0 \qquad (1\text{-}4)$$

where T = total kinetic energy of system

V = potential energy of system, including both strain energy and potential of any conservative external forces

W_{nc} = work done by nonconservative forces acting on system, including damping and any arbitrary external loads

δ = variation taken during indicated time interval

Hamilton's principle states that the variation of the kinetic and potential energy plus the variation of the work done by the nonconservative forces considered during any time interval t_1 to t_2 must equal zero. The application of this principle leads directly to the equations of motion for any given system. The process differs from the virtual-work analysis in that the inertia and elastic forces are not explicitly involved; the variations of the kinetic- and potential-energy terms, respectively, are utilized instead. This formulation thus has the advantage of dealing only with purely scalar energy quantities, whereas the forces and displacements used to represent corresponding effects in the virtual-work analysis are all vectorial in character even though the work terms themselves are scalar.

   It is of interest to note that Hamilton's principle can also be applied to statics problems. In this case, the kinetic-energy term T vanishes, and the remaining terms in the integrands of Eq. (1-4) are invariant with time; thus the equation reduces to

$$\delta(V - W_{nc}) = 0 \qquad (1\text{-}5)$$

which is the well-known principle of minimum potential energy, so widely used in static analyses.

Summary

It has been shown that the equations of motion of a dynamic system can be formulated by any one of three distinct procedures. The most straightforward approach is to establish directly the dynamic equilibrium of all forces acting in the system, taking account of inertial effects by means of d'Alembert's principle. In more complex systems, however, especially those involving mass and elasticity distributed over finite regions, a direct vectorial equilibration may be difficult, and work or energy formulations which involve only scalar quantities may be more convenient. The most direct of these procedures is based on the principle of virtual displacements, in which the forces acting on the system are evaluated explicitly but the equations of motion are derived by consideration of the work done during appropriate virtual displacements. On the other hand, the alternative energy formulation, which is based on Hamilton's principle, makes no direct use of the inertial or conservative forces acting in the system; the effects of these forces are represented instead by variations of the kinetic and potential energy of the system. It must be recognized that all three procedures are completely equivalent and lead to identical equations of motion. The method to be used in any given case is largely a matter of convenience and personal preference; the choice generally will depend on the nature of the dynamic system under consideration.

1-6 ORGANIZATION OF THE TEXT

In this development of the theory of structural dynamics, attention will be focused in Part One on the treatment of systems having but a single degree of freedom (SDOF), i.e., systems for which the displacement can be represented by the amplitude of a single coordinate. This class of problem will be studied in detail for two reasons: (1) the behavior of many practical structures can be expressed in terms of a single coordinate, so that the SDOF solution provides an adequate final result; (2) in linear structures of more complex forms, the total response may be expressed as the sum of the responses of a series of SDOF systems. Thus the SDOF analysis technique provides the basis for the vast majority of deterministic structural-dynamic analyses.

Part Two deals with discrete-parameter multi-degree-of-freedom (MDOF) systems, i.e., systems for which the behavior can be expressed in terms of a limited number of coordinates. In the analysis of linearly elastic systems, procedures will be presented for evaluating the vibration properties, and then the mode-superposition method will be derived, by means of which the total response is expressed as the sum of the individual responses in the various modes of vibration. The response calculation of each individual mode will be seen to involve a typical SDOF analysis. This superposition procedure is not applicable to nonlinear systems, however, and a step-by-step integration procedure is presented for the solution of such problems.

Dynamic systems having continuously distributed properties will be considered in Part Three. Such systems have an infinite number of degrees of freedom, and their equations of motion are written in the form of partial differential equations. However, it will be shown that the mode-superposition procedure is still applicable and that practical solutions can be obtained even in these cases by considering only a limited number of vibration modes.

Parts One to Three are all concerned with deterministic analyses which provide the response history to any given dynamic loading. The probabilistic approach to dynamic analysis is presented in Part Four, starting with the fundamentals of probability theory and including the analysis of both SDOF and MDOF systems.

Often it is not possible to define the excitation of a particular dynamic system fully. In such cases, however, it may be possible to characterize the excitation in a probabilistic manner, which then makes it possible to predict response by probabilistic methods. Such results, of course, are just as valuable to the design engineer as those obtained by deterministic means and often more so, particularly when questionable assumptions have been made in order to make a deterministic analysis possible. Certainly one cannot hope, for example, to predict deterministically with any degree of accuracy the future dynamic response of (1) airplanes flying in stormy weather conditions, (2) ships sailing the rough seas, (3) buildings withstanding strong-motion earthquake excitation, (4) missile components subjected to high-noise-level environments, or (5) vehicles traveling over rough roads.

Since probability theory is the basis for nondeterministic analysis, certain fundamentals in this field of study are presented in Chap. 22. These fundamentals are then applied to the characterization of stochastic processes in Chap 23, which in turn are used to study random vibrations of linear SDOF systems in Chap. 24 and MDOF systems in Chap. 25.

Finally, Part Five deals with applications of structural-dynamics theory to problems of earthquake engineering. It is in such applications that structural-dynamics analysis finds its principal use in civil engineering practice. These basic methods are equally applicable, however, to wind-loading analysis for civil engineering structures or to problems arising in the aerospace industry, in naval architecture, in mechanical engineering, or in any structural system subjected to dynamic loads.

PART ONE

Single-Degree-of-Freedom Systems

FORMULATION OF THE EQUATION OF MOTION

2-1 COMPONENTS OF THE BASIC DYNAMIC SYSTEM

The essential physical properties of any linearly elastic structural system subjected to dynamic loads include its mass, its elastic properties (flexibility or stiffness), its energy-loss mechanism, or damping, and the external source of excitation or loading. In the simplest model of a SDOF system, each of these properties is assumed to be concentrated in a single physical element. A sketch of such a system is shown in Fig. 2-1a.

The entire mass m of this system is included in the rigid block. Rollers constrain this block so that it can move only in simple translation; thus the single displacement coordinate v completely defines its position. The elastic resistance to displacement is provided by the weightless spring of stiffness k, while the energy-loss mechanism is represented by the damper c. The external-loading mechanism producing the dynamic response of this system is the time-varying load $p(t)$.

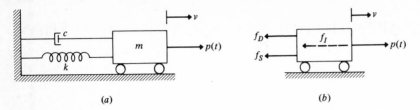

FIGURE 2-1
Idealized SDOF system: (a) basic components; (b) forces in equilibrium.

2-2   METHODS OF FORMULATION

Direct Equilibration

The equation of motion for the system of Fig. 2-1a can be derived by any of the three procedures discussed in Chap. 1. For this simple case, however, it is most easily formulated by directly expressing the equilibrium of all forces acting on the mass. As shown in Fig. 2-1b, the forces acting in the direction of the displacement degree of freedom include the applied load $p(t)$ and three forces resulting from the motion, inertia f_I, damping f_D, and the elastic spring force f_S. The equation of motion thus is merely an expression of the equilibrium of these forces, as follows:

$$f_I + f_D + f_S = p(t) \qquad (2\text{-}1)$$

Each of the forces represented on the left side of this equation is a function of the displacement v or of its derivatives; the positive sense of these forces has been chosen deliberately to correspond with the negative-displacement sense; thus they oppose the positive applied loads.

Consider first the elastic force. This is obviously given by the product of the spring stiffness and the displacement

$$f_S = kv \qquad (2\text{-}2a)$$

Similarly, by d'Alembert's principle, the inertia force is the product of the mass and the acceleration

$$f_I = m\ddot{v} \qquad (2\text{-}2b)$$

Finally, if a viscous damping mechanism is assumed, the damping force is the product of the damping constant c and the velocity

$$f_D = c\dot{v} \qquad (2\text{-}2c)$$

When Eqs. (2-2) are introduced into Eq. (2-1), the equation of motion of this SDOF system is found to be

$$m\ddot{v} + c\dot{v} + kv = p(t) \qquad (2\text{-}3)$$

Virtual-Work Analysis

It is instructive also to carry out a virtual-work formulation of this equation of motion. The forces acting on the mass are identified in Fig. 2-1b. If the mass is given a virtual displacement δv (the only displacement compatible with its constraints), these forces will each do work. The total work done by the system can then be written

$$-f_I\,\delta v - f_D\,\delta v - f_S\,\delta v + p(t)\,\delta v = 0 \qquad (2\text{-}4)$$

in which the negative signs result from the fact that the forces act opposite to the direction of the virtual displacement. Substituting Eqs. (2-2) into Eq. (2-4) and factoring out δv then leads to

$$[-m\ddot{v} - c\dot{v} - kv + p(t)]\,\delta v = 0 \qquad (2\text{-}5)$$

Since δv is nonzero, this can easily be put in the form of Eq. (2-3).

Application of Hamilton's Principle

To complete this presentation, the equation of motion of this system will be derived by use of Hamilton's principle [Eq. (1-4)]. The kinetic energy of the system, by definition, is given by

$$T = {}^1\!/_2 m\dot{v}^2 \qquad (2\text{-}6a)$$

while the potential energy, which represents merely the strain energy U of the spring, is given by

$$V = U = {}^1\!/_2 kv^2 \qquad (2\text{-}6b)$$

The nonconservative forces of the system of Fig. 2-1b are the damping force f_D and the applied load $p(t)$. The variation of the work done by these forces may be expressed

$$\delta W_{nc} = p(t)\,\delta v - c\dot{v}\,\delta v \qquad (2\text{-}6c)$$

which is equivalent to the virtual-work expression associated with these forces in Eq. (2-5). Introducing Eqs. (2-6) into Eq. (1-4), performing the variation of the first term, and rearranging leads to

$$\int_{t_1}^{t_2} [m\dot{v}\,\delta\dot{v} - c\dot{v}\,\delta v - kv\,\delta v + p(t)\,\delta v] = 0 \qquad (2\text{-}7)$$

Now the first term of Eq. (2-7) can be integrated by parts, as follows:

$$\int_{t_1}^{t_2} m\dot{v}\,\delta\dot{v}\,dt = m\dot{v}\,\delta v\,\Big|_{t_1}^{t_2} - \int_{t_1}^{t_2} m\ddot{v}\,\delta v\,dt \qquad (2\text{-}8)$$

noting that $\delta\dot{v} = d(\delta v)/dt$. But since it is assumed in Hamilton's principle that the variation δv vanishes at the limits of integration t_1 and t_2, the first term on the right side of Eq. (2-8) equals zero. Therefore if Eq. (2-8) is substituted into Eq. (2-7), the result may be written

$$\int_{t_1}^{t_2} [-m\ddot{v} - c\dot{v} - kv + p(t)] \, \delta v \, dt = 0 \qquad (2\text{-}9)$$

and since the variation δv is arbitrary, it is clear that Eq. (2-9) can be satisfied in general only if the expression in brackets vanishes. This can then be put in the form of Eq. (2-3).

This example shows how the same equation of motion can be derived by each of the three basic procedures. In this simple system the direct equilibration obviously would be preferred.

2-3 INFLUENCE OF GRAVITATIONAL FORCES

Consider now the system shown in Fig. 2-2a, which is the system of Fig. 2-1a rotated through 90° so that the force of gravity acts in the direction of the displacement. The system of forces acting on the mass in this case is shown in Fig. 2-2b, and when the expressions of Eqs. (2-2) are used, the equilibrium relationship can be written

$$m\ddot{v} + c\dot{v} + kv = p(t) + W \qquad (2\text{-}10)$$

where W is the weight of the rigid block.

However, if the total displacement v is expressed as the sum of the static displacement Δ_{st} caused by the weight W plus the additional dynamic displacement $\bar{v}$ as shown in Fig. 2-2c

$$v = \Delta_{st} + \bar{v} \qquad (2\text{-}11)$$

then the spring force may be written

$$f_S = kv = k\Delta_{st} + k\bar{v} \qquad (2\text{-}12)$$

Introducing Eq. (2-12) into (2-10)

$$m\ddot{v} + c\dot{v} + k\Delta_{st} + k\bar{v} = p(t) + W \qquad (2\text{-}13)$$

and noting that $k\Delta_{st} = W$ leads to

$$m\ddot{v} + c\dot{v} + k\bar{v} = p(t) \qquad (2\text{-}14)$$

Now by differentiating Eq. (2-11) and noting that Δ_{st} does not vary with time it is evident that $\dot{v} = \dot{\bar{v}}$, etc., so that Eq. (2-14) may be written

$$m\ddot{\bar{v}} + c\dot{\bar{v}} + k\bar{v} = p(t) \qquad (2\text{-}15)$$

Comparison of Eqs. (2-15) and (2-3) demonstrates that the equation of motion

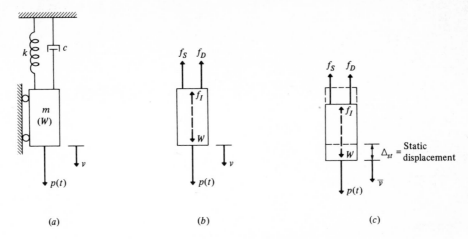

(a)   (b)   (c)

FIGURE 2-2
Influence of gravity on SDOF equilibrium.

expressed with reference to the static-equilibrium position of the dynamic system is not affected by gravity forces. For this reason displacements in all future discussions will be referenced from the static position, and the displacements which are determined will be the *dynamic* response. Therefore total deflections, stresses, etc., can be obtained only by adding the appropriate static quantities to the results of the dynamic analysis.

2-4   INFLUENCE OF SUPPORT EXCITATION

Dynamic stresses and deflections may be induced in a structure not only by a time-varying applied load, as indicated in Figs. 2-1 and 2-2, but also by motions of its support points. Important examples of such excitation are the motions of a building foundation caused by an earthquake or motions of the base of a piece of equipment due to vibrations of the building in which it is housed. A simplified model of the earthquake-excitation problem is shown in Fig. 2-3, in which the horizontal ground motion caused by the earthquake is indicated by the displacement v_g of the structure's base relative to the fixed reference axis.

The horizontal girder in this frame is assumed to be rigid and to include all the moving mass of the structure. The vertical columns are assumed to be weightless and inextensible in the vertical (axial) direction, and the resistance to girder displacement provided by each column is represented by its spring constant $k/2$. The mass thus has a single degree of freedom, v, which is associated with column flexure; the damper c provides a velocity-proportional resistance to this deformation.

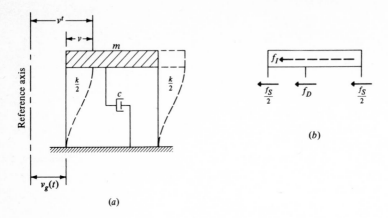

FIGURE 2-3
Influence of support excitation on SDOF equilibrium: (a) motion of system;
(b) equilibrium forces.

As shown in Fig. 2-3b, the equilibrium of forces for this system may be written

$$f_I + f_D + f_S = 0 \qquad (2\text{-}16)$$

in which the damping and elastic forces may be expressed as in Eqs. (2-2). However, the inertia force in this case is given by

$$f_I = m\ddot{v}_t \qquad (2\text{-}17)$$

where v_t represents the *total* displacement of the mass from the reference axis. Substituting for the inertia, damping, and elastic forces in Eq. (2-16) yields

$$m\ddot{v}^t + c\dot{v} + kv = 0 \qquad (2\text{-}18)$$

Before this equation can be solved all forces must be expressed in terms of a single variable, which can be accomplished by noting that the total motion of the mass can be expressed as the sum of the ground motion and the column distortion, i.e.,

$$v^t = v + v_g \qquad (2\text{-}19)$$

Expressing the inertia force in terms of the two acceleration components obtained by differentiation of Eq. (2-19) and substituting into Eq. (2-18) yields

$$m\ddot{v} + m\ddot{v}_g + c\dot{v} + kv = 0 \qquad (2\text{-}20)$$

or since the ground acceleration represents the specified dynamic input to the structure, the equation of motion may conveniently be written

$$m\ddot{v} + c\dot{v} + kv = -m\ddot{v}_g(t) \equiv p_{\text{eff}}(t) \qquad (2\text{-}21)$$

In this equation, $p_{\text{eff}}(t)$ denotes the effective support excitation loading; in other words, the structure responds to the ground acceleration $\ddot{v}_g(t)$ exactly as it would to an external load $p(t)$ equal to the product of the mass and the ground acceleration.

The negative sign in Eq. (2-21) indicates that the effective force opposes the direction of the ground acceleration; in practice this has little significance, inasmuch as the base input generally must be assumed to act in an arbitrary direction.

2-5 GENERALIZED SDOF SYSTEMS: RIGID-BODY ASSEMBLAGE

All the cases considered up to this point have been extremely simple because each of the physical properties—mass, damping, and elasticity—was represented by a single discrete element. However, the analysis of most real systems requires the use of more complicated idealizations, even when they can be considered to be SDOF structures. For the purposes of this discussion, it will be convenient to identify two classes of generalized SDOF structures: (1) assemblages of rigid bodies in which the elastic deformations are limited entirely to localized spring elements and (2) systems having distributed elasticity in which the deformations may be continuous throughout the structure or within some of its components. In both cases, the structure is forced to behave like a SDOF system by the assumption that displacements of only a single form or shape are permitted. For the rigid-body-assemblage class of structure discussed in this section, the limitation to a single displacement shape frequently is a consequence of the assemblage configuration; i.e., the rigid bodies are constrained by supports and hinges so that only one type of displacement is possible. In structures with distributed elasticity, considered in Sec. 2-7, the SDOF shape restriction is merely an assumption; the distributed elasticity actually permits an infinite variety of displacements to take place.

In formulating the equations of motion of the rigid-body assemblage, the elastic forces developed during the SDOF displacements can be expressed easily in terms of the displacement amplitude because each elastic element is a discrete spring subjected to a specified deformation. Similarly the damping forces can be expressed in terms of the specified velocities of the attachment points of the discrete dampers. On the other hand, the mass of the rigid bodies need not be localized, and distributed inertia forces generally will result from the assumed accelerations. However, for the purposes of dynamic analysis, it usually is most effective to treat the rigid-body inertia forces as though the mass and the mass moment of inertia were concentrated at the center of mass. The inertia-force resultants which are obtained thereby are entirely equivalent to the distributed inertia forces insofar as the assemblage behavior is concerned. (Similarly it is desirable to represent any distributed external loads acting on the rigid bodies by their force resultants.) The mass and mass moment of inertia of a uniform rod and of uniform plates of various shapes are summarized in Fig. 2-4 for convenient reference.

Uniform rod

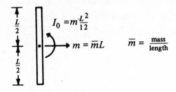

$$I_0 = m\frac{L^2}{12}$$

$m = \bar{m}L$ $\qquad \bar{m} = \frac{\text{mass}}{\text{length}}$

Uniform plates

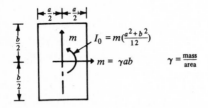

$$I_0 = m(\frac{a^2+b^2}{12})$$

$m = \gamma ab$ $\qquad \gamma = \frac{\text{mass}}{\text{area}}$

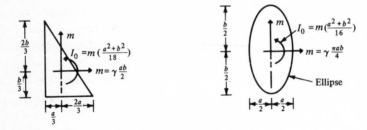

$$I_0 = m(\frac{a^2+b^2}{18})$$

$m = \gamma\frac{ab}{2}$

$$I_0 = m(\frac{a^2+b^2}{16})$$

$m = \gamma\frac{\pi ab}{4}$

Ellipse

FIGURE 2-4
Rigid-body mass and mass moment of inertia.

EXAMPLE E2-1   A representative example of a rigid-body assemblage, shown in Fig. E2-1, consists of two rigid bars connected by a hinge at B and supported by a pivot at A and a roller at C. Dynamic excitation is provided by a transverse load $p(x,t)$ varying linearly along the length of bar AB. In addition, a constant axial force N acts through the system, and its motion is constrained by discrete springs and dampers located as shown along the length of the bars. The mass is distributed uniformly through bar AB, and the weightless bar BC supports a point mass m_2.

Because the two bars are assumed rigid, this system has only a single degree of freedom, and its dynamic response can be expressed with a single equation of motion. This equation could be formulated by direct equilibration (the reader may find this a worthwhile exercise), but because of the complexity

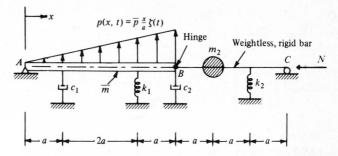

Example of a rigid-body-assemblage SDOF system.

of the system it is more convenient to use a work or energy formulation. A virtual-work analysis will be employed here, although Hamilton's principle would be equally effective.

For the form of displacement which may take place in this SDOF structure (Fig. E2-2) the hinge motion $Z(t)$ may be taken as the basic quantity and all other displacements expressed in terms of it; for example, $DD' = Z/4$, $EE' = 3Z/4$, $FF' = 2Z/3$, etc. The forces acting in the system (exclusive of the axial force N, which will be discussed later) are also shown in the figure. Each resisting force can be expressed in terms of Z or its time derivatives, as follows:

$$f_{S1} = k_1(EE') = k_1{}^3/_4 Z(t)$$

$$f_{S2} = k_2(GG') = k_2{}^1/_3 Z(t)$$

$$f_{D1} = c_1\left(\frac{d}{dt} DD'\right) = c_1{}^1/_4 \dot{Z}(t)$$

$$f_{D2} = c_2 \dot{Z}(t)$$

$$f_{I1} = m_1{}^1/_2 \ddot{Z}(t) = \overline{m}L^1/_2 \ddot{Z}(t) = 2a\overline{m}\ddot{Z}(t)$$

$$\mathfrak{M}_{I1} = I_0 \frac{1}{4a} \ddot{Z}(t) = \frac{\overline{m}L}{4a} \frac{L^2}{12} \ddot{Z}(t) = {}^4/_3 a^2 \overline{m}\ddot{Z}(t)$$

$$f_{I2} = m_2{}^2/_3 \ddot{Z}(t)$$

The externally applied load resultant is

$$p_1 = 8\overline{p}a\zeta(t)$$

In these expressions, $\overline{m}$ (or $\overline{p}$) denotes a reference value of mass (or force) per unit length and $\zeta(t)$ represents the dynamic load variation.

The equation of motion of this system may be established by equating to zero all work done by these forces during an arbitrary virtual displacement δZ. The virtual displacement through which the force components move is

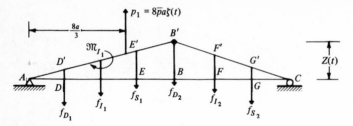

FIGURE E2-2
SDOF displacements and resultant forces.

proportional to Z, as indicated in Fig. E2-2. Thus the total virtual work may be written

$$\delta W = -k_1{}^3/_4 Z(t)^3/_4 \delta Z - k_2 \frac{Z(t)}{3} \frac{\delta Z}{3} - c_1 \frac{\dot{Z}(t)}{4} \frac{\delta Z}{4} - c_2 \dot{Z}(t)\, \delta Z$$

$$- 2a\bar{m}\ddot{Z}(t)\frac{\delta Z}{2} - {}^4/_3 a^2 \bar{m}\ddot{Z}(t)\frac{\delta Z}{4a} - m_2 \frac{2\ddot{Z}(t)}{3}{}^2/_3 \delta Z + 8\bar{p}a\zeta(t)^2/_3 \delta Z \qquad (a)$$

which when simplified becomes

$$\left[\left(a\bar{m} + \frac{a\bar{m}}{3} + {}^4/_9 m_2\right)\ddot{Z}(t) + \left(\frac{c_1}{16} + c_2\right)\dot{Z}(t)\right.$$

$$\left. + \left({}^9/_{16}k_1 + \frac{k_2}{9}\right)Z(t) - {}^{16}/_3\bar{p}a\zeta(t)\right]\delta Z = 0 \qquad (b)$$

Because the virtual displacement δZ is arbitrary, the term in square brackets must vanish; thus the final equation of motion becomes

$$({}^4/_3\bar{m}a + {}^4/_9 m_2)\ddot{Z}(t) + \left(\frac{c_1}{16} + c_2\right)\dot{Z}(t) + \left({}^9/_{16}k_1 + \frac{k_2}{9}\right)Z(t) = {}^{16}/_3\bar{p}a\zeta(t) \quad (c)$$

This may be written in the simplified form

$$m^*\ddot{Z}(t) + c^*\dot{Z}(t) + k^*Z(t) = p^*(t) \qquad \text{(2-22)}$$

if the new symbols are defined as

$$m^* = {}^4/_3\bar{m}a + {}^4/_9 m_2 \qquad c^* = {}^1/_{16}c_1 + c_2$$

$$k^* = {}^9/_{16}k_1 + {}^1/_9 k_2 \qquad p^*(t) = {}^{16}/_3\bar{p}a\zeta(t)$$

termed, respectively, the *generalized mass, damping, stiffness,* and *load* for this system; they have been evaluated with reference to the generalized coordinate $Z(t)$, which has been used here to define the displacements of the system.

Consider now the axial force N of Fig. E2-1. As may be seen in Fig. E2-3, the virtual work done by this force during the virtual displacement δZ is $N\,\delta e$.

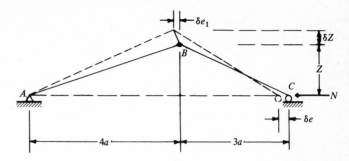

FIGURE E2-3
Displacement components in the direction of axial force.

The displacement δe is made up of two parts, δe_1 and δe_2, associated with the rotations of the two bars. Considering the influence of bar AB only, it is clear from the similar triangles shown in the figure (and assuming small deflections) that $\delta e_1 = (Z/4a)\,\delta Z$. Similarly $\delta e_2 = (Z/3a)\,\delta Z$, thus the total displacement is

$$\delta e_1 + \delta e_2 = \frac{7}{12}\frac{Z}{a}\,\delta Z$$

and the virtual work done by the axial force N is

$$\delta W_N = \frac{7}{12}\frac{NZ}{a}\,\delta Z \qquad (d)$$

Introducing Eq. (d) into Eq. (a) and carrying out simplifying operations similar to those which led to Eq. (c) show that only one term in the equation of motion is influenced by the axial force, the generalized stiffness. When the effect of the axial force in this system is included, the combined generalized stiffness $\bar{k}^*$ is reduced to

$$\bar{k}^* = k^* - \frac{7}{12}\frac{N}{a} = {}^{9}/_{16}k_1 + {}^{1}/_{9}k_2 - \frac{7}{12}\frac{N}{a} \qquad (e)$$

With this modified generalized-stiffness term, the equation of motion of the complete system of Fig. E2-4, including axial force, is given by an equation similar to Eq. (2-22).

It is of interest to note that the condition of zero generalized stiffness represents a neutral stability or critical buckling condition in the system. The value of axial force N_{cr} which would cause buckling of this structure thus can be found by equating $\bar{k}^*$ of Eq. (e) to zero:

$$0 = {}^{9}/_{16}k_1 + {}^{1}/_{9}k_2 - \frac{7}{12}\frac{N_{cr}}{a}$$

Thus
$$N_{cr} = ({}^{27}/_{28}k_1 + {}^{4}/_{21}k_2)a \qquad (f)$$

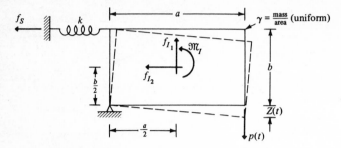

FIGURE E2-4
SDOF plate with dynamic forces.

In general, compressive axial forces tend to reduce the stiffness of a structural system, while tensile axial forces cause a corresponding increase of stiffness. Such loads can have a significant effect on the response of the structure to dynamic loads, and the resulting change of stiffness should always be evaluated to determine its importance in the given problem. It should be noted that *axial force* in this and in subsequent discussions refers to a force which acts parallel to the initial *undistorted* axis of the member; such a force is assumed not to change the direction of its line of action with the motion of the structure. ////

EXAMPLE E2-2 As a second example of the formulation of the equations of motion for a rigid-body assemblage, the system shown in Fig. E2-4 will be considered. The small-amplitude motion of this system can be characterized by the vertical displacement of the load point $Z(t)$, and all the system forces resisting this motion can be expressed in terms of it:

$$f_S = k\,\frac{b}{a}\,Z(t) \qquad\qquad f_{I1} = \gamma a b^1/_2 \ddot{Z}(t)$$

$$f_{I2} = \gamma a b\,\frac{b}{2a}\,\ddot{Z}(t) \qquad \mathfrak{M}_I = \gamma a b\,\frac{a^2 + b^2}{12}\,\frac{1}{a}\,\ddot{Z}(t)$$

The equation of motion for this simple system can be written directly by expressing the equilibrium of moments about the hinge:

$$f_S b + f_{I1}\,\frac{a}{2} + f_{I2}\,\frac{b}{2} + \mathfrak{M}_I = p(t)a$$

or dividing by the length a and substituting the above expressions for the forces:

$$\gamma a b\left[\frac{1}{12}\left(\frac{b^2}{a^2} + 1\right) + \frac{1}{4} + \frac{b^2}{4a^2}\right]\ddot{Z}(t) + k\,\frac{b^2}{a^2}\,Z(t) = p(t)$$

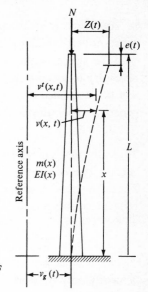

FIGURE 2-5
General structure treated as a SDOF system.

Finally, this may be written

$$m^*\ddot{Z}(t) + k^*Z(t) = p^*(t)$$

in which

$$m^* = \frac{\gamma ab}{3}\left(1 + \frac{b^2}{a^2}\right)$$

$$k^* = k\,\frac{b^2}{a^2} \qquad p^*(t) = p(t) \qquad ////$$

2-6 GENERALIZED SDOF SYSTEMS: DISTRIBUTED FLEXIBILITY

The example of Fig. E2-1 is a true SDOF system in spite of the complex interrelationships of its various components because the two rigid bars are supported so that only one type of displacement pattern is possible. If the bars could deform in flexure, the system would have an infinite number of degrees of freedom. A simple SDOF analysis could still be made, however, if it were assumed that only a single deflection pattern could be developed—including an appropriate flexural-deformation component.

As an illustration of this method of approximating SDOF behavior in a system actually having infinite degrees of freedom, consider the formulation of the equations of motion for the cantilever tower of Fig. 2-5. The essential properties of the tower are its flexural stiffness $EI(x)$ and its mass per unit of length $m(x)$. It is assumed to be subjected to an earthquake ground-motion excitation $v_g(t)$, and it supports a constant vertical load N applied at the top.

To approximate the motion of this system with a single degree of freedom, it is necessary to assume that it may deflect only in a single shape. The shape function will be designated $\psi(x)$, and the amplitude of the motion will be represented by the generalized coordinate $Z(t)$; thus

$$v(x,t) = \psi(x)Z(t) \qquad (2\text{-}23)$$

Typically the generalized coordinate is selected as the displacement of some convenient reference point in the system, such as the tip displacement in this tower. Then the shape function is the dimensionless ratio of the local displacement to this reference displacement:

$$\psi(x) = \frac{v(x,t)}{Z(t)} \qquad (2\text{-}24)$$

The equations of motion of this generalized system can be formulated conveniently only by work or energy principles, and Hamilton's principle will be used here to demonstrate the technique (although virtual work is equally applicable). The kinetic energy of the tower is given by

$$T = \frac{1}{2}\int_0^L m(x)[\dot{v}(x,t)]^2\,dx \qquad (2\text{-}25)$$

and the potential energy of the flexural deformations by

$$V_f = \frac{1}{2}\int_0^L EI(x)[v''(x,t)]^2\,dx \qquad (2\text{-}26)$$

in which v'' represents d^2v/dx^2.

To calculate the potential energy of the axial directed force (which is unchanged in direction or amplitude during the dynamic response and therefore is a conservative force) it is necessary to evaluate the vertical component of the motion of the top of the tower $e(t)$. By analogy with the development of Eq. (d) of Example E2-1, it can be shown that

$$e(t) = \frac{1}{2}\int_0^L [v'(x,t)]^2\,dx \qquad (2\text{-}27)$$

Thus the potential energy of the axial load N is given by

$$V_N = -\frac{N}{2}\int_0^L [v'(x,t)]^2\,dx \qquad (2\text{-}28)$$

where the negative sign results because the potential of the force N is reduced by the displacement $e(t)$. It should be noted, in passing, that if the axial force N varied with position along the tower (considering the dead weight of the tower itself, for example), it would be necessary to modify Eq. (2-28) only by including the axial-force expression under the integral sign.

In the system of Fig. 2-5, there are no directly applied dynamic loads, and damping has been neglected; thus there are no nonconservative forces to be considered. Consequently, Hamilton's principle takes the form

$$\int_{t_1}^{t_2} \delta(T - V)\, dt = 0$$

or, after introducing Eqs. (2-25) to (2-28) and performing the indicated variations,

$$\int_{t_1}^{t_2} \left[\int_0^L m(x)\ddot{v}^t(x,t)\, \delta\dot{v}_t\, dx - \int_0^L EI(x)v''(x,t)\, \delta v''\, dx \right. $$
$$\left. + N \int_0^L v'(x,t)\, \delta v'\, dx \right] = 0 \qquad (2\text{-}29)$$

Now noting the relationships

$$\dot{v}^t = \dot{v} + \dot{v}_g \qquad v'' = \psi'' Z \qquad v' = \psi' Z \qquad \dot{v} = \psi \dot{Z}$$
$$\delta\dot{v}^t = \delta\dot{v} \qquad \delta v'' = \psi''\, \delta Z \qquad \delta v' = \psi'\, \delta Z \qquad \delta\dot{v} = \psi\, \delta\dot{Z} \qquad (2\text{-}30)$$

and substituting into Eq. (2-29) leads to

$$\int_{t_1}^{t_2} \left[\dot{Z}\, \delta\dot{Z} \int_0^L m(x)\psi^2\, dx + \delta\dot{Z}\dot{v}_g(t) \int_0^L m(x)\psi\, dx \right.$$
$$\left. - Z\, \delta Z \int_0^L EI(x)(\psi'')^2\, dx + NZ\, \delta Z \int_0^L (\psi')^2\, dx \right] dt = 0 \qquad (2\text{-}31)$$

After integrating the first two terms by parts, this becomes

$$\int_{t_1}^{t_2} \left[m^*\ddot{Z} + k^*Z - k_G^*Z - p_{\text{eff}}^*(t) \right] \delta Z\, dt = 0 \qquad (2\text{-}32)$$

in which

$$m^* = \int_0^L m(x)\psi^2\, dx = \text{generalized mass}$$

$$k^* = \int_0^L EI(x)(\psi'')^2\, dx = \text{generalized stiffness}$$

$$k_G^* = N \int_0^L (\psi')^2\, dx = \text{generalized geometric stiffness} \qquad (2\text{-}33)$$

$$p_{\text{eff}}^*(t) = -\ddot{v}_g \int_0^L m(x)\psi\, dx = \text{generalized effective load}$$

But since the variation δZ is arbitrary, the term in brackets of Eq. (2-32) must vanish; hence the equation of motion finally may be written

$$m^*\ddot{Z}(t) + \bar{k}^*Z(t) = p_{\text{eff}}^*(t) \qquad (2\text{-}34)$$

in which

$$\bar{k}^* = k^* - k_G^* \qquad (2\text{-}35)$$

is the combined generalized stiffness of the system.

The critical buckling load can be calculated for this system by the same method used with Example E2-1: equating to zero the combined generalized stiffness. Thus

$$\bar{k}^* = k^* - k_G^* = \int_0^L EI(x)(\psi'')^2 \, dx - N_{cr} \int_0^L (\psi')^2 \, dx = 0$$

from which

$$N_{cr} = \frac{\int_0^L EI(x)(\psi'')^2 \, dx}{\int_0^L (\psi')^2 \, dx} \qquad (2\text{-}36)$$

This SDOF approximate analysis of the critical buckling load is called *Rayleigh's method*. The value determined for the critical load depends, of course, upon the assumed shape function.

EXAMPLE E2-3   To provide a numerical example of the formulation of the equations of motion for a SDOF system with distributed flexibility, it will be assumed that the tower of Fig. 2-5 has uniform flexural rigidity and uniform mass distribution along its length. Also, its deflected shape in free vibrations will be assumed as

$$\psi(x) = 1 - \cos\frac{\pi x}{2L} \qquad (a)$$

When Eqs. (2-33) are applied, the generalized mass and stiffness of the tower are

$$m^* = \int_0^L \bar{m}(\psi)^2 \, dx = \bar{m} \int_0^L \left(1 - \cos\frac{\pi x}{2L}\right)^2 dx = 0.228\bar{m}L \qquad (b)$$

$$k^* = \int_0^L EI(\psi'')^2 \, dx = EI \int_0^L \left(\frac{\pi^2}{4L^2}\cos\frac{\pi x}{2L}\right)^2 dx = \frac{\pi^4}{32}\frac{EI}{L^3} \qquad (c)$$

Also, if the tower is assumed to be subjected to a base excitation, the generalized force given by Eq. (2-33) is (neglecting the negative sign)

$$p_{\text{eff}}^*(t) = \ddot{v}_g(t) \int_0^L \bar{m}\psi \, dx = \bar{m}\ddot{v}_g(t) \int_0^L \left(1 - \cos\frac{\pi x}{2L}\right) dx = 0.364\bar{m}L\ddot{v}_g(t) \quad (d)$$

Hence, if the axial force is neglected, the equation of motion of the system, given by Eq. (2-34), is

$$0.228\bar{m}L\ddot{Z}(t) + \frac{\pi^4}{32}\frac{EI}{L^3} Z(t) = 0.364\bar{m}L\ddot{v}_g(t) \qquad (e)$$

With the axial force N considered, the generalized geometric stiffness of the tower, from Eq. (2-33), is

$$k_G^* = N \int_0^L (\psi')^2 \, dx = N \int_0^L \left(\frac{\pi}{2L} \sin \frac{\pi x}{2L} \right)^2 dx = \frac{N\pi^2}{8L} \qquad (f)$$

Combining this with Eq. (c) gives the combined generalized stiffness

$$\bar{k}^* = k^* - k_G^* = \frac{\pi^4 EI}{32L^3} - \frac{N\pi^2}{8L} \qquad (g)$$

Therefore the critical buckling load obtained by setting the combined stiffness to zero is, from Eq. (g),

$$N_{cr} = \frac{\pi^4 EI}{32L^3} \frac{8L}{\pi^2} = \frac{\pi^2}{4} \frac{EI}{L^2} \qquad (h)$$

This is the true buckling load for an end-loaded uniform cantilever column, because the assumed shape function of Eq. (a) is the true buckled shape. When Eq. (h) is substituted into Eq. (f), the geometric stiffness can be expressed conveniently as

$$k_G^* = \frac{\pi^4 EI}{32L^3} \frac{N}{N_{cr}} \qquad (i)$$

When this is used to modify Eq. (e), the equation of motion including axial-force effects becomes

$$0.228\bar{m}L\ddot{Z}(t) + \frac{\pi^4 EI}{32L^3} \left(1 - \frac{N}{N_{cr}} \right) Z(t) = 0.364\bar{m}L\ddot{v}_g(t) \qquad (j)$$

Of course, any other shape which satisfies the geometric boundary conditions of the structure might have been assumed for $\psi(x)$. For example, if the deflection were assumed to be of parabolic form

$$\psi(x) = \frac{x^2}{L^2} \qquad (k)$$

the generalized elastic stiffness would become

$$k^* = EI \int_0^L \left(\frac{2}{L^2} \right)^2 dx = \frac{4EI}{L^3}$$

and the generalized geometric stiffness would be

$$k_G^* = N \int_0^L \left(\frac{2x}{L^2} \right)^2 dx = \frac{4}{3} \frac{N}{L}$$

In this case the critical load obtained by setting $k^* = k_G^*$ becomes

$$N_{cr} = \frac{4EI}{L^3} \frac{3L}{4} = \frac{3EI}{L^2} \qquad (l)$$

which is 21 percent higher than the true value of Eq. (h).

As a matter of fact, the assumption of any shape other than the true buckled shape will require that additional external constraints act on the system to maintain its equilibrium. These external constraints represent a stiffening influence on the system; therefore the critical load computed by a Rayleigh analysis using any shape other than the true one must always be greater than the true critical load. Actually, it is apparent that the parabolic shape is not a good assumption for this structure, even though it satisfies the geometric boundary conditions, because the constant curvature of this shape implies that the moment is constant along its length. It is obvious here that the moment must vanish at the top of the column, and an assumed shape having zero curvature at the top would give much better results.   ////

2-7 EXPRESSIONS FOR GENERALIZED SYSTEM PROPERTIES

As might be inferred from the preceding examples, the equations of motion for any SDOF system, no matter how complex, can always be reduced to the form

$$m^*\ddot{Z}(t) + c^*\dot{Z}(t) + \bar{k}^*Z(t) = p^*(t)$$

in which $Z(t)$ is the single generalized coordinate expressing the motion of the system and the symbols with asterisks represent generalized physical properties corresponding to this coordinate. In general, the values of these generalized properties can be determined by application of either Hamilton's principle or the *principle of virtual displacements*. However, standardized forms of these expressions may be derived easily which are very convenient in practice.

Consider an arbitrary one-dimensional system (Fig. 2-6) assumed to displace only in the shape shown in Fig. 2-6a. Thus the displacements may be expressed in terms of the generalized coordinate $Z(t)$ as

$$v(x,t) = \psi(x)Z(t)$$

Then the generalized properties associated with this shape may be expressed as follows. For the mass distribution shown in Fig. 2-6b the generalized mass is

$$m^* = \int_0^L m(x)[\psi(x)]^2 \, dx + \sum m_i \psi_i^2 + \sum I_{0i}(\psi_i')^2 \qquad (2\text{-}37)$$

in which the effect of given rigid-body masses is represented by the summations and

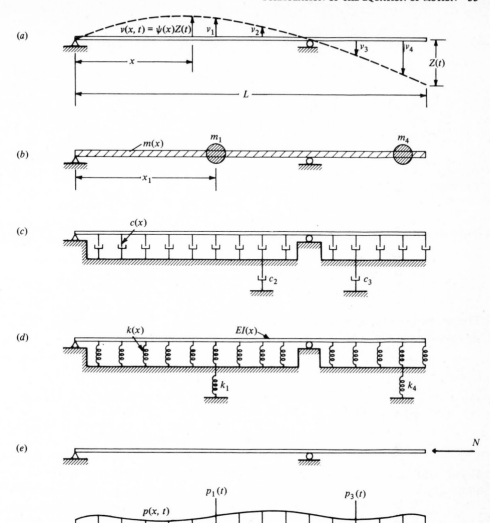

FIGURE 2-6
Properties of generalized SDOF system: (*a*) assumed shape; (*b*) mass properties; (*c*) dampling properties; (*d*) elastic properties; (*e*) axial loading; (*f*) applied loading.

ψ_i' represents the rotation of point i. The generalized damping resulting from the distributed foundation damping plus the local dampers c_i shown in Fig. 2-6c is given by

$$c^* = \int_0^L c(x)[\psi(x)]^2 \, dx + \sum c_i\psi_i^2 \qquad (2\text{-}38)$$

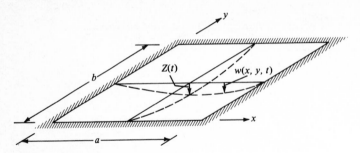

FIGURE 2-7
Two-dimensional slab treated as a SDOF system.

The generalized stiffness due to the elastic foundation, the flexural rigidity, and the local springs of Fig. 2-6d is given by

$$k^* = \int_0^L k(x)[\psi(x)]^2 \, dx + \int_0^L EI(x)[\psi''(x)]^2 \, dx + \sum k_i \psi_i^2 \qquad (2\text{-}39)$$

The geometric-stiffness term due to the axial force N (which does not change with time), shown in Fig. 2-6e, is

$$k_G^* = N \int_0^L [\psi'(x)]^2 \, dx \qquad (2\text{-}40)$$

For a more general case in which the axial force varied with position along the axis the expression would be

$$k_G^* = \int_0^L N(x)[\psi'(x)]^2 \, dx \qquad (2\text{-}40a)$$

Finally, the generalized force associated with the time-varying lateral loading of Fig. 2-6f is

$$p^*(t) = \int_0^L p(x,t)\psi(x) \, dx + \sum p_i \psi_i \qquad (2\text{-}41)$$

The vectorial nature of the force and displacement quantities contained in Eq. (2-41) must be carefully noted. Only displacement components in the direction of the applied loads can be included, and the sense of these must be assigned in accordance with the sense of the loads. In other words, Eq. (2-41) really represents the work done by the loads during a unit displacement of the generalized coordinate $Z(t)$.

The combined generalized stiffness $\bar{k}^*$ is given by $\bar{k}^* = k^* - k_G^*$, as mentioned earlier with Eq. (2-35).

These same generalized-coordinate concepts apply equally in the reduction of two-dimensional systems to a single degree of freedom. Consider, for example, the rectangular floor slab shown in Fig. 2-7. If the deflections of this slab are assumed to have the shape shown, and if the displacement amplitude at the middle is taken as the generalized coordinate, the displacements may be expressed

$$w(x,y,t) = \psi(x,y)Z(t) \qquad (2\text{-}42)$$

For a simply supported slab, the shape function might logically be of the form

$$\psi(x,y) = \sin\frac{\pi x}{a}\sin\frac{\pi y}{b} \qquad (2\text{-}43)$$

but any other reasonable shape consistent with the support conditions could be used.

The generalized properties of the system can then be calculated by expressions equivalent to those presented for the one-dimensional member but with the integrations carried out over the entire surface. For example, the generalized mass would be given by

$$m^* = \int_A m(x,y)[\psi(x,y)]^2\, dA + \sum m_i\psi_i^2 \qquad (2\text{-}44)$$

Corresponding expressions for generalized stiffness and generalized force, respectively, in a uniform plate-type structure are

$$k^* = D\int_A\left[\left(\frac{\partial^2\psi}{\partial x^2}+\frac{\partial^2\psi}{\partial y^2}\right)^2 - 2(1-v)\left(\frac{\partial^2\psi}{\partial x^2}\frac{\partial^2\psi}{\partial y^2}-\frac{\partial^2\psi}{\partial x\,\partial y}\right)\right]dA \qquad (2\text{-}45)$$

$$p^* = \int_A p(x,y)\psi(x,y)\, dA + \sum p_i\psi_i \qquad (2\text{-}46)$$

where $\quad D = Eh^3/12(1-v^2) = $ flexural rigidity of slab
$\quad p(x,y) = $ distributed loading on plate
$\quad v = $ Poisson's ratio
$\quad h = $ plate thickness

It also should be evident that the same procedures can easily be extended to three-dimensional systems, such as soil or concrete masses, by assuming an appropriate displacement function in three dimensions. However, the difficulty of selecting a suitable shape increases rapidly with the number of dimensions of the system, and the reliability of the results is reduced accordingly.

PROBLEMS

2-1 For the system shown in Fig. P2-1, determine the generalized physical properties m^*, c^*, k^*, and the generalized loading $p^*(t)$, all defined with respect to the displacement coordinate $Z(t)$. Express the results in terms of the given physical properties and dimensions.

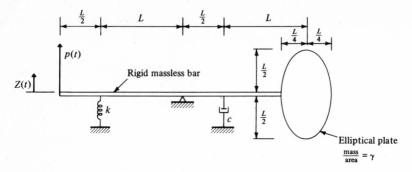

FIGURE P2-1

2-2 Repeat Prob. 2-1 for the structure shown in Fig. P2-2.

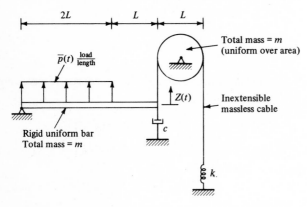

FIGURE P2-2

2-3 Repeat Prob. 2-1 for the structure shown in Fig. P2-3. (*Hint:* this system has only one dynamic degree of freedom because the springs completely control the relative motion of the two rigid bars.)

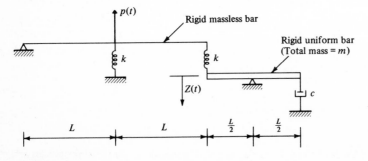

FIGURE P2-3

2-4   The column of Fig. P2-4 is to be treated as a SDOF system by defining its displaced shape as

$$\psi(x) = \frac{v(x,t)}{Z(t)} = \left(\frac{x}{L}\right)^2 \left(\frac{3}{2} - \frac{x}{2L}\right)$$

Denoting the uniformly distributed mass percent length by $\bar{m}$, the uniform stiffness by EI, and the uniformly distributed load per unit length by $\bar{p}(t)$, evaluate the generalized physical properties m^* and k^* and the generalized loading $p^*(t)$.

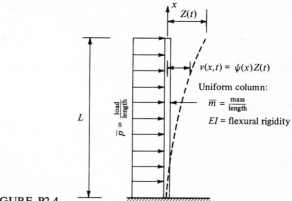

FIGURE P2-4

2-5   (a) If a downward load N is applied at the top of the column of Prob. 2-4, evaluate its combined generalized stiffness $\bar{k}^*$ using the same shape function $\psi(x)$.

(b) Repeat part (a) assuming that the axial force in the column varies linearly along its length as $N(x) = N(1 - x/L)$.

2-6   Assume that the uniform slab of Fig. 2-7 is square, with side length a, and is simply supported on all four edges.

(a) If its mass per unit area is γ and its flexural rigidity is D, determine its generalized properties m^* and k^* in terms of the central displacement coordinate $Z(t)$. Assume the displacement function is

$$\psi(x,y) = \sin\frac{\pi x}{a} \sin\frac{\pi y}{a}$$

(b) The uniformly distributed external loading per unit of area is $\bar{p}(t)$. Determine the generalized loading $p^*(t)$ based on the displacement function of part (a).

2-7   The outer diameters, height, and material properties of a conical concrete smokestack are shown in Fig. P2-5. Assuming a uniform wall thickness of 8 in and that the deflected shape is given by

$$\psi(x) = 1 - \frac{\cos \pi x}{2L}$$

compute the generalized mass m^* and stiffness k^* of the structure. Use Simpson's rule to evaluate the integrals, including in the summations the integrand values for the bottom, middle, and top sections. For example

$$m^* \doteq \frac{\Delta x}{3} (y_0 + 4y_1 + y_2)$$

where $y_i = m_i \psi_i^2$ evaluated at level "i."

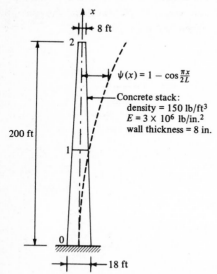

$\psi(x) = 1 - \cos\frac{\pi x}{2L}$

Concrete stack:
density = 150 lb/ft^3
$E = 3 \times 10^6$ lb/in.^{2}
wall thickness = 8 in.

8 ft

200 ft

18 ft

FIGURE P2-5

FREE-VIBRATION RESPONSE

3-1 SOLUTION OF THE EQUATION OF MOTION

It was shown in Chap. 2 that the equations of motion of any SDOF system can be reduced to the form

$$m^*\ddot{Z}(t) + c^*\dot{Z}(t) + \bar{k}^*Z(t) = p^*(t)$$

This is entirely equivalent to the equation of motion of a simple spring-mass system with damping, as shown in Fig. 3-1, which may be written

$$m\ddot{v}(t) + c\dot{v}(t) + kv(t) = p(t) \qquad (3\text{-}1)$$

Thus in the present discussion it will be convenient to use Eq. (3-1) and to visualize the response of this simple system. However, it should always be remembered that these results apply equally to the generalized-coordinate response of any complex system which has been represented as a SDOF system.

The solution of Eq. (3-1) will be obtained by considering first the homogeneous equation with the right side set equal to zero:

$$m\ddot{v}(t) + c\dot{v}(t) + kv(t) = 0 \qquad (3\text{-}2)$$

Motions taking place with the applied force set equal to zero are called *free vibrations*, and it is the free-vibration response of the system which we now wish to examine.

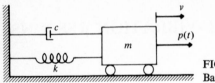

FIGURE 3-1
Basic SDOF system.

The solution of Eq. (3-2) is of the form

$$v(t) = Ge^{st} \qquad (3\text{-}3)$$

Substituting this into Eq. (3-2) leads to

$$(ms^2 + cs + k)Ge^{st} = 0 \qquad (3\text{-}4)$$

After dividing by mGe^{st} and introducing the notation

$$\omega^2 = \frac{k}{m} \qquad (3\text{-}5)$$

Eq. (3-4) becomes

$$s^2 + \frac{c}{m} s + \omega^2 = 0 \qquad (3\text{-}6)$$

The value of s which can be derived from this expression depends on the value of c; thus the type of motion represented by Eq. (3-3) will depend on the damping in the system.

3-2   UNDAMPED FREE VIBRATIONS

If the system is undamped, i.e., if $c = 0$, it is evident that the value of s given by Eq. (3-6) is

$$s = \pm i\omega \qquad (3\text{-}7)$$

Thus the response given by Eq. (3-3) is

$$v(t) = G_1 e^{i\omega t} + G_2 e^{-i\omega t} \qquad (3\text{-}8)$$

in which the two terms result from the two values of s and the constants G_1 and G_2 represent the (as yet) arbitrary amplitudes of the motion. Equation (3-8) can be put in a more convenient form by introducing Euler's equations

$$e^{\pm i\omega t} = \cos \omega t \pm i \sin \omega t \qquad (3\text{-}9)$$

The result may be written

$$v(t) = A \sin \omega t + B \cos \omega t \qquad (3\text{-}10)$$

in which the constants A and B may be expressed in terms of the initial conditions, i.e., the displacement $v(0)$ and velocity $\dot{v}(0)$ at time $t = 0$, which initiated the free

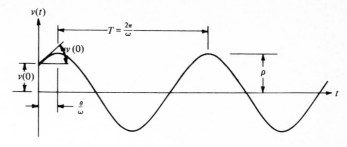

FIGURE 3-2
Undamped free-vibration response.

vibrations of the system. It is easily seen that $v(0) = B$ and $\dot{v}(0) = A\omega$; thus Eq. (3-10) becomes

$$v(t) = \frac{\dot{v}(0)}{\omega} \sin \omega t + v(0) \cos \omega t \qquad (3\text{-}11)$$

This solution represents a simple harmonic motion (SHM) and is portrayed graphically in Fig. 3-2. The quantity ω is the circular frequency or angular velocity of the motion; it is measured in radians per unit of time. The cyclic frequency f, which is usually referred to merely as the frequency of the motion, is given by

$$f = \frac{\omega}{2\pi} \qquad (3\text{-}12)$$

and its reciprocal is called the period T,

$$T = \frac{2\pi}{\omega} = \frac{1}{f} \qquad (3\text{-}13)$$

The motion represented by Eq. (3-11) also can be expressed in the form

$$v(t) = \rho \cos (\omega t - \theta) \qquad (3\text{-}14)$$

as may be noted from the Argand diagram or vector representation of Fig. 3-3. The response is given by the real part, or horizontal projection, of the two rotating vectors. Thus the amplitude of motion is given by the resultant

$$\rho = \sqrt{[v(0)]^2 + \left[\frac{\dot{v}(0)}{\omega}\right]^2} \qquad (3\text{-}15)$$

and the phase angle by

$$\theta = \tan^{-1} \frac{\dot{v}(0)}{\omega v(0)} \qquad (3\text{-}16)$$

It will be noted in Fig. 3-3 that the phase angle θ represents the angular distance by which the resultant motion lags behind the cosine term in the response.

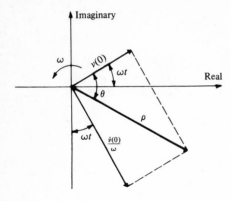

FIGURE 3-3
Rotating-vector representation of free
vibrations.

3-3 DAMPED FREE VIBRATIONS

If damping is present in the system, the solution of Eq. (3-6) which defines the response
is

$$s = -\frac{c}{2m} \pm \sqrt{\left(\frac{c}{2m}\right)^2 - \omega^2} \qquad (3\text{-}17)$$

Three types of motion are represented by this expression, according as the quantity
under the square-root sign is positive, negative, or zero. It is convenient to discuss
first the limiting case, when the radical vanishes; this is called the *critical-damping
condition*.

Critical Damping

If the radical in Eq. (3-17) is set equal to zero, it is evident that $c/2m = \omega$; thus the
critical damping value c_c is

$$c_c = 2m\omega \qquad (3\text{-}18)$$

Then the value of s in Eq. (3-17) is

$$s = -\frac{c}{2m} = -\omega \qquad (3\text{-}19)$$

and the response given by Eq. (3-3) is

$$v(t) = (G_1 + G_2 t)e^{-\omega t} \qquad (3\text{-}20)$$

in which the second term is multiplied by t because only a single value of s is available
in the solution, Eq. (3-19).

Introducing the initial conditions in Eq. (3-20) leads to the final form of the
critically damped response equation

$$v(t) = [v(0)(1 + \omega t) + \dot{v}(0)t]e^{-\omega t} \qquad (3\text{-}21)$$

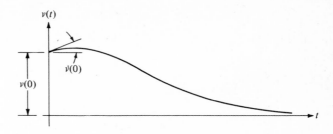

FIGURE 3-4
Free-vibration response with critical damping.

which is portrayed graphically in Fig. 3-4. It will be noted that the free response of a critically damped system does not include oscillation about the zero-deflection position; instead the displacement returns to zero in accordance with the exponential decay term of Eq. (3-21). One useful definition of the critically damped condition is that it is the smallest amount of damping for which no oscillation occurs in the free response.

Underdamped Systems

If the damping is less than critical, it is evident from Eq. (3-18) that $c < 2m\omega$ and thus that the radical in Eq. (3-17) must be negative. To evaluate the free-vibration response in this case, it is convenient to express the damping as a ratio ξ to the critical damping value; thus

$$\xi = \frac{c}{c_c} = \frac{c}{2m\omega} \qquad (3\text{-}22)$$

in which ξ is called the *damping ratio*. Introducing Eq. (3-22) into Eq. (3-17) leads to

$$s = -\xi\omega \pm \sqrt{(\xi\omega)^2 - \omega^2}$$

or changing the sign of the radical and introducing a new symbol ω_D gives

$$s = -\xi\omega \pm i\omega_D \qquad (3\text{-}23)$$

where

$$\omega_D = \omega\sqrt{1 - \xi^2} \qquad (3\text{-}24)$$

The quantity ω_D is called the *damped vibration frequency*; for damping ratios to be expected in typical structural systems ($\xi < 20$ percent) it differs very little from the undamped frequency, as may be noted in Eq. (3-24). To estimate the influence of damping on frequency, it is convenient to remember that a plot of the ratio of damped to undamped frequency ω_D/ω vs. the damping ratio ξ is a circle of unit radius, as shown in Fig. 3-5.

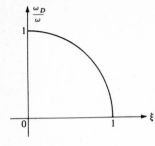

FIGURE 3-5
Relationship between damped frequency
and damping ratio.

The free-vibration response of an underdamped system can be evaluated by substituting Eq. (3-23) into Eq. (3-3); thus

$$v(t) = G_1 e^{-\xi\omega t + i\omega_D t} + G_2 e^{-\xi\omega t - i\omega_D t} = e^{-\xi\omega t}(G_1 e^{i\omega_D t} + G_2 e^{-i\omega_D t})$$

The term in parentheses represents simple harmonic motion [compare with Eq. (3-8)]; thus this expression can be written more conveniently as

$$v(t) = e^{-\xi\omega t}(A \sin \omega_D t + B \cos \omega_D t) \qquad (3\text{-}25)$$

Finally, when the initial conditions $v(0)$ and $\dot{v}(0)$ are introduced, the constants of Eq. (3-25) can be evaluated, giving

$$v(t) = e^{-\xi\omega t}\left[\frac{\dot{v}(0) + v(0)\xi\omega}{\omega_D} \sin \omega_D t + v(0) \cos \omega_D t\right] \qquad (3\text{-}26)$$

Alternatively, this response expression can be written in rotating-vector form:

$$v(t) = \rho e^{-\xi\omega t} \cos (\omega_D t - \theta) \qquad (3\text{-}27)$$

in which

$$\rho = \left\{\left[\frac{\dot{v}(0) + v(0)\xi\omega}{\omega_D}\right]^2 + [v(0)]^2\right\}^{1/2}$$

$$\theta = \tan^{-1} \frac{\dot{v}(0) + v(0)\xi\omega}{\omega_D v(0)} \qquad (3\text{-}28)$$

A plot of the response of an underdamped system to an initial displacement $v(0)$ but starting with zero velocity $[\dot{v}(0) = 0]$ (in other words releasing the mass from a stationary displaced position) is shown in Fig. 3-6. It is of interest to note that the underdamped system oscillates about the neutral position, with a constant circular frequency ω_D. The rotating-vector representation of Eq. (3-27) is equivalent to Fig. 3-3 except that the length of the vector diminishes exponentially as the response damps out.

The true damping characteristics of typical structural systems are very complex and difficult to define. However, it is common practice to express the damping of such

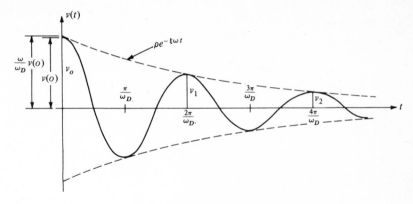

FIGURE 3-6
Free-vibration response of underdamped system.

real systems in terms of equivalent viscous-damping ratios ξ which show similar decay rates under free-vibration conditions. Therefore, let us now relate more fully the viscous-damping ratio ξ to the free-vibration response shown in Fig. 3-6.

Consider any two successive positive peaks shown in Figure 3-6, that is, v_n and v_{n+1}. From Eq. (3-27), the ratio of these two successive values is given by

$$\frac{v_n}{v_{n+1}} = \exp\left(2\pi\xi \frac{\omega}{\omega_D}\right) \qquad (3\text{-}29)$$

Taking the natural logarithm (ln) of both sides of Eq. (3-29) gives the logarithmic decrement δ

$$\delta \equiv \ln \frac{v_n}{v_{n+1}} = 2\pi\xi \frac{\omega}{\omega_D} \qquad (3\text{-}30)$$

or, with Eq. (3-24),

$$\delta = \frac{2\pi\xi}{\sqrt{1-\xi^2}} \qquad (3\text{-}31)$$

For low damping, Eq. (3-31) can be approximated by

$$\delta \doteq 2\pi\xi \qquad (3\text{-}32)$$

where the symbol $\doteq$ represents "approximately equal."

In such cases, Eq. (3-29) can be written as a series expansion

$$\frac{v_n}{v_{n+1}} = e^\delta \doteq e^{2\pi\xi} = 1 + 2\pi\xi + \frac{(2\pi\xi)^2}{2!} + \cdots \qquad (3\text{-}33)$$

For low values of ξ sufficient accuracy can be obtained by retaining only the first two terms in the series, in which case

$$\xi \doteq \frac{v_n - v_{n+1}}{2\pi v_{n+1}} \qquad (3\text{-}34)$$

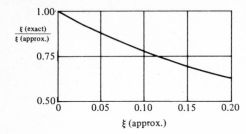

FIGURE 3-7
Damping-ratio correction factor [to be applied to result obtained from Eq. (3-34)].

To illustrate the accuracy of Eq. (3-34), the ratio of the exact value of ξ as given by Eq. (3-30) to the approximate value as given by Eq. (3-34) is plotted vs. the approximate value in Fig. 3-7. This graph permits one to correct the damping ratio obtained by the approximate method.

For lightly damped systems, greater accuracy can be obtained in evaluating the damping ratio by considering response peaks which are several cycles apart, say m cycles; then

$$\ln \frac{v_n}{v_{n+m}} = 2m\pi\xi \frac{\omega}{\omega_D} \qquad (3\text{-}35)$$

which can be simplified for very low damping to the approximate relation

$$\xi = \frac{v_n - v_{n+m}}{2m\pi v_{n+m}} \qquad (3\text{-}36)$$

When one is observing damped free vibrations experimentally, a convenient method for estimating the damping ratio is to count the number of cycles required to give a 50 percent reduction in amplitude. The relationship to be used in this case is presented graphically in Fig. 3-8. As a quick rule of thumb it is convenient to remember that for 10 percent critical damping, the amplitude is reduced by 50 percent in 1 cycle.

Overdamped Systems

Although structural systems having greater than critical damping are not encountered in normal conditions, it is useful to carry out the response analysis of an overdamped system to make this discussion complete. In this case $\xi > 1$, and Eq. (3-17) may be written

$$s = -\xi\omega \pm \omega\sqrt{\xi^2 - 1} = -\xi\omega \pm \hat{\omega} \qquad (3\text{-}37)$$

in which

$$\hat{\omega} = \omega\sqrt{\xi^2 - 1}$$

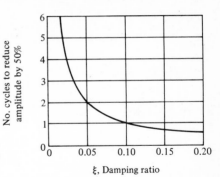

FIGURE 3-8
Damping ratio vs. number of cycles required to reduce amplitude by 50 percent.

Substituting Eq. (3-37) into Eq. (3-3) and simplifying leads eventually to

$$v(t) = e^{-\xi\omega t}(A \sinh \hat{\omega} t + B \cosh \hat{\omega} t) \qquad (3\text{-}38)$$

in which the constants A and B could be evaluated by consideration of the initial conditions. It will be noted from the form of Eq. (3-38) that the response of an over-damped system is not oscillatory; it is similar to the motion of the critically damped system of Fig. 3-3, but the return toward the neutral position is slowed as the damping ratio is increased.

EXAMPLE E3-1   A one-story building is idealized as a rigid girder supported by weightless columns, as shown in Fig. E3-1. In order to evaluate the dynamic properties of this structure, a free-vibration test is made, in which the roof system (rigid girder) is displaced laterally by a hydraulic jack and then released. During the jacking operation, it is observed that a force of 20 kips is required to displace the girder 0.20 in. After the instantaneous release of this initial displacement, the maximum displacement on the return swing is only 0.16 in and the period of this displacement cycle is $T = 1.40$ s.

From these data, the following dynamic behavior properties are to be determined:

1 Effective weight of the girder:

$$T = \frac{2\pi}{\omega} = 2\pi \sqrt{\frac{W}{kg}} = 1.40 \text{ s}$$

Hence $\qquad W = \left(\frac{1.40}{2\pi}\right)^2 kg = 0.0496 \frac{20}{0.2} 386 = 1{,}920 \text{ kips}$

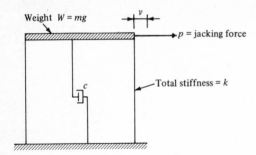

Weight $W = mg$

p = jacking force

Total stiffness = k

c

FIGURE E3-1
Vibration test of a simple building.

2   Frequency of vibration:

$$f = \frac{1}{T} = \frac{1}{1.40} = 0.714 \text{ Hz}$$

$$\omega = 2\pi f = 4.48 \text{ rad/s}$$

3   Damping properties:

Logarithmic decrement: $\delta = \ln \dfrac{0.20}{0.16} = 0.223$

Damping ratio: $\xi \doteq \dfrac{\delta}{2\pi} = 3.55\%$

Damping coefficient: $c = \xi c_c = \xi 2m\omega = 0.0355 \dfrac{2(1,920)}{386} 4.48$

$$= 1.584 \text{ kips/in} \cdot \text{s}$$

Damped frequency: $\omega_D = \omega\sqrt{1 - \xi^2} = \omega(0.999)^{1/2} \doteq \omega$

4   Amplitude after six cycles:

$$v_6 = \left(\frac{v_1}{v_0}\right)^6 v_0 = (^4/_5)^6(0.20) = 0.0524 \text{ in} \qquad ////$$

PROBLEMS

3-1   For the uniform cantilever tower of Example 2-3, the following expressions for the generalized mass and stiffness were determined:

$$m^* = 0.228\bar{m}L$$

$$k^* = \frac{\pi^4}{32} \frac{EI}{L^3}$$

Based on these expressions, compute the period of vibration for a concrete tower 200 ft high, with an outside diameter of 12 ft and wall thickness of 8 in, for which the following properties may be assumed:

$$\bar{m} = 110 \frac{\text{lb}}{\text{ft}^2} \text{sec}^2$$

$$EI = 165 \times 10^9 \text{ lb-ft}^2$$

3-2 Assuming that the tower of Prob. 3-1 supports an additional point weight of 400 kips at the top, determine the period of vibration (neglecting the geometric stiffness effect).

3-3 The weight W of the building of Fig. E3-1 is 200 kips and the building is set into free vibration by releasing it (at time $t = 0$) from a displacement of 1.20 in. If the maximum displacement on the return swing is 0.86 in at time $t = 0.64$ s, determine:
(a) the lateral spring stiffness k
(b) the damping ratio ξ
(c) the damping coefficient c

3-4 Assume that the mass and stiffness of the structure of Fig. 3-1 are as follows: $m = 2k \cdot s^2/\text{in}$, $k = 40k/\text{in}$. If the system is set into free vibration with the initial conditions $v(0) = 0.7$ in and $\dot{v}(0) = 5.6$ in/s, determine the displacement and velocity at $t = 1.0$ s, assuming:
(a) $c = 0$ (undamped system)
(b) $c = 2.8k \cdot s/\text{in}$

3-5 Assume that the mass and stiffness of the system of Fig. 3-1 are $m = 5k \cdot s^2/\text{in}$ and $k = 20k/\text{in}$, and that it is undamped. If the initial displacement is $v(0) = 1.8$ in, and the displacement at $t = 1.2$ s is also 1.8 in, determine:
(a) the displacement at $t = 2.4$ s
(b) the amplitude of free vibration ρ

4

RESPONSE TO HARMONIC LOADING

4-1 UNDAMPED SYSTEM

Complementary Solution

It will now be assumed that the system of Fig. 3-1 is subjected to a harmonically varying load $p(t)$ of amplitude p_0 and circular frequency $\bar{\omega}$. In this case the differential equation of motion becomes

$$m\ddot{v}(t) + c\dot{v}(t) + kv(t) = p_0 \sin \bar{\omega}t \qquad (4\text{-}1)$$

Before considering the general case with damping, it is instructive to examine the response of an undamped system to harmonic loading, for which the equation of motion becomes

$$m\ddot{v}(t) + kv(t) = p_0 \sin \bar{\omega}t \qquad (4\text{-}2)$$

The complementary solution of this equation is the free-vibration response of Eq. (3-10)

$$v_c(t) = A \sin \omega t + B \cos \omega t \qquad (4\text{-}3)$$

Particular Solution

The general solution includes also the particular solution, i.e., the specific behavior generated by the form of the dynamic loading. The response to the harmonic loading can be assumed to be harmonic and in phase with the loading; thus

$$v_p(t) = G \sin \bar{\omega}t \qquad (4\text{-}4)$$

in which the amplitude G is to be evaluated. Substituting Eq. (4-4) into Eq. (4-2) leads to

$$-m\bar{\omega}^2 G \sin \bar{\omega}t + kG \sin \bar{\omega}t = p_0 \sin \bar{\omega}t \qquad (4\text{-}5)$$

Dividing through by $\sin \bar{\omega}t$ (which is nonzero in general) and by k and noting that $k/m = \omega^2$ results (after some rearrangement) in

$$G\left(1 - \frac{\bar{\omega}^2}{\omega^2}\right) = \frac{p_0}{k} \qquad (4\text{-}6)$$

The amplitude of the response therefore becomes

$$G = \frac{p_0}{k} \frac{1}{1 - \beta^2} \qquad (4\text{-}7)$$

in which β represents the ratio of the applied load frequency to the natural free-vibration frequency; i.e.,

$$\beta \equiv \frac{\bar{\omega}}{\omega} \qquad (4\text{-}8)$$

General Solution

The general solution to the harmonic excitation of the undamped system is then given by the combination of the complementary solution and the particular solution, in which the value of G is given by Eq. (4-7); thus

$$v(t) = v_c(t) + v_p(t) = A \sin \omega t + B \cos \omega t + \frac{p_0}{k} \frac{1}{1 - \beta^2} \sin \bar{\omega}t \qquad (4\text{-}9)$$

In this equation, the values of A and B depend still on the conditions with which the response was initiated. For the system starting from rest, i.e., for "at rest" initial conditions $v(0) = \dot{v}(0) = 0$, it can easily be shown that the constants take the following values:

$$A = -\frac{p_0 \beta}{k} \frac{1}{1 - \beta^2} \qquad B = 0 \qquad (4\text{-}10)$$

Then the response given by Eq. (4-9) becomes

$$v(t) = \frac{p_0}{k} \frac{1}{1 - \beta^2} (\sin \bar{\omega}t - \beta \sin \omega t) \qquad (4\text{-}11)$$

where $p_0/k = v_{st}$ = static displacement, i.e., displacement which would be produced by the load p_0 applied statically

$1/(1 - \beta^2)$ = *magnification factor* (MF), representing dynamic amplification effect of harmonically applied load

$\sin \bar\omega t$ = response component at frequency of the applied load = *steady-state response*, directly related to the load

$\beta \sin \omega t$ = response component at natural vibration frequency = *free-vibration effect* induced by the initial conditions

Since in a practical case, damping will cause the last term to vanish eventually, it is termed the *transient response*. (For this hypothetical, undamped system, of course, this term would not vanish but would continue indefinitely.)

Response Ratio

A convenient measure of the influence of the dynamic character of the loading is provided by the ratio $R(t)$ of the dynamic response to the displacement that would be produced by the static application of the same load:

$$R(t) = \frac{v(t)}{v_{st}} = \frac{v(t)}{p_0/k} \qquad (4\text{-}12)$$

From Eq. (4-11) it is evident that the response ratio resulting from harmonic loading of an undamped system (starting from rest) is

$$R(t) = \frac{1}{1 - \beta^2} (\sin \bar\omega t - \beta \sin \omega t) \qquad (4\text{-}13)$$

It is informative to examine this response behavior in more detail by reference to Fig. 4-1. Figure 4-1a represents the steady-state component of the response. Figure 4-1b is the so-called transient response, the free-vibration motion initiated by the conditions at the start of the response. In this example it is assumed that $\beta = {}^2/_3$, that is, that the applied load frequency is two-thirds of the free-vibration frequency. The total response $R(t)$, the sum of these two terms, is shown in Fig. 4-1c. Two points are of interest: (1) the tendency for the two components to get in phase and then out again, which causes a *beating* effect in the response, and (2) the zero slope at time $t = 0$, showing that the velocity of the transient-response term is just sufficient to cancel the steady-state velocity and thus to satisfy the specified initial condition.

4-2 DAMPED SYSTEM

Returning to the equation of motion including damping, Eq. (4-1), dividing by m, and noting that $c/m = 2\xi\omega$ leads to

$$\ddot v(t) + 2\xi\omega\dot v(t) + \omega^2 v(t) = \frac{p_0}{m} \sin \bar\omega t \qquad (4\text{-}14)$$

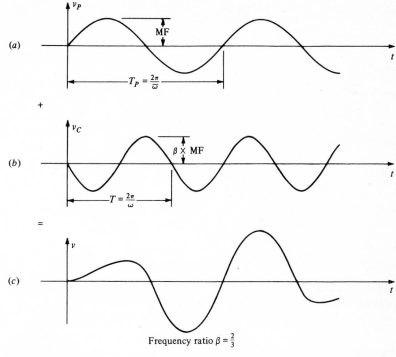

FIGURE 4-1
Response to harmonic load from at-rest initial conditions: (a) steady state; (b) transient; (c) total $R(t)$.

The complementary solution of this equation is the damped free-vibration response given by Eq. (3-25) (assuming that the structure is less than critically damped, as is the case for all practical structures):

$$v_c(t) = e^{-\xi\omega t}(A \sin \omega_D t + B \cos \omega_D t) \qquad (4\text{-}15)$$

The particular solution to this harmonic loading is of the form

$$v_p(t) = G_1 \sin \overline{\omega}t + G_2 \cos \overline{\omega}t \qquad (4\text{-}16)$$

in which the second term is required because, in general, the response of a damped system is not in phase with the loading.

Substituting Eq. (4-16) into Eq. (4-14) and separating the multiples of $\sin \overline{\omega}t$ from the multiples of $\cos \overline{\omega}t$ leads to

$$[-G_1\overline{\omega}^2 - G_2\overline{\omega}(2\xi\omega) + G_1\omega^2] \sin \overline{\omega}t = \frac{p_0}{m} \sin \overline{\omega}t \qquad (4\text{-}17a)$$

$$[-G_2\overline{\omega}^2 + G_1\overline{\omega}(2\xi\omega) + G_2\omega^2] \cos \overline{\omega}t = 0 \qquad (4\text{-}17b)$$

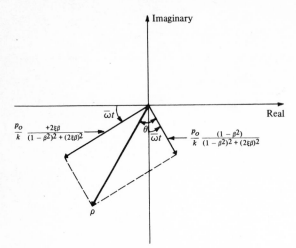

FIGURE 4-2
Steady-state displacement response.

These two relationships must be satisfied individually because the sine and cosine terms vanish at different times. Dividing both by ω^2, regrouping terms, and canceling out the trigonometric expressions yields

$$G_1(1 - \beta^2) - G_2(2\xi\beta) = \frac{p_0}{k} \qquad (4\text{-}18)$$

$$G_2(1 - \beta^2) + G_1(2\xi\beta) = 0$$

Solving these equations simultaneously then results in expressions for the response-amplitude factors:

$$G_1 = \frac{p_0}{k} \frac{1 - \beta^2}{(1 - \beta^2)^2 + (2\xi\beta)^2}$$

$$G_2 = \frac{p_0}{k} \frac{-2\xi\beta}{(1 - \beta^2)^2 + (2\xi\beta)^2} \qquad (4\text{-}19)$$

Introducing these expressions into the particular solution [Eq. (4-16)] and combining with the complementary solution finally yields the general solution:

$$v(t) = e^{-\xi\omega t}(A \sin \omega_D t + B \cos \omega_D t)$$

$$+ \frac{p_0}{k} \frac{1}{(1 - \beta^2)^2 + (2\xi\beta)^2} [(1 - \beta^2) \sin \bar{\omega}t - 2\xi\beta \cos \bar{\omega}t] \qquad (4\text{-}20)$$

The first term in Eq. (4-20) represents the transient response to the applied loading. The constants A and B could be evaluated for any given initial conditions, but this term damps out quickly and generally is of little interest; therefore its evaluation will not be pursued here. The second term in Eq. (4-20) is the steady-state response, at the frequency of the applied loading but out of phase with it. This steady-state displacement behavior can be interpreted most easily by plotting its two vectors in the Argand diagram shown in Fig. 4-2. The resultant ρ of the two vectors

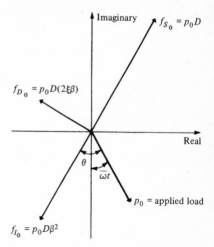

FIGURE 4-3
Force equilibrium in steady-state
response.

represents the amplitude of the steady-state response

$$\rho = \frac{p_0}{k} \left[(1 - \beta^2)^2 + (2\xi\beta)^2 \right]^{-1/2} \qquad (4\text{-}21)$$

and the phase angle θ by which the response lags behind the applied load is given by

$$\theta = \tan^{-1} \frac{2\xi\beta}{1 - \beta^2} \qquad (4\text{-}22)$$

where it is understood that the phase angle is limited to the range $0 < \theta < 180°$. Thus the steady-state response can also be written

$$v(t) = \rho \sin (\bar{\omega}t - \theta) \qquad (4\text{-}23)$$

The ratio of the resultant response amplitude to the static displacement which would be produced by the force p_0 will be called the *dynamic magnification factor D*; thus

$$D \equiv \frac{\rho}{p_0/k} = \left[(1 - \beta^2)^2 + (2\beta\xi)^2 \right]^{-1/2} \qquad (4\text{-}24)$$

It also is of interest to consider the balance of forces acting on the mass in this steady-state vibration condition. The force components are conveniently expressed in terms of the dynamic magnification factor and plotted in an Argand diagram (Fig. 4-3). Note that the elastic force acts in the direction opposite to the resultant displacement vector of Fig. 4-2. Similarly, the damping and inertia forces act in directions opposing the velocity and acceleration vectors, respectively. Finally, it is evident that the resultant of these resisting forces exactly balances the applied load p_0, as it must to maintain dynamic equilibrium.

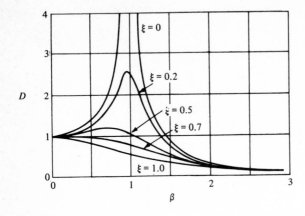

FIGURE 4-4
Variation of dynamic magnification factor with damping and frequency.

It may be seen from Eq. (4-24) that the dynamic magnification factor D varies with the frequency ratio β and the damping ratio ξ; plots of these relationships are shown in Fig. 4-4. The phase angle θ also varies with these same quantities, as is evident in Eq. (4-22) and as shown in the plots of Fig. 4-5.

EXAMPLE E4-1 A portable harmonic-loading machine provides an effective means for evaluating the dynamic properties of structures in the field. By operating the machine at two different frequencies and measuring the resulting structural-response amplitude and phase relationship in each case, it is possible to determine the mass, damping, and stiffness of a SDOF structure. In a test of this type on a single-story building, the shaking machine was operated at frequencies of $\bar{\omega}_1 = 16$ rad/s and $\bar{\omega}_2 = 25$ rad/s, with a force amplitude of 500 lb in each case. The response amplitudes and phase relationships measured in the two cases were

$$\rho_1 = 7.2 \times 10^{-3} \text{ in.} \qquad \theta_1 = 15° \qquad \cos \theta_1 = 0.966 \qquad \sin \theta_1 = 0.259$$
$$\rho_2 = 14.5 \times 10^{-3} \text{ in.} \qquad \theta_2 = 55° \qquad \cos \theta_2 = 0.574 \qquad \sin \theta_2 = 0.819$$

To evaluate the dynamic properties from these data, it is convenient to rewrite Eq. (4-21) as

$$\rho = \frac{p_0}{k} \frac{1}{1 - \beta^2} \left\{ \frac{1}{1 + [2\xi\beta/(1 - \beta^2)]^2} \right\}^{1/2} = \frac{p_0}{k} \frac{\cos \theta}{1 - \beta^2} \tag{a}$$

where the trigonometric function has been derived from Eq. (4-22). With further algebraic simplification this becomes

$$k(1 - \beta^2) = k - \bar{\omega}^2 m = \frac{p_0 \cos \theta}{\rho}$$

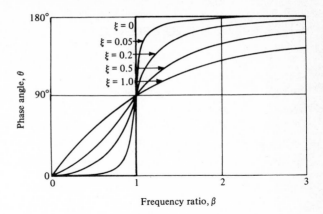

FIGURE 4-5
Variation of phase angle with damping and frequency.

Then introducing the two sets of test data leads to the matrix equation

$$\begin{bmatrix} 1 & -16^2 \\ 1 & -25^2 \end{bmatrix} \begin{bmatrix} k \\ m \end{bmatrix} = 500 \text{ lb} \begin{bmatrix} \dfrac{0.966}{7.2 \times 10^{-3}} \\ \dfrac{0.574}{14.5 \times 10^{-3}} \end{bmatrix}$$

which can be solved to give

$$k = 100 \times 10^3 \text{ lb/in} \qquad m = 128.5 \text{ lb s}^2/\text{in}$$

that is, $\qquad W = mg = 49.6 \times 10^3 \text{ lb}$

From these the natural frequency is

$$\omega = \sqrt{\frac{k}{m}} = 27.9 \text{ rad/s}$$

To determine the damping coefficient, two expressions for $\cos \theta$ can be derived from Eqs. (a) and (4-22). Equating these and solving for the damping ratio leads to

$$\xi = \frac{p_0 \sin \theta}{2\beta k \rho} = \frac{p_0 \sin \theta}{c_c \bar{\omega} \rho}$$

Thus with the data of the first test

$$c = \xi c_c = \frac{500(0.259)}{16(7.2 \times 10^{-3})} = 1{,}125 \text{ lb} \cdot \text{s/in}$$

and the same result (within slide-rule accuracy) is given by the data of the second test. The damping ratio therefore is

$$\xi = \frac{c}{2k/\omega} = \frac{1{,}125(27.9)}{200 \times 10^3} = 15.7\% \qquad ////$$

4-3 RESONANT RESPONSE

From Fig. 4-4 it may be noted that the peak steady-state response occurs at a frequency ratio near unity for lightly damped systems. The condition when the frequency ratio is unity, i.e., when the frequency of the applied load equals the natural vibration frequency, is called *resonance*. From Eq. (4-13) it is apparent that the steady-state response of an undamped system tends toward infinity at resonance. A more general result may be obtained from Eq. (4-24), which shows that for resonance ($\beta = 1$) the dynamic magnification factor is inversely proportional to the damping ratio:

$$D_{\beta=1} = \frac{1}{2\xi} \qquad (4\text{-}25)$$

However, although it is close to the maximum, this does not represent the maximum response for any damped system; the frequency ratio for maximum response may be found by differentiating Eq. (4-24), with respect to β and equating to zero. For practical structures having damping ratios $\xi < 1/\sqrt{2}$, the peak-response frequency is found to be

$$\beta_{\text{peak}} = \sqrt{1 - 2\xi^2} \qquad (4\text{-}26a)$$

and the corresponding peak response is

$$D_{\text{max}} = \frac{1}{2\xi\sqrt{1 - \xi^2}} \qquad (4\text{-}26b)$$

For reasonable amounts of damping, the difference between Eq. (4-26b) and the simpler Eq. (4-25) is negligible.

For a more complete understanding of the nature of the resonant response of a structure to harmonic loading, it is necessary to consider the general response equation (4-20), which includes the transient terms as well as the steady-state term. At the resonant exciting frequency ($\beta = 1$) this equation becomes

$$v(t) = e^{-\xi\omega t}(A \sin \omega_D t + B \cos \omega_D t) - \frac{p_0}{k}\frac{\cos \omega t}{2\xi} \qquad (4\text{-}27)$$

Assuming that the system starts from rest [$v(0) = \dot{v}(0) = 0$], the constants are

$$A = \frac{p_0}{k}\frac{\omega}{2\omega_D} = \frac{p_0}{k}\frac{1}{2\sqrt{1 - \xi^2}} \qquad B = \frac{p_0}{k}\frac{1}{2\xi} \qquad (4\text{-}28)$$

Thus Eq. (4-27) becomes

$$v(t) = \frac{1}{2\xi}\frac{p_0}{k}\left[e^{-\xi\omega t}\left(\frac{\xi}{\sqrt{1 - \xi^2}}\sin \omega_D t + \cos \omega_D t\right) - \cos \omega t\right] \qquad (4\text{-}29)$$

For the amounts of damping to be expected in a structural system, the sine term in this equation will contribute little to the response amplitude; moreover, the damped

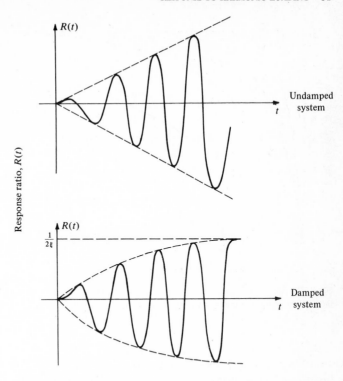

FIGURE 4-6
Response to resonant loading $\beta = 1$ for at-rest initial conditions.

frequency is nearly equal to the undamped frequency. Thus the response ratio in this case is approximately

$$R(t) = \frac{v(t)}{p_0/k} \doteq \frac{1}{2\xi}(e^{-\xi\omega t} - 1)\cos\omega t \qquad (4\text{-}30)$$

For zero damping, Eq. (4-29) becomes indeterminate, but when L'Hospital's rule is applied, the resonant response of an undamped system is

$$R(t) = \tfrac{1}{2}(\sin\omega t - \omega t\cos\omega t) \qquad (4\text{-}31)$$

Plots of Eqs. (4-30) and (4-31), shown in Fig. 4-6, show how the response builds up in cases of resonant excitation, with and without damping; in both cases, it is clear that the response builds up gradually. In the undamped system, the response continues to grow by the amount π for each cycle; thus it will eventually produce distress in the system unless the frequency is changed. On the other hand, the manner in which the damping limits the resonant-response amplitude is clear in the lower sketch. The number of cycles required for this damped resonant response to reach

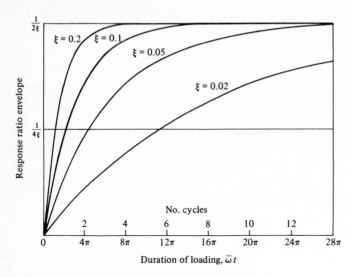

FIGURE 4-7
Rate of buildup of resonant response from rest.

essentially its peak amplitude depends on the amount of damping. Curves showing the buildup of the response envelope (dashed lines in Fig. 4-6), plotted for several values of damping as functions of the number of response cycles, are given in Fig. 4-7. Note how few cycles of excitation are required to reach nearly the full response amplitude.

4-4 ACCELEROMETERS AND DISPLACEMENT METERS

At this point it is convenient to discuss the fundamental principles on which the operation of an important class of dynamic measurement devices is based. These are seismic instruments, which consist essentially of a damped oscillator as shown in Fig. 4-8. The system is mounted in a housing which may be attached to the surface where the movement is to be measured; the response is measured in terms of the motion $v(t)$ of the mass relative to the housing.

The equation of motion of this system is given by

$$m\ddot{v} + c\dot{v} + kv = -m\ddot{v}_g(t) \equiv p_{\text{eff}}(t) \quad (2\text{-}21)$$

If the base on which the instrument is mounted is moving harmonically with an acceleration amplitude $\ddot{v}_g(t) = \ddot{v}_{g0} \sin \overline{\omega} t$, the effective loading of the mass is $p_{\text{eff}}(t) = -m\ddot{v}_{g0} \sin \overline{\omega} t$. The dynamic steady-state response of this system then has the

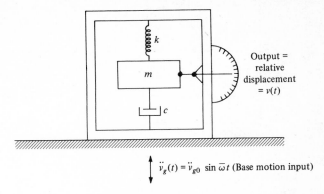

FIGURE 4-8
Schematic diagram of a typical seismometer.

amplitude given by Eq. (4-21),

$$\rho = \frac{m\ddot{v}_{g0}}{k} D \qquad (4\text{-}32)$$

in which D is given by Eq. (4-24) and is presented graphically in Fig. 4-4. Examination of this figure shows that for a damping ratio of $\xi = 0.7$ the value of D is nearly constant for the frequency range $0 < \beta < 0.6$. Thus it is clear from Eq. (4-32) that the response indicated by this instrument will be directly proportional to the base-acceleration amplitude for applied frequencies up to about six-tenths the natural frequency of the instrument. Hence this type of instrument when properly damped will serve effectively as an accelerometer for relatively low frequencies; its range of applicability will be increased by increasing its natural frequency relative to the exciting frequency, i.e., by increasing the stiffness of the spring. This is the basic principle of seismic accelerometers.

Consider now the response of this same instrument to a harmonic base displacement $v_g = v_{g0} \sin \overline{\omega}t$. In this case the corresponding acceleration is $\ddot{v}_g = -\overline{\omega}^2 v_{g0} \sin \overline{\omega}t$, and the effective loading is $p_{\text{eff}}(t) = m\overline{\omega}^2 v_{g0} \sin \overline{\omega}t$. By Eq. (4-21) the response amplitude then is

$$\rho = \frac{m\overline{\omega}^2 v_{g0}}{k} D = v_{g0}\beta^2 D \qquad (4\text{-}33)$$

A plot of the response function $\beta^2 D$ is presented in Fig. 4-9. In this case it is evident that $\beta^2 D$ is essentially constant at frequency ratios $\beta > 1$ for a damping ratio $\xi = 0.5$. Thus the response of the properly damped instrument is essentially proportional to the base-displacement amplitude for high-frequency base motions; i.e., it will serve as a displacement meter in measuring such motions. Its range of applicability for this purpose will be increased by reducing the natural frequency, i.e., by reducing the spring stiffness or increasing the mass.

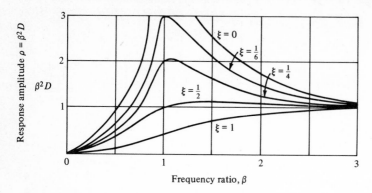

FIGURE 4-9
Response of seismometer to harmonic base displacement

4-5 VIBRATION ISOLATION

Although the subject of vibration isolation is too broad to be discussed extensively here, it is important to note that the basic principle is closely related to that of the seismic instrument. Two different classes of problem may be identified in which vibration isolation may be necessary: (1) operating equipment may generate oscillatory forces which could produce harmful vibrations in the supporting structure, or (2) sensitive instruments may be supported by a structure which is vibrating appreciably.

The first situation is illustrated in Fig. 4-10. A rotating machine produces an oscillatory vertical force $p_0 \sin \overline{\omega} t$ due to unbalance in its rotating parts. If the machine is mounted on a SDOF spring-damper support system, as shown, its steady-state displacement response is given by

$$v(t) = \frac{p_0}{k} D \sin (\overline{\omega} t - \theta) \qquad (4\text{-}34)$$

where D is defined by Eq. (4-24). Thus the force exerted against the base by the spring supports is

$$f_S = kv(t) = p_0 D \sin (\overline{\omega} t - \theta)$$

At the same time, the velocity of the motion relative to the base is

$$\dot{v}(t) = \frac{p_0}{k} D\overline{\omega} \cos (\overline{\omega} t - \theta)$$

which leads to a damping force on the base

$$f_D = c\dot{v}(t) = \frac{cp_0 D\overline{\omega}}{k} \cos (\overline{\omega} t - \theta) = 2\xi\beta p_0 D \cos (\overline{\omega} t - \theta)$$

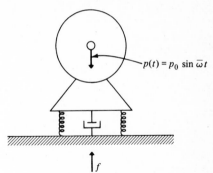

FIGURE 4-10
SDOF vibration-isolation system (applied loading).

Since this force is 90° out of phase with the spring force, it is evident that the amplitude of the base force f is

$$f_{max} = (f_{S,\,max}^2 + f_{D,\,max}^2)^{1/2} = p_0 D[1 + (2\xi\beta)^2]^{1/2}$$

The ratio of the maximum base force to the applied-force amplitude, which is known as the *transmissibility* (TR) of the support system, thus is given by

$$\text{TR} \equiv \frac{f_{max}}{p_0} = D\sqrt{1 + (2\xi\beta)^2} \qquad (4\text{-}35)$$

A plot of the transmissibility as a function of the frequency ratio and damping ratio is shown in Fig. 4-11. Although it is similar to Fig. 4-4, all curves pass through the same point at a frequency ratio of $\beta = \sqrt{2}$. This difference from Fig. 4-4 is due, of course, to the influence of the damping force. Because of this characteristic, it is

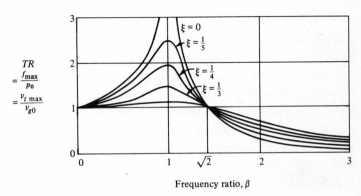

FIGURE 4-11
Vibration-transmissibility ratio (applied load or displacement).

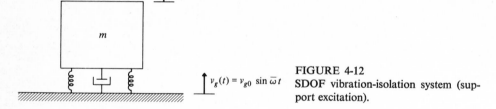

FIGURE 4-12
SDOF vibration-isolation system (support excitation).

evident that damping tends to reduce the effectiveness of a vibration-isolation system for frequencies greater than this critical ratio.

The second type of situation in which vibration isolation is important is illustrated in Fig. 4-12. The mass m to be isolated is supported by a spring-damper system on a foundation slab which is subjected to harmonic vertical motions. The displacement of the mass *relative* to the base then is given by [see Eq. (4-33)]

$$v(t) = v_{g0}\beta^2 D \sin(\bar{\omega}t - \theta) \qquad (4\text{-}36)$$

However, when the motion of the base is added vectorially it can be shown that the total motion of the mass is given by

$$v^t(t) = v_{g0}\sqrt{1 + (2\xi\beta)^2}\, D \sin(\bar{\omega}t - \bar{\theta}) \qquad (4\text{-}37)$$

in which the phase angle $\bar{\theta}$ is of no particular interest. Thus, if the transmissibility in this situation is defined as the ratio of the amplitude of motion of the mass to the base-motion amplitude, it can be seen that the expression for transmissibility is the same as that given by Eq. (4-35). This can be expressed mathematically as

$$\text{TR} \equiv \frac{v^t_{\max}}{v_{g0}} = D\sqrt{1 + (2\xi\beta)^2} \qquad (4\text{-}38)$$

and Fig. 4-11 serves to define the effectiveness of vibration-isolation systems for both basic SDOF isolation situations.

For the design of a vibration-isolation system, it is convenient to express the behavior of the system in terms of its isolation effectiveness rather than the transmissibility, where the effectiveness is defined as $1 - \text{TR}$. Also, when it is noted in Fig. 4-11 that an isolation system is effective only for frequency ratios $\beta > \sqrt{2}$ and that damping is undesirable in this range, it is evident that the isolation mounting should have very little damping. Thus it is acceptable to use the transmissibility expression for zero damping

$$\text{TR} = \frac{1}{\beta^2 - 1} \qquad 1 - \text{TR} = \frac{\beta^2 - 2}{\beta^2 - 1} \qquad (4\text{-}39)$$

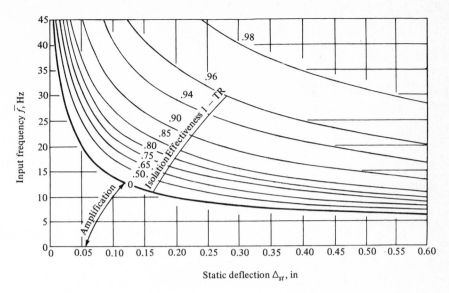

FIGURE 4-13
Vibration-isolation design chart.

in which it is understood that $\beta \geq \sqrt{2}$. Finally it may be noted that $\beta^2 = \bar{\omega}^2/\omega^2 = \bar{\omega}^2(m/k) = \bar{\omega}^2(W/kg) = \bar{\omega}^2(\Delta_{st}/g)$, where g is the acceleration of gravity and $\Delta_{st} = W/k$ is the deflection that the weight of the system to be isolated will produce on the vibration-mounting devices. Thus it is evident that the effectiveness of the mounting system can be expressed in terms of the frequency of the input motion $\bar{\omega}$ and this static-deflection value Δ_{st}. Solving Eq. (4-39) for the frequency ratio in terms of the isolation effectiveness leads to $\beta^2 = [2 - (1 - TR)]/[1 - (1 - TR)]$. This can now be expressed in terms of the input frequency ($\bar{f} = 2\pi\bar{\omega}$) and the static deflection:

$$\bar{f} = 3.13 \sqrt{\frac{1}{\Delta_{st}} \frac{2 - (1 - TR)}{1 - (1 - TR)}} \qquad (4\text{-}40)$$

where $\bar{f}$ is in Hertz and Δ_{st} is in inches. A plot of Eq. (4-40) is presented in Fig. 4-13. Knowing the frequency of the impressed excitation, one can determine directly from this graph the support-pad deflection Δ_{st} required to achieve any desired level of vibration isolation, assuming that the isolators have little damping. It also is apparent in this graph that the isolation system will have a deleterious effect if it is too stiff.

EXAMPLE E4-2   Deflections sometimes develop in concrete bridge girders due to creep, and if the bridge consists of a long series of identical spans, these deformations will cause a harmonic excitation in a vehicle traveling over the

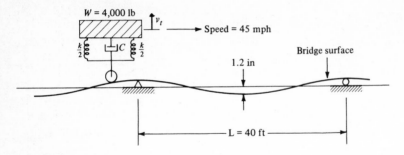

FIGURE E4-1
Idealized vehicle traveling over an uneven bridge deck.

bridge at constant speed. Of course, the springs and shock absorbers of the car are intended to provide a vibration-isolation system which will limit the vertical motions transmitted from the road to the occupants.

Figure E4-1 shows a highly idealized model of this type of system, in which the vehicle weight is 4,000 lb and its spring stiffness is defined by a test which showed that adding 100 lb caused a deflection of 0.08 in. The bridge profile is represented by a sine curve having a wavelength (girder span) of 40 ft and a (single) amplitude of 1.2 in. From these data it is desired to predict the steady-state vertical motions in the car when it is traveling at a speed of 45 mph, assuming that the damping is 40 percent of critical.

The transmissibility for this case is given by Eq. (4-38); hence the amplitude of vertical motion is

$$v^t_{max} = v_{g0} \left[\frac{1 + (2\xi\beta)^2}{(1 - \beta^2)^2 + (2\xi\beta)^2} \right]^{1/2}$$

When the car is traveling at 45 mph = 66 ft/s, the excitation period is

$$T_p = \frac{40 \text{ ft}}{66 \text{ ft/s}} = 0.606 \text{ s}$$

while the natural period of the vehicle is

$$T = \frac{2\pi}{\omega} = 2\pi \sqrt{\frac{w}{kg}} = 0.572 \text{ s}$$

Hence $\beta = T/T_p = 0.572/0.606 = 0.944$, and with $\xi = 0.4$ the response amplitude is

$$v^t_{max} = 1.2(1.642) = 1.97 \text{ in}$$

It also is of interest to note that if there were no damping in the vehicle ($\xi = 0$), the amplitude would be

$$v^t_{max} = v_{g0} \frac{1}{1 - \beta^2} = \frac{1.2}{0.11} = 10.9 \text{ in}$$

This is beyond the spring range, of course, and thus has little meaning, but it does demonstrate the important function of the shock absorbers in limiting the motions resulting from waviness of the road surface.   ////

EXAMPLE E4-3   A reciprocating machine weighing 20,000 lb is known to develop vertically oriented harmonic forces having an amplitude of 500 lb at its operating speed of 40 Hz. In order to limit the vibrations excited in the building in which this machine is to be installed, it is to be supported by a spring at each corner of its rectangular base. The designer wants to know what support spring stiffness will be required to limit to 80 lb the *total* harmonic force transmitted from the machine to the building.

The transmissibility in this case is 80/500 = 0.16; hence by Eq. (4-39)

$$\frac{1}{\text{TR}} = |\beta^2 - 1| = 6.25$$

from which

$$\beta^2 = 7.25 = \frac{\bar{\omega}^2 W}{kg}$$

Solving for the total spring stiffness gives

$$k = \frac{\bar{\omega}^2 W}{7.25g} = 451 \times 10^3 \text{ lb/in}$$

Thus the stiffness of each of the four support springs is

$$k/4 = {}^{451}/_4 = 113 \text{ kips/in}$$

It is of interest to note that the static deflection caused by the weight of the machine on these spring supports is

$$\Delta_{st} = {}^{20}/_{451} = 0.0444 \text{ in} \qquad ////$$

4-6 EVALUATION OF DAMPING IN SDOF SYSTEMS

In the foregoing discussion of response analysis in SDOF structures, it has been assumed that the physical properties of the system (mass, stiffness, and damping) are known. In most cases, the structural mass and stiffness can be evaluated rather easily, either by simple physical considerations or by generalized expressions such as Eqs. (2-37) and (2-39). On the other hand, the basic energy-loss mechanisms in practical structures are seldom fully understood; consequently it usually is not feasible to determine the damping coefficient by means of the corresponding generalized damping expression. For this reason, the damping in most structural systems must be evaluated directly by experimental methods. A brief survey of the principal procedures for evaluating damping from experimental measurements follows.

Free-Vibration Decay

Probably the simplest and most frequently used experimental method is measurement of the decay of free vibrations, as mentioned in Chap. 3. When a system has been set into free vibration by any means, the damping ratio can be determined from the ratio of two displacement amplitudes measured at an interval of m cycles. Thus if v_n is the amplitude of vibration at any time and v_{n+m} is the amplitude m cycles later, the damping ratio is given by

$$\xi = \frac{\delta_m}{2\pi m (\omega/\omega_D)} \doteq \frac{\delta_m}{2\pi m} \qquad (4\text{-}41)$$

where $\delta_m = \ln (v_n/v_{n+m})$ represents the logarithmic decrement and ω and ω_D are the undamped and damped frequencies, respectively. In most practical structures, the damping ratio is less than 0.2, so that the approximate form of Eq. (4-41), based on neglecting the change of frequency due to damping, is sufficiently accurate (the error in ξ is less than 2 percent). A major advantage of this free-vibration method is that equipment and instrumentation requirements are minimal; the vibrations can be initiated by any convenient method, and only the relative displacement amplitudes need be measured.

Resonant Amplification

The other principal techniques for evaluating damping are based on observations of steady-state harmonic response behavior and thus require a means of applying harmonic excitations to the structure at prescribed frequencies and amplitudes. With such equipment the frequency-response curve for the structure can be constructed by applying a harmonic load $p_0 \sin \overline{\omega} t$ at a closely spaced sequence of frequencies which span the resonance frequency and plotting the resulting displacement amplitudes as a function of the applied frequencies. A typical frequency-response curve for a moderately damped structure is shown in Fig. 4-14.

The dynamic magnification factor for any given frequency is the ratio of the response amplitude at that frequency to the zero-frequency (static) response. It was shown earlier, by Eq. (4-25), that the damping ratio is closely related to the dynamic magnification factor at resonance. When the static-response and the resonant-response amplitude are denoted by ρ_0 and $\rho_{\beta=1}$, respectively, the damping ratio is given by

$$\xi = \frac{1}{2} \frac{\rho_0}{\rho_{\beta=1}} \qquad (4\text{-}42)$$

In practice, however, it is difficult to apply the exact resonance frequency, and it is more convenient to determine the maximum response amplitude $\rho_{\max}$ which occurs

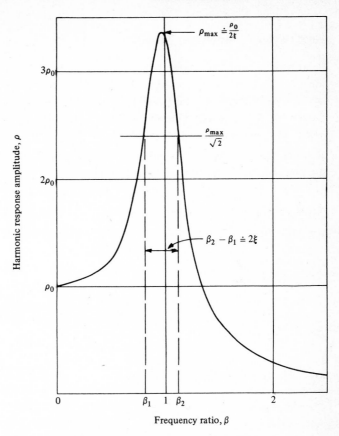

FIGURE 4-14
Frequency-response curve for moderately damped system.

at a slightly lower frequency. In this case, it is evident from Eq. (4-26b) that the damping ratio may be evaluated as follows:

$$\xi = \frac{1}{2} \frac{\rho_0}{\rho_{max}} \frac{\omega}{\omega_D} \doteq \frac{1}{2} \frac{\rho_0}{\rho_{max}} \qquad (4\text{-}43)$$

The error involved in the simpler expression again results from neglecting the difference between the damped and undamped frequencies and is unimportant in ordinary structures.

This method of damping analysis requires only simple instrumentation, capable of measuring relative displacement amplitudes; however, the evaluation of the static displacement may present a problem because many types of loading systems cannot be operated at zero frequency.

Half-Power (Bandwidth) Method

It is evident from the general harmonic-response expression [Eq. (4-21)] that the shape of the entire frequency-response wave is controlled by the amount of damping in the system; therefore, it is possible to derive the damping ratio from many different properties of the curve. One of the most convenient of these is the bandwidth, or half-power, method, in which the damping ratio is determined from the frequencies at which the response is reduced to $(1/\sqrt{2})\rho_{\beta=1}$, that is, at the frequencies for which the power input is half the input at resonance.

The values of these half-power frequencies can be determined by setting the response amplitude in Eq. (4-21) equal to $1/\sqrt{2}$ times the resonant amplitude derived from Eq. (4-42), that is,

$$\frac{1}{\sqrt{2}}\frac{\rho_0}{2\xi} = \rho_0\left[\frac{1}{(1-\beta^2)^2 + (2\xi\beta)^2}\right]^{1/2}$$

or, squaring both sides,

$$\frac{1}{8\xi^2} = \frac{1}{(1-\beta^2)^2 + (2\xi\beta)^2}$$

Solving for the frequency ratio then gives

$$\beta^2 = 1 - 2\xi^2 \pm 2\xi\sqrt{1+\xi^2}$$

from which (neglecting ξ^2 in the square-root term) the two half-power frequencies are

$$\beta_1{}^2 \doteq 1 - 2\xi - 2\xi^2 \qquad \beta_1 \doteq 1 - \xi - \xi^2$$
$$\beta_2{}^2 \doteq 1 + 2\xi - 2\xi^2 \qquad \beta_2 \doteq 1 + \xi - \xi^2$$

Hence, the damping ratio is given by half the difference between these half-power frequencies:

$$\xi \doteq {}^1\!/_2(\beta_2 - \beta_1) \qquad (4\text{-}44)$$

This method of evaluating the damping ratio also is illustrated with the typical frequency-response curve of Fig. 4-14. A horizontal line has been drawn across the curve at $1/\sqrt{2}$ times the resonant-response value; the difference between the frequencies at which this line intersects the response curve is equal to twice the damping ratio. It is evident that this technique avoids the need for the static response; however, it does require that the response curve be plotted accurately in the half-power range and at resonance.

EXAMPLE E4-4 Data from a frequency-response test of a SDOF system have been plotted in Fig. E4-2. The pertinent data for evaluating the damping ratio are also shown. The sequence of steps in the analysis after the curve was plotted were as follows:

1 Determine peak response $= 5.67 \times 10^{-2}$ in

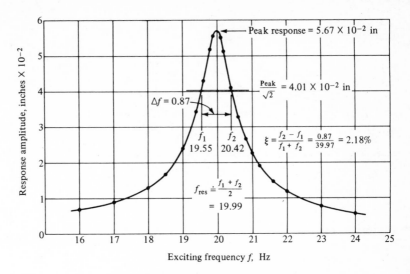

FIGURE E4-2
Frequency-response experiment to determine damping ratio.

2   Construct line at peak/$\sqrt{2}$ = 4.01 × 10^{-2} in

3   Determine the two frequencies at which this line cuts the response curve: f_1 = 19.55, f_2 = 20.42

4   The damping ratio then is given by

$$\xi = \frac{\Delta\beta}{2} = \frac{f_2 - f_1}{f_2 + f_1} = 2.18\% \qquad ////$$

Energy Loss per Cycle (Resonance Testing)

If instrumentation is available to measure the phase relationship between the input force and the resulting displacements, the damping can be evaluated from tests run only at resonance and there is no need to construct the frequency-response curve. The procedure involves establishing resonance by adjusting the input frequency until the response is 90° out of phase with the applied loading. Then the applied load is exactly balanced by the damping force, so that if the relationship between the applied load and the resulting displacements is plotted for one loading cycle as shown in Fig. 4-15, the result can be interpreted as the damping-force–displacement diagram.

If the structure has linear viscous damping, the curve will be an ellipse, as shown by the dashed line in Fig. 4-15. In this case, the damping coefficient can be determined directly from the ratio of the maximum damping force to the maximum velocity:

$$c = \frac{f_{D,\,max}}{\dot{v}_{max}} = \frac{p_0}{\omega\rho} \qquad (4\text{-}45)$$

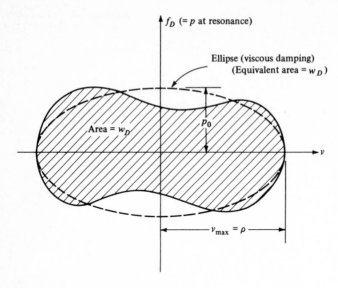

FIGURE 4-15
Actual and equivalent damping energy per cycle.

where it is noted that the maximum velocity is given by the product of frequency and displacement amplitude. If the damping is not linear viscous, the shape of the force-displacement diagram will not be elliptical; a curve like the solid line in Fig. 4-15 might have been obtained, for example. In this case, an equivalent viscous damping coefficient can be defined which would cause the same energy loss per cycle as in the observed force-displacement diagram. In other words, the equivalent viscous damper is associated with the elliptical force-displacement diagram having the same area and maximum displacements as the actual force-displacement diagram. In this sense, the dashed-line curve in Fig. 4-15 is equivalent to the solid-line curve. Then the equivalent applied-force amplitude is given by

$$p_0 = \frac{w_D}{\pi \rho}$$

where w_D is the area under the force-displacement diagram, i.e., energy loss per cycle. Substituting this into Eq. (4-45) leads to an expression for the equivalent viscous-damping coefficient in terms of the energy loss per cycle:

$$c_{eq} = \frac{w_D}{\pi \omega \rho^2} \qquad (4\text{-}46)$$

In most cases, it is more convenient to define the damping in terms of the critical damping ratio than as a damping coefficient. For this purpose, it is necessary to define also a measure of the critical damping coefficient of the structure, and this

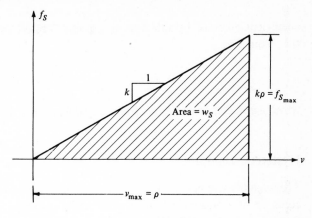

FIGURE 4-16
Elastic stiffness and strain energy.

can be expressed in terms of the mass and frequency [as in Eq. (3-18)] or in the alternate form involving stiffness and frequency:

$$c_c = \frac{2k}{\omega} \qquad (4\text{-}47)$$

This latter expression is more convenient here because the stiffness of the structure can be measured by the same instrumentation used to measure the damping-energy loss per cycle, merely by operating the system very slowly at essentially static conditions. The static-force–displacement diagram obtained in this way will be of the form shown in Fig. 4-16 if the structure is linearly elastic, and the stiffness is represented by the slope of the curve. Alternatively, the stiffness may be expressed by the area under the force-displacement diagram, w_s, as follows:

$$k = \frac{2w_s}{\rho^2} \qquad (4\text{-}48)$$

Thus the damping ratio can be obtained by combining Eqs. (4-46) to (4-48):

$$\xi = \frac{c}{c_c} = \frac{w_D}{4\pi w_s} \qquad (4\text{-}49)$$

The damping ratio defined by Eq. (4-49) appears to be independent of frequency: it depends directly on the ratio of damping-energy loss per cycle to the strain energy stored at maximum displacement. However, for any given viscous-damping mechanism, the energy loss in the system will be proportional to the harmonic frequency, and thus the damping ratio will be too. Alternatively, when the damping ratio has been evaluated from a resonance test, the corresponding viscous-damping coefficient obtained by substituting Eq. (4-46) into (4-49) is inversely proportional to the frequency:

$$c_{eq} = \xi \frac{4w_s}{\omega \rho^2} \qquad (4\text{-}50)$$

which again demonstrates the frequency dependence of the viscous-damping behavior.

Hysteretic Damping

Although the viscous-damping mechanism leads to a convenient form of the structural equation of motion, the results of experiments seldom correspond closely with this type of energy-loss behavior. In many practical cases, the equivalent viscous-damper concept defined in terms of energy loss per cycle provides a reasonable approximation of experimental results. However, the essential frequency dependence of the viscous mechanism mentioned above is at variance with a great deal of test evidence, much of which indicates that the damping forces are nearly independent of the test frequency.

A mathematical model which has this property of frequency independence is provided by the concept of *hysteretic* damping, which may be defined as a damping force in phase with the velocity but proportional to the displacements. This force-displacement relationship may be expressed as

$$f_D = \zeta k |v| \frac{\dot{v}}{|\dot{v}|} \qquad (4\text{-}51)$$

where ζ is the hysteretic-damping coefficient which defines the damping forces as a fraction of the elastic-stiffness forces. The force-displacement diagram for hysteretic damping during a cycle of harmonic displacement is depicted in Fig. 4-17. It will be noted that the damping resistance is similar to the linear elastic forces during displacements of increasing magnitudes but that the sense of the damping force reverses when the displacements diminish. The hysteretic energy loss per cycle given by this mechanism is

$$w_D = 2\zeta k \rho^2 \qquad (4\text{-}52)$$

If this hysteretic energy loss is assumed to be represented by an equivalent viscous damper, the equivalent viscous-damping ratio still is given by Eq. (4-49). In other words, Eq. (4-49) can be used to express the damping ratio of a structure regardless of the actual internal-energy-loss mechanism. However, if the specific hysteretic-damping coefficient corresponding to a given test is to be determined, this may be expressed in terms of the damping ratio [by substituting Eqs. (4-52) and (4-48) into Eq. (4-49)] as

$$\zeta = \pi \xi \qquad (4\text{-}53)$$

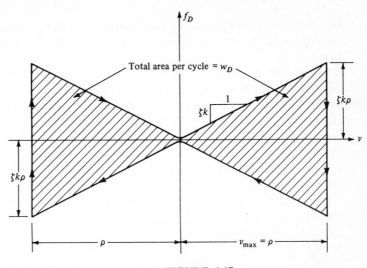

FIGURE 4-17
Hysteretic damping force vs. displacement.

Thus it is clear that the hysteretic-damping coefficient is independent of the frequency at which the test was run, in contrast with the frequency dependence of the viscous-damping coefficient shown by Eq. (4-50).

PROBLEMS

4-1   Consider the basic structure of Fig. 3-1 with zero damping and subjected to harmonic excitation at the frequency ratio $\beta = 0.8$. Including both steady-state and transient effects, plot the response ratio $R(t)$. Evaluate the response at increments $\bar{\omega}\Delta t = 80°$ and continue the analysis for 10 increments.

4-2   Consider the basic system of Fig. 3-1 with the following properties: $m = 2k \cdot s^2/\text{in}$ and $k = 20k/\text{in}$. If this system is subjected to resonant harmonic loading ($\bar{\omega} = \omega$) starting from "at rest" conditions, determine the value of the response ratio $R(t)$ after four cycles ($\bar{\omega}t = 8\pi$), assuming:
(a) $c = 0$ [use Eq. (4-31)]
(b) $c = 0.5k \cdot s/\text{in}$ [use Eq. (4-30)]
(c) $c = 2.0k \cdot s/\text{in}$ [use Eq. (4-30)]

4-3   Consider the same vehicle and bridge structure of Example E4-2, except with the girder spans reduced to $L = 36$ ft. Determine:
(a) the vehicle speed required to induce resonance in the vehicle spring system.
(b) the total amplitude of vertical motion v_{max}^t at resonance.
(c) the total amplitude of vertical motion v_{max}^t at the speed of 45 mph.

4-4   A control console containing delicate instrumentation is to be located on the floor of a test laboratory where it has been determined that the floor slab is vibrating vertically with an amplitude of 0.03 in at 20 Hz. If the weight of the console is 800 lb, determine the stiffness of the vibration isolation system required to reduce the vertical-motion amplitude of the console to 0.005 in.

4-5 A sieving machine weighs 6,500 lb, and when operating at full capacity, it exerts a harmonic force on its supports of 700 lb amplitude at 12 Hz. After mounting the machine on spring-type vibration isolators, it was found that the harmonic force exerted on the supports had been reduced to a 50-lb amplitude. Determine the spring stiffness k of the isolation system.

4-6 The structure of Fig. P4-1a can be idealized by the equivalent system of Fig. P4-1b. In order to determine the values of c and k for this mathematical model, the concrete column was subjected to a harmonic load test as shown in Fig. P4-1c. When operating at a test frequency of $\bar{\omega} = 10$ rads/s, the force-deflection (hysteresis) curve of Fig. P4-1d was obtained. From this data:

(a) determine the stiffness k.

(b) assuming a viscous damping mechanism, determine the apparent viscous damping ratio ξ and damping coefficient c.

(c) assuming a hysteretic damping mechanism, determine the apparent hysteretic damping factor ζ.

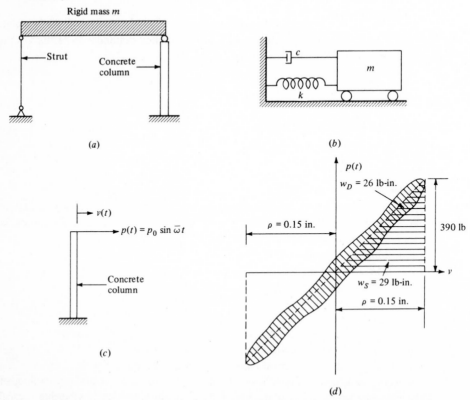

(a)

(b)

(c)

(d)

FIGURE P4-1

4-7 Suppose that the test of Prob. 4-6 were repeated, using a test frequency $\bar{\omega} = 20$ rads/s, and that the force-deflection curve (Fig. P4-1d) was found to be unchanged. In this case:

(*a*) determine the apparent viscous damping values ξ and c.

(*b*) determine the apparent hysteretic damping factor ζ.

(*c*) Based on these two tests ($\bar{\omega} = 10$ and $\bar{\omega} = 20$ rads/s), which type of damping mechanism appears more reasonable—viscous or hysteretic?

4-8 If the damping of the system of Prob. 4-6 actually were provided by a viscous damper as indicated in Fig. P4-1b, what would be the value of w_D obtained in a test performed at $\bar{\omega} = 20$ rads/s?

5

RESPONSE TO PERIODIC LOADINGS

5-1 FOURIER SERIES EXPRESSION OF THE LOADING

Equations expressing the response of a SDOF system to any harmonic applied loading were developed in Chap. 4. It will now be shown how these same response expressions can be used to evaluate the response of a SDOF system to any periodic loading. To treat the periodic loading, it is necessary only to express it in Fourier series form; the response to each term of the series is then merely the response to a harmonic loading, and by the principle of superposition the total response is the sum of the responses to the separate load terms.

Consider any periodic loading, such as that shown in Fig. 5-1. Such a periodic function may be expressed by the Fourier series

$$p(t) = a_0 + \sum_{n=1}^{\infty} a_n \cos \frac{2\pi n}{T_p} t + \sum_{n=1}^{\infty} b_n \sin \frac{2\pi n}{T_p} t \qquad (5\text{-}1)$$

in which T_p represents the period of the load function and the coefficients can be evaluated from the following expressions:

$$a_0 = \frac{1}{T_p} \int_0^{T_p} p(t)\, dt$$

$$a_n = \frac{2}{T_p} \int_0^{T_p} p(t) \cos \frac{2\pi n}{T_p} t \qquad (5\text{-}2)$$

$$b_n = \frac{2}{T_p} \int_0^{T_p} p(t) \sin \frac{2\pi n}{T_p} t$$

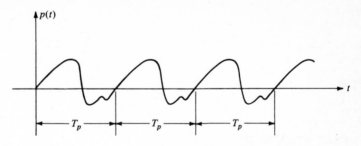

FIGURE 5-1
Arbitrary periodic loading.

5-2 RESPONSE TO THE FOURIER SERIES LOADING

When an arbitrary periodic loading has been expressed in the Fourier series form of
Eq. (5-1), it is apparent that it consists of a constant load (the average load value
represented by the coefficient a_0) plus a series of harmonic loads of frequencies $\bar{\omega}_n$
and amplitudes a_n and b_n. The steady-state response produced in an undamped SDOF
structure by each sine term of the harmonic loading series is given by an expression
of the form of Eq. (4-11), but omitting the transient term, as follows:

Sine terms:
$$v_n(t) = \frac{b_n}{k} \frac{1}{1 - \beta_n^{2}} \sin n\bar{\omega}_1 t \qquad (5\text{-}3a)$$

where
$$\beta_n = \frac{\bar{\omega}_n}{\omega} = \frac{nT}{T_p} = \frac{n\bar{\omega}_1}{\omega}$$

Similarly, the steady-state response to each cosine term of the series is given by:

Cosine terms:
$$v_n(t) = \frac{a_n}{k} \frac{1}{1 - \beta_n^{2}} \cos n\bar{\omega}_1 t \qquad (5\text{-}3b)$$

Finally, the steady-state response to the constant load component is merely the static
deflection, i.e.,

Constant:
$$v_0 = \frac{a_0}{k} \qquad (5\text{-}3c)$$

The total periodic response of the undamped structure can therefore be expressed
as the sum of the individual response expressions for all terms of the loading series,
as follows:

$$v(t) = \frac{1}{k}\left[a_0 + \sum_{n=1}^{\infty} \frac{1}{1 - \beta_n^{2}} (a_n \cos n\bar{\omega}_1 t + b_n \sin n\bar{\omega}_1 t) \right] \qquad (5\text{-}3)$$

where the load-amplitude coefficients are given by the Fourier series expressions of
Eq. (5-2).

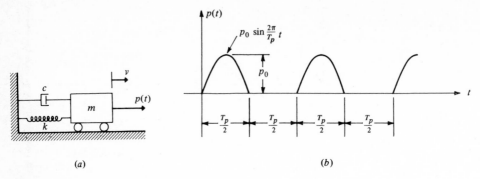

FIGURE E5-1
Example analysis of response to periodic loading: (a) SDOF system; (b) periodic loading.

To take account of damping in evaluating the response of a SDOF structure to a periodic loading, it is necessary merely to substitute the damped-harmonic-response expressions of the form of Eq. (4-20) for the undamped-response expressions used above. In this case, the total steady-state response is given by

$$v(t) = \frac{1}{k} \left(a_0 + \sum_{n=1}^{\infty} \frac{1}{(1 - \beta_n{}^2)^2 + (2\xi\beta_n)^2} \right.$$

$$\times \{ [a_n 2\xi\beta_n + b_n(1 - \beta_n{}^2)] \sin n\bar{\omega}_1 t$$

$$\left. + [a_n(1 - \beta_n{}^2) - b_n 2\xi\beta_n] \cos n\bar{\omega}_1 t \} \right) \qquad (5\text{-}4)$$

EXAMPLE E5-1 As an example of the response analysis of a periodically loaded structure, consider the system and loading shown in Fig. E5-1. The loading in this case consists of the positive portion of a simple sine function. The Fourier coefficients defining this loading are found by applying Eqs. (5-2):

$$a_0 = \frac{1}{T_p} \int_0^{T_p/2} p_0 \sin \frac{2\pi t}{T_p} \, dt = \frac{p_0}{\pi}$$

$$a_n = \frac{2}{T_p} \int_0^{T_p/2} p_0 \sin \frac{2\pi t}{T_p} \cos \frac{2\pi n t}{T_p} \, dt = \begin{cases} 0 & n = \text{odd} \\ \dfrac{p_0}{\pi} \dfrac{2}{1 - n^2} & n = \text{even} \end{cases}$$

$$b_n = \frac{2}{T_p} \int_0^{T_p/2} p_0 \sin \frac{2\pi t}{T_p} \sin \frac{2\pi n t}{T_p} \, dt = \begin{cases} \dfrac{p_0}{2} & n = 1 \\ 0 & n > 1 \end{cases}$$

Thus substituting these in Eq. (5-1) leads to the following series expression for the periodic loading:

$$p(t) = \frac{p_0}{\pi}\left(1 + \frac{\pi}{2}\sin \bar{\omega}_1 t - \frac{2}{3}\cos 2\bar{\omega}_1 t\right.$$

$$\left. - \frac{2}{15}\cos 4\bar{\omega}_1 t - \frac{2}{35}\cos 6\bar{\omega}_1 t - \cdots\right) \qquad (a)$$

in which

$$\bar{\omega}_1 = \frac{2\pi}{T_p}$$

If it is now assumed that the structure of Fig. E5-1 is undamped, and if, for example, the period of the loading is taken at four-thirds the period of vibration of the structure, i.e.,

$$\frac{T_p}{T} = \frac{4}{3} \qquad \frac{\bar{\omega}_1}{\omega} \equiv \beta_1 = \frac{3}{4} \qquad \frac{2\bar{\omega}_1}{\omega} \equiv \beta_2 = \frac{3}{2} \qquad (b)$$

the steady-state response of the structure will be given by an equation of the form of Eq. (5-3). Introducing numerical values for the load coefficients and frequency ratios from Eqs. (a) and (b) leads finally to

$$v(t) = \frac{p_0}{k\pi}\left(1 + \frac{8\pi}{7}\sin \bar{\omega}_1 t + \frac{8}{15}\cos 2\bar{\omega}_1 t + \frac{1}{60}\cos 4\bar{\omega}_1 t + \cdots\right) \qquad (c)$$

If the structure were damped, the analysis would proceed similarly, using Eq. (5-4) in place of Eq. (5-3). ////

5-3 EXPONENTIAL FORM OF FOURIER SERIES SOLUTION

The Fourier series expressions of Eqs. (5-1) and (5-2) can also be written in exponential form by substituting for the trigonometric functions the corresponding exponential terms given by Euler's equation, namely,

$$\sin x = -\frac{1}{2}i(e^{-ix} - e^{-ix}) \qquad \cos x = \frac{1}{2}(e^{ix} + e^{-ix}) \qquad (5-5)$$

The result is

$$p(t) = \sum_{n=-\infty}^{\infty} c_n \exp(in\bar{\omega}_1 t) \qquad (5-6)$$

where

$$c_n = \frac{1}{T_p}\int_0^{T_p} p(t)\exp(-in\bar{\omega}_1 t)\,dt \qquad (5-7)$$

In Eq. (5-6) it should be noted that for each positive value of n, say $n = +m$, there is a corresponding $n = -m$. The two terms $\exp(im\bar{\omega}_1 t)$ and $\exp(-im\bar{\omega}_1 t)$

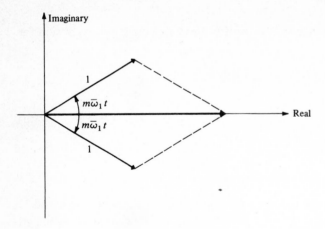

FIGURE 5-2
Vector representation of exponential load term.

may be thought of as unit vectors rotating in counterclockwise and clockwise directions, respectively, with the angular velocity $m\bar{\omega}_1$, as shown in Fig. 5-2. The imaginary components of such vector pairs always cancel each other. It should also be noted in Eq. (5-7) that c_{+m} is the complex conjugate of c_{-m}; thus all imaginary terms in the equation cancel, as they should if $p(t)$ is a real loading function.

Having expressed the arbitrary periodic loading in the exponential Fourier series form, by means of Eqs. (5-6) and (5-7), it is desirable also to write the equations defining the response to harmonic loading in exponential form. Again only the steady-state response will be considered, it being assumed that the periodic loading has continued long enough for the starting transient to have died out. Introducing the unit complex forcing function $\exp(i\bar{\omega}t)$ into the equation of motion (4-1) leads to

$$m\ddot{v}(t) + c\dot{v}(t) + kv(t) = \exp(i\bar{\omega}t) \qquad (5\text{-}8)$$

which has a steady-state solution of the form

$$v(t) = H(\bar{\omega}) \exp(i\bar{\omega}t) \qquad (5\text{-}9)$$

When Eq. (5-9) is substituted into Eq. (5-8), it is found that the function $H(\bar{\omega})$, which will be designated henceforth as the *complex-frequency-response function*, takes the form

$$H(\bar{\omega}) = \frac{1}{-\bar{\omega}^2 m + i\bar{\omega}c + k} \qquad (5\text{-}10)$$

Upon introducing expressions for the frequency ratio β and the damping ratio ζ this becomes

$$H(\bar{\omega}) = \frac{1}{k(-\beta^2 + 2i\beta\zeta + 1)} \qquad (5\text{-}11)$$

Consequently, the complex frequency response to a forcing frequency $\bar{\omega}_n = n\bar{\omega}_1$ will be

$$H(n\bar{\omega}_1) = \frac{1}{k(-n^2\beta_1{}^2 + 2in\beta_1\xi + 1)} \qquad (5\text{-}12)$$

where $\beta_1 = \bar{\omega}_1/\omega$. From the form of Eq. (5-10) it can be seen that $H(n\bar{\omega}_1)$ is the complex conjugate of $H(-n\bar{\omega}_1)$. Hence it is possible to express the steady-state response of the SDOF system to the forcing function which represents each term of the Fourier series. From the principle of superposition, it follows that the total steady-state response of the system to any periodic forcing function can be written

$$v(t) = \sum_{n=-\infty}^{\infty} H(n\bar{\omega}_1)c_n \exp(in\bar{\omega}_1 t) \qquad (5\text{-}13)$$

The advantage in simplicity of the exponential form of the periodic-response analysis is apparent when Eq. (5-13) is compared with the equivalent trigonometric-series expression of Eq. (5-4).

PROBLEMS

5-1 Express the periodic loading shown in Fig. P5-1 as a Fourier series. Thus, determine the coefficients a_n and b_n by means of Eqs. (5-2) for the periodic loading given by

$$p(t) = p_0 \sin\frac{3\pi}{T_p}t \qquad (0 < t < 2\pi)$$

$$p(t) = 0 \qquad (2\pi < t < 3\pi)$$

Then write the loading in the series form of Eq. (5-1).

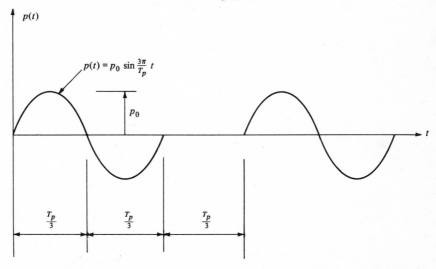

FIGURE P5-1

5-2 Repeat Prob. 5-1 for the periodic loading shown in Fig. P5-2.

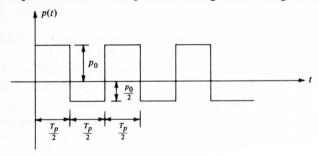

FIGURE P5-2

5-3 Solve the problem of Example E5-1, assuming that the structure is 10 percent critically damped.

5-4 Construct an Argand diagram similar to that of Fig. 4-3, showing to scale the applied load vector and the steady-state inertia, damping, and elastic resisting-force vectors. Assume the structure has 15 percent critical damping and is subjected to the harmonic loading term $p_n(t) = a_n \cos \bar{\omega}_n t$, where $\bar{\omega}_n = {}^6/_5 \omega$ (i.e., $\beta_n = {}^6/_5$). Construct the diagram for the time when $\bar{\omega}_n t = \pi/4$.

5-5 The periodic loading of Fig. P5-3 can be expressed by the sine series

$$p(t) = \sum_{n=1}^{\infty} b_n \sin \bar{\omega}_n t$$

where

$$b_n = -\frac{2p_0}{n\pi} (-1)^n$$

Plot the steady-state response of the structure of Fig. E5-1a to this loading for one full period, considering only the first four terms of the series and evaluating at time increments given by $\bar{\omega}_1 \Delta t = 30°$.

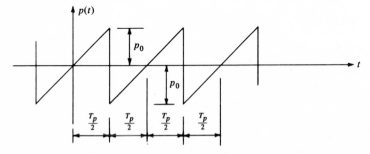

FIGURE P5-3

6

RESPONSE TO IMPULSIVE LOADS

6-1 GENERAL NATURE OF IMPULSIVE LOADS

Another special class of dynamic loading of the SDOF system will now be considered, the impulsive load. Such a load consists of a single principal impulse, as illustrated by the example in Fig. 6-1, and generally is of relatively short duration. Impulsive or shock loads frequently are of great importance in the design of certain classes of structural systems, e.g., vehicles such as trucks or automobiles or traveling cranes. Damping has much less importance in controlling the maximum response of a structure to impulsive loads than for periodic and harmonic loads. The maximum response to an impulsive load will be reached in a very short time, before the damping forces can absorb much energy from the structure. For this reason only the undamped response to impulsive loads will be considered in this section.

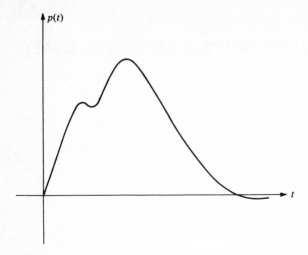

FIGURE 6-1
Arbitrary impulsive loading.

6-2 SINE-WAVE IMPULSE

For impulsive loads which can be expressed by simple analytical functions, closed-form solutions of the equations of motion can be obtained. As an example of this type of load, consider the sine-wave impulse shown in Fig. 6-2. The response will be divided into two phases, as shown, corresponding to the interval during which the load acts, followed by the free-vibration phase.

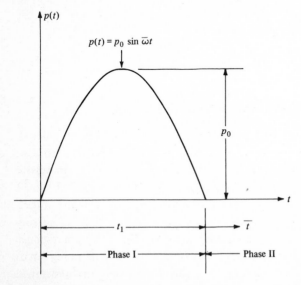

FIGURE 6-2
Half-sine-wave impulse.

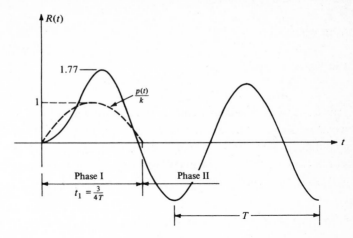

FIGURE 6-3
Response ratio due to sine pulse ($t_1 = {}^3/_4 T$).

PHASE I   During this phase the structure is subjected to harmonic loading, starting from rest. The undamped response, including the transient as well as the steady-state term, is given by [Eq. (4-11)]:

For $0 \leq t \leq t_1$:
$$v(t) = \frac{p_0}{k} \frac{1}{1 - \beta^2} (\sin \bar{\omega} t - \beta \sin \omega t) \qquad (6\text{-}1)$$

PHASE II   The free-vibration motion which occurs during this phase depends on the displacement $v(t_1)$ and velocity $\dot{v}(t_1)$ existing at the end of phase I, and may be expressed as follows [see Eq. (3-11)]:

For $\bar{t} = t - t_1 \geq 0$:
$$v(\bar{t}) = \frac{\dot{v}(t_1)}{\omega} \sin \omega \bar{t} + v(t_1) \cos \omega \bar{t} \qquad (6\text{-}2)$$

in which the new time variable $\bar{t} = t - t_1$ has been introduced for convenience.

The magnitude of dynamic response which results from this impulsive load depends on the ratio of the load duration to the period of vibration of the structure. The response ratio $R(t) = v(t)/(p_0/k)$ for $t_1/T = {}^3/_4$ is shown in Fig. 6-3. Also plotted for comparison is $p(t)/k$, which has a peak value of unity on the response-ratio scale.

In general, the maximum response produced by the impulsive load rather than the complete history is of most interest to the structural engineer. The time when the peak response occurs can be determined by differentiating Eq. (6-1) with respect to time and equating to zero; thus

$$\frac{dv(t)}{dt} = 0 = \frac{p_0}{k} \frac{1}{1 - \beta^2} (\bar{\omega} \cos \bar{\omega} t - \bar{\omega} \cos \omega t)$$

from which

$$\cos \bar{\omega}t = \cos \omega t$$

and hence

$$\bar{\omega}t = 2\pi n \pm \omega t \qquad n = 0, \pm 1, 2, 3, \ldots \qquad (6\text{-}3)$$

This expression is valid, of course, only so long as $\bar{\omega}t \leq \pi$, that is, if the maximum response occurs while the impulsive load is acting. For the most interesting loading condition, where the load frequency approaches the free-vibration frequency, i.e., where $\bar{\omega} \to \omega$, the time of maximum response will be given by substituting $n = 1$ and using the negative sign in Eq. (6-3), which leads to

$$\bar{\omega}t = \frac{2\pi}{1 + (\omega/\bar{\omega})} \qquad (6\text{-}4)$$

The maximum-response amplitude can then be obtained by introducing Eq. (6-4) into Eq. (6-1); the result is valid only if $\bar{\omega}t \leq 1$, which will be the case if $\beta < 1$, that is, $\bar{\omega} < \omega$.

For $\beta > 1$ $(\bar{\omega} > \omega)$ the maximum response occurs during the free-vibration phase (phase II). The initial displacement and velocity for this phase are given by introducing $\bar{\omega}t_1 = \pi$ into Eq. (6-1):

$$v(t_1) = \frac{p_0}{k} \frac{1}{1 - \beta^2} \left(0 - \beta \sin \frac{\pi}{\beta} \right)$$
$$\dot{v}(t_1) = \frac{p_0}{k} \frac{\bar{\omega}}{1 - \beta^2} \left(-1 - \cos \frac{\pi}{\beta} \right) \qquad (6\text{-}5)$$

The amplitude of this free-vibration motion is then given by Eq. (3-15), i.e.,

$$\rho = \left\{ \left[\frac{\dot{v}(t_1)}{\omega} \right]^2 + [v(t_1)]^2 \right\}^{1/2} = \frac{p_0/k}{1 - \beta^2} \beta \left(2 + 2\cos \frac{\pi}{\beta} \right)^{1/2}$$

Hence the dynamic magnification factor for this condition is

For $\beta > 1, t > t_1$: $\qquad\qquad D = \frac{v_{max}}{p_0/k} = \frac{2\beta}{1 - \beta^2} \cos \frac{\pi}{2\beta} \qquad (6\text{-}6)$

EXAMPLE E6-1 As an example of the maximum-response analysis for a *long*-duration sine impulse, where the maximum occurs while the load is acting, consider the case where $\beta = {}^2/_3 (\bar{\omega} = {}^2/_3\omega$ or $t_1 = {}^3/_4 T)$. For this case Eq. (6-4) gives

$$\bar{\omega}t = \frac{2\pi}{1 + {}^3/_2} = {}^4/_5\pi$$

and with this value substituted in Eq. (6-1), the dynamic magnification factor is

$$D = \frac{1}{1 - \frac{4}{9}} \left(\sin \frac{4}{5}\pi - \frac{2}{3} \sin \frac{6}{5}\pi\right) = 1.77$$

A specific example of a *short*-duration impulse, where the peak response occurs during the free-vibration phase, is the case where $\beta = \frac{4}{3}$ ($\overline{\omega} = \frac{4}{3}\omega$ or $t_1 = \frac{3}{8}T$). In this case the dynamic magnification factor is found from Eq. (6-6) to be

$$D = \frac{2(\frac{4}{3})}{1 - \frac{16}{9}} \cos \frac{\pi}{2(\frac{4}{3})} = 1.31$$

With similar procedures, the maximum response to a resonant impulsive loading ($\beta = 1$) can be found from Eq. (4-31), which is the resonance equation. In this case the maximum response occurs at the end of the impulse, $\overline{\omega}t = \pi$, and the dynamic magnification factor is

$$D = \frac{\pi}{2} = 1.57 \qquad \text{////}$$

6-3 RECTANGULAR IMPULSE

The second case of the analysis of the response to an impulsive load will concern the rectangular impulse shown in Fig. 6-4. Again the response will be divided into the loading phase and the subsequent free-vibration phase.

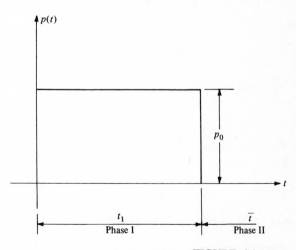

FIGURE 6-4
Rectangular impulse.

PHASE I The suddenly applied load which remains constant during phase I is called a *step loading*. The particular solution for a step loading is simply the static deflection which it would produce:

$$v_p = \frac{p_0}{k} \qquad (6\text{-}7a)$$

From this result, the general solution in which the free-vibration constants have been evaluated to satisfy the at-rest initial conditions is easily found to be:

For $0 \leq t \leq t_1$:
$$v(t) = \frac{p_0}{k}(1 - \cos \omega t) \qquad (6\text{-}7b)$$

PHASE II The free vibration during phase II is again given by Eq. (6-2):

For $\bar{t} = t - t_1 \geq 0$:
$$v(t) = \frac{\dot{v}(t_1)}{\omega} \sin \omega \bar{t} + v(t_1) \cos \omega \bar{t} \qquad (6\text{-}8)$$

For this rectangular impulse, it is evident that the maximum response will always occur in phase I if $t_1 \geq T/2$ and that the dynamic magnification factor D in this case is 2. For shorter-duration loadings, the maximum response will occur during the free vibration of phase II, and the response amplitude will be given by Eq. (3-15):

$$\rho = v_{\max} = \sqrt{\left[\frac{\dot{v}(t_1)}{\omega}\right]^2 + [v(t_1)]^2} \qquad (6\text{-}9)$$

With $\dot{v}(t) = p_0\omega/k \sin \omega t$ and $\omega = 2\pi/T$ this becomes

$$v_{\max} = \frac{p_0}{k}\left[\left(1 - 2\cos\frac{2\pi}{T}t_1 + \cos^2\frac{2\pi}{T}t_1\right) + \sin^2\frac{2\pi}{T}t_1\right]^{1/2}$$

$$= \frac{p_0}{k}\left[2\left(1 - \cos\frac{2\pi}{T}t_1\right)\right]^{1/2}$$

from which

$$D \equiv \frac{v_{\max}}{p_0/k} = 2\sin\frac{\pi t_1}{T} \qquad \frac{t_1}{T} \leq \frac{1}{2} \qquad (6\text{-}10)$$

Thus the dynamic magnification factor varies as a sine function of the load pulse-length ratio t_1/T for ratios less than $^1/_2$.

6-4 TRIANGULAR IMPULSE

The last impulse loading to be analyzed in detail is the decreasing triangular impulse shown in Fig. 6-5.

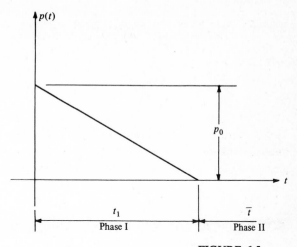

FIGURE 6-5
Triangular impulse.

PHASE I The loading during this phase is $p_0(1 - t/t_1)$, and it is easily demonstrated that the particular solution to this loading is

$$v_p(t) = \frac{p_0}{k}\left(1 - \frac{t}{t_1}\right) \qquad (6\text{-}11)$$

If zero initial conditions are assumed, the free-vibration constants in the general solution can be evaluated, leading to

$$v(t) = \frac{p_0}{k}\left(\frac{\sin \omega t}{\omega t_1} - \cos \omega t - \frac{t}{t_1} + 1\right) \qquad (6\text{-}12)$$

PHASE II Evaluating Eq. (6-12) and its first derivative at the end of phase I ($t = t_1$) gives

$$v(t_1) = \frac{p_0}{k}\left(\frac{\sin \omega t_1}{\omega t_1} - \cos \omega t_1\right)$$

$$\dot{v}(t_1) = \frac{p_0\omega}{k}\left(\frac{\cos \omega t_1}{\omega t_1} + \sin \omega t_1 - \frac{1}{\omega t_1}\right) \qquad (6\text{-}13)$$

which can be substituted into Eq. (6-2) to obtain the free-vibration response in phase II.

The maximum values of these response functions are found, as for the other examples, by evaluating them for the times at which the zero-velocity condition is achieved. For loadings of very short duration ($t_1/T < 0.4$) the maximum response

Table 6-1 DYNAMIC MAGNIFICATION FACTOR FOR TRIANGULAR IMPULSE LOADING

t_1/T	0.20	0.40	0.50	0.75	1.00	1.50	2.00
D	0.60	1.05	1.19	1.38	1.53	1.68	1.76

occurs during the free vibrations of phase II; otherwise it occurs during the loading interval (phase I). Values of the dynamic magnification factor $D = v_{max}/(p_0/k)$ computed for various loading durations are presented in Table 6-1.

6-5 SHOCK OR RESPONSE SPECTRA

In the expressions derived above the maximum response produced in an undamped SDOF structure by a given form of impulsive loading depends only on the ratio of the impulse duration to the natural period of the structure t_1/T. Thus, it is convenient to plot the dynamic magnification factor D as a function of t_1/T for various forms of impulsive loading. For example, the data presented in Table 6-1 have been plotted as one curve of Fig. 6-6. Similar plots, corresponding to other forms of impulsive load,

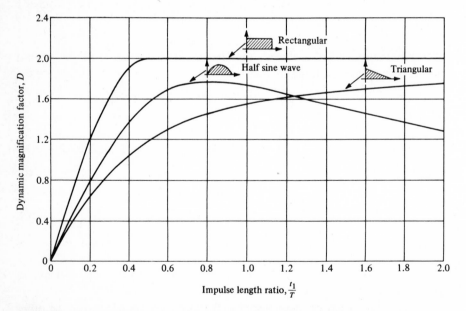

FIGURE 6-6
Displacement-response spectra (shock spectra) for three types of impulse.

are presented in the other curves of this figure; these are known as the *displacement-response spectra*, or merely the *response spectra*, of the impulsive loads. Generally plots like these can be used to predict with necessary engineering accuracy the maximum effect to be expected from a given type of impulsive loading acting on a simple structure.

These response spectra also serve to indicate the response of the structure to an acceleration impulse applied to its base. If the applied base acceleration is $\ddot{v}_g(t)$, it produces an effective impulsive loading $p_{eff}(t) = -m\ddot{v}_g(t)$ [see Eq. (2-21)]. If the maximum base acceleration is denoted by $\ddot{v}_{g0}$, the maximum effective impulsive load is $p_{0,\,eff} = -m\ddot{v}_{g0}$. The dynamic magnification factor thus becomes

$$D = \left| \frac{v_{max}}{m\ddot{v}_{g0}/k} \right| \qquad (6\text{-}14)$$

in which only the absolute magnitude of the response generally is of interest. Alternatively this may be written

$$D = \left| \frac{\ddot{v}^t_{max}}{\ddot{v}_{g0}} \right| \qquad (6\text{-}15)$$

where $\ddot{v}^t_{max}$ is the maximum total acceleration of the mass. This follows from the fact that in an undamped system the product of the mass and the total acceleration must be equal in magnitude to the elastic spring force kv_{max}. Accordingly it is evident that the response spectrum plots of Fig. 6-6 may be used to predict the maximum acceleration response of the mass m to an impulsive base acceleration as well as the maximum displacement response to impulsive loads. When used for this purpose, the plots are generally referred to as *shock spectra*.

EXAMPLE E6-2   As an example of the use of shock spectra in evaluating the maximum response of a SDOF structure to an impulsive load, consider the system shown in Fig. E6-1, which represents a single-story building subjected to a blast load. For the given weight and column stiffness of this structure, the period of vibration is

$$T = \frac{2\pi}{\omega} = 2\pi \sqrt{\frac{W}{kg}} = 2\pi \sqrt{\frac{600}{10,000(386)}} = 0.079 \text{ s}$$

The impulse-length ratio thus is

$$\frac{t_1}{T} = \frac{0.05}{0.079} = 0.63$$

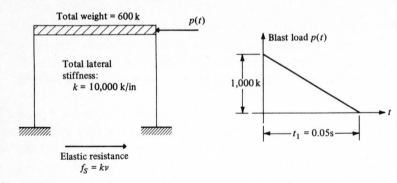

FIGURE E6-1
SDOF building subjected to blast load.

and from Fig. 6-6 the dynamic magnification factor is $D = 1.33$. Thus the maximum displacement will be

$$v_{\max} = D \frac{p_0}{k} = 1.33 \frac{1,000}{10,000} = 0.133 \text{ in}$$

and the maximum elastic forces which will develop are

$$f_{S,\max} = k v_{\max} = 10,000(0.133) = 1,330 \text{ kips}$$

If the blast-pressure impulse had been only one-tenth as long ($t_1 = 0.005$ s), the dynamic magnification factor for this impulse-length ratio ($t_1/T = 0.063$) would have been only $D = 0.44$ and hence the elastic resisting forces would have been only $f_S = 440$ kips. Thus for an impulse of very short duration, a large part of the applied load is resisted by the inertia of the structure, and the stresses produced are much smaller than those due to the longer loading.

$////$

6-6 APPROXIMATE ANALYSIS OF IMPULSIVE-LOAD RESPONSE

From a study of the response spectra presented in Fig. 6-6 and similar spectra for other forms of loadings, two general conclusions may be drawn concerning the response of structures to impulsive loadings:

1 For long-duration loadings, for example, $t_1/T > 1$, the dynamic magnification factor depends principally on the rate of increase of the load to its maximum value. A step loading of sufficient duration produces a magnification factor of 2; a very gradual increase causes a magnification factor of 1.

2 For short-duration loads, for example, $t_1/T < {}^1/_4$, the maximum displacement amplitude v_{max} depends principally upon the magnitude of the applied impulse $I = \int_0^{t_1} p(t)\,dt$ and is not strongly influenced by the form of the loading impulse. The dynamic magnification factor D, however, is quite dependent upon the form of loading because it is proportional to the ratio of impulse area to peak-load amplitude, as may be noted by comparing the curves of Fig. 6-6 in the short-period range. Thus v_{max} is the more significant measure of response.

A convenient approximate procedure for evaluating the maximum response to a short-duration impulsive load, which represents a mathematical expression of this second conclusion, may be derived as follows. The impulse-momentum relationship for the mass m may be written

$$m\,\Delta\dot{v} = \int_0^{t_1} [p(t) - kv(t)]\,dt \qquad (6\text{-}16)$$

in which $\Delta\dot{v}$ represents the change of velocity produced by the loading. In this expression it may be observed that for small values of t_1 the displacement developed during the loading $v(t_1)$ is of the order of $(t_1)^2$ while the velocity change $\Delta\dot{v}$ is of the order of t_1. Thus since the impulse is also of the order of t_1, the elastic force term $kv(t)$ vanishes from the expression as t_1 approaches zero and is negligibly small for short-duration loadings.

On this basis, the approximate relationship may be used:

$$m\,\Delta\dot{v} \doteq \int_0^{t_1} p(t)\,dt \qquad (6\text{-}17)$$

or

$$\Delta\dot{v} = \frac{1}{m}\int_0^{t_1} p(t)\,dt \qquad (6\text{-}18)$$

The response after the termination of the loading is a free vibration

$$v(\bar{t}) = \frac{\dot{v}(t_1)}{\omega}\sin\omega\bar{t} + v(t_1)\cos\omega\bar{t}$$

in which $\bar{t} = t - t_1$. But since the displacement term $v(t_1)$ is negligibly small and the velocity $\dot{v}(t_1) = \Delta\dot{v}$, the following approximate relationship may be used:

$$v(\bar{t}) \doteq \frac{1}{m\omega}\left(\int_0^{t_1} p(t)\,dt\right)\sin\omega\bar{t} \qquad (6\text{-}19)$$

EXAMPLE E6-3 As an example of the use of this approximate formula, consider the response of the structure shown in Fig. E6-2 to the impulsive load-

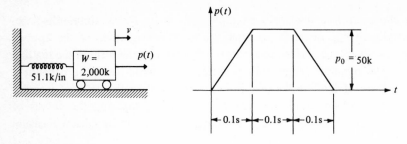

FIGURE E6-2
Approximate impulse-response analysis.

ing indicated. In this case, $\omega = \sqrt{kg/W} = 3.14$ rad/s and $\int_0^{t_1} p(t)\,dt = 10$ kip · s. The response then is approximately

$$v(\bar{t}) = \frac{10(386)}{2,000(3.14)} \sin \omega \bar{t}$$

in which the acceleration of gravity is taken as $g = 386$ in/s². The maximum response results when $\sin \omega \bar{t} = 1$, that is,

$$v_{max} \doteq 0.614 \text{ in}$$

The maximum elastic force developed in the spring, which is of major concern to the structural engineer, is

$$f_{S,\,max} = kv_{max} = 51.1(0.614) = 31.4 \text{ kips}$$

Since the period of vibration of this system is $T = 2\pi/\omega = 2$ s, for this short-duration loading $(t_1/T = 0.15)$ the approximate analysis may be expected to be quite reliable. In fact, the maximum response determined by direct integration of the equation of motion is 0.604 in, and so the error in the approximate result is less than 2 percent. ////

PROBLEMS

6-1 Consider the basic dynamic system of Fig. 3-1 with the following properties: $W = 600$ lb $(m = W/g)$ and $k = 1,000$ lb/in. Assume that it is subjected to a half sine-wave impulse (Fig. 6-2) of amplitude $p_0 = 500$ lb and duration $t_1 = 0.15$ s. Determine:

(a) The time at which the maximum response will occur.

(b) The maximum spring force produced by this loading; check this result with that obtained by use of Fig. 6-6.

6-2 A triangular impulse that increases linearly from zero to the peak value is expressed as $p(t) = p_0(t/t_1)$ $(0 < t < t_1)$.

(a) Derive an expression for the response of a SDOF structure to this loading, starting from "at rest" conditions.

(b) Determine the maximum response ratio

$$\left(R_{max} = \frac{v_{max}}{p_0/k} \right)$$

resulting from this loading if $t_1 = 3\pi/\omega$.

6-3 A quarter cosine-wave impulse is expressed as

$$p(t) = p_0 \cos \bar{\omega} t \left(0 < t < \frac{\pi}{2\bar{\omega}} \right)$$

(a) Derive an expression for the response to this impulse, starting from rest.

(b) Determine the maximum response ratio

$$R_{max} = \frac{v_{max}}{p_0/k} \text{ if } \bar{\omega} = \omega$$

6-4 The basic SDOF system of Fig. 3-1, having the following properties, $k = 20$ k/in and $m = 4$ k·s^2/in, is subjected to a triangular impulse of the form of Fig. 6-5 with $p_0 = 15$ k and $t_1 = 0.15T$.

(a) Using the shock spectra of Fig. 6-6, determine the maximum spring force $f_{S\,max}$.

(b) Using Eq. (6-19), compute approximately the maximum displacement and spring force; compare with the result of part a.

6-5 The water tank of Fig. P6-1a can be treated as a SDOF structure with the following properties: $m = 4$ k·s^2/in, $k = 40$ k/in. As a result of an explosion, the tank is subjected to the dynamic-load history shown in Fig. P6-1b. Compute approximately the maximum overturning moment $\mathfrak{M}_0$ at the base of the tower using Eq. (6-14) and evaluating the impulse integral by means of Simpson's rule:

$$\int p \, dt = \frac{\Delta t}{3} (p_0 + 4p_1 + 2p_2 + 4p_3 + p_4)$$

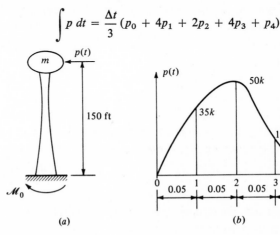

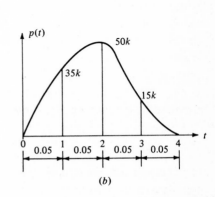

(a)   (b)

FIGURE P6-1

7

RESPONSE TO GENERAL DYNAMIC LOADING

7-1 DUHAMEL INTEGRAL FOR AN UNDAMPED SYSTEM

The procedure described in Chap. 6 for approximating the response of a structure to a short-duration impulse may be used as the basis for developing a formula for evaluating response to a general dynamic loading. Consider the arbitrary general loading $p(t)$ shown in Fig. 7-1, specifically the intensity of loading $p(\tau)$ acting at time $t = \tau$. This loading acting during the short interval of time $d\tau$ produces a short-duration impulse $p(\tau)\, d\tau$ on the structure, and Eq. (6-19) can be used to evaluate the response to this impulse. It should be noted carefully that although the procedure is only approximate for impulses of finite duration, it becomes exact as the duration of loading approaches zero. Thus for the differential time interval $d\tau$, the response produced by the loading $p(\tau)$ is *exactly* (for $t > \tau$)

$$dv(t) = \frac{p(\tau)\, d\tau}{m\omega} \sin \omega(t - \tau) \qquad (7\text{-}1)$$

In this expression, the term $dv(t)$ represents the differential response to the differential impulse over the entire response history for $t > \tau$; it is not the change of v during a time interval dt.

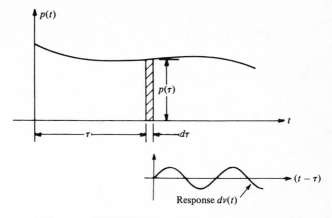

FIGURE 7-1
Derivation of the Duhamel integral (undamped).

The entire loading history may be considered to consist of a succession of such short impulses, each producing its own differential response of the form of Eq. (7-1). For this linearly elastic system, then, the total response can be obtained by summing all the differential responses developed during the loading history, that is, by integrating Eq. (7-1) as follows:

$$v(t) = \frac{1}{m\omega} \int_0^t p(\tau) \sin \omega(t - \tau)\, d\tau \qquad (7\text{-}2)$$

Equation (7-2) is generally known as the *Duhamel integral* for an undamped system. It may be used to evaluate the response of an undamped SDOF system to any form of dynamic loading $p(t)$, although in the case of arbitrary loadings the evaluation will have to be performed numerically.

Equation (7-2) may also be expressed in the form

$$v(t) = \int_0^t p(\tau) h(t - \tau)\, d\tau \qquad (7\text{-}3)$$

where the new symbol has the definition

$$h(t - \tau) \equiv \frac{1}{m\omega} \sin \omega(t - \tau) \qquad (7\text{-}4)$$

Equation (7-3) is called the *convolution integral*; computing the response of a structure to an arbitrary loading using this integral is known as obtaining the response through the time domain. The function $h(t - \tau)$ is generally referred to as the unit-impulse response (defined in this case for an undamped system), because it expresses the response of the system to an impulse of unit magnitude applied at time $t = \tau$.

In Eq. (7-2) it has been tacitly assumed that the loading was initiated at time $t = 0$ and that the structure was at rest at that time. For any other specified initial conditions, $v(0) \neq 0$ and $\dot{v}(0) \neq 0$, an additional free-vibration response must be added to this solution; thus, in general,

$$v(t) = \frac{\dot{v}(0)}{\omega} \sin \omega t + v(0) \cos \omega t + \frac{1}{m\omega} \int_0^t p(\tau) \sin \omega(t - \tau) \, d\tau \qquad (7\text{-}5)$$

7-2   NUMERICAL EVALUATION OF THE DUHAMEL INTEGRAL FOR AN UNDAMPED SYSTEM

If the applied-loading function is integrable, the dynamic response of the structure can be evaluated by the formal integration of Eq. (7-2) or (7-5). In many practical cases, however, the loading is known only from experimental data, and the response must be evaluated by numerical processes. For such analyses it is useful to note the trigonometric identity, $\sin (\omega t - \omega \tau) = \sin \omega t \cos \omega \tau - \cos \omega t \sin \omega \tau$, and to write Eq. (7-2) in the form (zero initial conditions being assumed)

$$v(t) = \sin \omega t \, \frac{1}{m\omega} \int_0^t p(\tau) \cos \omega \tau \, d\tau - \cos \omega t \, \frac{1}{m\omega} \int_0^t p(\tau) \sin \omega \tau \, d\tau$$

or
$$v(t) = \bar{A}(t) \sin \omega t - \bar{B}(t) \cos \omega t \qquad (7\text{-}6)$$

where
$$\bar{A}(t) = \frac{1}{m\omega} \int_0^t p(\tau) \cos \omega \tau \, d\tau$$

$$\bar{B}(t) = \frac{1}{m\omega} \int_0^t p(\tau) \sin \omega \tau \, d\tau \qquad (7\text{-}7)$$

The numerical integration of the Duhamel integral thus requires the evaluation of the integrals $\bar{A}(t)$ and $\bar{B}(t)$ numerically. Consider, for example, the first of these; the function to be integrated is depicted graphically in Fig. 7-2. For convenience of numerical calculation, the function has been evaluated at equal time increments $\Delta\tau$, successive values of the function being identified by appropriate subscripts. The value of the integral can then be obtained approximately by summing these ordinates multiplied by appropriate weighting factors. Expressed mathematically, this is

$$\bar{A}(t) = \frac{1}{m\omega} \int_0^t y(\tau) \, d\tau \doteq \frac{\Delta\tau}{m\omega} \frac{1}{\zeta} \sum_\zeta^{\bar{A}} (t) \qquad (7\text{-}8)$$

in which $y(\tau) = p(\tau) \cos \omega \tau$ and $(1/\zeta) \sum_\zeta^{\bar{A}}$ represents the numerical summation process, the specific form of which depends on the order of the integration approx-

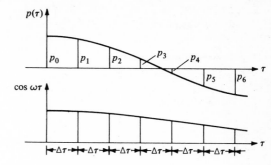

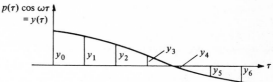

FIGURE 7-2
Formulation of numerical summation
process for Duhamel integral.

imation being used. For three elementary approximation procedures, the summations
are performed as follows:

Simple summation ($\zeta = 1$):

$$\sum_{1}^{\bar{A}} (t) = y_0 + y_1 + y_2 + \cdots + y_{N-1} \qquad (7\text{-}9a)$$

Trapezoidal rule ($\zeta = 2$):

$$\sum_{2}^{\bar{A}} (t) = y_0 + 2y_1 + 2y_2 + \cdots + 2y_{N-1} + y_N \qquad (7\text{-}9b)$$

Simpson's rule ($\zeta = 3$):

$$\sum_{3}^{\bar{A}} (t) = y_0 + 4y_1 + 2y_2 + \cdots + 4y_{N-1} + y_N \qquad (7\text{-}9c)$$

where $N = t/\Delta\tau$ must be an even number for Simpson's rule.

Using any of the summation processes of Eq. (7-9) with Eq. (7-8) leads to an
approximation of the integral for the specific time t under consideration. Generally,
however, the entire history of response is required rather than merely the displacement at some specific time; in other words, the response must be evaluated successively
at a sequence of times $t_1, t_2, \ldots$, where the interval between these times is $\Delta\tau$ (or
$2\,\Delta\tau$ if Simpson's rule is used). To provide this complete response history it is more
convenient to express the summations of Eq. (7-9) in incremental form:

Simple summation ($\zeta = 1$):

$$\sum_{1}^{\bar{A}} (t) = \sum_{1}^{\bar{A}} (t - \Delta\tau) + p(t - \Delta\tau) \cos \omega(t - \Delta\tau) \qquad (7\text{-}10a)$$

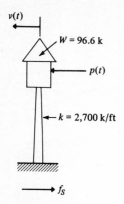

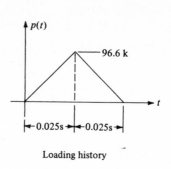

Loading history

FIGURE E7-1
Water tower subjected to blast load.

Trapezoidal rule ($\zeta = 2$):

$$\sum_{2}^{\bar{A}}(t) = \sum_{2}^{\bar{A}}(t - \Delta\tau) + \left[p(t - \Delta\tau) \cos \omega(t - \Delta\tau) + p(t) \cos \omega t \right] \qquad (7\text{-}10b)$$

Simpson's rule ($\zeta = 3$):

$$\sum_{3}^{\bar{A}}(t) = \sum_{3}^{\bar{A}}(t - 2\Delta\tau) + \left[p(t - 2\Delta\tau) \cos \omega(t - 2\Delta\tau) \right.$$
$$\left. + 4p(t - \Delta\tau) \cos \omega(t - \Delta\tau) + p(t) \cos \omega t \right] \qquad (7\text{-}10c)$$

in which $\sum_{\zeta}^{\bar{A}}(t - \Delta\tau)$ represents the value of the summation determined at the preceding time $t - \Delta\tau$.

The evaluation of the term $\bar{B}(t)$ can be carried out in exactly the same way, that is,

$$\bar{B}(t) = \frac{\Delta\tau}{m\omega} \frac{1}{\zeta} \sum_{\zeta}^{B} \qquad (7\text{-}11)$$

in which $\sum_{\zeta}^{B}(t)$ can be evaluated by expressions identical to Eqs. (7-10) but with sine functions replacing the cosine functions. Substituting Eqs. (7-8) and (7-11) into Eq. (7-6) leads to the final response equation for an undamped system:

$$v(t) = \frac{\Delta\tau}{m\omega} \frac{1}{\zeta} \left[\sum_{\zeta}^{\bar{A}}(t) \sin \omega t - \sum_{\zeta}^{B}(t) \cos \omega t \right] \qquad (7\text{-}12)$$

EXAMPLE E7-1 The dynamic response of a water tower subjected to a blast loading has been calculated to illustrate the numerical evaluation of the Duhamel integral. The idealization of the structure and of the blast loading are shown in Fig. E7-1. For this system, the vibration frequency and period are

$$\omega = \sqrt{\frac{kg}{W}} = \sqrt{\frac{2,700(32.3)}{96.6}} = 30 \text{ rad/s} \qquad T = \frac{2\pi}{\omega} = 0.209 \text{ s}$$

The time increment used in the numerical integration was $\Delta\tau = 0.005$ s, which corresponds to an angular increment in free vibrations of $\omega \Delta\tau = 0.15$ rad (probably a longer increment would have given equally satisfactory results). In this undamped analysis, the Simpson's rule summation was used; hence the factor $\zeta = 3$ was used in Eqs. (7-10) to (7-12).

A hand solution of the first 10 steps of the undamped response is presented in a convenient tabular format in Table E7-1 (see p. 106). The operations performed in each column are generally apparent from the labels at the top. $\Delta\bar{A}$ and $\Delta\bar{B}$ represent the summing of column 7 (or column 12) by groups of three terms, as indicated by the braces. Column 17 is the term in square brackets of Eq. (7-12), and the displacements given in column 18 were obtained by multiplying column 17 by $G = \Delta\tau/m\omega\zeta$. The forces in the last column are given by $f_S = kv(t)$. It should be noted that this is slide-rule work, so that the final results, which involve differences of large numbers, are rather rough.

Since the blast loading terminates at the end of these 10 time steps, the values of $\bar{A}$ and $\bar{B}$ remain constant after this time. If these constant values of the integrals are designated $\bar{A}^*$ and $\bar{B}^*$, the free vibrations which follow the blast loading are given by [see Eq. (7-6)]

$$v(t) = \bar{A}^* \sin \omega t - \bar{B}^* \cos \omega t$$

and the amplitude of motion is $v_{\max} = [(\bar{A}^*)^2 + (\bar{B}^*)^2]^{1/2}$.

The Duhamel integral could easily have been evaluated by formal integration for this simple form of loading, but the advantage of the numerical procedure is that it can be applied to any arbitrary loading history, even where the loads have been determined by experiment and cannot be expressed analytically.   ////

7-3   RESPONSE OF DAMPED SYSTEMS

The derivation of the Duhamel integral equation which expresses the response of a damped system to a general dynamic loading is entirely equivalent to the undamped analysis except that the free-vibration response initiated by the differential load impulse $p(\tau) \, d\tau$ is subjected to exponential decay. Thus setting $v(0) = 0$ and letting $\dot{v}(0) = [p(\tau) \, d\tau]/m$ in Eq. (3-26) leads to

$$dv(t) = e^{-\xi\omega(t-\tau)}\left[\frac{p(\tau) \, d\tau}{m\omega_D} \sin \omega_D(t - \tau)\right] \qquad t > \tau \qquad (7\text{-}13)$$

in which the exponential decay begins as soon as the load is applied—at time $t = \tau$.

Table E7-1 NUMERICAL DUHAMEL INTEGRAL ANALYSIS OF STRUCTURE IN FIG. E7-1 (Undamped)

$$\omega = \sqrt{\frac{kg}{W}} = 30 \text{ rad/s} \qquad \Delta\tau = 0.005 \text{ s} \qquad \omega\Delta\tau = 0.15 \text{ rad} = 8.59° \qquad G \equiv \frac{\Delta\tau}{m\omega\xi} = 1.852 \times 10^{-5} \qquad k = 2{,}700 \text{ kips/f}^2$$

				Evaluation of $\bar{A}$					Evaluation of $\bar{B}$									
τ, s (1)	$p(\tau)$, kips (2)	$\sin\omega\tau$ (3)	$\cos\omega\tau$ (4)	(2)× (4) (5)	Mult. (6)	(5)× (6) (7)	$\Delta\bar{A}$ (8)	$\bar{A}$ (9)	(2)× (3) (10)	Mult. (11)	(10)× (11) (12)	$\Delta\bar{B}$ (13)	$\bar{B}$ (14)	(9)× (3) (15)	(14)× (4) (16)	(15)− (16) (17)	$v(\tau)$ $G\times$ (17), ft (18)	$f_s(\tau)$ $k\times$ (18), kips (19)
0.000	0	0	1.000	0	1	0		0	0	1	0		0	0	0	0	0	0
0.005	19.32	0.150	0.989	19.09	4	76.4	113.3		2.9	4	11.6	23.0						
0.010	38.64	0.295	0.955	36.9	1	36.9		113.3	11.4	1	11.4		23.0	33.4	22.0	11.4	0.0002	0.57
0.015	57.96	0.435	0.901	52.1	4	208.4	309.1		25.2	4	100.8	155.8						
0.020	77.28	0.564	0.826	63.8	1	63.8		422.4	43.6	1	43.6		178.8	238.2	147.8	90.4	0.0017	4.52
0.025	96.60	0.677	0.736	71.1	4	284.4	396.2		67.5	4	270.0	374.1						
0.030	77.28	0.783	0.622	48.0	1	48.0		818.6	60.5	1	60.5		552.9	641.0	344.2	296.8	0.0055	14.84
0.035	57.96	0.870	0.493	28.55	4	114.2	176.2		50.5	4	202.0	298.5						
0.040	38.64	0.932	0.363	14.03	1	14.0		994.8	36.0	1	36.0		851.4	928	309.6	618	0.0114	30.9
0.045	19.32	0.976	0.220	4.25	4	17.0	31.0		18.9	4	75.6	111.6						
0.050	0	0.997	0.0715	0	1	0		1,025.8	0	1	0		963.0	1,022	69	953	0.0176	47.6

Summing these differential response terms over the entire loading interval then results in

$$v(t) = \frac{1}{m\omega_D} \int_0^t p(\tau) e^{-\xi\omega(t-\tau)} \sin \omega_D(t - \tau) \, d\tau \qquad (7\text{-}14)$$

which is the damped-response equivalent of Eq. (7-2).

Comparing Eq. (7-14) with the convolution integral of Eq. (7-3) shows that the unit-impulse response for a damped system is given by

$$h(t - \tau) = \frac{1}{m\omega_D} e^{-\xi\omega(t-\tau)} \sin \omega_D(t - \tau) \qquad (7\text{-}15)$$

For numerical evaluation of the damped-system response, Eq. (7-14) may be written in a form similar to Eq. (7-6):

$$v(t) = A(t) \sin \omega_D t - B(t) \cos \omega_D t \qquad (7\text{-}16)$$

where, in this case,

$$A(t) = \frac{1}{m\omega_D} \int_0^t p(\tau) \frac{e^{\xi\omega\tau}}{e^{\xi\omega t}} \cos \omega_D\tau \, d\tau$$

$$B(t) = \frac{1}{m\omega_D} \int_0^t p(t) \frac{e^{\xi\omega\tau}}{e^{\xi\omega t}} \sin \omega_D\tau \, d\tau \qquad (7\text{-}17)$$

These integrals can be evaluated by an incremental summation process equivalent to that used previously but taking account of the exponential decay in the process. The first integral is given by

$$A(t) \doteq \frac{\Delta\tau}{m\omega_D} \frac{1}{\zeta} \sum_\zeta^A (t) \qquad (7\text{-}18)$$

in which the summations can be expressed for the different processes considered before as follows:

Simple summation ($\zeta = 1$):

$$\sum_1^A (t) = \left[\sum_1^A (t - \Delta\tau) + p(t - \Delta\tau) \cos \omega_D(t - \Delta\tau) \right] \exp(-\xi\omega \, \Delta\tau) \qquad (7\text{-}19a)$$

Trapezoidal rule ($\zeta = 2$):

$$\sum_2^A (t) = \left[\sum_2^A (t - \Delta\tau) + p(t - \Delta\tau) \cos \omega_D(t - \Delta\tau) \right] \exp(-\xi\omega \, \Delta\tau)$$
$$+ p(t) \cos \omega_D t \qquad (7\text{-}19b)$$

Simpson's rule ($\zeta = 3$):

$$\sum_3^A (t) = \left[\sum_3^A (t - 2\Delta\tau) + p(t - 2\Delta\tau) \cos \omega_D(t - 2\Delta\tau) \right] \exp(-\xi\omega \, 2\Delta\tau)$$
$$+ 4p(t - \Delta\tau) \cos \omega_D(t - \Delta\tau) \exp(-\xi\omega \, \Delta\tau) + p(t) \cos \omega_D t \qquad (7\text{-}19c)$$

The $B(t)$ term is given by similar expressions involving the sine functions.

The accuracy of the solution to be expected from any of these numerical processes depends, of course, upon the length of the time interval $\Delta\tau$. In general, this must be selected short enough for both the load function and trigonometric functions to be well defined; $\Delta\tau \leq T/10$ is a common rule of thumb which usually provides satisfactory results. The accuracy and the computational effort increase with the order of the summation process; as a general rule, the increased accuracy of the Simpson's rule procedure justifies its use despite its greater numerical complexity.

EXAMPLE E7-2   To demonstrate how damping may be included conveniently in the numerical evaluation of the Duhamel integral, the response analysis of the system of Fig. E7-1 has been repeated using a damping ratio $\xi = 5$ percent. Again the integrals were evaluated by Simpson's rule summation; hence Eq. (7-19c) (and its sine-function counterpart) were used. For this lightly damped system, the damped frequency was taken to be the same as the undamped frequency.

A hand solution of the first 10 steps of the damped response is presented in Table E7-2. The exponential-decay factors, which represent the damping effect, have been combined with the Simpson's rule multipliers for convenience. The factor $\exp\left(-\xi\omega\,2\Delta\tau\right)$ acts on the existing value of the integral A (or B) plus the first part of the new increment; hence these are added together in column 7 (or 12) before multiplying by the decay factor. Thus the three terms making up the new $A(t)$ are column 8 $(t - 2\Delta\tau)$ plus column 8 $(t - \Delta\tau)$ plus column 5 (t). Similarly

$$B(t) = \text{col. } 13\ (t - 2\Delta\tau) + \text{col. } 13\ (t - \Delta\tau) + \text{col. } 5\ (t)$$

Since the rest of the analysis is completely equivalent to that required for the undamped system, damping leads to only a slight increase in the computational effort.

A plot of the elastic-force histories in the damped and undamped cases calculated by digital computer for 46 time steps is presented in Fig. E7-2. It is apparent here that damping has little effect during the first part of the response but causes a noticeable reduction in the maximum response and continuing reductions thereafter.   ////

7-4   RESPONSE ANALYSIS THROUGH THE FREQUENCY DOMAIN

Although the time-domain analysis described above is completely general and can be used to evaluate the response of any linear SDOF system to any arbitrary input, it sometimes is more convenient to perform the analysis in the frequency domain.

Table E7-2 NUMERICAL DUHAMEL INTEGRAL ANALYSIS INCLUDING DAMPING

Problem data in Fig. E7.1 and Table E7.1; with 5% damping, multipliers are $\exp(-2\zeta\omega\,\Delta\tau) = \exp(0.015) = 0.985$ and $4\exp(-\zeta\omega\,\Delta\tau) = 4\exp(-0.0075) = 3.97$

τ, s (1)	p(τ), kips (2)	sin ωτ (3)	cos ωτ (4)	(2)×(4) (5)	Mult. (6)	(5)+A (7)	(6)×(5) or (7) (8)	A (9)	(2)×(3) (10)	Mult. (11)	(10)+B (12)	(11)×(10) or (12) (13)	B (14)	(9)×(3) (15)	(14)×(4) (16)	(15)−(16) (17)	v(τ) G×(17), ft (18)	f_s(τ) k×(18), kips (19)
				Evaluation of A					Evaluation of B									
0.000	0	0.000	1.000	0	0.985	0	0	0	0	0.985	0	0	0	0	0	0	0	0
0.005	19.32	0.150	0.989	19.1	3.97	…	75.7	…	2.9	3.97	…	11.5	…	…	…	…	…	…
0.010	38.64	0.295	0.955	36.9	0.985	149.5	147.3	112.6	11.4	0.985	34.3	33.8	22.9	33.2	21.9	11.5	0.0002	0.56
0.015	57.96	0.435	0.901	52.1	3.97	…	207	…	25.2	3.97	…	100	…	…	…	…	…	…
0.020	77.28	0.564	0.826	63.8	0.985	482	475	418	43.6	0.985	221.0	218	177.4	236	147	89	0.0016	4.45
0.025	96.60	0.677	0.736	71.1	3.97	…	282	…	67.5	3.97	…	268	…	…	…	…	…	…
0.030	77.28	0.783	0.622	48.0	0.985	853	840	805	60.5	0.985	606	596	546	630	340	290	0.0054	14.50
0.035	57.96	0.870	0.493	28.6	3.97	…	113.2	…	50.5	3.97	…	200	…	…	…	…	…	…
0.040	38.64	0.932	0.363	14.0	0.985	981	966	967	36.0	0.985	868	854	832	900	302	598	0.0111	29.9
0.045	19.32	0.976	0.220	4.25	3.97	…	16.9	…	18.9	3.97	…	75	…	…	…	…	…	…
0.050	0	0.997	0.0715	0	0.985	…	…	983	0	0.985	…	…	929	981	66	915	0.0169	45.7

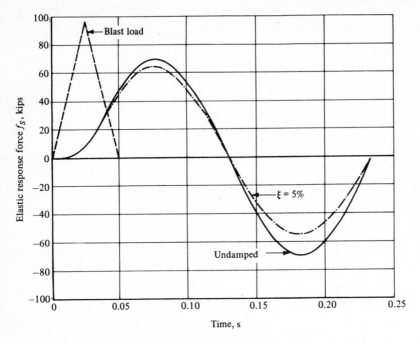

FIGURE E7-2
Response of water tower to blast load.

The frequency-domain approach is similar in concept to the periodic-load-analysis procedure presented in Chap. 5. Both these methods involve expressing the applied loading in terms of harmonic components, evaluating the response of the structure to each component, and then superposing the harmonic responses to obtain the total structural response. However, to apply the periodic-load technique to arbitrary loadings, it obviously is necessary to extend the Fourier series concept to the representation of nonperiodic functions. In this development, it will be convenient to use the concise exponential form of the Fourier series expressions, as given by Eqs. (5-6) and (5-7).

Consider, for example, the arbitrary nonperiodic loading shown in Fig. 7-3. If this function were represented by a Fourier series, the coefficients c_n obtained by integrating Eq. (5-7) over the interval $0 < t < T_p$ would actually define the periodic function shown in the figure by the dashed as well as the solid lines. However, it is apparent that the spurious repetitive loadings could be eliminated by extending the loading period to infinity. Thus it is necessary to reformulate the Fourier series expression so that it extends over an infinite time range. Toward this end, it will be

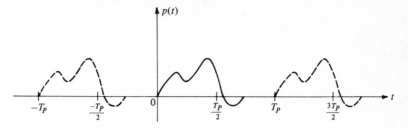

FIGURE 7-3
Arbitrary loading represented by Fourier series.

convenient first to rewrite Eqs. (5-6) and (5-7) in slightly modified form, using notation defined as follows:

$$\frac{1}{T_p} = \frac{\bar{\omega}_1}{2\pi} \equiv \frac{\Delta\bar{\omega}}{2\pi}$$

$$n\bar{\omega}_1 = n\,\Delta\bar{\omega} \equiv \bar{\omega}_n$$

$$c_n \equiv \frac{1}{T_p} c(\bar{\omega}_n) = \frac{\Delta\bar{\omega}}{2\pi} c(\bar{\omega}_n)$$

With these new symbols, the Fourier series equations (5-6) and (5-7) become

$$p(t) = \frac{\Delta\bar{\omega}}{2\pi} \sum_{n=-\infty}^{\infty} c(\bar{\omega}_n) \exp(i\bar{\omega}_n t) \qquad (7\text{-}20)$$

$$c(\bar{\omega}_n) = T_p c_n = \int_{t=-T_p/2}^{t=T_p/2} p(t) \exp(-i\bar{\omega}_n t)\, dt \qquad (7\text{-}21)$$

where advantage has been taken of the fact that the limits of the integral are arbitrary so long that they span one complete loading period.

Now if the loading period is extended to infinity ($T_p \rightarrow \infty$), the frequency increment becomes an infinitesimal ($\Delta\bar{\omega} \rightarrow d\bar{\omega}$) and the discrete frequencies $\bar{\omega}_n$ become a continuous function $\bar{\omega}$. Thus in the limit, the Fourier series expression of Eq. (7-20) becomes the following *Fourier integral*:

$$p(t) = \frac{1}{2\pi} \int_{\bar{\omega}=-\infty}^{\infty} c(\bar{\omega}) \exp(i\bar{\omega}t)\, d\bar{\omega} \qquad (7\text{-}22)$$

in which the harmonic-amplitude function is given by

$$c(\bar{\omega}) = \int_{t=-\infty}^{\infty} p(t) \exp(-i\bar{\omega}t)\, dt \qquad (7\text{-}23)$$

The two integrals of Eqs. (7-22) and (7-23) are known as a *Fourier transform pair* because the time function can be derived from the frequency function or vice versa by equivalent processes. A necessary condition for the existence of the Fourier transform is that the integral

$$\int_{-\infty}^{\infty} |p(t)| \, dt$$

be finite.

By analogy with the Fourier series expression of Eq. (7-20), the Fourier integral of Eq. (7-22) may be interpreted as representing an arbitrary loading as an infinite sum of harmonic components, where $(1/2\pi)c(\bar{\omega})$ defines the amplitude per unit of $\bar{\omega}$ of the load component at frequency $\bar{\omega}$. Multiplying this by the complex-frequency-response function $H(\bar{\omega})$ therefore yields the amplitude per unit of $\bar{\omega}$ of the response component at frequency $\bar{\omega}$. Hence the total response can be obtained by summing these response components over the entire frequency range. Expressing this concept mathematically leads to the basic equation for the analysis of response through the frequency domain:

$$v(t) = \frac{1}{2\pi} \int_{\bar{\omega}=-\infty}^{\infty} H(\bar{\omega})c(\bar{\omega}) \exp(i\bar{\omega}t) \, d\bar{\omega} \qquad (7\text{-}24)$$

To apply this frequency-domain procedure, it is necessary to evaluate the harmonic components $c(\bar{\omega})$ of the given loading by means of Eq. (7-23), and to make use of the complex-frequency-response function for a SDOF structure given by Eq. (5-11).

EXAMPLE E7-3   As an example of a response analysis through the frequency domain, consider the rectangular impulse loading of Fig. 6-4: $p(t) = p_0$ in the interval $0 < t < t_1$, with zero loading otherwise. The Fourier transform [Eq. (7-23)] of this load function is

$$c(\bar{\omega}) = \frac{p_0}{-i\bar{\omega}} \left[\exp(-i\bar{\omega}t) - 1 \right] \qquad (a)$$

Substituting this load expression together with the complex-frequency-response expression of Eq. (5-11) into Eq. (7-24) leads to the response in integral form:

$$v(t) = \frac{i\bar{\omega}_D}{2\pi k} \left[\int_{-\infty}^{\infty} \frac{e^{-i\omega\beta(t_1-t)}}{\beta(\beta-\gamma_1)(\beta-\gamma_2)} \, d\beta - \int_{-\infty}^{\infty} \frac{e^{i\omega\beta t}}{\beta(\beta-\gamma_1)(\beta-\gamma_2)} \, d\beta \right] \qquad (b)$$

where $\qquad \gamma_1 = \xi i + \sqrt{1-\xi^2} \qquad \gamma_2 = \xi i - \sqrt{1-\xi^2} \qquad (c)$

The two integrals of Eq. (*b*) can readily be determined by contour integration in the complex β plane, giving

$$
v(t) = \begin{cases} 0 & t < 0 \\[2mm] \dfrac{p_0}{k}\left[1 - e^{-\xi\omega t}\left(\cos \omega_D t + \dfrac{\xi}{\sqrt{1 - \xi^2}} \sin \omega_D t\right)\right] & 0 < t < t_1 \\[4mm] \dfrac{p_0}{k}\, e^{-\xi\omega(t-t_1)} \end{cases}
$$

$$
\times \left\{\left[e^{-\xi\omega t_1}\left(\sin \omega_D t_1 - \dfrac{\xi}{\sqrt{1 - \xi^2}} \cos \omega_D t_1\right)\right.\right.
$$

$$
\left.+ \dfrac{\xi}{\sqrt{1 - \xi^2}}\right] \sin \omega_D(t - t_1) \tag{d}
$$

$$
+ \left[1 - e^{-\xi\omega t_1}\left(\cos \omega_D t_1 + \dfrac{\xi}{\sqrt{1 - \xi^2}} \sin \omega_D t_1\right)\right]
$$

$$
\times \cos \omega_D(t - t_1)\Bigg\} \qquad t > t_1
$$

These results are equivalent to those obtained previously by analysis through the time domain [Eqs. (6-7*b*) and (6-8)] except that damping is included in the present analysis.   ////

7-5   NUMERICAL ANALYSIS IN THE FREQUENCY DOMAIN

The formal application of the frequency-domain-analysis procedure, as illustrated in the foregoing example, is limited to cases for which the Fourier transforms of the applied-load functions are available, and even in these cases the evaluation of the resulting integrals may be a tedious process. Thus to make practical use of the method, it is necessary to formulate it in terms of a numerical-analysis procedure. The numerical formulation may be divided conveniently into two phases: (1) derivation of discrete Fourier transform (DFT) expressions which correspond to the integral expressions of Eqs. (7-22) and (7-23) and (2) development of an efficient numerical technique for evaluating the DFTs. Each of these phases will be discussed briefly here.

Discrete Fourier Transforms

The first step in deriving the expressions is to assume that the loading is periodic of period T_p. This constitutes an approximation in the treatment of an arbitrary general loading but is necessary in order to replace the infinite time integral of Eq. (7-23) by a finite sum. The selection of the load period also serves to define the lowest frequency that may be considered in the analysis; thus

$$\bar{\omega}_1 = \Delta\bar{\omega} = \frac{2\pi}{T_p}$$

The load period is then divided into N equal time increments Δt, and the load is defined for the discrete times $t_m = m\,\Delta t$. When use is made of these relationships, the exponential term in Eq. (7-20) can be written

$$\exp{(i\bar{\omega}_n t_m)} = \exp{(in\Delta\bar{\omega}m\Delta t)} = \exp{\left(2\pi i\,\frac{nm}{N}\right)}$$

Accordingly, Eq. (7-20) takes the discrete form

$$p(t_m) = \frac{\Delta\bar{\omega}}{2\pi} \sum_{n=0}^{N-1} c(\bar{\omega}_n) \exp{\left(2\pi i\,\frac{nm}{N}\right)} \qquad (7\text{-}25)$$

in which the highest frequency to be considered has been arbitrarily set at $(N-1)\Delta\bar{\omega}$.

The corresponding discrete expression for the amplitude function $c(\bar{\omega}_n)$ can be obtained by merely substituting the sum of a finite series of discrete terms for the integral of Eq. (7-21), with the following result:

$$c(\bar{\omega}_n) = \Delta t \sum_{m=0}^{N-1} p(t_m) \exp{\left(-2\pi i\,\frac{nm}{N}\right)} \qquad (7\text{-}26)$$

Equations (7-25) and (7-26) are the DFT pairs which correspond to the continuous transforms of Eqs. (7-22) and (7-23). When using the discrete transforms, it is important to remember that they are based on the assumption that the loading is periodic. To minimize errors in the analysis of nonperiodic loads, the load period may be extended by the inclusion of a significant interval of zero load in the period T_p; the resulting load history would thus appear like that in Fig. 7-3.

Fast Fourier Transform Analysis

The evaluation of the sums involved in the two DFT equations is greatly simplified by the fact that the exponential functions involved are harmonic and extend over a range of N^2. Only discrete values of m and n are employed in the exponentials, and full advantage may be taken of the resulting duplication of values when the DFT sums are formed. Although it would be necessary to go into details of computer coding to discuss the optimum formulation of a DFT analysis fully, which is beyond the scope of this brief summary, it will be worthwhile to demonstrate something of the concept underlying the Fast Fourier Transform (FFT) technique. This relatively new computer procedure is so efficient and powerful that it has made the frequency-domain approach computationally competitive with traditional time-domain analyses and thus is revolutionizing the field of structural dynamics.

For the purpose of this discussion, either one of the DFT pairs may be represented by

$$B_m = \sum_{n=0}^{N-1} A_n W_N^{nm} \qquad (7\text{-}27)$$

where

$$W_N = e^{2\pi i/N} \qquad (7\text{-}28)$$

The evaluation of the sum will be most efficient if the number of time increments N into which the load period T_p is divided is a power M of 2, that is,

$$N = M^2$$

In this case, the integers m and n can be expressed in binary form as

$$m = m_0 + 2m_1 + 4m_2 + \cdots + 2^{M-1}m_{M-1} \qquad (7\text{-}29)$$

$$n = n_0 + 2n_1 + 4n_2 + \cdots + 2^{M-1}n_{M-1}$$

where the values of the coefficients m_j and n_j are either 0 or 1. With this binary notation, Eq. (7-27) can be written

$$B(m) = \sum_{n_0=0}^{1} \sum_{n_1=0}^{1} \sum_{n_2=0}^{1} \cdots \sum_{n_{M-1}=0}^{1} A(n)W_N^{(m_0+2m_1+4m_2+\cdots)(n_0+2n_1+\cdots)} \qquad (7\text{-}30)$$

The advantage of binary notation will be illustrated by considering the very simple case where the load period is divided into only eight time increments, that is, $N = 8$, $M = 3$. In this case, Eq. (7-30) becomes

$$B(m) = \sum_{n_0=0}^{1} \sum_{n_1=0}^{1} \sum_{n_2=0}^{1} A(n)W_8^{(m_0+2m_1+4m_2)(n_0+2n_1+4n_2)} \qquad (7\text{-}31)$$

However, the exponential term here can be written

$$W_8^{mn} = W_8^{8(m_1n_2+2m_2n_2+m_2n_1)} W_8^{4n_2m_0} W_8^{2n_1(2m_1+m_0)} W_8^{n_0(4m_2+2m_1+m_0)}$$

Moreover, the first term on the right side is unity because

$$W_8^{8(\text{integer})} = \exp\left[2\pi i(^8/_8)(\text{integer})\right] \equiv 1$$

Consequently, only the remaining three terms need be considered in the summation. These may be treated conveniently in sequence, introducing new notation to denote the successive stages of the summation process. Thus the first stage will be designated

$$A_1(m_0, n_1, n_0) = \sum_{n_2=0}^{1} A(n_2, n_1, n_0)W_8^{n_2m_0} \qquad (7\text{-}32a)$$

Similarly, the second stage will be denoted

$$A_2(m_0,m_1,n_0) = \sum_{n_1=0}^{1} A_1(m_0,n_1,n_0)W_8^{2n_1(2m_1+m_0)} \qquad (7\text{-}32b)$$

while the third stage (the final stage, where $M = 3$) is

$$B(m_0,m_1,m_2) = \sum_{n_0=0}^{1} A_2(m_0,m_1,n_0)W_8^{n_0(4m_2+2m_1+m_0)} \qquad (7\text{-}32c)$$

This process is particularly efficient because the results of one stage are used immediately in the next stage (thus minimizing storage requirements) and also because the exponential takes the value of unity in the first term of each summation. Further savings result from the harmonic nature of the exponential, that is, $W_8^0 = -W_8^4$, $W_8^1 = -W_8^5$, $W_8^2 = -W_8^6$, etc. The reductions in computational effort which result from this formulation are enormous when the time interval is divided into a large number of increments; for example, when $N = 1{,}024$, the FFT algorithm requires only about $^1/_2$ percent of the computer effort involved in the direct evaluation of Eq. (7-27).

PROBLEMS

7·1 The undamped SDOF system of Fig. P7-1a is subjected to the half sine-wave loading of Fig. P7-1b. Calculate the spring force history $f_s(t)$ for the time $0 < t < 0.6\,\text{s}$ by numerical evaluation of the Duhamel integral with $\Delta\tau = 0.1$ s using:

(a) Simple summation ($\zeta = 1$)
(b) Trapezoidal rule ($\zeta = 2$)
(c) Simpson's rule ($\zeta = 3$)

Compare these results with those obtained with Eq. (6-1) evaluated at the same 0.1 s time increments.

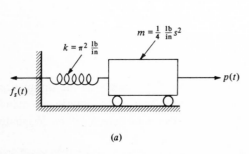

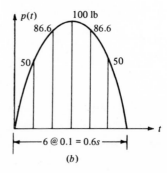

(a) (b)

FIGURE P7-1

7-2 Solve Example E7-1 using the trapezoidal rule.

7-3 Solve Example E7-2 using the trapezoidal rule.

7-4 The SDOF frame of Fig. P7-2*a* is subjected to the blast loading history shown in Fig. P7-2*b*. Compute the displacement history for the time $0 < t < 0.72$ s by numerical evaluation of the Duhamel integral using Simpson's rule with $\Delta\tau = 0.12$ s.

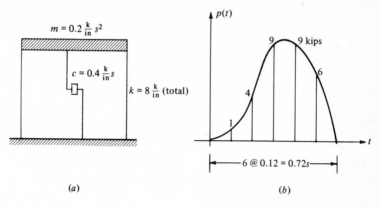

(*a*)   (*b*)

FIGURE P7-2

8

ANALYSIS OF NONLINEAR STRUCTURAL RESPONSE

8-1 ANALYSIS PROCEDURE

In the analysis of linear structures subjected to arbitrary dynamic loadings, the Duhamel integral or the frequency-domain analysis described in Chap. 7 generally provides the most convenient solution technique. However, it must be emphasized that since the principle of superposition was employed in the derivation of both these procedures, they may be employed *only* with linear systems, that is, systems for which the properties remain constant during the response. On the other hand, there are many important classes of structural-dynamics problems which cannot be assumed to be linear, e.g., the response of a building to an earthquake motion severe enough to cause serious damage. Consequently it is necessary to develop another method of analysis suitable for use with nonlinear systems.

Probably the most powerful technique for nonlinear analysis is the step-by-step integration procedure. In this approach, the response is evaluated for a series of short time increments Δt, generally taken of equal length for computational convenience. The condition of dynamic equilibrium is established at the beginning and end of each interval, and the motion of the system during the time increment is evaluated approximately on the basis of an assumed response mechanism (generally ignoring the lack of equilibrium which may develop during the interval). The nonlinear nature of the system is accounted for by calculating new properties appropriate to the current deformed state at the beginning of each time increment. The complete response is obtained by using the velocity and displacement computed at the end of one computational interval as the initial conditions for the next interval; thus the process may be continued step by step from the initiation of loading to any desired time, approximating the nonlinear behavior as a sequence of successively changing linear systems.

8-2 INCREMENTAL EQUATION OF EQUILIBRIUM

The structure to be considered in this discussion is the SDOF system shown in Fig. 8-1a. The properties of the system, m, k, c, and $p(t)$, may represent generalized quantities, as discussed in Sec. 2-5, instead of the simple localized effects implied in the sketch. The forces acting on the mass of the system are indicated in Fig. 8-1b, and the general nonlinear characteristics of the spring and damping forces are shown in Fig. 8-1c and d, respectively, while an arbitrary applied loading is sketched in Fig. 8-1e.

At any instant of time t the equilibrium of forces acting on the mass m requires

$$f_I(t) + f_D(t) + f_S(t) = p(t) \qquad (8\text{-}1a)$$

while a short time Δt later the equation would be

$$f_I(t + \Delta t) + f_D(t + \Delta t) + f_S(t + \Delta t) = p(t + \Delta t) \qquad (8\text{-}1b)$$

Subtracting Eq. (8-1a) from Eq. (8-1b) then yields the incremental form of the equation of motion for the time interval t:

$$\Delta f_I(t) + \Delta f_D(t) + \Delta f_S(t) = \Delta p(t) \qquad (8\text{-}2)$$

The incremental forces in this equation may be expressed as follows:

$$\Delta f_I(t) = f_I(t + \Delta t) - f_I(t) = m\,\Delta\ddot{v}(t)$$
$$\Delta f_D(t) = f_D(t + \Delta t) - f_D(t) = c(t)\,\Delta\dot{v}(t)$$
$$\Delta f_S(t) = f_S(t + \Delta t) - f_S(t) = k(t)\,\Delta v(t) \qquad (8\text{-}3)$$
$$\Delta p(t) = p(t + \Delta t) - p(t)$$

in which it is tacitly assumed that the mass remains constant, and where the terms $c(t)$ and $k(t)$ represent the damping and stiffness properties corresponding to the velocity and displacement existing during the time interval, as indicated in Fig. 8-1c and d, respectively. In practice, the secant slopes indicated could be evaluated only by iteration because the velocity and displacement at the end of the time increment

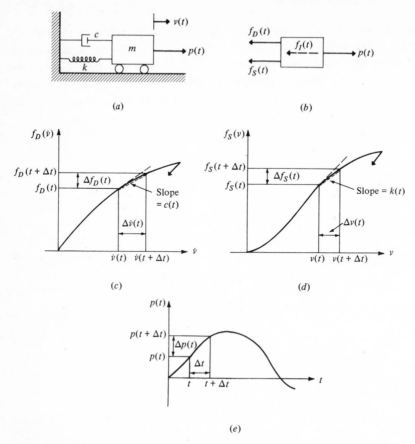

FIGURE 8-1
Definition of a nonlinear dynamic system: (a) basic SDOF structure; (b) force equilibrium; (c) nonlinear damping; (d) nonlinear stiffness; (e) applied load.

depend on these properties; for this reason the tangent slopes defined at the beginning of the time intervals frequently are used instead:

$$c(t) \doteq \left(\frac{df_D}{d\dot{v}}\right)_t \qquad k(t) \doteq \left(\frac{df_S}{dv}\right)_t \qquad (8\text{-}4)$$

Substituting the force expressions of Eqs. (8-3) into Eq. (8-2) leads to the final form of the incremental equilibrium equations for time t:

$$m\,\Delta\ddot{v}(t) + c(t)\,\Delta\dot{v}(t) + k(t)\,\Delta v(t) = \Delta p(t) \qquad (8\text{-}5)$$

The material properties considered in this type of analysis may include any form of nonlinearity. Thus, there is no necessity for the spring force f_S to be dependent only on the displacement, as in a nonlinear elastic material. A nonlinear hysteretic material may equally well be specified, in which the force depends on the past history of deformation as well as the current value of displacement. The only requirement is that the stiffness properties be completely defined by the past as well as the current state of deformation. Moreover, it is evident that the implicit assumption of a constant mass is arbitrary: it also could be represented as a time-varying quantity.

8-3 STEP-BY-STEP INTEGRATION

Many procedures are available for the numerical integration of Eq. (8-5). The technique employed here is simple in concept but has been found to yield excellent results with relatively little computational effort. The basic assumption of the process is that the acceleration varies linearly during each time increment while the properties of the system remain constant during this interval. The motion of the mass during the time interval is indicated in graphical form in Fig. 8-2, together with equations for the assumed linear variation of the acceleration and the corresponding quadratic and cubic variations of the velocity and displacement, respectively. Evaluating these latter expressions at the end of the interval ($\tau \equiv \Delta t$) leads to the following equations for the increments of velocity and displacement:

$$\Delta \dot{v}(t) = \ddot{v}(t)\, \Delta t + \Delta \ddot{v}(t)\, \frac{\Delta t}{2} \tag{8-6a}$$

$$\Delta v(t) = \dot{v}(t)\, \Delta t + \ddot{v}(t)\, \frac{\Delta t^2}{2} + \Delta \ddot{v}(t)\, \frac{\Delta t^2}{6} \tag{8-6b}$$

Now it will be convenient to use the incremental displacement as the basic variable of the analysis; hence Eq. (8-6a) is solved for the incremental acceleration, and this expression is substituted into Eq. (8-6b) to obtain

$$\Delta \ddot{v}(t) = \frac{6}{\Delta t^2}\, \Delta v(t) - \frac{6}{\Delta t}\, \dot{v}(t) - 3\ddot{v}(t) \tag{8-7a}$$

$$\Delta \dot{v}(t) = \frac{3}{\Delta t}\, \Delta v(t) - 3\dot{v}(t) - \frac{\Delta t}{2}\, \ddot{v}(t) \tag{8-7b}$$

Substituting Eqs. (8-7) into Eq. (8-5) leads to the following form of the equation of motion:

$$m\left[\frac{6}{\Delta t^2}\, \Delta v(t) - \frac{6}{\Delta t}\, \dot{v}(t) - 3\ddot{v}(t) \right] + c(t)\left[\frac{3}{\Delta t}\, \Delta v(t) - 3\dot{v}(t) - \frac{\Delta t}{2}\, \ddot{v}(t) \right]$$
$$+ k(t)\, \Delta v(t) = \Delta p(t)$$

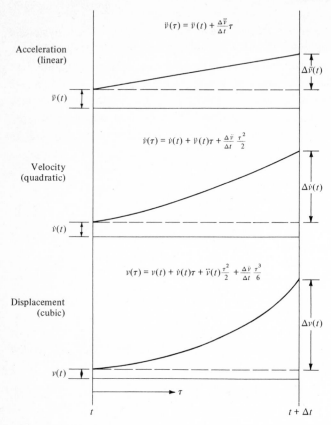

FIGURE 8-2
Motion of system during time increment (based on linear acceleration).

Finally transferring all terms associated with the known initial conditions to the right-hand side gives

$$\tilde{k}(t)\, \Delta v(t) = \Delta \tilde{p}(t) \qquad (8\text{-}8)$$

in which

$$\tilde{k}(t) = k(t) + \frac{6}{\Delta t^2}\, m + \frac{3}{\Delta t}\, c(t) \qquad (8\text{-}9a)$$

$$\Delta \tilde{p}(t) = \Delta p(t) + m\left[\frac{6}{\Delta t}\, \dot{v}(t) + 3\ddot{v}(t)\right] + c(t)\left[3\dot{v}(t) + \frac{\Delta t}{2}\, \ddot{v}(t)\right] \qquad (8\text{-}9b)$$

It will be noted that Eq. (8-8) is equivalent to a static incremental-equilibrium relationship, and may be solved for the incremental displacement by dividing the incremental load by the stiffness. The dynamic behavior is accounted for by the inclusion of inertial and damping effects in the effective-load and stiffness terms.

After Eq. (8-8) is solved for the displacement increment, this value is substituted into Eq. (8-7b) to obtain the incremental velocity. The initial conditions for the next time step then result from the addition of these incremental values to the velocity and displacement at the beginning of the time step.

This numerical-analysis procedure includes two significant approximations: (1) that the acceleration varies linearly and (2) that the damping and stiffness properties remain constant during the time step. In general, neither of these assumptions is entirely correct, even though the errors are small if the time step is short. Therefore errors generally will arise in the incremental-equilibrium relationship which might tend to accumulate from step to step, and this accumulation should be avoided by imposing the total-equilibrium condition at each step of the analysis. This may be accomplished conveniently by expressing the accelerations at the beginning of the time step in terms of the total external load minus the total damping and elastic forces.

8-4 SUMMARY OF PROCEDURE

For any given time increment, the analysis procedure consists of the following operations:

1 Initial velocity and displacement values $\dot{v}(t)$ and $v(t)$ are known, either from values at the end of the preceding increment or as initial conditions of the problem.

2 With these values and the specified nonlinear properties of the structure, the damping $c(t)$ and stiffness $k(t)$ for the interval, as well as current values of the damping $f_D(t)$ and elastic $f_S(t)$ forces are found, e.g., from Fig. 8-1c and d.

3 The initial acceleration is given by

$$\ddot{v}(t) = \frac{1}{m} \left[p(t) - f_D(t) - f_S(t) \right] \qquad (8\text{-}10)$$

This is merely a rearrangement of the equation of equilibrium for time t.

4 The effective load increment $\Delta \tilde{p}(t)$ and effective stiffness $\tilde{k}(t)$ are computed from Eqs. (8-9).

5 The displacement increment is given by Eq. (8-8), and with it the velocity increment is found from Eq. (8-7b).

6 Finally the velocity and displacement at the end of the increment are obtained from

$$\dot{v}(t + \Delta t) = \dot{v}(t) + \Delta \dot{v}(t)$$
$$v(t + \Delta t) = v(t) + \Delta v(t) \qquad (8\text{-}11)$$

When step 6 has been completed, the analysis for this time increment is finished, and the entire process may be repeated for the next time interval. Obviously the process can be carried out consecutively for any desired number of time increments; thus the complete response history can be evaluated for any SDOF system having any prescribed nonlinear properties. Linear systems also can be treated by the same process, of course; in this case the damping and stiffness properties remain constant so that the analysis procedure is somewhat simpler.

As with any numerical-integration process, the accuracy of this step-by-step method will depend on the length of the time increment Δt. Three factors must be considered in the selection of this interval: (1) the rate of variation of the applied loading $p(t)$, (2) the complexity of the nonlinear damping and stiffness properties, and (3) the period T of vibration of the structure. The time increment must be short enough to permit the reliable representation of all these factors, the last one being associated with the free-vibration behavior of the system. In general, the material-property variation is not a critical factor; if a significant sudden change takes place, as in the yielding of an elastoplastic spring, a special subdivided time increment may be introduced to treat this effect accurately. Also the time increment required to approximate the significant dynamic aspects of the loading adequately may be estimated with little difficulty.

Thus, if the load history is relatively simple, the choice of the time interval will depend essentially on the period of vibration of the structure. This linear-acceleration method is only *conditionally* stable and will give a divergent solution if the time increment is greater than about half the vibration period. However, the increment must be considerably shorter than this to provide reasonable accuracy, so that instability is not a practical problem. In general, an increment-period ratio $\Delta t/T \leq {}^1/_{10}$ is a good rule of thumb for obtaining reliable results. If there is any doubt about the adequacy of a given solution, a second analysis can be made halving the time increment; if the response is not changed appreciably in the second analysis, it may be assumed that the errors introduced by the numerical integration are negligible.

EXAMPLE E8-1 To demonstrate a hand-solution technique for applying the linear-acceleration step-by-step method described above, the response of the elastoplastic SDOF frame shown in Fig. E8-1 to the loading history indicated has been calculated. A time step of 0.1 s has been used for this analysis, which is longer than desirable for good accuracy but will be adequate for the present purpose.

In this structure, the damping coefficient has been assumed to remain constant; hence the nonlinearity results only from the change of stiffness as

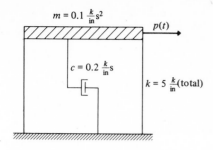

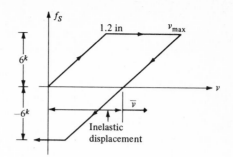

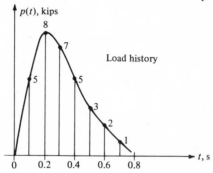

FIGURE E8-1
Elastoplastic frame and dynamic loading.

yielding takes place. The effective stiffness thus may be expressed [see Eq. (8-9a)] as

$$\tilde{k}(t) = k(t) + \frac{6}{(0.1)^2} m + \frac{3}{0.1} c = 66 + k(t)$$

where $k(t)$ is either 5 kips/in or zero, according as the frame is elastic or yielding. Also the effective incremental loading is given by [see Eq. (8.9b)]

$$\Delta \tilde{p}(t) = \Delta p(t) + \left(\frac{6m}{0.1} + 3c\right) \dot{v} + \left(3m + \frac{0.1}{2} c\right) \ddot{v} = \Delta p(t) + 6.6\dot{v} + 0.31\ddot{v}$$

The velocity increment given by Eq. (8-7b) becomes

$$\Delta \dot{v} = 30\Delta v - 3\dot{v} - 0.05\ddot{v}$$

A convenient tabular arrangement for the hand calculations is shown in Table E8-1 (see p. 126).

For this elastoplastic system, the response behavior changes drastically as the yielding starts and stops, and to obtain best accuracy it would be desirable

Table E8-1 NONLINEAR RESPONSE ANALYSIS: LINEAR ACCELERATION STEP-BY-STEP METHOD

Structure and loading in Fig. E8-1

t, s (1)	p, kips (2)	v, in. (3)	$\dot v$, in./s (4)	f_S $5\bar v$† (5)	f_D $0.2\dot v$ (6)	f_I $(2)-(5)$ (7)	$\ddot v$ $\times 10$ (8)	Δp (9)	$6.6\dot v$ (10)	$0.31\bar v$ (11)	$\Delta\bar p$ $(9)+(10)+(11)$ (12)	k (13)	$\tilde k$ $66+(13)$ (14)	Δv $(12)\div(14)$ (15)	$30\,\Delta v$ (16)	$3\dot v$ (17)	$0.05\bar v$ (18)	$\Delta\ddot v$ $(16)-(17)-(18)$ (19)
0.0	0				0	0	0	5	0	0	0	5	71	0.070	2.11	0	0	2.11
0.1	5	0.070	2.11	0.35	0.42	4.23	42.3	3	13.92	13.12	30.04	5	71	0.423	12.68	6.33	2.11	4.24
0.2	8	0.493	6.35	2.46	1.27	4.27	42.7	−1	41.90	13.25	54.15	0‡	71	0.763	22.88	19.06	2.14	1.68
0.3	7	1.256	8.03	6	1.61	−0.61	−6.1	−2	53.02	−1.89	49.13	0	66	0.744	22.33	24.08	−0.30	−1.45
0.4	5	2.000	6.58	6	1.32	−2.32	−23.2	−2	43.43	−7.19	34.24	5	66	0.519	15.57	19.74	−1.16	−3.01
0.5	3	2.519	3.57	6	0.71	−3.71	−37.1	−1	23.56	−11.50	11.06	5	71	0.168	5.02	10.72	−1.85	−3.85
0.6	2	2.687	−0.28		−0.06	−3.94	−39.4	−1	−1.85	−12.22	−15.07	5	71	−0.212	−6.36	−0.84	−1.97	−3.55
0.7	1	2.475	−3.83		−0.77	−2.17	−21.7	−1	−25.28	−6.73	−33.01		71	−0.465	−11.49	−13.95	−1.08	−1.38
0.8	0	2.010	−5.21	4.94	−1.04	−1.58	−15.8	0	−34.39	−4.90	−39.29			−0.554	−16.21	−15.63	−0.79	0.21
0.9	0	1.456	−5.00	−0.16	−1.00	+1.16	11.6	0	−33.00	3.60	−29.40			−0.414	−12.42	−15.00	0.58	2.00
1.0	0	1.042	−3.00															

† $\bar v = v - v_i$, where v_i = inelastic displ. = $v_{\max} - 1.2$ in.; ‡ $k = 0$ while frame is yielding.

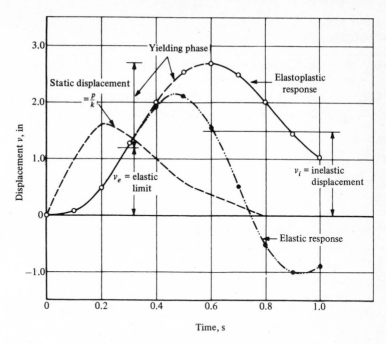

FIGURE E8-2
Comparison of elastoplastic with elastic response (frame of Fig. E8-1).

to divide each time increment involving such a change of phase into two sub-increments. The properties then would be constant during the sub-increments, and the analysis would be quite precise; however, an iterative procedure would be required to establish the lengths of the subincrements. In the present analysis, this refinement has not been used. The initial stiffness has been assumed to act during the entire increment, and thus significant errors may have arisen during the phase transitions.

The dynamic elastoplastic response calculated in Table E8-1 is plotted in Fig. E8-2, with the response during the yielding phase shown as a dashed line. Also plotted for comparison is the linear elastic response obtained by a similar step-by-step analysis but with $\tilde{k} = 71$ and $f_S = 5v$ throughout the calculations. The effect of the plastic yielding shows up clearly in this comparison; the permanent set (the position about which the subsequent free vibrations of the nonlinear system occur) amounts to about 1.49 in. Also shown to indicate the character of the loading is the static displacement p/k, that is, the deflection which would have occurred in the elastic structure if there had been no damping and inertia effects. ////

PROBLEMS

8-1 Solve the linear elastic response of Prob. 7-4 by step-by-step integration, using the linear acceleration method.

8-2 Solve Prob. 8-1, assuming an elastoplastic force-displacement relation for the columns and a yield force level of 8 kips, as shown in Fig. P8-1a.

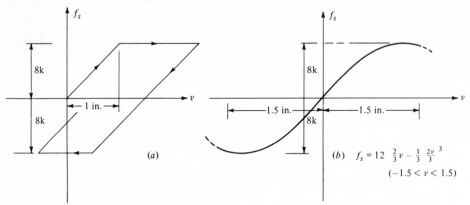

(a)

(b) $f_s = 12 \left[\frac{2}{3} v - \frac{1}{3} \frac{2v}{3} \right]^3$

$(-1.5 < v < 1.5)$

FIGURE P8-1

8-3 Solve Prob. 8-1, assuming the nonlinear elastic force-displacement relation, $f_s = 12[^2/_3 v - {^1/_3}(2v/3)^3]$, which is sketched in Fig. P8-1b (f_s is in kips, v in inches).

8-4 Solve Prob. 8-1, replacing the viscous damper by a hysteretic damping device that results in a damping force given by Eq. (5-51), where $\zeta k = 4$ k/in, i.e.,

$$f_D = 4|v| \frac{\dot{v}}{|\dot{v}|}$$

9

VIBRATION ANALYSIS BY RAYLEIGH'S METHOD

9-1 BASIS OF THE METHOD

In the preceding discussions of the analysis of response to dynamic loadings it is evident that the vibration frequency or period of a SDOF system has a controlling influence on its dynamic behavior. For this reason it is useful to develop a simple method of evaluating the vibration frequency for SDOF systems. Probably the most useful procedure in general is Rayleigh's method, described in this section.

By definition, the undamped frequency of vibration of a SDOF system is given by Eq. (3-5), that is,

$$\omega = \sqrt{\frac{k}{m}} \qquad (9\text{-}1)$$

This expression applies directly to a simple spring-mass system with spring stiffness k and mass m. It can also be applied to any structure which can be represented as a SDOF system through the use of an assumed displacement shape ψ. In this case the quantities in Eq. (9-1) represent the generalized stiffness k^* and generalized mass m^*, defined by Eqs. (2-37) and (2-39).

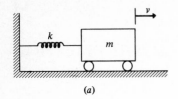

(a)

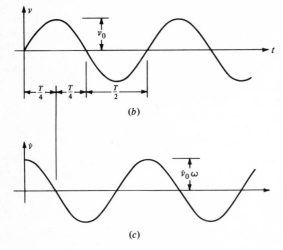

(b)

(c)

FIGURE 9-1
Free vibration of undamped SDOF structure: (a) SDOF structure; (b) displacement; (c) velocity.

Although this generalized-coordinate concept may be used to determine approximately the vibration frequency of any structure, it is desirable to examine the frequency-analysis problem from another point of view, originated by Lord Rayleigh. The basic concept in the Rayleigh method is the principle of conservation of energy; the energy in a freely vibrating system must remain constant if no damping forces act to absorb it. Consider the free-vibration motion of the undamped spring-mass system shown in Fig. 9-1a. With an appropriate choice of time origin, the displacement can be expressed (Fig. 9-1b) by

$$v = v_0 \sin \omega t \qquad (9\text{-}2a)$$

and the velocity (Fig. 9-1c) by

$$\dot{v} = v_0 \omega \cos \omega t \qquad (9\text{-}2b)$$

The potential energy of this system is represented entirely by the strain energy of the spring:

$$V = {}^1\!/_2 k v^2 = {}^1\!/_2 k v_0{}^2 \sin^2 \omega t \qquad (9\text{-}3a)$$

while the kinetic energy of the mass is

$$T = {}^1\!/_2 m \dot{v}^2 = {}^1\!/_2 m v_0{}^2 \omega^2 \cos^2 \omega t \qquad (9\text{-}3b)$$

Now considering the time $t = T/4 = \pi/2\omega$, it is clear from Fig. 9-1 [or from Eqs.

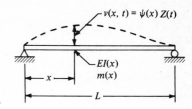

FIGURE 9-2
Vibration of a nonuniform beam.

(9-3)] that the kinetic energy is zero and that the potential energy has reached its maximum value:

$$V_{max} = {}^1\!/_2 k v_0{}^2 \qquad (9\text{-}4a)$$

Similarly, at the time $t = T/2 = \pi/\omega$, the potential energy vanishes and the kinetic energy is maximum

$$T_{max} = {}^1\!/_2 m v_0{}^2 \omega^2 \qquad (9\text{-}4b)$$

Hence, if the total energy in the vibrating system remains constant (as it must in undamped free vibration), it is apparent that the maximum kinetic energy must equal the maximum potential energy, $V_{max} = T_{max}$; that is,

$${}^1\!/_2 k v_0{}^2 = {}^1\!/_2 m v_0{}^2 \omega^2$$

from which

$$\omega^2 = \frac{k}{m}$$

This, of course, is the same frequency expression which was cited earlier; in this case it has been derived by the Rayleigh concept of equating expressions for the maximum strain energy and kinetic energy.

9-2   APPROXIMATE ANALYSIS OF A GENERAL SYSTEM

There is no advantage to be gained from the application of Rayleigh's method to vibration analysis of a spring-mass system as described above; its principal use is for the approximate frequency analysis of a system having many degrees of freedom. Consider, for example, the nonuniform simple beam shown in Fig. 9-2. This beam actually has an infinite number of degrees of freedom; that is, it can displace in an infinite variety of displacement patterns. To apply the Rayleigh procedure, it is necessary to assume the shape which the beam will take in its fundamental mode of vibration. As explained in Chap. 2 in the discussion of assumed shape functions, this assumption may be expressed by Eq. (2-29), or noting the harmonic variation of the generalized coordinate in free vibrations

$$v(x,t) = \psi(x) Z_0 \sin \omega t \qquad (9\text{-}5)$$

in which $\psi(x)$ is the shape function, which represents the ratio of the displacement at any point x to the reference displacement or generalized coordinate $Z(t)$. Equation (9-5) expresses the assumption that the shape of the vibrating beam does not change with time; only the amplitude of motion varies, and it varies harmonically in a free-vibration condition.

The assumption of the shape function $\psi(x)$ effectively reduces the beam to a SDOF system. Thus the frequency of vibration can be found by equating the maximum strain energy developed during the motion to the maximum kinetic energy. The strain energy of this flexural system is given by

$$V = \frac{1}{2} \int_0^L EI(x) \left(\frac{\partial^2 v}{\partial x^2}\right)^2 dx \qquad (9\text{-}6)$$

Thus, substituting the assumed shape function of Eq. (9-5) and letting the displacement amplitude take its maximum value leads to

$$V_{max} = \frac{1}{2}Z_0^2 \int_0^L EI(x)[\psi''(x)]^2 \, dx \qquad (9\text{-}7)$$

The kinetic energy of the nonuniformly distributed mass is

$$T = \frac{1}{2} \int_0^L m(x)(\dot{v})^2 \, dx \qquad (9\text{-}8)$$

Thus, when Eq. (9-5) is differentiated with respect to time to obtain the velocity and the amplitude is allowed to reach its maximum,

$$T_{max} = \frac{1}{2}Z_0^2\omega^2 \int_0^L m(x)[\psi(x)]^2 \, dx \qquad (9\text{-}9)$$

Finally, after equating the maximum potential energy to the maximum kinetic energy, the frequency is found to be

$$\omega^2 = \frac{\displaystyle\int_0^L EI(x)[\psi''(x)]^2 \, dx}{\displaystyle\int_0^L m(x)[\psi(x)]^2 \, dx} \qquad (9\text{-}10)$$

At this point, it may be noted that the numerator of Eq. (9-10) is merely the generalized stiffness of the beam k^* for this assumed displacement shape [see Eq. (2-39)] while the denominator is its generalized mass m^* [see Eq. (2-37)]. Thus Rayleigh's method leads directly to the generalized form of Eq. (9-1), as is to be expected since it employs the same generalized-coordinate concept to reduce the system to a single degree of freedom.

9-3   SELECTION OF THE VIBRATION SHAPE

The accuracy of the vibration frequency obtained by Rayleigh's method depends entirely on the shape function $\psi(x)$, which is assumed to represent the vibration-mode shape. In principal, any shape may be selected which satisfies the geometric boundary conditions of the beam, that is, which is consistent with the specified support conditions. However, any shape other than the true vibration shape would require the action of additional external constraints to maintain equilibrium; these extra constraints would stiffen the system, adding to its strain energy, and thus would cause an increase in the computed frequency. Consequently, it may be recognized that the true vibration shape will yield the lowest frequency obtainable by Rayleigh's method, and in choosing between approximate results given by this method the lowest frequency is always the best approximation.

> EXAMPLE E9-1   To illustrate this point, assume that the beam of Fig. 9-2 has uniform mass $\overline{m}$ and stiffness. As a first approximation for the frequency analysis, assume that the vibration shape is parabolic: $\psi(x) = (x/L)(x/L - 1)$. Then, $\psi''(x) = 2/L^2$, and
>
> $$V_{max} = {}^1\!/_2 Z_0{}^2 EI \int_0^L \left(\frac{2}{L^2}\right)^2 dx = {}^1\!/_2 Z_0{}^2 \frac{4EI}{L^3}$$
>
> while
>
> $$T_{max} = {}^1\!/_2 Z_0{}^2 \omega^2 \overline{m} \int_0^L \left[\frac{x}{L}\left(\frac{x}{L} - 1\right)\right]^2 dx = {}^1\!/_2 Z_0{}^2 \omega^2 \frac{\overline{m}L}{30}$$
>
> from which
>
> $$\omega^2 = \frac{V_{max}}{(1/\omega^2)T_{max}} = \frac{120EI}{\overline{m}L^4}$$
>
> If the shape were assumed to be a sine curve, $\psi(x) = \sin(\pi x/L)$, the same type of analysis would lead to the result
>
> $$\omega^2 = \frac{EI\pi^4/2L^3}{mL/2} = \pi^4 \frac{EI}{\overline{m}L^4}$$

This second frequency is significantly less than the first (actually almost 20 percent less); thus it is a much better approximation. As a matter of fact, it is the exact answer because the assumed sine-curve shape is the true vibration shape of a uniform simple beam. The first assumption should not be expected to lead to very good results; the assumed parabolic shape implies a uniform bending moment along the span which does not correspond to the simple end-support conditions. It is a *valid* shape, since it satisfies the geometric requirements of zero end displacements, but is not a realistic assumption.   ////

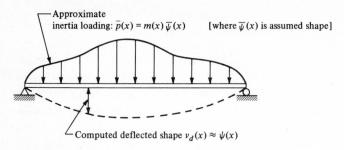

FIGURE 9-3
Deflected shape resulting from inertia load of assumed shape.

The question now arises how a reasonable deflected shape can be selected in order to ensure good results with Rayleigh's method (or the equivalent generalized-coordinate approach described earlier). The concept to be used in selecting the vibration shape is that the displacements in free vibrations result from the application of inertia forces and that the inertia forces (which are the product of mass and acceleration) are proportional to the mass distribution and to the displacement amplitude. Thus, the correct vibration shape $\psi_c(x)$ is that deflected shape resulting from a loading $p_c(x)$ proportional to $m(x)\psi_c(x)$. Of course, it is not possible to guess the exact shape $\psi_c(x)$, but the deflection shape computed from the loading $\bar{p}(x) = m(x)\bar{\psi}(x)$ [as shown in Fig. 9-3, where $\bar{\psi}(x)$ is any reasonable approximation of the true shape] will provide extremely good accuracy in the solution.

In general, the evaluation of the generalized coordinate shape on the basis of an assumed shape in this fashion involves more computational effort than is necessary in an approximate analysis. The Rayleigh procedure will give good accuracy with a considerably less refined approach than this. One common assumption is that the inertia loading $\bar{p}(x)$ (see Fig. 9-3) is merely the weight of the beam, that is, $\bar{p}(x) = m(x)g$, where $m(x)$ is the mass distribution and g is the acceleration of gravity. The frequency then is evaluated on the basis of the deflected shape $v_d(x)$ resulting from this dead-weight load. The maximum strain energy can be found very simply in this case from the fact that stored energy must be equal to the work done on the system by the applied loading:

$$V_{\max} = \frac{1}{2} \int_0^L \bar{p}(x) v_d(x)\, dx = \frac{1}{2} g Z_0 \int_0^L m(x) \psi(x)\, dx \qquad (9\text{-}11)$$

The kinetic energy is given still by Eq. (9-7), in which $\psi(x) = v_d(x)/Z_0$ is the shape function computed from the dead load. Thus the frequency found by equating the strain and kinetic-energy expressions is

$$\omega^2 = \frac{g}{Z_0} \frac{\displaystyle\int_0^L m(x)\psi(x)\, dx}{\displaystyle\int_0^L m(x)[\psi(x)]^2\, dx} = g \frac{\displaystyle\int_0^L m(x) v_d(x)\, dx}{\displaystyle\int_0^L m(x)[v_d(x)]^2\, dx} \qquad (9\text{-}12)$$

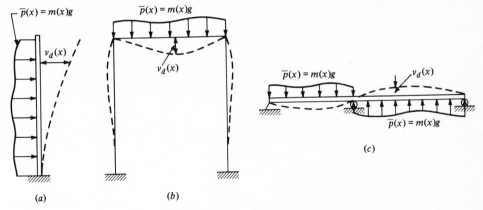

FIGURE 9-4
Assumed shapes resulting from dead loads.

Equation (9-12) is commonly used for the approximate frequency analysis of any type of system. It should be noted that the reference amplitude Z_0 must be included in the expression if the shape is defined by the dimensionless shape function $\psi(x)$, but it is not involved if the actual dead-load deflections are used.

The loading $\bar{p}(x)$ used to calculate the dead-weight deflection $v_d(x)$ in Eq. (9-12) is actually a gravitational loading only in cases where the principal vibratory motion is in the vertical direction. For a structure like the vertical cantilever of Fig. 9-4a, in which the principal motion is horizontal, the loading must be applied laterally, as shown in this figure; in effect it is assumed that gravity acts horizontally for this purpose. An appropriate deflected shape to approximate the *symmetrical* vibration frequency of the frame of Fig. 9-4b could be obtained by applying a vertical gravity load, as shown. However, the fundamental vibrations of this type of structure will generally be in the horizontal direction; to obtain a shape $\psi(x)$ for approximating the lateral vibration frequency, the gravity forces should be applied *laterally*. Furthermore, in the fundamental mode of vibration of the two-span beam shown in Fig. 9-4c, the two spans will deflect in opposite directions. Thus, to obtain a deflected shape for this case, the gravitational forces should be applied in opposite directions. A considerably higher vibration frequency would be obtained from the deflected shape resulting from downward loads acting in both spans.

The reader must be cautioned, however, against spending too much time in computing deflected shapes which will yield extremely accurate results. The principal value of the Rayleigh method is in providing a simple and reliable approximation to the natural frequency. Almost any reasonable shape assumption will give useful results.

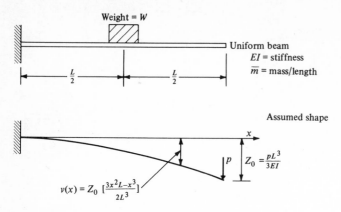

FIGURE E9-1
Rayleigh method analysis of beam vibration frequency.

EXAMPLE E9-2   The use of the Rayleigh method to compute the vibration frequency of a practical system will be illustrated by the analysis of the uniform cantilever beam supporting a weight at midspan, shown in Fig. E9-1. For this study, the vibration shape has been taken to be that produced by a load applied to the end of the cantilever, as shown in the figure. The resulting deflected shape is

$$v(x) = \frac{pL^3}{3EI} \frac{3x^2L - x^3}{2L^3} \equiv Z_0\psi(x)$$

The maximum potential energy of the beam can be found in this case from

$$V_{max} = {}^1\!/_2 pZ_0 = \frac{1}{2} \frac{3EI}{L^3} Z_0{}^2$$

where Z_0 is the deflection under the load and p has been expressed in terms of this end deflection.

The maximum kinetic energy of the beam can be calculated in two parts, considering separately the beam and the supported weight:

Beam:
$$T_{max}^B = \frac{\omega^2}{2} \int_0^L \bar{m} \, v^2 \, dx = \frac{\bar{m}}{2} \omega^2 Z_0{}^2 \int_0^L [\psi(x)]^2 \, dx$$

$$= \frac{33}{140} \frac{\bar{m}L}{2} \omega^2 Z_0{}^2$$

Weight:
$$T_{max}^W = \frac{W}{2g} \omega^2 \left[v\left(x = \frac{L}{2}\right) \right]^2 = \frac{W}{2g} \omega^2 ({}^5\!/_{16}Z_0)^2$$

$$= \frac{25}{256} \frac{W}{2g} \omega^2 Z_0{}^2$$

Hence the total kinetic energy is

$$T_{\max} = \left(\frac{33}{140} + \frac{25}{256}\frac{W}{\overline{m}Lg}\right)\frac{\overline{m}L}{2}\omega^2 Z_0^2$$

and equating the maximum kinetic- and potential-energy expressions leads to the frequency equation

$$\omega^2 = \frac{3}{\dfrac{33}{140} + \dfrac{25}{256}\dfrac{W}{\overline{m}Lg}}\frac{EI}{\overline{m}L^4} \qquad\qquad ////$$

9-4   IMPROVED RAYLEIGH METHOD

The idea of using a deflected shape resulting from an inertia loading in a Rayleigh analysis, as described above, can be applied systematically to develop improved versions of the procedure. The standard analysis involves the arbitrary selection of a deflected shape which satisfies the geometric boundary conditions of the structure. For the purposes of this discussion, this initially selected shape will be identified with the superscript zero:

$$v^{(0)}(x,t) = \psi^{(0)}(x)Z_0^{(0)}\sin\omega t \qquad (9\text{-}13)$$

The maximum potential and kinetic energy associated with this shape are then given by

$$V_{\max} = \frac{1}{2}\int_0^L EI(x)\left(\frac{\partial^2 v^{(0)}}{\partial x^2}\right)^2 dx = \frac{Z_0^{(0)2}}{2}\int_0^L EI(x)(\psi''^{(0)})^2\,dx \qquad (9\text{-}14)$$

$$T_{\max} = \frac{1}{2}\int_0^L m(x)(\dot{v}^{(0)})^2\,dx = \frac{Z_0^{(0)2}}{2}\omega^2\int_0^L m(x)(\psi^{(0)})^2\,dx \qquad (9\text{-}15)$$

Method R_{00}   The standard Rayleigh frequency expression, designated as method R_{00}, is

$$\omega^2 = \frac{\displaystyle\int_0^L EI(x)(\psi''^{(0)})^2\,dx}{\displaystyle\int_0^L m(x)(\psi^{(0)})^2\,dx} \qquad (9\text{-}16)$$

However, a better approximation of the frequency can be obtained by computing the potential energy from the work done in deflecting the structure by the inertia force associated with the assumed deflection. The distributed inertia force is (at the time of maximum displacement)

$$p^{(0)}(x) = \omega^2 m(x)v^{(0)} = Z_0^{(0)}\omega^2 m(x)\psi^{(0)} \qquad (9\text{-}17)$$

The deflection produced by this loading may be written

$$v^{(1)} = \omega^2 \frac{v^{(1)}}{\omega^2} = \omega^2 \psi^{(1)} \frac{Z_0^{(1)}}{\omega^2} \equiv \omega^2 \psi^{(1)} \bar{Z}_0^{(1)} \qquad (9\text{-}18)$$

in which ω^2 is the unknown frequency. It may be looked upon as a proportionality factor in both Eqs. (9-17) and (9-18); it is not combined into the expression because its value is not known. The potential energy of the strain produced by this loading is given by

$$V_{max} = \frac{1}{2} \int_0^L p^{(0)} v^{(1)} \, dx = \frac{Z_0^{(0)} \bar{Z}_0^{(1)}}{2} \omega^4 \int m(x) \psi^{(0)} \psi^{(1)} \, dx \qquad (9\text{-}19)$$

Method R_{01}   Equating this expression for the potential energy to the kinetic energy given by the originally assumed shape [Eq. (9-15)] leads to the improved Rayleigh frequency expression, here designated as method R_{01}:

$$\omega^2 = \frac{Z_0^{(0)}}{\bar{Z}_0^{(1)}} \frac{\displaystyle\int_0^L m(x)(\psi^{(0)})^2 \, dx}{\displaystyle\int_0^L m(x) \psi^{(0)} \psi^{(1)} \, dx} \qquad (9\text{-}20)$$

This expression often is recommended in preference to Eq. (9-16) because it avoids the differentiation operation required in the standard formula. In general, curvatures $\psi''(x)$ associated with an assumed deflected shape will be much less accurate than the shape function $\psi(x)$, and thus Eq. (9-20), which involves no derivatives, will give improved accuracy.

However, a still better approximation can be obtained with relatively little additional effort by computing the kinetic energy from the calculated shape $v^{(1)}$ rather than from the initial shape $v^{(0)}$. In this case the result is

$$T_{max} = \frac{1}{2} \int_0^L m(x)(\dot{v}^{(1)})^2 \, dx = \tfrac{1}{2}\omega^6 (\bar{Z}^{(1)})^2 \int_0^L m(x)(\psi^{(1)})^2 \, dx \qquad (9\text{-}21)$$

Method R_{11}   Equating this to the strain energy of Eq. (9-19) leads to the further improved result (here designated as method R_{11}):

$$\omega^2 = \frac{Z_0^{(0)}}{\bar{Z}_0^{(1)}} \frac{\displaystyle\int_0^L m(x) \psi^{(0)} \psi^{(1)} \, dx}{\displaystyle\int_0^L m(x)(\psi^{(1)})^2 \, dx} \qquad (9\text{-}22)$$

Further improvement could be made by continuing the process another step, that is, by using the inertia loading associated with $\psi^{(1)}$ to calculate a new shape $\psi^{(2)}$. In fact, as will be shown later, the process will eventually converge to the exact vibration shape if it is carried through enough cycles and therefore will yield the exact fre-

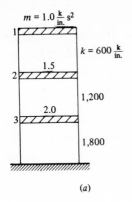

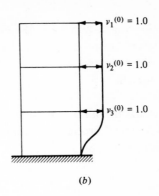

(a)

(b)

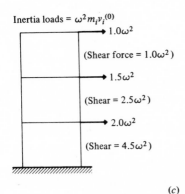

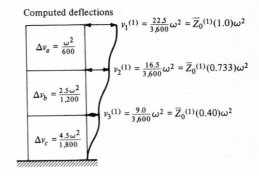

(c)

FIGURE E9-2
Frame for Rayleigh method frequency analysis: (a) mass and stiffness values;
(b) initial assumed shape; (c) deflections resulting from initial inertia forces.

quency. However, for practical use of the Rayleigh method there is no need to go beyond the improved procedure represented by Eq. (9-22). Also, it should be noted that the generalized-coordinate amplitudes $Z_0{}^{(0)}$ and $\bar{Z}_0{}^{(1)}$ in Eqs. (9-20) and (9-22) are arbitrary and can be set to unity if the shape functions $\psi^{(0)}$ and $\psi^{(1)}$ are defined appropriately. However, it is advisable to leave the generalized coordinates in the equations to show that the relative amplitude of $v^{(0)}$ and $v^{(1)}$ is a factor in computing the frequency.

EXAMPLE E9-3   The two improved versions of the Rayleigh method will be demonstrated and compared with the standard method in carrying out the frequency analysis of the three-story frame shown in Fig. E9-2a. The mass of this frame is lumped in the girders, with values as shown, and the columns are assumed to be weightless. Also, the girders are assumed to be rigid, so that the columns in each story act as simple lateral springs with stiffness coefficients as indicated.

STANDARD METHOD R_{00} In order to demonstrate the effectiveness of the improvement procedures, a poor choice will be deliberately assumed for the initial vibration shape for the frame. This shape consists of equal displacements for the three stories, as shown in Fig. E9-2b; thus

$$v_1{}^{(0)} = v_2{}^{(0)} = v_3{}^{(0)} = 1.0 = Z_0{}^{(0)}\psi_i{}^{(0)} \qquad \text{where } \psi_i{}^{(0)} = Z_0{}^{(0)} = 1.0$$

From this shape, the maximum kinetic energy is given by

$$T_{\max}^{(0)} = {}^1\!/_2 \sum m_i(v_i{}^{(0)})^2 = {}^1\!/_2\omega^2 Z_0{}^{(0)2} \sum m_i(\psi_i{}^{(0)})^2 = {}^1\!/_2\omega^2(4.5)$$

The maximum potential energy depends on the relative story-to-story deformations Δv_i and is given by

$$V_{\max}^{(0)} = {}^1\!/_2 \sum k_i(\Delta v_i{}^{(0)})^2 = {}^1\!/_2 Z_0{}^{(0)2} \sum k_i(\Delta \psi_i{}^{(0)})^2 = {}^1\!/_2(1{,}800)$$

Hence, when the potential and kinetic energies are equated, the frequency is

$$\omega^2 = \frac{1{,}800}{4.5} = 400 \qquad \omega = 20 \text{ rad/s}$$

IMPROVED METHOD R_{01} The assumption that the structure behaves as though the columns were rigid above the first story clearly is not reasonable for this frame and can be expected to give a gross overestimate of the frequency. However, using the inertia forces associated with these initial deflections to calculate an improved shape, in accordance with the improved method R_{01}, leads to much better results.

The inertia loads of the initial shape and the deflections they produce are shown in Fig. E9-2c. The deflections can easily be calculated because the deformation Δv_i in each story is given by the story shear divided by the story stiffness. The maximum potential energy of this new shape $v_i{}^{(1)}$ may be found as follows:

$$V_{\max}^{(1)} = {}^1\!/_2 \sum p_i{}^{(0)} v_i{}^{(1)} = \frac{\omega^4}{2} \overline{Z}_0{}^{(1)} \sum m_i \psi_i{}^{(0)} \psi_i{}^{(1)} = \frac{\omega^4}{2} \overline{Z}_0{}^{(1)}(2.90)$$

When this is equated to the kinetic energy found previously, the frequency is

$$\omega^2 = \frac{1}{\overline{Z}_0{}^{(1)}} \frac{4.50}{2.90} = \frac{1}{22.5/3{,}600} \frac{4.5}{2.9} = 248 \qquad \omega = 15.73 \text{ rad/s}$$

It is apparent that this much smaller frequency represents a great improvement over the result obtained by the standard method R_{00}.

IMPROVED METHOD R_{11} Still better results can be obtained by using the improved shape $\psi_1^{(1)}$ in calculating the kinetic as well as the potential energy. Thus the maximum kinetic energy becomes

$$T_{max}^{(1)} = \frac{\omega^2}{2}(\bar{Z}_0^{(1)})^2 \sum m_i(\psi_i^{(1)})^2 = \frac{\omega^6}{2}\left(\frac{22.5}{3{,}600}\right)^2 \quad (2.124)$$

Hence, equating this to the improved potential-energy expression leads to the frequency value

$$\omega^2 = \frac{1}{\bar{Z}_0^{(1)}}\frac{2.90}{2.124} = \frac{3{,}600}{22.5}\frac{2.90}{2.124} = 218 \qquad \omega = 14.76 \text{ rad/s}$$

This is quite close to the exact first-mode frequency for this structure, $\omega_1 = 14.5$ rad/s, as will be derived in Chap. 11.

It is interesting to note that method R_{11} gives the same result here as would be given by Eq. (9-12), where the deflections due to a lateral gravity acceleration are the basis of the analysis. This is because the inertia forces associated with *equal* story displacements are equivalent to the lateral gravity forces. However, if a more reasonable estimate had been made of the initial shape (rather than equal story deflections), the improved method R_{11} would have given a better result than Eq. (9-12). ////

PROBLEMS

9-1 By Rayleigh's method, compute the period of vibration of the uniform beam supporting a central mass m_1 shown in Fig. P9-1. For the assumed shape, use the deflection produced by a central load p; i.e., $v(x) = px(3L^2 - 4x^2)/48EI$. Consider the cases: (*a*) $m_1 = 0$, and (*b*) $m_1 = 3\bar{m}L$.

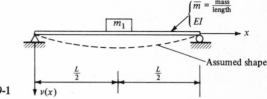

FIGURE P9-1

9-2 (*a*) Determine the period of vibration of the frame shown in Fig. P9-2, assuming the girder to be rigid and the deflected shape of the columns to be that due to a lateral load p acting on the girder $v(x) = p(2L^3 - 3L^2x + x^3)/12EI$;

(b) What fraction of the total column weight assumed lumped with the girder weight will give the same period of vibration as was found in part (a)?

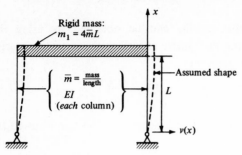

Rigid mass:
$m_1 = 4\bar{m}L$

$\bar{m} = \frac{\text{mass}}{\text{length}}$
EI
(each column)

Assumed shape

L

x

$v(x)$

FIGURE P9-2

9-3 The shear building of Fig. P9-3 has its entire mass lumped in the rigid girders. For the given mass and stiffness properties, and assuming a linear initial shape (as shown), evaluate the period of vibration by:
(a) Rayleigh method R_{00}
(b) Rayleigh method R_{01}
(c) Rayleigh method R_{11}

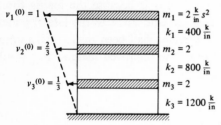

$v_1{}^{(0)} = 1$

$v_2{}^{(0)} = \frac{2}{3}$

$v_3{}^{(0)} = \frac{1}{3}$

$m_1 = 2 \frac{k}{in} s^2$
$k_1 = 400 \frac{k}{in}$
$m_2 = 2$
$k_2 = 800 \frac{k}{in}$
$m_3 = 2$
$k_3 = 1200 \frac{k}{in}$

FIGURE P9-3

9-4 Repeat Prob. 9-3 if the building properties are $m_1 = 1$, $m_2 = 2$, $m_3 = 3 \text{ k} \cdot \text{s}^2/\text{in}$ and $k_1 = k_2 = k_3 = 800 \text{ k/in}$.

PART TWO

Multi-Degree-of-Freedom Systems

FORMULATION OF THE MDOF
EQUATIONS OF MOTION

10-1 SELECTION OF THE DEGREES OF FREEDOM

The discussion presented in Part One has demonstrated that any structure can be represented as a SDOF system the dynamic response of which can be evaluated by the solution of a single differential equation of motion. If the physical properties of the system are such that its motion can be described by a single coordinate and no other motion is possible, then it actually is a SDOF system and the solution of the equation provides the exact dynamic response. On the other hand, if the structure actually has more than one possible mode of displacement and it is reduced mathematically to a SDOF approximation by assuming its deformed shape, the solution of the equation of motion is only an approximation of the true dynamic behavior.

The quality of the result obtained with a SDOF approximation depends on many factors, principally the spatial distribution and time variation of the loading and the stiffness and mass properties of the structure. If the physical properties of the system constrain it to move most easily with the assumed shape, and if the loading is such as to excite a significant response in this shape, the SDOF solution will probably be a good approximation; otherwise, the true behavior may bear little resemblance to the computed response. One of the greatest disadvantages of the SDOF approximation is that it is difficult to assess the reliability of the results obtained from it.

In general, the dynamic response of a structure cannot be described adequately by a SDOF model; usually the response includes time variations of the displacement shape as well as its amplitude. Such behavior can be described only in terms of more than one displacement coordinate; that is, the motion must be represented by more than one degree of freedom. As noted in Chap. 1, the degrees of freedom in a discrete-parameter system may be taken as the displacement amplitudes of certain selected points in the structure, or they may be generalized coordinates representing the amplitudes of a specified set of displacement patterns. In the present discussion, the former approach will be adopted; this includes both the finite-element and the lumped-mass type of idealization. The generalized-coordinate procedure will be discussed at the end of Part Two (Sec. 16-2).

In this development of the equations of motion of a general MDOF system, it will be convenient to refer to the general simple beam shown in Fig. 10-1 as a typical example. The discussion applies equally to any type of structure, but the visualization of the physical factors involved in evaluating all the forces acting is simplified for this type of structure.

The motion of this structure will be assumed to be defined by the displacements of a set of discrete points on the beam: $v_1(t)$, $v_2(t)$, ..., $v_i(t)$, ..., $v_N(t)$. In principle, these points may be located arbitrarily on the structure; in practice, they should be associated with any specific features of the physical properties which may be significant and should be distributed so as to provide a good definition of the deflected shape. The number of degrees of freedom (displacement components) to be considered is left to the discretion of the analyst; greater numbers provide better approximations of the true dynamic behavior, but in many cases excellent results can be obtained with only two or three degrees of freedom. In the beam of Fig. 10-1 only one displacement component has been associated with each nodal point on the beam. It should be noted, however, that several displacement components could be identified with each point; e.g., the rotation $\partial v/\partial x$ and longitudinal motions might be used as additional degrees of freedom at each point.

10-2 DYNAMIC-EQUILIBRIUM CONDITION

The equation of motion of the system of Fig. 10-1 can be formulated by expressing the equilibrium of the effective forces associated with each of its degrees of freedom. In general four types of forces will be involved at any point i: the externally applied load $p_i(t)$ and the forces resulting from the motion, that is, inertia f_{Ii}, damping f_{Di},

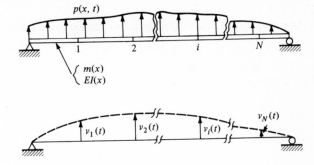

FIGURE 10-1
Discretization of a general beam-type structure.

and elastic f_{Si}. Thus for each of the several degrees of freedom the dynamic equilibrium may be expressed as

$$
\begin{aligned}
f_{I1} + f_{D1} + f_{S1} &= p_1(t) \\
f_{I2} + f_{D2} + f_{S2} &= p_2(t) \qquad (10\text{-}1) \\
f_{I3} + f_{D3} + f_{S3} &= p_3(t)
\end{aligned}
$$

$$\cdots\cdots\cdots\cdots\cdots$$

or when the force vectors are represented in matrix form,

$$\mathbf{f}_I + \mathbf{f}_D + \mathbf{f}_S = \mathbf{p}(t) \qquad (10\text{-}2)$$

which is the MDOF equivalent of the SDOF equation (2-1).

Each of the resisting forces is expressed most conveniently by means of an appropriate set of influence coefficients. Consider, for example, the elastic-force component developed at point 1; this depends in general upon the displacement components developed at all points of the structure:

$$f_{S1} = k_{11}v_1 + k_{12}v_2 + k_{13}v_3 + \cdots + k_{1N}v_N \qquad (10\text{-}3a)$$

Similarly, the elastic force corresponding to the degree of freedom v_2 is

$$f_{S2} = k_{21}v_1 + k_{22}v_2 + k_{23}v_3 + \cdots + k_{2N}v_N \qquad (10\text{-}3b)$$

and, in general,

$$f_{Si} = k_{i1}v_1 + k_{i2}v_2 + k_{i3}v_3 + \cdots + k_{iN}v_N \qquad (10\text{-}3c)$$

In these expressions it has been tacitly assumed that the structural behavior is linear, so that the principle of superposition applies. The coefficients k_{ij} are called *stiffness influence coefficients*, defined as follows:

$$
\begin{aligned}
k_{ij} = \ &\text{force corresponding to coordinate } i \text{ due to} \\
&\text{a unit } \textit{displacement} \text{ of coordinate } j \qquad (10\text{-}4)
\end{aligned}
$$

In matrix form, the complete set of elastic-force relationships may be written

$$
\begin{bmatrix} f_{S1} \\ f_{S2} \\ \cdot \\ f_{Si} \\ \cdot \end{bmatrix} = \begin{bmatrix} k_{11} & k_{12} & k_{13} & \cdots & k_{1i} & \cdots & k_{1N} \\ k_{21} & k_{22} & k_{23} & \cdots & k_{2i} & \cdots & k_{2N} \\ \cdots\cdots\cdots\cdots\cdots\cdots\cdots\cdots\cdots\cdots\cdots \\ k_{i1} & k_{i2} & k_{i3} & \cdots & k_{ii} & \cdots & k_{iN} \\ \cdots\cdots\cdots\cdots\cdots\cdots\cdots\cdots\cdots\cdots\cdots \end{bmatrix} \begin{bmatrix} v_1 \\ v_2 \\ \cdot \\ v_i \\ \cdot \end{bmatrix} \tag{10-5}
$$

or, symbolically,

$$
\mathbf{f}_S = \mathbf{k}\mathbf{v} \tag{10-6}
$$

in which the matrix of stiffness coefficients **k** is called the *stiffness matrix* of the structure (for the specified set of displacement coordinates) and **v** is the displacement vector representing the displaced shape of the structure.

If it is assumed that the damping depends on the velocity, that is, viscous type, the damping forces corresponding to the selected degrees of freedom may be expressed by means of damping influence coefficients in similar fashion. By analogy with Eq. (10-5), the complete set of damping forces is given by

$$
\begin{bmatrix} f_{D1} \\ f_{D2} \\ \cdot \\ f_{Di} \\ \cdot \end{bmatrix} = \begin{bmatrix} c_{11} & c_{12} & c_{13} & \cdots & c_{1i} & \cdots & c_{1N} \\ c_{21} & c_{22} & c_{23} & \cdots & c_{2i} & \cdots & c_{2N} \\ \cdots\cdots\cdots\cdots\cdots\cdots\cdots\cdots\cdots\cdots\cdots \\ c_{i1} & c_{i2} & c_{i3} & \cdots & c_{ii} & \cdots & c_{iN} \\ \cdots\cdots\cdots\cdots\cdots\cdots\cdots\cdots\cdots\cdots\cdots \end{bmatrix} \begin{bmatrix} \dot{v}_1 \\ \dot{v}_2 \\ \cdot \\ \dot{v}_i \\ \cdot \end{bmatrix} \tag{10-7}
$$

in which $\dot{v}_i$ represents the time rate of change (velocity) of the i displacement co-ordinate and the coefficients c_{ij} are called *damping influence coefficients*. The definition of these coefficients is exactly parallel to Eq. (10-4):

$$
c_{ij} = \text{force corresponding to coordinate } i \text{ due to unit}
$$
$$
\textit{velocity of coordinate } j \tag{10-8}
$$

Symbolically, Eq. (10-7) may be written

$$
\mathbf{f}_D = \mathbf{c}\dot{\mathbf{v}} \tag{10-9}
$$

in which the matrix of damping coefficients **c** is called the *damping matrix* of the structure (for the specified degrees of freedom) and $\dot{\mathbf{v}}$ is the velocity vector.

The inertia forces may also be expressed by a set of influence coefficients called the *mass coefficients*. These represent the relationship between the accelerations of the degrees of freedom and the resulting inertia forces; by analogy with Eq. (10-5), the inertia forces may be expressed as

$$
\begin{bmatrix} f_{I1} \\ f_{I2} \\ \cdot \\ f_{Ii} \\ \cdot \end{bmatrix} = \begin{bmatrix} m_{11} & m_{12} & m_{13} & \cdots & m_{1i} & \cdots & m_{1N} \\ m_{21} & m_{22} & m_{23} & \cdots & m_{2i} & \cdots & m_{2N} \\ \cdots\cdots\cdots\cdots\cdots\cdots\cdots\cdots\cdots\cdots\cdots \\ m_{i1} & m_{i2} & m_{i3} & \cdots & m_{ii} & \cdots & m_{iN} \\ \cdots\cdots\cdots\cdots\cdots\cdots\cdots\cdots\cdots\cdots\cdots \end{bmatrix} \begin{bmatrix} \ddot{v}_1 \\ \ddot{v}_2 \\ \cdot \\ \ddot{v}_i \\ \cdot \end{bmatrix} \tag{10-10}
$$

where $\ddot{v}_i$ is the acceleration of the i displacement coordinate and the coefficients m_{ij} are the *mass influence coefficients*, defined as follows:

$$m_{ij} = \text{force corresponding to coordinate } i \text{ due to}$$
$$\text{unit } acceleration \text{ of coordinate } j \qquad (10\text{-}11)$$

Symbolically, Eq. (10-10) may be written

$$\mathbf{f}_I = \mathbf{m}\ddot{\mathbf{v}} \qquad (10\text{-}12)$$

in which the matrix of mass coefficients $\mathbf{m}$ is called the *mass matrix* of the structure and $\ddot{\mathbf{v}}$ is its acceleration vector, both defined for the specified set of displacement coordinates.

Substituting Eqs. (10-6), (10-9), and (10-12) into Eq. (10-2) gives the complete dynamic equilibrium of the structure, considering all degrees of freedom:

$$\mathbf{m}\ddot{\mathbf{v}} + \mathbf{c}\dot{\mathbf{v}} + \mathbf{k}\mathbf{v} = \mathbf{p}(t) \qquad (10\text{-}13)$$

This equation is the MDOF equivalent of Eq. (2-3); each term of the SDOF equation is represented by a matrix in Eq. (10-13), the order of the matrix corresponding to the number of degrees of freedom used in describing the displacements of the structure. Thus, Eq. (10-13) expresses the N equations of motion which serve to define the response of the MDOF system.

10-3 AXIAL-FORCE EFFECTS

It was observed in the discussion of SDOF systems that axial forces or any load which may tend to cause buckling of a structure may have a significant effect on the stiffness of the structure. Similar effects may be observed in MDOF systems; the force component acting parallel to the original axis of the members leads to additional load components which act in the direction (and sense) of the nodal displacements and which will be denoted by f_G. When these forces are included, the dynamic-equilibrium expression, Eq. (10-2), becomes

$$\mathbf{f}_I + \mathbf{f}_D + \mathbf{f}_S - \mathbf{f}_G = \mathbf{p}(t) \qquad (10\text{-}14)$$

in which the negative sign results from the fact that the forces $\mathbf{f}_G$ are assumed to contribute to the deflection rather than oppose it.

These forces resulting from axial loads depend on the displacements of the structure and may be expressed by influence coefficients, called the *geometric-stiffness coefficients*, as follows:

$$
\begin{bmatrix} f_{G_1} \\ f_{G_2} \\ \cdot \\ f_{G_i} \\ \cdot \end{bmatrix} =
\begin{bmatrix}
k_{G_{11}} & k_{G_{12}} & k_{G_{13}} & \cdots & k_{G_{1i}} & \cdots & k_{G_{1N}} \\
k_{G_{21}} & k_{G_{22}} & k_{G_{23}} & \cdots & k_{G_{2i}} & \cdots & k_{G_{2N}} \\
\multicolumn{7}{c}{\cdots\cdots\cdots\cdots\cdots\cdots\cdots\cdots\cdots} \\
k_{G_{i1}} & k_{G_{i2}} & k_{G_{i3}} & \cdots & k_{G_{ii}} & \cdots & k_{G_{iN}} \\
\multicolumn{7}{c}{\cdots\cdots\cdots\cdots\cdots\cdots\cdots\cdots\cdots}
\end{bmatrix}
\begin{bmatrix} v_1 \\ v_2 \\ \cdot \\ v_i \\ \cdot \end{bmatrix} \qquad (10\text{-}15)
$$

in which the geometric-stiffness influence coefficients $k_{G_{ij}}$ have the following definition:

$$k_{G_{ij}} = \text{force corresponding to coordinate } i \text{ due to unit}$$
$$\text{displacement of coordinate } j \text{ and resulting from}$$
$$\text{axial-force components in the structure} \qquad (10\text{-}16)$$

Symbolically Eq. (10-15) may be written

$$\mathbf{f}_G = \mathbf{k}_G \mathbf{v} \qquad (10\text{-}17)$$

where $\mathbf{k}_G$ is called the *geometric-stiffness matrix* of the structure.

When this expression is introduced, the equation of dynamic equilibrium of the structure [given by Eq. (10-13) without axial-force effects] becomes

$$\mathbf{m}\ddot{\mathbf{v}} + \mathbf{c}\dot{\mathbf{v}} + \mathbf{k}\mathbf{v} - \mathbf{k}_G\mathbf{v} = \mathbf{p}(t) \qquad (10\text{-}18)$$

or when it is noted that both the elastic stiffness and the geometric stiffness are multiplied by the displacement vector, the combined stiffness effect can be expressed by a single symbol and Eq. (10-18) written

$$\mathbf{m}\ddot{\mathbf{v}} + \mathbf{c}\dot{\mathbf{v}} + \bar{\mathbf{k}}\mathbf{v} = \mathbf{p}(t) \qquad (10\text{-}19)$$

in which

$$\bar{\mathbf{k}} = \mathbf{k} - \mathbf{k}_G \qquad (10\text{-}20)$$

is called the combined stiffness matrix, which includes both elastic and geometric effects. The dynamic properties of the structure are expressed completely by the four influence-coefficient matrices of Eq. (10-18), while the dynamic loading is fully defined by the load vector. The evaluation of these physical-property matrices and the evaluation of the load vector resulting from externally applied forces will be discussed in detail in the following chapter. The effective-load vector resulting from support excitation will be discussed in connection with earthquake-response analysis in Chap. 27.

EVALUATION OF
STRUCTURAL-PROPERTY MATRICES

11-1 ELASTIC PROPERTIES

Flexibility

Before discussing the elastic-stiffness matrix expressed in Eq. (10-5), it will be useful to define the inverse relationship, the flexibility. The definition of a flexibility influence coefficient $\tilde{f}_{ij}$ is

$$\tilde{f}_{ij} = \text{deflection of coordinate } i \text{ due to unit load}$$
$$\text{applied to coordinate } j \qquad (11\text{-}1)$$

For the simple beam shown in Fig. 11-1, the physical significance of some of the flexibility influence coefficients associated with a set of vertical-displacement degrees of freedom is illustrated. Horizontal or rotational degrees of freedom might also have been considered, in which case it would have been necessary to use the corresponding horizontal or rotational unit loads in defining the complete set of influence coefficients; however, it will be convenient to restrict the present discussion to the vertical motions.

The evaluation of the flexibility influence coefficients for any given system is a standard problem of static structural analysis; any desired method of analysis may

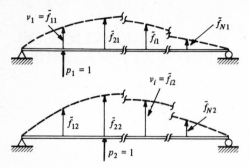

FIGURE 11-1
Definition of flexibility influence coefficients.

be used to compute these deflections resulting from the applied unit loads. When the complete set of influence coefficients has been determined, they are used to calculate the displacement vector resulting from any combination of the applied loads. For example, the deflection at point 1 due to any combination of loads may be expressed

$$v_1 = \tilde{f}_{11}p_1 + \tilde{f}_{12}p_2 + \tilde{f}_{13}p_3 + \cdots + \tilde{f}_{1N}p_N \qquad (11\text{-}2)$$

Since similar expressions can be written for each displacement component, the complete set of displacements is expressed

$$\begin{bmatrix} v_1 \\ v_2 \\ \cdot \\ v_i \\ \cdot \end{bmatrix} = \begin{bmatrix} \tilde{f}_{11} & \tilde{f}_{12} & \tilde{f}_{13} & \cdots & \tilde{f}_{1i} & \cdots & \tilde{f}_{1N} \\ \tilde{f}_{21} & \tilde{f}_{22} & \tilde{f}_{23} & \cdots & \tilde{f}_{2i} & \cdots & \tilde{f}_{2N} \\ \cdots\cdots\cdots\cdots\cdots\cdots\cdots\cdots\cdots \\ \tilde{f}_{i1} & \tilde{f}_{i2} & \tilde{f}_{i3} & \cdots & \tilde{f}_{ii} & \cdots & \tilde{f}_{iN} \\ \cdots\cdots\cdots\cdots\cdots\cdots\cdots\cdots\cdots \end{bmatrix} \begin{bmatrix} p_1 \\ p_2 \\ \cdot \\ p_i \\ \cdot \end{bmatrix} \qquad (11\text{-}3)$$

or symbolically

$$\mathbf{v} = \tilde{\mathbf{f}}\mathbf{p} \qquad (11\text{-}4)$$

in which the matrix of flexibility influence coefficients $\tilde{\mathbf{f}}$ is called the *flexibility matrix* of the structure.

In Eq. (11-4) the deflections are expressed in terms of the vector of externally applied loads $\mathbf{p}$, which are considered positive when acting in the same sense as the positive displacements. The deflection may also be expressed in terms of the elastic forces $\mathbf{f}_S$ which *resist* the deflections and which are considered positive when acting opposite to the positive displacements. Obviously by statics $\mathbf{f}_S = \mathbf{p}$, and Eq. (11-4) may be revised to read

$$\mathbf{v} = \tilde{\mathbf{f}}\mathbf{f}_S \qquad (11\text{-}5)$$

Stiffness

The physical meaning of the stiffness influence coefficients defined in Eq. (10-4) is illustrated for a few degrees of freedom in Fig. 11-2; they represent the forces developed in the structure when a unit displacement corresponding to one degree of freedom

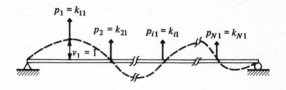

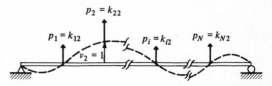

FIGURE 11-2
Definition of stiffness influence
coefficients.

is introduced and no other nodal displacements are permitted. It should be noted
that the stiffness influence coefficients in Fig. 11-2 are numerically equal to the applied
forces required to maintain the specified displacement condition. They are positive
when the sense of the applied force corresponds to a positive displacement, and
negative otherwise.

Basic Structural Concepts

Strain energy The strain energy stored in any structure may be expressed conveniently in terms of either the flexibility or the stiffness matrix. The strain energy
U is equal to the work done in distorting the system; thus

$$U = \frac{1}{2}\sum_{i=1}^{N} p_i v_i = {}^1/_2 \mathbf{p}^T \mathbf{v} \qquad (11\text{-}6)$$

where $\mathbf{p}^T$ represents the transpose of $\mathbf{p}$. By substituting Eq. (11-4) this becomes

$$U = {}^1/_2 \mathbf{p}^T \tilde{\mathbf{f}} \mathbf{p} \qquad (11\text{-}7)$$

Alternatively, transposing Eq. (11-6) and substituting Eq. (10-6) leads to the second
strain-energy expression (note that $\mathbf{p} = \mathbf{f}_S$):

$$U = {}^1/_2 \mathbf{v}^T \mathbf{k} \mathbf{v} \qquad (11\text{-}8)$$

Finally, when it is noted that the strain energy stored in a stable structure during any
distortion must always be positive, it is evident that

$$\mathbf{v}^T \mathbf{k} \mathbf{v} > 0 \qquad \text{and} \qquad \mathbf{p}^T \mathbf{f} \mathbf{p} > 0 \qquad (11\text{-}9)$$

Matrices which satisfy this condition, where $\mathbf{v}$ or $\mathbf{p}$ is any arbitrary nonzero vector,
are said to be *positive definite*; positive definite matrices (and consequently the
flexibility and stiffness matrices) are nonsingular and can be inverted.

Load system a:

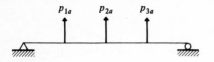

Load system b:

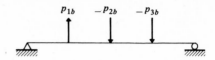

Deflections a:

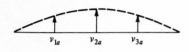

Deflections b:

FIGURE 11-3
Two independent load systems and resulting deflections.

Inverting the stiffness matrix and premultiplying both sides of Eq. (10-6) by the inverse leads to

$$\mathbf{k}^{-1}\mathbf{f}_S = \mathbf{v}$$

which upon comparison with Eq. (11-5) demonstrates that the flexibility matrix is the inverse of the stiffness matrix:

$$\mathbf{k}^{-1} = \tilde{\mathbf{f}} \qquad (11\text{-}10)$$

In practice, the evaluation of stiffness coefficients by direct application of the definition, as implied by Eq. (11-2), may be a tedious computational problem. In many cases, the most convenient procedure for obtaining the stiffness matrix is direct evaluation of the flexibility coefficients and inversion of the flexibility matrix.

Betti's law A property which is very important in structural-dynamics analysis can be derived by applying two sets of loads to a structure in reverse sequence and comparing expressions for the work done in the two cases. Consider, for example, the two different load systems and their resulting displacements shown in Fig. 11-3. If the loads a are applied first followed by loads b, the work done will be as follows:

Case 1:

Loads a: $\qquad W_{aa} = {}^1\!/_2 \sum p_{ia}v_{ia} = {}^1\!/_2 \mathbf{p}_a^T \mathbf{v}_a$

Loads b: $W_{bb} + W_{ab} = {}^1\!/_2 \mathbf{p}_b^T \mathbf{v}_b + \mathbf{p}_a^T \mathbf{v}_b$

Total: $\qquad W_1 = W_{aa} + W_{bb} + W_{ab} = {}^1\!/_2 \mathbf{p}_a^T \mathbf{v}_a + {}^1\!/_2 \mathbf{p}_b^T \mathbf{v}_b + \mathbf{p}_a^T \mathbf{v}_b \qquad (11\text{-}11)$

Note that the work done by loads a during the application of loads b is not multiplied by ${}^1\!/_2$; they act at their full value during the entire displacement $\mathbf{v}_b$. Now if the loads are applied in reverse sequence, the work done is:

Case 2:

Loads b: $\qquad W_{bb} = {}^{1}/_{2}\mathbf{p}_b^T\mathbf{v}_b$

Loads a: $W_{aa} + W_{ba} = {}^{1}/_{2}\mathbf{p}_a^T\mathbf{v}_a + \mathbf{p}_b^T\mathbf{v}_a$

Total: $\qquad W_2 = W_{bb} + W_{aa} + W_{ba} = \tfrac{1}{2}\mathbf{p}_b^T\mathbf{v}_b + \tfrac{1}{2}\mathbf{p}_a^T\mathbf{v}_a + \mathbf{p}_b^T\mathbf{v}_a \qquad (11\text{-}12)$

The deformation of the structure is independent of the loading sequence, however; therefore the strain energy and hence also the work done by the loads is the same in both these cases; that is, $W_1 = W_2$. From a comparison of Eqs. (11-11) and (11-12) it may be concluded that $W_{ab} = W_{ba}$; thus

$$\mathbf{p}_a^T\mathbf{v}_b = \mathbf{p}_b^T\mathbf{v}_a \qquad (11\text{-}13)$$

Equation (11-13) is an expression of *Betti's law*; it states that the work done by one set of loads on the deflection due to a second set of loads is equal to the work of the second set of loads acting on the deflections due to the first.

If Eq. (11-4) is written for the two sets of forces and displacements and substituted into both sides of Eq. (11-13):

$$\mathbf{p}_a^T\tilde{\mathbf{f}}\mathbf{p}_b = \mathbf{p}_b^T\tilde{\mathbf{f}}\mathbf{p}_a$$

it is evident that

$$\tilde{\mathbf{f}} = \tilde{\mathbf{f}}^T \qquad (11\text{-}14)$$

Thus the flexibility matrix must be symmetric; that is, $\tilde{f}_{ij} = \tilde{f}_{ji}$. This is an expression of Maxwell's law of reciprocal deflections. Substituting similarly with Eq. (10-6) (and noting that $\mathbf{p} = \mathbf{f}_S$) leads to

$$\mathbf{k} = \mathbf{k}^T \qquad (11\text{-}15)$$

That is, the stiffness matrix also is symmetric.

Finite-Element Stiffness

In principle, the flexibility or stiffness coefficients associated with any prescribed set of nodal displacements can be obtained by direct application of their definitions. In practice, however, the finite-element concept, described in Chap. 1, frequently provides the most convenient means for evaluating the elastic properties. By this approach the structure is assumed to be divided into a system of discrete elements which are interconnected only at a finite number of nodal points. The properties of the complete structure are then found by evaluating the properties of the individual finite elements and superposing them appropriately.

The problem of defining the stiffness properties of any structure is thus reduced basically to the evaluation of the stiffness of a typical element. Consider, for example, the nonuniform straight-beam segment shown in Fig. 11-4. The two nodal points by which this type of element can be assembled into a structure are located at its

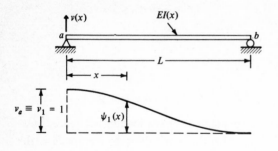

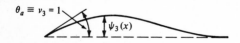

FIGURE 11-4
Beam deflections due to unit nodal
displacements at left end.

ends, and if only transverse plane displacements are considered, it has two degrees of
freedom at each node, vertical translation and rotation. The deflected shapes resulting
from applying a unit displacement of each type at the left end of the element while
constraining the other three nodal displacements are shown in Fig. 11-4. These displacement functions could be any arbitrary shapes which satisfy nodal and internal
continuity requirements, but they generally are assumed to be the shapes developed
in a *uniform* beam subjected to these nodal displacements. These are cubic hermitian
polynomials which may be expressed as

$$\psi_1(x) = 1 - 3\left(\frac{x}{L}\right)^2 + 2\left(\frac{x}{L}\right)^3 \quad (11\text{-}16a)$$

$$\psi_3(x) = x\left(1 - \frac{x}{L}\right)^2 \quad (11\text{-}16b)$$

The equivalent shape functions for displacements applied at the right end are

$$\psi_2(x) = 3\left(\frac{x}{L}\right)^2 - 2\left(\frac{x}{L}\right)^3 \quad (11\text{-}16c)$$

$$\psi_4(x) = \frac{x^2}{L}\left(\frac{x}{L} - 1\right) \quad (11\text{-}16d)$$

With these four interpolation functions, the deflected shape of the element can now
be expressed in terms of its nodal displacements:

$$v(x) = \psi_1(x)v_1 + \psi_2(x)v_2 + \psi_3(x)v_3 + \psi_4(x)v_4 \quad (11\text{-}17)$$

where the numbered degrees of freedom are related to those shown in Fig. 11-4 as
follows:

$$\begin{bmatrix} v_1 \\ v_2 \\ v_3 \\ v_4 \end{bmatrix} \equiv \begin{bmatrix} v_a \\ v_b \\ \theta_a \\ \theta_b \end{bmatrix} \quad (11\text{-}17a)$$

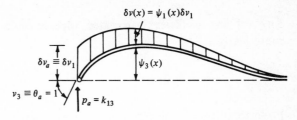

FIGURE 11-5
Beam subjected to real rotation and virtual translation of node.

It should be noted that both rotations and translations are represented as basic nodal degrees of freedom v_i.

By definition, the stiffness coefficients of the element represent the nodal forces due to unit nodal displacements. The nodal forces associated with any nodal-displacement component can be determined by the principle of virtual displacements, as described in Sec. 2-5. Consider, for example, the stiffness coefficient k_{13} for the beam element of Fig. 11-4, that is, the vertical force developed at end a due to a unit rotation applied at that point.

This force component can be evaluated by introducing a virtual vertical displacement of end a, as shown in Fig. 11-5, while the unit rotation is applied as shown, and equating the work done by the external forces to the work done on the internal forces: $W_E = W_I$. In this case, the external work is done only by the vertical-force component at a because the virtual displacements of all other nodal components vanish; thus

$$W_E = \delta v_a p_a = \delta v_1 k_{13} \qquad (11\text{-}18)$$

The internal virtual work is done by the internal moments associated with $\theta_a = 1$ acting on the virtual curvatures, which are $\partial^2/\partial x^2[\delta v(x)] = \psi_1''(x)\,\delta v_1$ (neglecting the effects of shear distortion). However, the internal moments due to $\theta_a = 1$ may be expressed as

$$\mathcal{M}(x) = EI(x)\psi_3''(x)$$

Thus the internal work is given by

$$W_I = \delta v_1 \int_0^L EI(x)\psi_1''(x)\psi_3''(x)\,dx \qquad (11\text{-}19)$$

When the work expressions of Eqs. (11-18) and (11-19) are equated, the expression for this stiffness coefficient is

$$k_{13} = \int_0^L EI(x)\psi_1''(x)\psi_3''(x)\,dx \qquad (11\text{-}20)$$

Any stiffness coefficient associated with beam flexure therefore may be written equivalently as

$$k_{ij} = \int_0^L EI(x)\psi_i''(x)\psi_j''(x)\, dx \qquad (11\text{-}21)$$

From the form of this expression, the symmetry of the stiffness matrix is evident; that is, $k_{ij} = k_{ji}$. Its equivalence to the corresponding term in Eq. (2-39) for $i = j$ should be noted.

For the special case of a uniform beam segment, the stiffness matrix resulting from Eq. (11-21) when the interpolation functions of Eqs. (11-16) are used may be expressed by

$$\begin{bmatrix} f_{S1} \\ f_{S2} \\ f_{S3} \\ f_{S4} \end{bmatrix} = \frac{2EI}{L^3} \begin{bmatrix} 6 & -6 & 3L & 3L \\ -6 & 6 & -3L & -3L \\ 3L & -3L & 2L^2 & L^2 \\ 3L & -3L & L^2 & 2L^2 \end{bmatrix} \begin{bmatrix} v_1 \\ v_2 \\ v_3 \\ v_4 \end{bmatrix} \qquad (11\text{-}22)$$

where the nodal displacements **v** are defined by Eq. (11-17a) and $\mathbf{f}_S$ is the corresponding vector of nodal forces. These stiffness coefficients are the exact values for a uniform beam without shear distortion because the interpolation functions used in Eq. (11-21) are the true shapes for this case. If the stiffness of the beam is not uniform, applying these shape functions in Eq. (11-21) will provide only an approximation to the true stiffness, but the final result for the complete beam will be very good if it is divided into a sufficient number of finite elements.

As mentioned earlier, when the stiffness coefficients of all the finite elements in a structure have been evaluated, the stiffness of the complete structure can be obtained by merely adding the element stiffness coefficients appropriately; this is called the *direct stiffness method*. In effect, any stiffness coefficient k_{ij} of the complete structure can be obtained by adding together the corresponding stiffness coefficients of the elements associated with those nodal points. Thus if elements m, n, and p were all attached to nodal point i of the complete structure, the structure stiffness coefficient for this point would be

$$k_{ii} = \hat{k}_{ii}{}^{(m)} + \hat{k}_{ii}{}^{(n)} + \hat{k}_{ii}{}^{(p)} \qquad (11\text{-}23)$$

in which the superscripts identify the individual elements. Before the element stiffnesses can be superposed in this fashion, they must be expressed in a common global-coordinate system which applies to the entire structure. The hats are placed over each element stiffness symbol in Eq. (11-23) to indicate that they have been transformed from their local-coordinate form [for example, Eq. (11-22)] to the global coordinates.

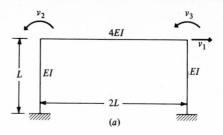

(a)

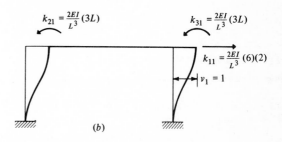

(b)

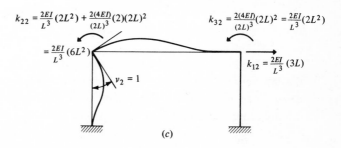

(c)

FIGURE E11-1
Analysis of frame stiffness coefficients: (a) frame properties and degrees of freedom; (b) forces due to displacement $v_1 = 1$; (c) forces due to rotation $v_2 = 1$.

EXAMPLE E11-1 The evaluation of the structural stiffness matrix is a basic operation of the matrix-displacement method of static structural analysis; although a general discussion of this subject is beyond the scope of this structural-dynamics text, it may be useful to apply the procedure to a simple frame structure in order to demonstrate how the element stiffness coefficients of Eq. (11-22) may be used.

Consider the structure of Fig. E11-1a. If it is assumed that the members do not distort axially, this frame has the three joint degrees of freedom shown. The corresponding stiffness coefficients can be evaluated by successively applying a unit displacement to each degree of freedom while constraining the other

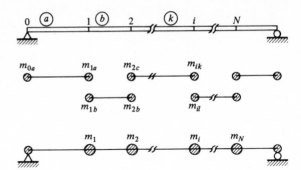

FIGURE 11-6
Lumping of mass at beam nodes.

two and determining the forces developed in each member by the coefficients of Eq. (11-22).

When the sidesway displacement shown in Fig. E11-1*b* is applied, it is clear that only the vertical members are deformed; their end forces are given by elements 1, 3, and 4 in the first column of the stiffness matrix of Eq. (11-22). It will be noted that the structure coefficient k_{11} receives a contribution from each column.

Considering the joint rotation shown in Fig. E11-1*c*, both the girder and the left vertical contribute to the structure coefficient k_{22}, the contributions being given by element 3 of column 3 in the stiffness matrix of Eq. (11-22) (taking proper account of the girder properties, of course). Only the left vertical contributes to k_{12} and only the girder to k_{32}. The structure stiffness coefficients due to the right-joint rotation are analogous to these.

The structure stiffness matrix finally obtained by assembling all these coefficients is

$$\begin{bmatrix} f_{S1} \\ f_{S2} \\ f_{S3} \end{bmatrix} = \frac{2EI}{L^3} \begin{bmatrix} 12 & 3L & 3L \\ 3L & 6L^2 & 2L^2 \\ 3L & 2L^2 & 6L^2 \end{bmatrix} \begin{bmatrix} v_1 \\ v_2 \\ v_3 \end{bmatrix} \qquad \text{////}$$

11-2 MASS PROPERTIES

Lumped-Mass Matrix

The simplest procedure for defining the mass properties of any structure is to assume that the entire mass is concentrated at the points at which the translational displacements are defined. The usual procedure for defining the point mass to be located at each node is to assume that the structure is divided into segments, the nodes serving as connection points. Figure 11-6 illustrates the procedure for a beam-type structure. The mass of each segment is assumed to be concentrated in point masses at each of its nodes, the distribution of the segment mass to these points being determined by statics. The total mass concentrated at any node of the complete structure then is the sum of the nodal contributions from all the segments attached to that node. In the beam system of Fig. 11-6 there are two segments contributing to each node; for example, $m_1 = m_{1a} + m_{1b}$.

For a system in which only translational degrees of freedom are defined, the lumped-mass matrix has a diagonal form; for the system of Fig. 11-6 it would be written

$$\mathbf{m} = \begin{bmatrix} m_1 & 0 & 0 & \cdots & 0 & \cdots & 0 \\ 0 & m_2 & 0 & \cdots & 0 & \cdots & 0 \\ 0 & 0 & m_3 & \cdots & 0 & \cdots & 0 \\ \cdots\cdots\cdots\cdots\cdots\cdots\cdots\cdots\cdots\cdots \\ 0 & 0 & 0 & \cdots & m_i & \cdots & 0 \\ \cdots\cdots\cdots\cdots\cdots\cdots\cdots\cdots\cdots\cdots \\ 0 & 0 & 0 & \cdots & 0 & \cdots & m_N \end{bmatrix} \tag{11-24}$$

in which there are as many terms as there are degrees of freedom. The off-diagonal terms m_{ij} of this matrix vanish because an acceleration of any mass point produces an inertia force at that point *only*. The inertia force at i due to a unit acceleration of point i is obviously equal to the mass concentrated at that point; thus the mass influence coefficient $m_{ii} = m_i$ in a lumped-mass system.

If more than one translational degree of freedom is specified at any nodal point, the same point mass will be associated with each degree of freedom. On the other hand, the mass associated with any rotational degree of freedom will be zero because of the assumption that the mass is lumped in points which have no rotational inertia. (Of course, if a rigid mass having a finite rotational inertia is associated with a rotational degree of freedom, the diagonal mass coefficient for that degree of freedom would be the rotational inertia of the mass.) Thus the lumped-mass matrix is a diagonal matrix which will include zero diagonal elements for the rotational degrees of freedom, in general.

Consistent-Mass Matrix

Making use of the finite-element concept, it is possible to evaluate mass influence coefficients for each element of a structure by a procedure similar to the analysis of element stiffness coefficients. Consider, for example, the nonuniform beam segment shown in Fig. 11-7, which may be assumed to be the same as that of Fig. 11-4. The degrees of freedom of the segment are the translation and rotation at each end, and it will be assumed that the displacements within the span are defined by the same interpolation functions $\psi_i(x)$ used in deriving the element stiffness.

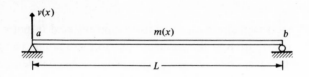

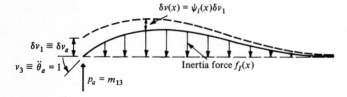

FIGURE 11-7
Node subjected to real angular acceleration and virtual translation.

If the beam were subjected to a unit angular acceleration of the left end, $\ddot{v}_3 = \ddot{\theta}_a = 1$, accelerations would be developed along its length, as follows:

$$\ddot{v}(x) = \psi_3(x)\ddot{v}_3 \qquad (11\text{-}25)$$

which can be obtained by taking the second time derivative of Eq. (11-17). By d'Alembert's principle, the inertia force resisting this acceleration is

$$f_I(x) = m(x)\ddot{v}(x) = m(x)\psi_3(x)\ddot{v}_3 \qquad (11\text{-}26)$$

Now the mass influence coefficients associated with this acceleration are defined as the nodal inertia forces which it produces; these can be evaluated from the distributed inertia force of Eq. (11-26) by the principle of virtual displacements. For example, the vertical force at the left end can be evaluated by introducing a vertical virtual displacement and equating the work done by the external nodal force p_a to the work done on the distributed inertial forces $f_I(x)$. Thus

$$p_a\, \delta v_a = \int_0^L f_I(x)\, \delta v(x)\, dx$$

Expressing the internal virtual displacement in terms of the interpolation function and substituting Eq. (11-26) leads finally to

$$m_{13} = \int_0^L m(x)\psi_1(x)\psi_3(x)\, dx \qquad (11\text{-}27)$$

It should be noted in Fig. 11-7 that the mass influence coefficient represents the inertia force opposing the acceleration, but that it is numerically equal to the external force producing the acceleration.

From Eq. (11-27) it is evident that any mass influence coefficient m_{ij} of an arbitrary beam segment can be evaluated by the equivalent expression

$$m_{ij} = \int_0^L m(x)\psi_i(x)\psi_j(x)\,dx \qquad (11\text{-}28)$$

The symmetric form of this equation shows that the mass matrix (like the stiffness matrix) is symmetric; that is, $m_{ij} = m_{ji}$; also it may be noted that this expression is equivalent to the corresponding term of Eq. (2-37) in the case where $i = j$. When the mass coefficients are computed in this way, using the same interpolation functions which are used for calculating the stiffness coefficients, the result is called the *consistent-mass matrix*. In general, the cubic hermitian polynomials of Eqs. (11-16) are used for evaluating the mass coefficients of any straight beam segment. In the special case of a beam with uniformly distributed mass the results are

$$\begin{bmatrix} f_{I1} \\ f_{I2} \\ f_{I3} \\ f_{I4} \end{bmatrix} = \frac{\bar{m}L}{420} \begin{bmatrix} 156 & 54 & 22L & -13L \\ 54 & 156 & 13L & -22L \\ +22L & 13L & 4L^2 & -3L^2 \\ -13L & -22L & -3L^2 & 4L^2 \end{bmatrix} \begin{bmatrix} \ddot{v}_1 \\ \ddot{v}_2 \\ \ddot{v}_3 \\ \ddot{v}_4 \end{bmatrix} \qquad (11\text{-}29)$$

When the mass coefficients of the elements of a structure have been evaluated, the mass matrix of the complete element assemblage can be developed by exactly the same type of superposition procedure as that described for developing the stiffness matrix from the element stiffness [Eq. (11-23)]. The resulting mass matrix in general will have the same configuration, that is, arrangement of nonzero terms, as the stiffness matrix.

The dynamic analysis of a consistent-mass system generally requires considerably more computational effort than a lumped-mass system does, for two reasons: (1) the lumped-mass matrix is diagonal, while the consistent-mass matrix has many off-diagonal terms (leading to what is called *mass coupling*); (2) the rotational degrees of freedom can be eliminated from a lumped-mass analysis (by static condensation, explained later), whereas all rotational and translational degrees of freedom must be included in a consistent-mass analysis.

EXAMPLE E11-2 The structure of Example E11-1, shown again in Fig. E11-2a, will be used to illustrate the evaluation of the structural mass matrix. First the lumped-mass procedure is used: half the mass of each member is lumped at the ends of the members, as shown in Fig. E11-2b. The sum of the four contributions at the girder level then acts in the sidesway degree of freedom m_{11}; no mass coefficients are associated with the other degrees of freedom because these point masses have no rotational inertia.

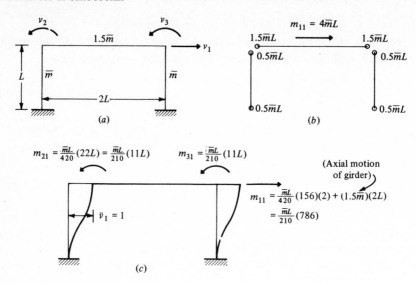

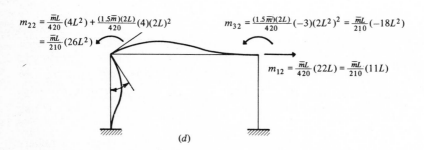

FIGURE E11-2
Analysis of lumped- and consistent-mass matrices: (a) uniform mass in members; (b) lumping of mass at member ends; (c) forces due to acceleration $\ddot{v}_1 = 1$ (consistent); (d) forces due to acceleration $\ddot{v}_2 = 1$ (consistent).

The consistent-mass matrix is obtained by applying unit accelerations to each degree of freedom in succession while constraining the others and determining the resulting inertia forces from the coefficients of Eq. (11-29). Considering first the sideway acceleration, as shown in Fig. E11-2c, it must be noted that the coefficients of Eq. (11-29) account only for the transverse inertia of the columns. The inertia of the girder due to the acceleration parallel to its axis must be added as a rigid-body mass $(3\bar{m}L)$, as shown.

The joint rotational acceleration induces only accelerations transverse to the members, and the resulting girder and column contributions are given by

Eq. (11-29), as shown in Fig. E11-2d. The final mass matrices, from the lumped-and consistent-mass formulations, are

$$\mathbf{m} = \frac{\overline{m}L}{210} \begin{bmatrix} 840 & 0 & 0 \\ 0 & 0 & 0 \\ 0 & 0 & 0 \end{bmatrix} \qquad \mathbf{m} = \frac{\overline{m}L}{210} \begin{bmatrix} 786 & 11L & 11L \\ 11L & 26L^2 & -18L^2 \\ 11L & -18L^2 & 26L^2 \end{bmatrix} \qquad ////$$
$$\text{Lumped} \qquad\qquad\qquad \text{Consistent}$$

11-3 DAMPING PROPERTIES

If the various damping forces acting on a structure could be determined quantitatively, the finite-element concept could be used again to define the damping coefficients of the system. For example, the coefficient for any element might be of the form [compare with Eq. (2-38)]

$$c_{ij} = \int_0^L c(x)\psi_i(x)\psi_j(x)\, dx \qquad (11\text{-}30)$$

in which $c(x)$ represents a distributed viscous-damping property. After the element damping influence coefficients were determined, the damping matrix of the complete structure could be obtained by a superposition process equivalent to the direct stiffness method. In practice, however, evaluation of the damping property $c(x)$ (or any other specific damping property) is impracticable. For this reason, the damping is generally expressed in terms of damping ratios established from experiments on similar structures rather than by means of an explicit damping matrix $\mathbf{c}$. If an explicit expression of the damping matrix is needed, it generally will be computed from the specified damping ratios, as described in Chap. 13.

11-4 EXTERNAL LOADING

If the dynamic loading acting on a structure consists of concentrated forces corresponding with the displacement coordinates, the load vector of Eq. (10-2) can be written directly. In general, however, the load is applied at other points as well as the nodes and may include distributed loadings. In this case, the load terms in Eq. (10-2) are generalized forces associated with the corresponding displacement components. Two procedures which can be applied in the evaluation of these generalized forces are described in the following paragraphs.

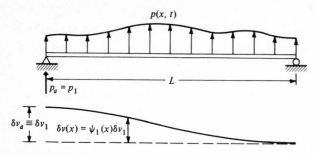

FIGURE 11-8
Virtual nodal translation of a laterally loaded beam.

Static Resultants

The most direct means of determining the effective nodal forces generated by loads distributed between the nodes is by application of the principles of simple statics; in other words, the nodal forces are defined as a set of concentrated loads which are statically equivalent to the distributed loading. In effect, the analysis is made as though the actual loading were applied to the structure through a series of simple beams supported at the nodal points. The reactive forces developed at the supports then become the concentrated nodal forces acting on the structure. In this type of analysis it is evident that generalized forces will be developed corresponding only to the translational degrees of freedom; the rotational nodal forces will be zero unless external moments are applied directly to the joints.

Consistent Nodal Loads

A second procedure which can be used to evaluate nodal forces corresponding to all nodal degrees of freedom can be developed from the finite-element concept. This procedure employs the principle of virtual displacements in the same way as in evaluating the consistent-mass matrix, and the generalized nodal forces which are derived are called the *consistent* nodal loads. Consider the same beam segment as in the consistent-mass analysis but subjected to the externally applied dynamic loading shown in Fig. 11-8. When a virtual displacement δv_1 is applied, as shown in the sketch, and external and internal work are equated, the generalized force corresponding to v_1 is

$$p_1(t) = \int_0^L p(x,t)\psi_1(x)\,dx \qquad (11\text{-}31)$$

Thus, the element generalized loads can be expressed in general as

$$p_i(t) = \int_0^L p(x,t)\psi_i(x)\,dx \qquad (11\text{-}32)$$

The generalized load p_3 corresponding to $v_3 = \theta_a$ is an external moment applied at point a. The positive sense of the generalized loads corresponds to the positive coordinate axes. The equivalence of Eq. (11-32) to Eq. (2-11) should be noted.

For the loads to be properly called consistent, the interpolation functions $\psi_i(x)$ used in Eq. (11-32) must be the same as those used to define the element stiffness coefficients. If linear interpolation functions

$$\psi_1(x) = 1 - \frac{x}{L} \qquad \psi_2(x) = \frac{x}{L} \qquad (11\text{-}33)$$

were used instead, Eq. (11-32) would provide the static nodal resultants; in general this is the easiest way to compute the statically equivalent loads.

In some cases, the applied loading may have the special form

$$p(x,t) = \chi(x)\zeta(t) \qquad (11\text{-}34)$$

that is, the form of load distribution $\chi(x)$ does not change with time; only its amplitude changes. In this case the generalized force becomes

$$p_i(t) = \zeta(t) \int_0^L \chi(x)\psi_i(x)\,dx \qquad (11\text{-}34a)$$

which shows that the generalized force has the same time variation as the applied loading; the integral indicates the extent to which the load participates in developing the generalized force.

When the generalized forces acting on each element have been evaluated by Eq. (11-32), the total effective load acting at the nodes of the assembled structure can be obtained by a superposition procedure equivalent to the direct stiffness process.

11-5   GEOMETRIC STIFFNESS

Linear Approximation

The geometric-stiffness property represents the tendency toward buckling induced in a structure by axially directed load components; thus it depends not only on the configuration of the structure but also on its condition of loading. In this discussion, it is assumed that the forces tending to cause buckling are constant during the dynamic loading; thus they are assumed to result from an independent static loading and are not significantly affected by the dynamic response of the structure. (When these forces do vary significantly with time, they result in a time-varying stiffness property, and analysis procedures based on superposition are not valid for such a nonlinear system.)

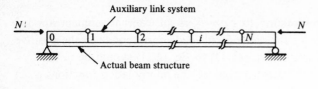

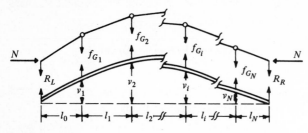

FIGURE 11-9
Idealization of axial-load mechanism in beam.

In general, two different levels of approximation can be established for the evaluation of geometric-stiffness properties, more or less in parallel with the preceding discussions for mass matrices and load vectors. The simplest approximation is conveniently derived from the physical model illustrated in Fig. 11-9, in which it is assumed that all axial forces are acting in an auxiliary structure consisting of rigid bar segments connected by hinges. The hinges are located at points where the transverse-displacement degrees of freedom of the actual beam are identified, and they are attached to the main beam by links which transmit transverse forces but no axial-force components.

When the actual beam is deflected by any form of loading, the auxiliary link system is forced to deflect equally, as shown in the sketch. As a result of these deflections and the axial forces in the auxiliary system, forces will be developed in the links coupling it to the main beam. In other words, the resistance of the main beam will be required to stabilize the auxiliary system.

The forces required for equilibrium in a typical segment i of the auxiliary system are shown in Fig. 11-10. The transverse force components f_{Gi} and f_{Gj} depend on the value of the axial-force component in the segment N_i and on the slope of the segment. They are assumed to be positive when they act in the positive-displacement sense on the main beam. In matrix form, these forces may be expressed

$$\begin{bmatrix} f_{Gi} \\ f_{Gj} \end{bmatrix} = \frac{N_i}{l_i} \begin{bmatrix} 1 & -1 \\ -1 & 1 \end{bmatrix} \begin{bmatrix} v_i \\ v_j \end{bmatrix} \qquad (11\text{-}35)$$

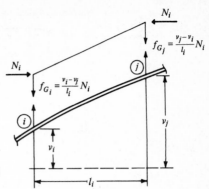

FIGURE 11-10
Equilibrium forces due to axial load in
auxiliary link.

By combining expressions of this type for all segments the transverse forces
due to axial loads can be written for the beam structure of Fig. 11-9 as

$$
\begin{bmatrix} f_{G1} \\ f_{G2} \\ \cdot \\ f_{Gi} \\ \cdot \\ f_{GN} \end{bmatrix}
=
\begin{bmatrix}
\dfrac{N_0}{l_0}+\dfrac{N_1}{l_1} & -\dfrac{N_1}{l_1} & 0 & \cdots & 0 & \cdots \\
-\dfrac{N_1}{l_1} & \dfrac{N_1}{l_1}+\dfrac{N_2}{l_2} & -\dfrac{N_2}{l_2} & \cdots & 0 & \cdots \\
\cdots\cdots\cdots\cdots\cdots\cdots\cdots\cdots\cdots\cdots\cdots\cdots\cdots\cdots \\
0 & 0 & 0 & \cdots & \dfrac{N_{i-1}}{l_{i-1}}+\dfrac{N_i}{l_i} & \cdots \\
\cdots\cdots\cdots\cdots\cdots\cdots\cdots\cdots\cdots\cdots\cdots\cdots \\
0 & 0 & 0 & \cdots & 0 & \cdots
\end{bmatrix}
\begin{bmatrix} v_1 \\ v_2 \\ \cdot \\ v_i \\ \cdot \\ v_N \end{bmatrix}
\qquad (11\text{-}36)
$$

in which it will be noted that magnitude of the axial force may change from segment
to segment; for the loading shown in Fig. 11-9 all axial forces would be the same,
and the term N could be factored from the matrix.

Symbolically, Eq. (11-36) may be expressed

$$\mathbf{f}_G = \mathbf{k}_G \mathbf{v} \qquad (11\text{-}37)$$

where the square symmetric matrix $\mathbf{k}_G$ is called the *geometric-stiffness matrix* of the
structure. For this linear approximation of a beam system, the matrix has a tri-
diagonal form, as may be seen in Eq. (11-36), with contributions from two adjacent
elements making up the diagonal terms and a single element providing each off-
diagonal, or coupling, term.

Consistent Geometric Stiffness

The finite-element concept can be used to obtain a higher-order approximation of the
geometric stiffness, as demonstrated for the other physical properties. Consider the
same beam element used previously but now subjected to distributed axial loads

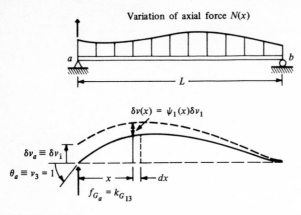

FIGURE 11-11
Axially loaded beam with real rotation and virtual translation of node.

which result in an arbitrary variation of axial force $N(x)$, as shown in Fig. 11-11. In the lower sketch, the beam is shown subjected to a unit rotation of the left end $v_3 = 1$. By definition, the nodal forces associated with this displacement component are the corresponding geometric-stiffness influence coefficients; for example, k_{G13} is the vertical force developed at the left end.

These coefficients may be evaluated by application of virtual displacements and equating the internal and external work components. The virtual displacement δv_1 required to determine k_{G13} is shown in the sketch. The external virtual work in this case is

$$W_E = f_{Ga}\,\delta v_a = k_{G13}\,\delta v_1 \qquad (11\text{-}38)$$

in which it will be noted that the positive sense of the geometric-stiffness coefficient corresponds with the positive displacements. To develop an expression for the internal virtual work, it is necessary to consider a differential segment of length dx, taken from the system of Fig. 11-11 and shown enlarged in Fig. 11-12. The work done in this segment by the axial force $N(x)$ during the virtual displacement is

$$dW_I = N(x)\,d(\delta e) \qquad (11\text{-}39)$$

where $d(\delta e)$ represents the distance the forces acting on this differential segment move toward each other. By similar triangles it may be seen in the sketch that

$$d(\delta e) = \frac{dv}{dx}\,d(\delta v)$$

Interchanging the differentiation and variation symbols on the right side gives

$$d(\delta e) = \frac{dv}{dx}\,\delta\left(\frac{dv}{dx}\,dx\right)$$

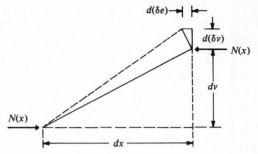

FIGURE 11-12
Differential segment of deformed beam
of Fig. 11-11.

and hence introducing this into Eq. (11-39) leads to

$$dW_I = N(x) \frac{dv}{dx} \delta\left(\frac{dv}{dx}\right) dx$$

Expressing the lateral displacements in terms of interpolation functions and integrating finally gives

$$W_I = \delta v_1 \int_0^L N(x) \frac{d\psi_3(x)}{dx} \frac{d\psi_1(x)}{dx} dx \qquad (11\text{-}40)$$

Hence, by equating internal to external work, this geometric-stiffness coefficient is found to be

$$k_{G13} = \int_0^L N(x)\psi_3'(x)\psi_1'(x) \, dx \qquad (11\text{-}41)$$

or in general the element geometric-stiffness influence coefficients are

$$k_{Gij} = \int_0^L N(x)\psi_i'(x)\psi_j'(x) \, dx \qquad (11\text{-}42)$$

The equivalence of this equation to Eq. (2-40a) should be noted; also its symmetry is apparent, that is, $k_{Gij} = k_{Gji}$.

If the hermitian interpolation functions [Eqs. (11-16)] are used in deriving the geometric-stiffness coefficients, the result is called the *consistent geometric-stiffness matrix*. In the special case where the axial force is constant through the length of the element, the consistent geometric-stiffness matrix is

$$\begin{bmatrix} f_{G1} \\ f_{G2} \\ f_{G3} \\ f_{G4} \end{bmatrix} = \frac{N}{30L} \begin{bmatrix} 36 & -36 & 3L & 3L \\ -36 & 36 & -3L & -3L \\ 3L & -3L & 4L^2 & -L^2 \\ 3L & -3L & -L^2 & 4L^2 \end{bmatrix} \begin{bmatrix} v_1 \\ v_2 \\ v_3 \\ v_4 \end{bmatrix} \qquad (11\text{-}43)$$

On the other hand, if linear-interpolation functions [Eq. (11-33)] are used in Eq. (11-42), and if the axial force is constant through the element, its geometric stiffness will be as derived earlier in Eq. (11-35).

The assembly of the element geometric-stiffness coefficients to obtain the structure geometric-stiffness matrix can be carried out exactly as for the elastic-stiffness matrix, and the result will have a similar configuration (positions of the non-zero terms). Thus the consistent geometric-stiffness matrix represents rotational as well as translational degrees of freedom, whereas the linear approximation [Eq. (11-35)] is concerned only with the translational displacements. However, either type of relationship may be represented symbolically by Eq. (11-37).

11-6 Choice of Property Formulation

In the preceding discussion, two different levels of approximation have been considered for the evaluation of the mass, geometric-stiffness, and external-load properties: (1) an elementary approach taking account only of the translational degrees of freedom of the structure and (2) a "consistent" approach, which accounts for the rotational as well as translational displacements. The elementary approach is considerably easier to apply; not only are the element properties defined more simply but the number of coordinates to be considered in the analysis is much less for a given structural assemblage. In principle, the consistent approach should lead to greater accuracy in the results, but in practice the improvement is often slight. Apparently the rotational degrees of freedom are much less significant in the analysis than the translational terms. The principal advantage of the consistent approach is that all the energy contributions to the response of the structure are evaluated in a consistent manner, which makes it possible to draw certain conclusions regarding bounds on the vibration frequency; however, this advantage seldom outweighs the additional effort required.

The elementary lumped-mass approach presents a special problem when the elastic-stiffness matrix has been formulated by the finite-element approach or by any other procedure which includes the rotational degrees of freedom in the matrix. If the evaluation of all the other properties has excluded these degrees of freedom, it is necessary to exclude them also from the stiffness matrix before the equations of motion can be written.

The process of eliminating these unwanted degrees of freedom from the stiffness matrix is called *static condensation*. For the purpose of this discussion, assume that the rotational and translational degrees of freedom have been segregated, so that Eq. (10-5) can be written in partitioned form

$$
\begin{bmatrix} \mathbf{k}_{tt} & \mathbf{k}_{t\theta} \\ \mathbf{k}_{\theta t} & \mathbf{k}_{\theta\theta} \end{bmatrix} \begin{bmatrix} \mathbf{v}_t \\ \mathbf{v}_\theta \end{bmatrix} = \begin{bmatrix} \mathbf{f}_{St} \\ \mathbf{f}_{S\theta} \end{bmatrix} = \begin{bmatrix} \mathbf{f}_{St} \\ \mathbf{0} \end{bmatrix} \qquad (11\text{-}44)
$$

where $\mathbf{v}_t$ represents the translations and $\mathbf{v}_\theta$ the rotations, with corresponding subscripts to identify the submatrices of stiffness coefficients. Now, if none of the other force vectors acting in the structure include any rotational components, it is evident that the elastic rotational forces also must vanish, that is, $\mathbf{f}_{S\theta} = \mathbf{0}$. When this static constraint is introduced into Eq. (11-44), it is possible to express the rotational displacements in terms of the translations by means of the second submatrix equation, with the result

$$\mathbf{v}_\theta = -\mathbf{k}_{\theta\theta}^{-1}\mathbf{k}_{\theta t}\mathbf{v}_t \qquad (11\text{-}45)$$

Substituting this into the first of the submatrix equations of Eq. (11-44) leads to

$$(\mathbf{k}_{tt} - \mathbf{k}_{t\theta}\mathbf{k}_{\theta\theta}^{-1}\mathbf{k}_{\theta t})\mathbf{v}_t = \mathbf{f}_{St}$$

or

$$\mathbf{k}_t\mathbf{v}_t = \mathbf{f}_{St} \qquad (11\text{-}46)$$

where

$$\mathbf{k}_t = \mathbf{k}_{tt} - \mathbf{k}_{t\theta}\mathbf{k}_{\theta\theta}^{-1}\mathbf{k}_{\theta t} \qquad (11\text{-}47)$$

is the translational elastic stiffness. This stiffness matrix is suitable for use with the other elementary property expressions; in other words, it is the type of stiffness matrix implied in Fig. 11-2.

EXAMPLE E11-3 To demonstrate the use of the static-condensation procedure, the two rotational degrees of freedom will be eliminated from the stiffness matrix evaluated in Example E11-1. The resulting condensed stiffness matrix will retain only the translational degree of freedom of the frame and thus will be compatible with the lumped-mass matrix derived in Example E11-2.

The stiffness submatrix associated with the rotational degrees of freedom of Example E11-1 is

$$\mathbf{k}_{\theta\theta} = \frac{2EI}{L^3}\begin{bmatrix} 6L^2 & 2L^2 \\ 2L^2 & 6L^2 \end{bmatrix} = \frac{4EI}{L}\begin{bmatrix} 3 & 1 \\ 1 & 3 \end{bmatrix}$$

and its inverse is

$$\mathbf{k}_{\theta\theta}^{-1} = \frac{L}{32EI}\begin{bmatrix} 3 & -1 \\ -1 & 3 \end{bmatrix}$$

When this is used in Eq. (11-45), the rotational degrees of freedom can be expressed in terms of the translation:

$$\begin{bmatrix} v_2 \\ v_3 \end{bmatrix} = -\frac{L}{32EI}\begin{bmatrix} 3 & -1 \\ -1 & 3 \end{bmatrix}\frac{2EI}{L^3}\begin{bmatrix} 3L \\ 3L \end{bmatrix}v_1 = -\frac{3}{8L}\begin{bmatrix} 1 \\ 1 \end{bmatrix}v_1$$

The condensed stiffness given by Eq. (11-47) then is

$$\mathbf{k}_t = \frac{2EI}{L^3}\left(12 - \begin{bmatrix} 3L & 3L \end{bmatrix}\begin{bmatrix} \dfrac{3}{8L} \\ \dfrac{3}{8L} \end{bmatrix} \right) = \frac{2EI}{L^3}\frac{39}{4} \qquad ////$$

PROBLEMS

11-1 Using the hermitian polynomials, Eq. (11-16), as shape functions $\psi_i(x)$, evaluate by means of Eq. (11-21) the finite-element stiffness coefficient k_{23} for a beam having the following variation of flexural rigidity: $EI(x) = EI_0(1 + x/L)$.

11-2 Making use of Eq. (11-28), compute the consistent mass coefficient m_{23} for a beam with the following nonuniform mass distribution: $m(x) = \bar{m}(1 + x/L)$. Assume the shape functions of Eq. (11-16) and evaluate the integral by Simpson's rule, dividing the beam into four segments of equal length.

11-3 The distributed load applied to a certain beam may be expressed as

$$p(x,t) = \bar{p}\left(2 + \frac{x}{L}\right)\sin \bar{\omega}t$$

Making use of Eq. (11-34a), write an expression for the time variation of the consistent load component $p_2(t)$ based on the shape function of Eq. (11-16).

11-4 Using Eq. (11-42), evaluate the consistent geometric stiffness coefficient k_{G24} for a beam having the following distribution of axial force: $N(x) = N_0(2 - x/L)$. Make use of the shape functions of Eq. (11-16) and evaluate the integral by Simpson's rule using $\Delta x = L/4$.

11-5 The plane frame of Fig. P11-1 is formed of uniform members, with the properties of each as shown. Assemble the stiffness matrix defined for the three DOFs indicated, evaluating the member stiffness coefficients by means of Eq. (11-22).

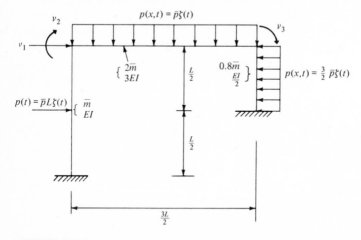

FIGURE P11-1

11-6 Assemble the mass matrix for the structure of Prob. 11-5, evaluating the individual member mass coefficients by means of Eq. (11-29).

11-7 Assemble the load vector for the structure of Prob. 11-5, evaluating the individual member nodal loads by means of Eq. (11-32).

11-8 For a plane frame of the same general form as that of Prob. 11-5, but having different member lengths and physical properties, the stiffness and lumped mass matrices are as follows:

$$\mathbf{k} = \frac{EI}{L^3} \begin{bmatrix} 20 & -10L & -5L \\ -10L & 15L^2 & 8L^2 \\ -5L & 8L^2 & 12L^2 \end{bmatrix} \qquad \mathbf{m} = \bar{m}L \begin{bmatrix} 30 & 0 & 0 \\ 0 & 0 & 0 \\ 0 & 0 & 0 \end{bmatrix}$$

(a) Using static condensation, eliminate the two rotational degrees of freedom from the stiffness matrix.

(b) Using the condensed stiffness matrix, write the SDOF equation of motion for undamped free vibrations.

12

UNDAMPED FREE VIBRATIONS

12-1 ANALYSIS OF VIBRATION FREQUENCIES

The equations of motion for a freely vibrating undamped system can be obtained by omitting the damping matrix and applied-loads vector from Eq. (10-13):

$$\mathbf{m\ddot{v} + kv = 0} \qquad (12\text{-}1)$$

in which $\mathbf{0}$ is a zero vector. The problem of vibration analysis consists of determining the conditions under which Eq. (12-1) will permit motions to occur. By analogy with the behavior of SDOF systems, it will be assumed that the free-vibration motion is simple harmonic, which may be expressed for a MDOF system as

$$\mathbf{v}(t) = \hat{\mathbf{v}} \sin{(\omega t + \theta)} \qquad (12\text{-}2)$$

In this expression $\hat{\mathbf{v}}$ represents the shape of the system (which does not change with time; only the amplitude varies) and θ is a phase angle. When the second time derivative of Eq. (12-2) is taken, the accelerations in free vibration are

$$\mathbf{\ddot{v}} = -\omega^2 \hat{\mathbf{v}} \sin{(\omega t + \theta)} = -\omega^2 \mathbf{v} \qquad (12\text{-}3)$$

Substituting Eqs. (12-2) and (12-3) into Eq. (12-1) gives

$$-\omega^2 \mathbf{m}\hat{\mathbf{v}} \sin(\omega t + \theta) + \mathbf{k}\hat{\mathbf{v}} \sin(\omega t + \theta) = \mathbf{0}$$

which (since the sine term is arbitrary and may be omitted) may be written

$$[\mathbf{k} - \omega^2 \mathbf{m}]\hat{\mathbf{v}} = \mathbf{0} \qquad (12\text{-}4)$$

Now it can be shown by Cramer's rule that the solution of this set of simultaneous equations is of the form

$$\hat{\mathbf{v}} = \frac{\mathbf{0}}{\|\mathbf{k} - \omega^2 \mathbf{m}\|} \qquad (12\text{-}5)$$

Hence a nontrivial solution is possible only when the denominator determinant vanishes. In other words, finite-amplitude free vibrations are possible only when

$$\|\mathbf{k} - \omega^2 \mathbf{m}\| = 0 \qquad (12\text{-}6)$$

Equation (12-6) is called the *frequency equation* of the system. Expanding the determinant will give an algebraic equation of the Nth degree in the frequency parameter ω^2 for a system having N degrees of freedom. The N roots of this equation $(\omega_1^2, \omega_2^2, \omega_3^2, \ldots, \omega_N^2)$ represent the frequencies of the N modes of vibration which are possible in the system. The mode having the lowest frequency is called the first mode, the next higher frequency is the second mode, etc. The vector made up of the entire set of modal frequencies, arranged in sequence, will be called the *frequency vector* $\boldsymbol{\omega}$:

$$\boldsymbol{\omega} = \begin{bmatrix} \omega_1 \\ \omega_2 \\ \omega_3 \\ \vdots \\ \omega_N \end{bmatrix} \qquad (12\text{-}7)$$

It can be shown that for the real, symmetric, positive definite mass and stiffness matrices which pertain to stable structural systems, all roots of the frequency equation will be real and positive.

EXAMPLE E12-1   The analysis of vibration frequencies by the solution of the determinantal equation (12-6) will be demonstrated with reference to the structure of Fig. E12-1, the same frame for which an approximation of the fundamental frequency was obtained by the Rayleigh method in Example E9-3. The stiffness matrix for this frame can be determined by applying a unit displacement to each story in succession and evaluating the resulting story forces, as shown in the figure. Because the girders are assumed to be rigid, the story forces

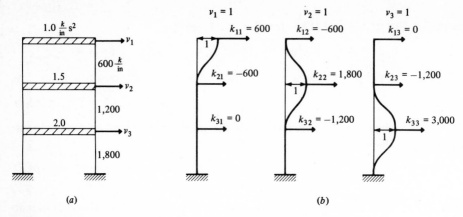

(a) **(b)**

FIGURE E12-1
Frame used in example of vibration analysis: (a) structural system; (b) stiffness
influence coefficients.

can easily be determined here by merely adding the sidesway stiffnesses of the
appropriate stories.

The mass and stiffness matrices for this frame thus are

$$\mathbf{m} = (1 \text{ kip} \cdot \text{s}^2/\text{in}) \begin{bmatrix} 1.0 & 0 & 0 \\ 0 & 1.5 & 0 \\ 0 & 0 & 2.0 \end{bmatrix} \qquad \mathbf{k} = (600 \text{ kips/in}) \begin{bmatrix} 1 & -1 & 0 \\ -1 & 3 & -2 \\ 0 & -2 & 5 \end{bmatrix}$$

from which

$$\mathbf{k} - \omega^2 \mathbf{m} = (600 \text{ kips/in}) \begin{bmatrix} 1 - B & -1 & 0 \\ -1 & 3 - 1.5B & -2 \\ 0 & -2 & 5 - 2B \end{bmatrix} \qquad (a)$$

where

$$B \equiv \frac{\omega^2}{600}$$

The frequencies of the frame are given by the condition that $\Delta = 0$, where
Δ is the determinant of the square matrix in Eq. (a). Evaluating this determinant,
simplifying, and equating to zero leads to the cubic equation

$$B^3 - 5.5B^2 + 7.5B - 2 = 0$$

The three roots of this equation may be solved directly or obtained by trial
and error; their values are $B_1 = 0.351$, $B_2 = 1.61$, $B_3 = 3.54$. Hence the
frequencies are

$$\begin{bmatrix} \omega_1{}^2 \\ \omega_2{}^2 \\ \omega_3{}^2 \end{bmatrix} = \begin{bmatrix} 210 \\ 966 \\ 2,124 \end{bmatrix} \qquad \begin{bmatrix} \omega_1 \\ \omega_2 \\ \omega_3 \end{bmatrix} = \begin{bmatrix} 14.5 \\ 31.1 \\ 46.1 \end{bmatrix} \text{ rad/s} \qquad ////$$

12-2 ANALYSIS OF VIBRATION MODE SHAPES

When the frequencies of vibration have been determined from Eq. (12-6), the equations of motion [Eq. (12-4)] may be expressed as

$$\tilde{\mathbf{E}}^{(n)}\hat{\mathbf{v}}_n = 0 \qquad (12\text{-}8)$$

in which

$$\tilde{\mathbf{E}}^{(n)} = \mathbf{k} - \omega_n{}^2\mathbf{m} \qquad (12\text{-}9)$$

Thus $\tilde{\mathbf{E}}^{(n)}$ represents the matrix obtained by subtracting $\omega_n{}^2\mathbf{m}$ from the stiffness matrix; since it depends on the frequency, it is different for each mode. Equation (12-8) is satisfied identically because the frequencies were evaluated from this condition; therefore the *amplitude* of the vibrations is indeterminate. However, the *shape* of the vibrating system can be determined by solving for all the displacements in terms of any one coordinate.

For this purpose it wlll be assumed that the first element of the displacement vector has a unit amplitude, that is,

$$\begin{bmatrix} \hat{v}_{1n} \\ \hat{v}_{2n} \\ \hat{v}_{3n} \\ \vdots \\ \hat{v}_{Nn} \end{bmatrix} = \begin{bmatrix} 1 \\ \hat{v}_{2n} \\ \hat{v}_{3n} \\ \vdots \\ \hat{v}_{Nn} \end{bmatrix} \qquad (12\text{-}10)$$

In expanded form, Eq. (12-8) may then be written

$$\begin{bmatrix} e_{11}{}^{(n)} & e_{12}{}^{(n)} & e_{13}{}^{(n)} & \cdots & e_{1N}{}^{(n)} \\ e_{21}{}^{(n)} & e_{22}{}^{(n)} & e_{23}{}^{(n)} & \cdots & e_{2N}{}^{(n)} \\ e_{31}{}^{(n)} & e_{32}{}^{(n)} & e_{33}{}^{(n)} & \cdots & e_{3N}{}^{(n)} \\ \vdots & \vdots & \vdots & & \vdots \\ e_{N1}{}^{(n)} & e_{N2}{}^{(n)} & e_{N3}{}^{(n)} & \cdots & e_{NN}{}^{(n)} \end{bmatrix} \begin{bmatrix} 1 \\ \hat{v}_{2n} \\ \hat{v}_{3n} \\ \vdots \\ \hat{v}_{Nn} \end{bmatrix} = \begin{bmatrix} 0 \\ 0 \\ 0 \\ \vdots \\ 0 \end{bmatrix} \qquad (12\text{-}11)$$

in which partitioning is indicated to correspond with the as yet unknown displacement amplitudes. For convenience, Eq. (12-11) will be expressed symbolically as

$$\begin{bmatrix} e_{11}^{(n)} & \tilde{\mathbf{E}}_{10}^{(n)} \\ \tilde{\mathbf{E}}_{01}^{(n)} & \tilde{\mathbf{E}}_{00}^{(n)} \end{bmatrix} \begin{bmatrix} 1 \\ \hat{\mathbf{v}}_{0n} \end{bmatrix} = \begin{bmatrix} 0 \\ 0 \end{bmatrix} \qquad (12\text{-}11a)$$

from which

$$\tilde{\mathbf{E}}_{01}{}^{(n)} + \tilde{\mathbf{E}}_{00}{}^{(n)}\hat{\mathbf{v}}_{0n} = 0 \qquad (12\text{-}12)$$

as well as

$$e_{11}{}^{(n)} + \tilde{\mathbf{E}}_{10}{}^{(n)}\hat{\mathbf{v}}_{0n} = 0 \qquad (12\text{-}13)$$

Equation (12-12) can be solved simultaneously for the displacement amplitudes

$$\hat{\mathbf{v}}_{0n} = -(\tilde{\mathbf{E}}_{00}{}^{(n)})^{-1}\tilde{\mathbf{E}}_{01}{}^{(n)} \qquad (12\text{-}14)$$

but Eq. (12-13) is redundant; the redundancy corresponds to the fact that Eq. (12-7) is satisfied identically. The displacement vector obtained in Eq. (12-14) must satisfy Eq. (12-13), however, and this condition provides a useful check on the accuracy of the solution. It should be noted that it is not always wise to let the first element of the displacement vector be unity; numerical accuracy will be improved if the unit element is associated with one of the larger displacement amplitudes. The same solution process can be employed in any case, however, by merely rearranging the order of the rows and columns of $\tilde{\mathbf{E}}^{(n)}$ appropriately.

The displacement amplitudes obtained from Eq. (12-14) together with the unit amplitude of the first component constitute the displacement vector associated with the nth mode of vibration. For convenience the vector is usually expressed in dimensionless form by dividing all the components by one reference component (usually the largest). The resulting vector is called the nth mode shape $\boldsymbol{\phi}_n$; thus

$$\boldsymbol{\phi}_n = \begin{bmatrix} \phi_{1n} \\ \phi_{2n} \\ \phi_{3n} \\ \vdots \\ \phi_{Nn} \end{bmatrix} \equiv \frac{1}{\hat{v}_{kn}} \begin{bmatrix} 1 \\ \hat{v}_{2n} \\ \hat{v}_{3n} \\ \vdots \\ \hat{v}_{Nn} \end{bmatrix} \qquad (12\text{-}15)$$

in which $\hat{v}_{kn}$ is the reference component.

The shape of each of the N modes of vibration can be found by this same process; the square matrix made up of the N mode shapes will be represented by $\boldsymbol{\Phi}$; thus

$$\boldsymbol{\Phi} = \begin{bmatrix} \boldsymbol{\phi}_1 & \boldsymbol{\phi}_2 & \boldsymbol{\phi}_3 & \cdots & \boldsymbol{\phi}_N \end{bmatrix} = \begin{bmatrix} \phi_{11} & \phi_{12} & \cdots & \phi_{1N} \\ \phi_{21} & \phi_{22} & \cdots & \phi_{2N} \\ \phi_{31} & \phi_{32} & \cdots & \phi_{3N} \\ \phi_{41} & \phi_{42} & \cdots & \phi_{4N} \\ \cdots\cdots\cdots\cdots\cdots\cdots \\ \phi_{N1} & \phi_{N2} & \cdots & \phi_{NN} \end{bmatrix} \qquad (12\text{-}16)$$

It should be noted that the vibration analysis of a structural system is a form of characteristic-value, or eigenvalue, problem of matrix-algebra theory. The frequency-squared terms are the eigenvalues, and the mode shapes are the eigenvectors. A brief discussion of the reduction of the equation of motion in free vibrations to standard eigenproblem form is presented in Chap. 14.

EXAMPLE E12-2   The analysis of vibration mode shapes by means of Eq. (12-14) will be demonstrated by applying it to the structure of Fig. E12-1. The vibration matrix for this structure was derived in Example E12-1, and when the second and third rows of this matrix are used, Eq. (12-14) may be expressed as

$$\begin{bmatrix} \phi_{2n} \\ \phi_{3n} \end{bmatrix} = -\begin{bmatrix} 3 - 1.5B_n & -2 \\ -2 & 5 - 2B_n \end{bmatrix}^{-1} \begin{bmatrix} -1 \\ 0 \end{bmatrix}$$

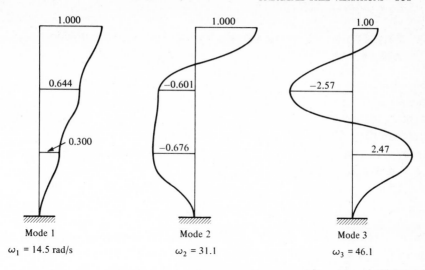

FIGURE E12-2
Vibration properties for the frame of Fig. E12-1.

Thus the mode shapes can be found by introducing the values of B_n computed in Example E12-1, inverting, and multiplying as indicated. The calculations for the three mode shapes of this system follow.

Mode 1:

$$B_1 = 0.35 \qquad \tilde{\mathbf{E}}_{00}^{(1)} = \begin{bmatrix} 2.475 & -2 \\ -2 & 4.300 \end{bmatrix} \qquad (\tilde{\mathbf{E}}_{00}^{(1)})^{-1} = \frac{1}{6.68} \begin{bmatrix} 4.300 & 2 \\ 2 & 2.475 \end{bmatrix}$$

$$\begin{bmatrix} \phi_{21} \\ \phi_{31} \end{bmatrix} = \frac{1}{6.68} \begin{bmatrix} 4.300 \\ 2.000 \end{bmatrix} = \begin{bmatrix} 0.644 \\ 0.300 \end{bmatrix}$$

Mode 2:

$$B_2 = 1.61 \qquad \tilde{\mathbf{E}}_{00}^{(2)} = \begin{bmatrix} 0.585 & -2 \\ -2 & 1.780 \end{bmatrix} \qquad (\tilde{\mathbf{E}}_{00}^{(2)})^{-1} = -\frac{1}{2.959} \begin{bmatrix} 1.780 & 2 \\ 2 & 0.585 \end{bmatrix}$$

$$\begin{bmatrix} \phi_{22} \\ \phi_{32} \end{bmatrix} = -\frac{1}{2.959} \begin{bmatrix} 1.780 \\ 2.000 \end{bmatrix} = -\begin{bmatrix} 0.601 \\ 0.676 \end{bmatrix}$$

Mode 3:

$$B_3 = 3.54 \qquad \tilde{\mathbf{E}}_{00}^{(3)} = \begin{bmatrix} -2.31 & -2 \\ -2 & -2.08 \end{bmatrix} \qquad (\tilde{\mathbf{E}}_{00}^{(3)})^{-1} = \frac{1}{0.81} \begin{bmatrix} -2.08 & 2 \\ 2 & -2.31 \end{bmatrix}$$

$$\begin{bmatrix} \phi_{23} \\ \phi_{33} \end{bmatrix} = \frac{1}{0.81} \begin{bmatrix} -2.08 \\ 2.00 \end{bmatrix} = \begin{bmatrix} -2.57 \\ 2.47 \end{bmatrix}$$

Of course, the displacement of mass a in each mode has been assumed to be unity. The three mode shapes for this structure are sketched in Fig. E12-2. ////

12-3 FLEXIBILITY FORMULATION OF VIBRATION ANALYSIS

The preceding discussion of vibration analysis was based on a stiffness-matrix formulation of the equations of motion. In many cases it may be more convenient to express the elastic properties of the structure by means of the flexibility matrix rather than the stiffness. Equation (12-4) can be converted readily into the flexibility form by multiplying by $(1/\omega^2)\tilde{\mathbf{f}}$, where the flexibility matrix $\tilde{\mathbf{f}}$ is the inverse of the stiffness matrix $\mathbf{k}$. The result is

$$\left[\frac{1}{\omega^2}\mathbf{I} - \tilde{\mathbf{f}}\mathbf{m}\right]\hat{\mathbf{v}} = \mathbf{0} \qquad (12\text{-}17)$$

in which $\mathbf{I}$ represents an identity matrix of order N. As before, this set of homogeneous equations can have a nonzero solution only if the determinant of the square matrix vanishes; thus the frequency equation in this case is

$$\left\|\frac{1}{\omega^2}\mathbf{I} - \tilde{\mathbf{f}}\mathbf{m}\right\| = 0 \qquad (12\text{-}18)$$

Evaluation of the roots of this equation can be carried out as for Eq. (12-6); similarly the mode shape corresponding to each frequency can be evaluated as before. The only basic difference between the solutions is that the roots of Eq. (12-18) represent the reciprocals of the frequency-squared values rather than the frequency squared.

It should be noted that the matrix product $\tilde{\mathbf{f}}\mathbf{m}$ in Eq. (12-18) is not symmetrical, in general, even though the mass and flexibility matrices are both symmetric. In digital-computer analyses of the eigenvalue problem it may be desirable to retain the symmetry of the matrix being solved; techniques for obtaining a symmetric form of the flexibility eigenvalue equation are presented in Chap. 14.

12-4 INFLUENCE OF AXIAL FORCES

Free Vibrations

The vibration mode shapes and frequencies of a structure which is subjected to a constant axial-force loading can be evaluated in exactly the same way as for a system without axial-force effects. In this case the geometric stiffness must be included in the equations of motion; thus Eq. (12-1) takes the form

$$\mathbf{m}\ddot{\mathbf{v}} + \mathbf{k}\mathbf{v} - \mathbf{k}_G\mathbf{v} = \mathbf{m}\ddot{\mathbf{v}} + \bar{\mathbf{k}}\mathbf{v} = \mathbf{0} \qquad (12\text{-}19)$$

and the frequency equation becomes

$$\|\bar{\mathbf{k}} - \omega^2\mathbf{m}\| = 0 \qquad (12\text{-}20)$$

In the mode-shape and frequency analysis, it is necessary only to substitute the combined stiffness $\bar{\mathbf{k}}$ for the elastic stiffness $\mathbf{k}$; otherwise the analysis is as described before. For any given condition of axial loading, the geometric-stiffness matrix (and therefore the combined stiffness) can be evaluated numerically. The effect of a compressive axial-force system is to reduce the effective stiffness of the structure; thus the frequencies of vibration are reduced; in addition the mode shapes generally are modified by the axial loads.

Buckling Load

If the frequency of vibration is zero, the inertia forces in Eq. (12-19) vanish and the equations of equilibrium become

$$\mathbf{kv} - \mathbf{k}_G\mathbf{v} = 0 \qquad (12\text{-}21)$$

The conditions under which a nonzero displacement vector is possible in this case constitute the static buckling condition; in other words, a useful definition of buckling is the condition in which the vibration frequency becomes zero. In order to evaluate the critical buckling loading of the structure, it is convenient to express the geometric stiffness in terms of a reference loading multiplied by a load factor λ_G. Thus

$$\mathbf{k}_G = \lambda_G\mathbf{k}_{G0} \qquad (12\text{-}22)$$

in which the element geometric-stiffness coefficients from which $\mathbf{k}_{G0}$ is formed are given by

$$k_{G_{ij}} = \int_0^L N_0(x)\psi_i'(x)\psi_j'(x)\,dx \qquad (12\text{-}23)$$

In this expression $N_0(x)$ is the reference axial loading in the element. The loading of the structure therefore is proportional to the parameter λ_G; its relative distribution, however, is constant. Substituting Eq. (12-22) into Eq. (12-21) leads to the eigenvalue equation

$$[\mathbf{k} - \lambda_G\mathbf{k}_{G0}]\hat{\mathbf{v}} = 0 \qquad (12\text{-}24)$$

A nontrivial solution of this set of equations can be obtained only under the condition

$$\|\mathbf{k} - \lambda_G\mathbf{k}_{G0}\| = 0 \qquad (12\text{-}25)$$

which represents the buckling condition for the structure. The roots of this equation represent the values of the axial-load vector λ_G at which buckling will occur. The buckling mode shapes can be evaluated exactly like the vibration mode shapes. In practice, only the first buckling load and mode shape have any real significance; buckling in the higher modes generally is of little practical importance because the system will have failed when the load exceeds the lowest critical load.

Buckling with Harmonic Excitation

Although the concept has found little application in practice, it is at least of academic interest to note that a range of different "buckling" loads can be defined for a harmonically excited structure, just as a range of different vibration frequencies exists in an axially loaded structure. Suppose the structure is subjected to a harmonic excitation at the frequency $\bar{\omega}$; that is, assume that an applied-load vector of the following form is acting

$$\mathbf{p}(t) = \mathbf{p}_0 \sin \bar{\omega}t \qquad (12\text{-}26)$$

where $\bar{\omega}$ is the applied-load frequency. The undamped equations of equilibrium in this case become [from Eq. (10-18)]:

$$\mathbf{m}\ddot{\mathbf{v}} + \mathbf{k}\mathbf{v} - \mathbf{k}_G\mathbf{v} = \mathbf{p}_0 \sin \bar{\omega}t \qquad (12\text{-}27)$$

The steady-state response will then take place at the applied-load frequency,

$$\mathbf{v}(t) = \hat{\mathbf{v}} \sin \bar{\omega}t \qquad (12\text{-}28a)$$

and the accelerations become

$$\ddot{\mathbf{v}}(t) = -\bar{\omega}^2\hat{\mathbf{v}} \sin \bar{\omega}t \qquad (12\text{-}28b)$$

Introducing Eqs. (12-28) into Eq. (12-27) gives (after dividing by $\sin \bar{\omega}t$):

$$-\bar{\omega}^2\mathbf{m}\hat{\mathbf{v}} + \mathbf{k}\hat{\mathbf{v}} - \mathbf{k}_G\hat{\mathbf{v}} = \mathbf{p}_0 \qquad (12\text{-}29)$$

The symbol $\bar{\mathbf{k}}$ will be used to represent the *dynamic stiffness* of the system, where $\bar{\mathbf{k}}$ is defined as

$$\bar{\mathbf{k}} \equiv \mathbf{k} - \bar{\omega}^2\mathbf{m} \qquad (12\text{-}30a)$$

Substituting this into Eq. (12-29) and expressing the geometric stiffness in terms of the load factor λ_G leads to

$$[\bar{\mathbf{k}} - \lambda_G\mathbf{k}_{G0}]\hat{\mathbf{v}} = \mathbf{p}_0 \qquad (12\text{-}30b)$$

If the amplitude of the applied-load vector in this equation is allowed to approach zero, it is apparent by comparison with Eq. (12-5) that a nonzero response is still possible if the determinant of the square matrix is zero. Thus the condition

$$\|\bar{\mathbf{k}} - \lambda_G\mathbf{k}_{G0}\| = 0 \qquad (12\text{-}31)$$

defines the buckling condition for the harmonically excited structure.

When the applied load is allowed to vanish, Eq. (12-30b) may be written

$$[\mathbf{k} - \omega^2\mathbf{m} - \lambda_G\mathbf{k}_{G0}]\hat{\mathbf{v}} = 0 \qquad (12\text{-}32)$$

Now it is apparent that an infinite variety of combinations of buckling loads λ_G and frequencies ω^2 will satisfy this eigenvalue equation. For any given "buckling" load specified by a prescribed λ_G, the corresponding frequency of vibration can be found

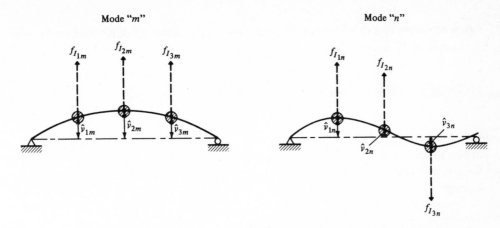

FIGURE 12-1
Vibration mode shapes and resulting inertia forces.

from Eq. (12-20). Similarly, for any given frequency of vibration ω^2, the corresponding buckling loading is defined by Eq. (12-31). It is interesting to note that a zero-axial-load condition causes "buckling" at the unstressed natural-vibration frequency according to this definition.

12-5 ORTHOGONALITY CONDITIONS

Basic Conditions

The free-vibration mode shapes ϕ_n have certain special properties which are very useful in structural-dynamics analyses. These properties, which are called *orthogonality relationships*, can be demonstrated by application of Betti's law. Consider, for example, two different modes of vibration of a structural system, as shown in Fig. 12-1. For convenience, the structure has been shown as a lumped-mass system, but the following analysis applies equally well to a consistent-mass idealization.

The equations of motion for a system in free vibration, Eq. (12-4), may be rewritten

$$\mathbf{k}\hat{\mathbf{v}}_n = \omega_n{}^2\mathbf{m}\hat{\mathbf{v}}_n \qquad (12\text{-}33)$$

in which the right-hand side represents the applied-inertia-load vector $-\mathbf{f}_I$ and the left-hand side is the elastic-resisting-force vector $\mathbf{f}_S$. Thus the free-vibration motion may be considered to involve deflections produced by inertia forces acting as applied loads, as shown in Fig. 12-1. On this basis, the two vibration modes shown in the figure represent two different applied-load systems and their resulting displacements; consequently Betti's law may be applied as follows:

$$-\mathbf{f}_{Im}{}^T\hat{\mathbf{v}}_n = -\mathbf{f}_{In}{}^T\hat{\mathbf{v}}_m$$

Introducing the inertia-force expression used in Eq. (12-33) gives

$$\omega_m{}^2\hat{\mathbf{v}}_m{}^T\mathbf{m}\hat{\mathbf{v}}_n = \omega_n{}^2\hat{\mathbf{v}}_n{}^T\mathbf{m}\hat{\mathbf{v}}_m \qquad (12\text{-}34)$$

where the rules of transposing matrix products have been observed, taking account of the symmetry of $\mathbf{m}$. When it is noted that the matrix products in Eq. (12-34) are scalars and can be transposed arbitrarily, it is evident that the equation may be written

$$(\omega_m{}^2 - \omega_n{}^2)\hat{\mathbf{v}}_m{}^T\mathbf{m}\hat{\mathbf{v}}_n = 0 \qquad (12\text{-}35)$$

Subject to the condition that the two mode frequencies are not the same, this gives the first orthogonality condition

$$\hat{\mathbf{v}}_m{}^T\mathbf{m}\hat{\mathbf{v}}_n = 0 \qquad \omega_m \neq \omega_n \qquad (12\text{-}36)$$

A second orthogonality condition can be derived directly from this by premultiplying Eq. (12-33) by $\hat{\mathbf{v}}_m{}^T$; thus

$$\hat{\mathbf{v}}_m{}^T\mathbf{k}\hat{\mathbf{v}}_n = \omega_n{}^2\hat{\mathbf{v}}_m{}^T\mathbf{m}\mathbf{v}_n$$

When Eq. (12-36) is applied to the right-hand side, it is clear that

$$\hat{\mathbf{v}}_m{}^T\mathbf{k}\hat{\mathbf{v}}_n = 0 \qquad \omega_m \neq \omega_n \qquad (12\text{-}37)$$

which shows that the vibrating shapes are orthogonal with respect to the stiffness matrix as well as with respect to the mass.

In general, it is convenient to express the orthogonality conditions in terms of the dimensionless mode-shape vectors $\boldsymbol{\phi}_n$ rather than for the arbitrary amplitudes $\hat{\mathbf{v}}_n$. Equations (12-36) and (12-37) are obviously equally valid when divided by any reference displacement value; thus the orthogonality conditions become

$$\boldsymbol{\phi}_m{}^T\mathbf{m}\boldsymbol{\phi}_n = 0 \qquad m \neq n \qquad (12\text{-}38a)$$

$$\boldsymbol{\phi}_m{}^T\mathbf{k}\boldsymbol{\phi}_n = 0 \qquad m \neq n \qquad (12\text{-}38b)$$

For systems in which no two modes have the same frequency, the orthogonality conditions apply to any two different modes, as indicated in Eqs. (12-38); they do *not* apply to two modes having the same frequency.

Additional Relationships

A complete family of additional orthogonality relationships can be derived directly from Eq. (12-33) by successive multiplications. In order to obtain the results in terms of mode-shape vectors, it is convenient to divide both sides of Eq. (12-33) by a reference amplitude, which gives the equivalent expression

$$\mathbf{k}\boldsymbol{\phi}_n = \omega_n{}^2\mathbf{m}\boldsymbol{\phi}_n \qquad (12\text{-}39)$$

Premultiplying this by $\phi_m{}^T \mathbf{km}^{-1}$ leads to

$$\phi_m{}^T \mathbf{km}^{-1}\mathbf{k}\phi_n = \omega_n{}^2 \phi_m{}^T \mathbf{k}\phi_n$$

from which [using Eq. (12-38b)]

$$\phi_m{}^T \mathbf{km}^{-1}\mathbf{k}\phi_n = 0 \qquad (12\text{-}40)$$

Premultiplying Eq. (12-39) by $\phi_m{}^T \mathbf{km}^{-1}\mathbf{km}^{-1}$ leads to

$$\phi_m{}^T \mathbf{km}^{-1}\mathbf{km}^{-1}\mathbf{k}\phi_n = \omega_n{}^2 \phi_m{}^T \mathbf{km}^{-1}\mathbf{k}\phi_n$$

from which [using Eq. (12-40)]

$$\phi_m{}^T \mathbf{km}^{-1}\mathbf{km}^{-1}\mathbf{k}\phi_n = 0 \qquad (12\text{-}41)$$

By proceeding similarly any number of orthogonality relationships of this type can be developed.

The first of a second series of relationships can be derived by premultiplying Eq. (12-39) by $(1/\omega_n{}^2)\phi_m{}^T \mathbf{m}\tilde{\mathbf{f}}$, with the result

$$\frac{1}{\omega_n{}^2} \phi_m{}^T \mathbf{m}\phi_n = \phi_m{}^T \mathbf{m}\tilde{\mathbf{f}}\mathbf{m}\phi_n$$

from which [using Eq. (12-38a)]

$$\phi_m{}^T \mathbf{m}\tilde{\mathbf{f}}\mathbf{m}\phi_n = 0 \qquad (12\text{-}42)$$

Premultiplying Eq. (12-39) by $(1/\omega_n{}^2)\phi_m{}^T \mathbf{m}\tilde{\mathbf{f}}\mathbf{m}\tilde{\mathbf{f}}$ then gives

$$\frac{1}{\omega_n^2} \phi_m{}^T \mathbf{m}\tilde{\mathbf{f}}\mathbf{m}\phi_n = \phi_m{}^T \mathbf{m}\tilde{\mathbf{f}}\mathbf{m}\tilde{\mathbf{f}}\mathbf{m}\phi_n = 0 \qquad (12\text{-}43)$$

Again the series can be extended indefinitely by similar operations.

Both complete families of orthogonality relationships, including the two basic relationships, can be compactly expressed as

$$\phi_m{}^T \mathbf{m}[\mathbf{m}^{-1}\mathbf{k}]^b \phi_n = 0 \qquad -\infty < b < \infty \qquad (12\text{-}44)$$

The two basic relationships, Eqs. (12-38a) and (12-38b), are given by exponents $b = 0$ and $b = +1$ in Eq. (12-44), respectively.

Normalizing

It was noted earlier that the vibration mode amplitudes obtained from the eigen-problem solution are arbitrary; any amplitude will satisfy the basic frequency equation (12-4), and only the resulting shapes are uniquely defined. In the analysis process described above, the amplitude of one degree of freedom (the first, actually) has been set to unity, and the other displacements have been determined relative to this reference value. This is called *normalizing* the mode shapes with respect to the specified reference coordinate.

Other normalizing procedures also are frequently used; e.g., in many computer programs the shapes are normalized relative to the maximum displacement value in each mode rather than with respect to any particular coordinate. Thus, the maximum value in each modal vector is unity, which provides convenient numbers for use in subsequent calculations. The normalizing procedure most often used in computer programs for structural-vibration analysis, however, involves adjusting each modal amplitude to the amplitude $\hat{\phi}_n$ which satisfies the condition

$$\hat{\phi}_n^T \mathbf{m} \hat{\phi}_n = 1 \qquad (12\text{-}45)$$

This can be accomplished by computing the scalar factor

$$\hat{v}_n^T \mathbf{m} \hat{v}_n = \hat{M}_n \qquad (12\text{-}46)$$

where $\hat{v}_n$ represents an arbitrarily determined modal amplitude, and then computing the normalized mode shapes as follows:

$$\hat{\phi}_n = \hat{v}_n \hat{M}_n^{-1/2} \qquad (12\text{-}47)$$

By simple substitution it is easy to show that this gives the required result.

A consequence of this type of normalizing, together with the modal orthogonality relationships relative to the mass matrix [Eq. (12-38a)], is that

$$\hat{\Phi}^T \mathbf{m} \hat{\Phi} = \mathbf{I} \qquad (12\text{-}48)$$

where $\hat{\Phi}$ is the complete set of N normalized mode shapes and $\mathbf{I}$ is an $N \times N$ identity matrix. The mode shapes normalized in this fashion are said to be *orthonormal* relative to the mass matrix. Although the use of the orthonormalized mode shapes is convenient in the development of digital-computer programs for structural-dynamic analyses, it has no particular merit when the calculations are to be done by hand. For that reason, no specific normalizing procedure is assumed in the discussions which follow.

EXAMPLE E12-3 The modal orthogonality properties and the orthonormalizing procedure will be demonstrated with the mode shapes calculated in Example E12-2. The normalizing factors obtained by applying Eq. (12-46) to these shapes are given in this lumped-mass case by

$$\hat{M}_n = \sum_{i=1}^{3} \phi_{in}^2 m_i$$

Their values are

$$\hat{M}_1 = 1.801 \qquad \hat{M}_2 = 2.455 \qquad \hat{M}_3 = 23.10$$

Dividing the respective mode shapes by the square root of these factors then leads to the orthonormalized mode-shape matrix

$$\hat{\Phi} = \begin{bmatrix} 0.745 & 0.638 & 0.208 \\ 0.480 & -0.384 & -0.535 \\ 0.223 & -0.432 & 0.514 \end{bmatrix}$$

Finally, performing the multiplication of Eq. (12-48) gives

$$\hat{\Phi}^T \mathbf{m} \hat{\Phi} = \begin{bmatrix} 1.000 & 0.006 & 0.000 \\ 0.006 & 1.000 & -0.003 \\ 0.000 & -0.003 & 0.998 \end{bmatrix}$$

The slight difference between this result and the desired identity matrix is due to round-off error in the mode-shape calculations and in forming this product since all calculations were done by slide rule.   ////

PROBLEMS

12-1 The properties of a three-story shear building in which it is assumed that the entire mass is lumped in the rigid girders are shown in Prob. 9-3.
   (a) By solving the determinantal equation, evaluate the undamped vibration frequencies of this structure.
   (b) On the basis of the computed frequencies, evaluate the corresponding vibration mode shapes, normalizing them to unity at the top story.
   (c) Demonstrate numerically that the computed mode shapes satisfy the orthogonality conditions with respect to mass and stiffness.

12-2 Repeat Prob. 12-1 for the mass and stiffness properties given in Prob. 9-4.

12-3 Two identical uniform beams are arranged, as shown in isometric view in Fig. P12-1, to support a piece of equipment weighing 3 kips. The flexural rigidity and weight per foot of the beams is shown. Assuming the distributed mass of each beam to be lumped half at its center and 1/4 at each end, compute the two frequencies and mode shapes in terms of the coordinates v_1 and v_2. [*Hint:* Note that the central deflection of a uniform beam with central load is $PL^3/48EI$. Use the flexibility formulation of the determinantal solution method, Eq. (12-18).]

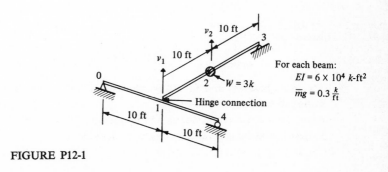

FIGURE P12-1

12-4 A rigid rectangular slab is supported by three columns rigidly attached to the slab and at the base (as shown in Fig. P12-2).

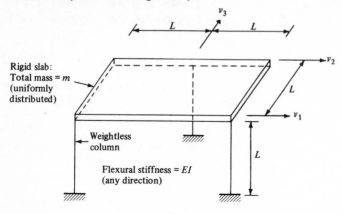

FIGURE P12-2

(*a*) Evaluate the mass and stiffness matrices for this system (in terms of *m*, *EI*, and *L*), considering the three displacement coordinates shown. (*Hint:* Apply a unit displacement or acceleration corresponding to each coordinate and evaluate the forces acting in each coordinate required for equilibrium.)

(*b*) Compute the frequencies and mode shapes of this system, normalizing the mode shapes so that either v_2 or v_3 is unity.

12-5 Repeat Prob. 12-4 using the rotation and translation (parallel and perpendicular to the symmetry axis) of the center of mass as the displacement coordinates.

12-6 A rigid bar is supported by a weightless column as shown in Fig. P12-3.

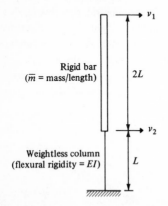

FIGURE P12-3

(*a*) Evaluate the mass and flexibility matrices of this system defined for the two coordinates shown.

(*b*) Compute the two mode shapes and frequencies of the system. Normalize the mode shapes so that the generalized mass for each mode is unity; i.e., so that $M_1 = M_2 = 1$.

ANALYSIS OF DYNAMIC RESPONSE

13-1 NORMAL COORDINATES

In the preceding discussion of any arbitrary N-DOF system, the displaced position was defined by the N components of the displacement vector $\mathbf{v}$. However, for dynamic-response analysis of linear systems, a much more useful representation of the displacements is provided by the free-vibration mode shapes. These shapes constitute N independent displacement patterns, the amplitudes of which may serve as generalized coordinates to express any form of displacement. The mode shapes thus serve the same purpose as the trigonometric functions in a Fourier series, and they are advantageous for the same reasons—because of their orthogonality properties and because they describe the displacements efficiently so that good approximations can be made with few terms.

Consider, for example, the cantilever column shown in Fig. 13-1, for which the deflected shape is defined by translational displacement coordinates at three levels. Any displacement vector $\mathbf{v}$ for this structure can be developed by superposing suitable amplitudes of the three modes of vibration, as shown. For any modal component $\hat{\mathbf{v}}_n$, the displacements are given by the mode-shape vector $\boldsymbol{\phi}_n$ multiplied by the modal amplitude Y_n; thus

$$\hat{\mathbf{v}}_n = \boldsymbol{\phi}_n Y_n \qquad (13\text{-}1)$$

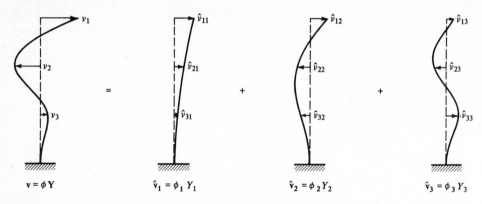

FIGURE 13-1
Representing deflections as sum of modal components.

The total displacement is then obtained as the sum of the modal components,

$$\mathbf{v} = \boldsymbol{\phi}_1 Y_1 + \boldsymbol{\phi}_2 Y_2 + \cdots + \boldsymbol{\phi}_N Y_N = \sum_{n=1}^{N} \boldsymbol{\phi}_n Y_n$$

or, in matrix notation,

$$\mathbf{v} = \boldsymbol{\Phi} \mathbf{Y} \qquad (13\text{-}2)$$

In this equation it is apparent that the mode-shape matrix $\boldsymbol{\Phi}$ serves to transform from the generalized coordinates $\mathbf{Y}$ to the geometric coordinates $\mathbf{v}$. These mode-amplitude generalized coordinates are called the *normal coordinates* of the structure.

Because the mode-shape matrix $\boldsymbol{\Phi}$ for a system with N degrees of freedom consists of N independent modal vectors, it is nonsingular and can be inverted. Thus, it is always possible to solve Eq. (13-2) directly for the normal-coordinate amplitudes $\mathbf{Y}$ associated with any given displacement vector $\mathbf{v}$. However, the orthogonality property makes it unnecessary to solve any simultaneous equations in evaluating $\mathbf{Y}$. To evaluate any arbitrary normal coordinate Y_n, Eq. (13-2) can be multiplied by the product of the transpose of the corresponding modal vector and the mass matrix $\boldsymbol{\phi}_n^T \mathbf{m}$; thus

$$\boldsymbol{\phi}_n^T \mathbf{m} \mathbf{v} = \boldsymbol{\phi}_n^T \mathbf{m} \boldsymbol{\Phi} \mathbf{Y} \qquad (13\text{-}3)$$

The right-hand side of this equation can be expanded to give

$$\boldsymbol{\phi}_n^T \mathbf{m} \boldsymbol{\Phi} \mathbf{Y} = \boldsymbol{\phi}_n^T \mathbf{m} \boldsymbol{\phi}_1 Y_1 + \boldsymbol{\phi}_n^T \mathbf{m} \boldsymbol{\phi}_2 Y_2 + \cdots + \boldsymbol{\phi}_n^T \mathbf{m} \boldsymbol{\phi}_N Y_N \qquad (13\text{-}4)$$

However, all terms of this series vanish except that corresponding to $\boldsymbol{\phi}_n$ because of the orthogonality property with respect to mass; thus introducing this one term on the right side of Eq. (13-3) gives

$$\boldsymbol{\phi}_n^T \mathbf{m} \mathbf{v} = \boldsymbol{\phi}_n^T \mathbf{m} \boldsymbol{\phi}_n Y_n$$

from which

$$Y_n = \frac{\boldsymbol{\phi}_n^T \mathbf{m} \mathbf{v}}{\boldsymbol{\phi}_n^T \mathbf{m} \boldsymbol{\phi}_n} \qquad (13\text{-}5)$$

Of course, each of the normal coordinates is given by an expression of this type. The equivalence of this analysis process to the standard derivation of expressions [such as Eqs. (5-2)] for evaluating Fourier series coefficients should be noted.

13-2 UNCOUPLED EQUATIONS OF MOTION: UNDAMPED

The orthogonality properties of the normal coordinates now may be used to simplify the equations of motion of the MDOF system. In general form these equations are given by Eq. (10-13) [or its equivalent Eq. (10-19) if axial forces are present]; for the undamped system they become

$$\mathbf{m\ddot{v}} + \mathbf{kv} = \mathbf{p}(t) \qquad (13\text{-}6)$$

Introducing Eq. (13-2) and its second time derivative $\mathbf{\ddot{v}} = \mathbf{\Phi\ddot{Y}}$ (noting that the mode shapes do not change with time) leads to

$$\mathbf{m\Phi\ddot{Y}} + \mathbf{k\Phi Y} = \mathbf{p}(t) \qquad (13\text{-}7)$$

If Eq. (13-7) is premultiplied by the transpose of the nth mode-shape vector $\boldsymbol{\phi}_n^T$, it becomes

$$\boldsymbol{\phi}_n^T \mathbf{m\Phi\ddot{Y}} + \boldsymbol{\phi}_n^T \mathbf{k\Phi Y} = \boldsymbol{\phi}_n^T \mathbf{p}(t) \qquad (13\text{-}8)$$

but if the two terms on the left-hand side are expanded as shown in Eq. (13-4), all terms except the nth will vanish because of the mode-shape orthogonality properties; hence the result is

$$\boldsymbol{\phi}_n^T \mathbf{m}\boldsymbol{\phi}_n \ddot{Y}_n + \boldsymbol{\phi}_n^T \mathbf{k}\boldsymbol{\phi}_n Y_n = \boldsymbol{\phi}_n^T \mathbf{p}(t) \qquad (13\text{-}9)$$

Now new symbols will be defined as follows:

$$M_n \equiv \boldsymbol{\phi}_n^T \mathbf{m}\boldsymbol{\phi}_n \qquad (13\text{-}10a)$$

$$K_n \equiv \boldsymbol{\phi}_n^T \mathbf{k}\boldsymbol{\phi}_n \qquad (13\text{-}10b)$$

$$P_n(t) \equiv \boldsymbol{\phi}_n^T \mathbf{p}(t) \qquad (13\text{-}10c)$$

which are called the normal-coordinate generalized mass, generalized stiffness, and generalized load for mode n, respectively. With them Eq. (13-9) can be written

$$M_n \ddot{Y}_n + K_n Y_n = P_n(t) \qquad (13\text{-}11)$$

which is a SDOF equation of motion for mode n. If Eq. (12-39), $\mathbf{k}\boldsymbol{\phi}_n = \omega_n^2 \mathbf{m}\boldsymbol{\phi}_n$, is multiplied on both sides by $\boldsymbol{\phi}_n^T$, the generalized stiffness for mode m is related to the generalized mass by the frequency of vibration

$$\boldsymbol{\phi}_n^T \mathbf{k}\boldsymbol{\phi}_n = \omega_n^2 \boldsymbol{\phi}_n^T \mathbf{m}\boldsymbol{\phi}_n$$

or

$$K_n = \omega_n^2 M_n \qquad (13\text{-}10d)$$

(Capital letters are used to denote all normal-coordinate properties.)

The procedure described above can be used to obtain an independent SDOF equation for each mode of vibration of the structure. Thus the use of the normal coordinates serves to transform the equations of motion from a set of N simultaneous differential equations, which are coupled by the off-diagonal terms in the mass and stiffness matrices, to a set of N independent normal-coordinate equations. The dynamic response therefore can be obtained by solving separately for the response of each normal (modal) coordinate and then superposing these by Eq. (13-2) to obtain the response in the original coordinates. This procedure is called the *mode-super-position method*.

13-3 UNCOUPLED EQUATIONS OF MOTION: DAMPED

Derivation of the Equations

Now it is of interest to examine the conditions under which this normal-coordinate transformation will also serve to uncouple the damped equations of motion. These equations [Eq. (10-13)] are

$$\mathbf{m}\ddot{\mathbf{v}} + \mathbf{c}\dot{\mathbf{v}} + \mathbf{k}\mathbf{v} = \mathbf{p}(t)$$

Introducing the normal-coordinate expression of Eq. (13-2) and its time derivatives and premultiplying by the transpose of the nth mode-shape vector ϕ_n^T leads to

$$\phi_n^T\mathbf{m}\mathbf{\Phi}\ddot{\mathbf{Y}} + \phi_n^T\mathbf{c}\mathbf{\Phi}\dot{\mathbf{Y}} + \phi_n^T\mathbf{k}\mathbf{\Phi}\mathbf{Y} = \phi_n^T\mathbf{p}(t) \qquad (13\text{-}12)$$

It was noted above that the orthogonality conditions

$$\phi_m^T\mathbf{m}\phi_n = 0$$
$$\phi_m^T\mathbf{k}\phi_n = 0 \qquad m \neq n$$

cause all components except the nth mode term in the mass and stiffness expressions of Eq. (13-12) to vanish. A similar reduction will apply to the damping expression if it is *assumed* that the corresponding orthogonality condition applies to the damping matrix; that is, assume that

$$\phi_m^T\mathbf{c}\phi_n = 0 \qquad m \neq n \qquad (13\text{-}13)$$

In this case Eq. (13-12) may be written

$$M_n\ddot{Y}_n + C_n\dot{Y}_n + K_nY_n = P_n(t) \qquad (13\text{-}14a)$$

or alternatively

$$\ddot{Y}_n + 2\xi_n\omega_n\dot{Y}_n + \omega_n^2Y_n = \frac{P_n(t)}{M_n} \qquad (13\text{-}14b)$$

in which

$$M_n = \phi_n^T\mathbf{m}\phi_n \qquad\qquad C_n = \phi_n^T\mathbf{c}\phi_n = 2\xi_n\omega_nM_n$$
$$K_n = \phi_n^T\mathbf{k}\phi_n = \omega_n^2M_n \qquad P_n(t) = \phi_n^T\mathbf{p}(t) \qquad (13\text{-}15)$$

The normal-coordinate generalized mass, stiffness, and load for the damped system are identical to those for the undamped system [Eq. (13-10)]. The generalized damping for mode n, which is given by Eq. (13-15), is of equivalent form. The right-hand term in this equation constitutes a definition of the nth-mode damping ratio ξ_n, because the other factors in the expression are known. As noted earlier, it generally is much more convenient and physically reasonable to define the damping by the damping ratio for each mode than it is to try to evaluate the coefficients of the damping matrix $\mathbf{c}$.

Conditions for Damping Orthogonality

In this derivation of the normal-coordinate equations of motion, it has been assumed that the normal-coordinate transformation serves to uncouple the damping forces in the same way that it uncouples the inertia and elastic forces. The vibration mode shapes in the damped system will then be the same as the undamped mode shapes. It is now useful to consider the conditions under which this uncoupling will occur, that is, the form of damping matrix to which Eq. (13-13) applies.

Rayleigh showed that a damping matrix of the form

$$\mathbf{c} = a_0\mathbf{m} + a_1\mathbf{k} \qquad (13\text{-}16)$$

in which a_0 and a_1 are arbitrary proportionality factors, will satisfy the orthogonality condition, Eq. (13-13). This is readily demonstrated by applying the orthogonality operation on both sides of Eq. (13-16); thus it is evident that a damping matrix proportional to the mass and/or stiffness matrices will permit uncoupling the equations of motion. However, it was demonstrated earlier that an infinite number of matrices formed from the mass and stiffness matrices also satisfy the orthogonality condition [Eq. (12-42)]; thus the damping matrix can also be made up of combinations of these. In general, then, the orthogonal damping matrix may be of the form

$$\mathbf{c} = \mathbf{m} \sum_b a_b[\mathbf{m}^{-1}\mathbf{k}]^b \equiv \sum_b \mathbf{c}_b \qquad (13\text{-}17)$$

in which as many terms may be included as desired.

Rayleigh damping [Eq. (13-16)] obviously is contained in Eq. (13-17); however, by including additional terms in this equation it is possible to obtain a greater degree of control over the modal damping ratios resulting from the damping matrix. With this type of damping matrix it is possible to compute the damping influence coefficients necessary to provide a decoupled system having any desired damping ratios in any specified number of modes. For each mode n, the generalized damping is given by Eq. (13-15):

$$C_n = \boldsymbol{\phi}_n^T\mathbf{c}\boldsymbol{\phi}_n = 2\xi_n\omega_n M_n$$

But if $\mathbf{c}$ is given by Eq. (13-17), the contribution of term b of the series to the generalized damping is

$$C_{nb} = \phi_n{}^T \mathbf{c}_b \phi_n = a_b \phi_n{}^T \mathbf{m}[\mathbf{m}^{-1}\mathbf{k}]^b \phi_n \qquad (13\text{-}18)$$

Now if Eq. (12-39) $(\mathbf{k}\phi_n = \omega_n{}^2 \mathbf{m}\phi_n)$ is premultiplied on both sides by $\phi_n{}^T \mathbf{k}\mathbf{m}^{-1}$, the result is

$$\phi_n{}^T \mathbf{k}\mathbf{m}^{-1}\mathbf{k}\phi_n = \omega_n{}^2 \phi_n{}^T \mathbf{k}\phi_n = \omega_n{}^4 M_n \qquad (13\text{-}19)$$

By operations equivalent to this, it can be shown that

$$\phi_n{}^T \mathbf{m}[\mathbf{m}^{-1}\mathbf{k}]^b \phi_n = \omega_n{}^{2b} M_n \qquad (13\text{-}20)$$

and consequently

$$C_{nb} = a_b \omega_n{}^{2b} M_n \qquad (13\text{-}21)$$

On this basis, the damping matrix associated with any mode n is

$$C_n = \sum_b C_{nb} = \sum_b a_b \omega_n{}^{2b} M_n = 2\xi_n \omega_n M_n \qquad (13\text{-}22)$$

from which

$$\xi_n = \frac{1}{2\omega_n} \sum_b a_b \omega_n{}^{2b} \qquad (13\text{-}23)$$

Equation (13-23) provides the means for evaluating the constants a_b to give the desired damping ratios in any specified number of modes. As many terms must be included in the series as there are specified modal damping ratios; then the constants can be determined from the resulting set of simultaneous equations. In principle, the values of b can lie anywhere in the range $-\infty < b < \infty$, but in practice it is desirable to select values as near to zero as possible. For example, to evaluate the coefficients to provide for three specified damping ratios, the equations resulting from Eq. (13-22) would be

$$\begin{bmatrix} \xi_1 \\ \xi_2 \\ \xi_2 \end{bmatrix} = \frac{1}{2} \begin{bmatrix} \dfrac{1}{\omega_1{}^3} & \dfrac{1}{\omega_1} & \omega_1 \\ \dfrac{1}{\omega_2{}^3} & \dfrac{1}{\omega_2} & \omega_2 \\ \dfrac{1}{\omega_3{}^3} & \dfrac{1}{\omega_3} & \omega_3 \end{bmatrix} \begin{bmatrix} a_{-1} \\ a_0 \\ a_1 \end{bmatrix} \qquad (13\text{-}24)$$

In general, the corresponding relationship may be written symbolically as

$$\xi = \tfrac{1}{2}\mathbf{Q}\mathbf{a} \qquad (13\text{-}25)$$

where $\mathbf{Q}$ is a square matrix involving different powers of the modal frequencies. Equation (13-25) can then be solved for the coefficients $\mathbf{a}$

$$\mathbf{a} = 2\mathbf{Q}^{-1}\xi \qquad (13\text{-}26)$$

and finally the damping matrix can be obtained from Eq. (13-17).

It is of interest to note in Eq. (13-23) [or (13-24)] that when the damping matrix is proportional to the mass matrix ($\mathbf{c} = a_0\mathbf{m}$; that is, $b = 0$), the damping ratio is inversely proportional to the frequency of vibration; thus the higher modes of a structure will have very little damping. Similarly, where the damping is proportional to the stiffness matrix ($\mathbf{c} = a_1\mathbf{k}$; that is, $b = 1$), the damping ratio is directly proportional to the frequency; and the higher modes of the structure will be very heavily damped.

A second method is available for evaluating the damping matrix associated with any given set of modal damping ratios. In principle, the procedure can be explained by considering the complete diagonal matrix of generalized damping coefficients, which may be obtained by pre- and postmultiplying the damping matrix by the mode-shape matrix:

$$\mathbf{C} = \mathbf{\Phi}^T\mathbf{c}\mathbf{\Phi} = 2\begin{bmatrix} \xi_1\omega_1 M_1 & 0 & 0 & \cdots \\ 0 & \xi_2\omega_2 M_2 & 0 & \cdots \\ 0 & 0 & \xi_3\omega_3 M_3 & \cdots \\ \cdots\cdots\cdots\cdots\cdots\cdots\cdots\cdots\cdots \end{bmatrix} \quad (13\text{-}27)$$

It is evident from this equation that the damping matrix can be obtained by pre- and postmultiplying $\mathbf{C}$ by the inverse of the mode-shape matrix or its transpose:

$$[\mathbf{\Phi}^T]^{-1}\mathbf{C}\mathbf{\Phi}^{-1} = [\mathbf{\Phi}^T]^{-1}\mathbf{\Phi}^T\mathbf{c}\mathbf{\Phi}\mathbf{\Phi}^{-1} = \mathbf{c} \quad (13\text{-}28)$$

Thus for any specified set of modal damping ratios ξ_n the generalized damping coefficients $\mathbf{C}$ can be evaluated, as indicated in Eq. (13-27), and then the damping matrix $\mathbf{c}$ evaluated as in Eq. (13-28).

In practice, however, this is not a very convenient procedure because the inversion of the mode-shape matrix is a large computational job. Instead it is useful to take advantage of the orthogonality properties of the mode shapes relative to the mass matrix. The diagonal generalized-mass matrix of the system is obtained by pre- and postmultiplying the mass matrix by the complete mode-shape matrix:

$$\mathbf{M} = \mathbf{\Phi}^T\mathbf{m}\mathbf{\Phi} \quad (13\text{-}29)$$

Premultiplying this by the inverse of the generalized-mass matrix then gives

$$\mathbf{I} = \mathbf{M}^{-1}\mathbf{M} = [\mathbf{M}^{-1}\mathbf{\Phi}^T\mathbf{m}]\mathbf{\Phi} = \mathbf{\Phi}^{-1}\mathbf{\Phi} \quad (13\text{-}30)$$

from which it is evident that the mode-shape-matrix inverse is

$$\mathbf{\Phi}^{-1} = \mathbf{M}^{-1}\mathbf{\Phi}^T\mathbf{m} \quad (13\text{-}31)$$

The damping matrix now is given by substituting Eq. (13-31) into Eq. (13-28):

$$\mathbf{c} = [\mathbf{m}\mathbf{\Phi}\mathbf{M}^{-1}]\mathbf{C}[\mathbf{M}^{-1}\mathbf{\Phi}^T\mathbf{m}] \quad (13\text{-}32)$$

Since $c_n = 2\xi_n\omega_n M_n$, the elements of the diagonal matrix obtained as the product of the three central diagonal matrices in Eq. (13-32) are given by

$$\zeta_n \equiv \frac{2\xi_n\omega_n}{M_n} \qquad (13\text{-}33)$$

and Eq. (13-32) may be written

$$\mathbf{c} = \mathbf{m}\mathbf{\Phi}\boldsymbol{\zeta}\mathbf{\Phi}^T\mathbf{m} \qquad (13\text{-}34)$$

where $\boldsymbol{\zeta}$ is the diagonal matrix of elements ζ_n. In practice it is more convenient to note that each modal damping ratio provides an independent contribution to the damping matrix, as follows:

$$\mathbf{c}_n = \mathbf{m}\boldsymbol{\phi}_n\zeta_n\boldsymbol{\phi}_n^T\mathbf{m} \qquad (13\text{-}35)$$

Thus the total damping matrix is obtained as the sum of the modal contributions

$$\mathbf{c} = \sum_{n=1}^{N} \mathbf{c}_n = \mathbf{m}\left[\sum_{n=1}^{N} \boldsymbol{\phi}_n\zeta_n\boldsymbol{\phi}_n^T\right]\mathbf{m} \qquad (13\text{-}36)$$

By substituting from Eq. (13-33) this may be written

$$\mathbf{c} = \mathbf{m}\left[\sum_{n=1}^{N} \frac{2\xi_n\omega_n}{M_n} \boldsymbol{\phi}_n\boldsymbol{\phi}_n^T\right]\mathbf{m} \qquad (13\text{-}37)$$

In this equation the contribution to the damping matrix from each mode is proportional to the modal damping ratio; thus any undamped mode will contribute nothing to the damping matrix. In other words, only those modes specifically included in the formation of the damping matrix will have any damping; all other modes will be undamped.

At this point, it is well to consider under what circumstances it may be desirable to evaluate the elements of a damping matrix explicitly, as by Eq. (13-17) or (13-37). It has been noted that the modal damping ratios are the most effective measures of the damping in the system when the analysis is to be carried out by the mode-superposition method. Hence the damping matrix will be needed in explicit form primarily when the dynamic response is to be obtained by some other analysis procedure, e.g., step-by-step integration of a nonlinear system.

Damping Coupling

In the foregoing paragraphs, it has been emphasized that where the damping matrix of the structure is of a form which satisfies the modal orthogonality conditions, the transformation to the undamped modal coordinates leads to a set of uncoupled equations. Since the response of the system can then be obtained by superposing the responses given by these SDOF equations, this decoupling is a major advantage of the normal coordinates. It was mentioned earlier, however, that these coordinates have another major advantage which can be equally important: the essential dynamic response often is associated with the lowest few modal coordinates, which means that a good approximation to the response can be often obtained with a drastically reduced number of coordinates.

Where the dynamic response is contained in only a few of the lower modes, it clearly will be advantageous to apply the normal-coordinate transformation, even with structures for which the damping matrix does not satisfy the orthogonality condition. In this case, the generalized damping matrix will not be diagonal; that is, the modal equations will be coupled by the generalized damping forces. Consequently, the response must be obtained by integrating these equations simultaneously rather than individually. However, this integration can be carried out by step-by-step methods (see Chap. 15), and certainly it is more efficient to perform the integration for a few coupled normal-coordinate equations than for the original coupled-equation system.

An alternative procedure would be to solve the complex eigenproblem (which results when the damping matrix is of general form) and then to obtain an uncoupled set of equations by transforming to the damped modal[1] coordinates. However, the evaluation of the damped mode shapes requires much more computation than the undamped eigenproblem solution does; the problem is of order $2N$ for a system with N degrees of freedom because a phase angle must be evaluated for each degree of freedom as well as its relative amplitude. For this reason the use of the undamped mode shapes generally is more efficient.

13-4 SUMMARY OF THE MODE-SUPERPOSITION PROCEDURE

The normal-coordinate transformation, which serves to change the set of N coupled equations of motion of a MDOF system into a set of N uncoupled equations, is the basis of the mode-superposition method of dynamic analysis. This method can be used to evaluate the dynamic response of any linear structure for which the displacements have been expressed in terms of a set of N discrete coordinates and where the damping can be expressed by modal damping ratios. The procedure consists of the following steps.

STEP 1: EQUATIONS OF MOTION   For this class of system, the equations of motion may be expressed [Eqs. (10-13)] as

$$\mathbf{m}\ddot{\mathbf{v}} + \mathbf{c}\dot{\mathbf{v}} + \mathbf{k}\mathbf{v} = \mathbf{p}(t)$$

---

[1] Such an approach is described in *Dynamics of Structures* by W. C. Hurty and M. F. Rubinstein, Prentice-Hall, Inc., Englewood Cliffs, N.J., 1964, chap. 9.

STEP 2: MODE SHAPE AND FREQUENCY ANALYSIS   For undamped, free vibrations, this matrix equation can be reduced to the eigenvalue equation [Eq. (12-4)]:

$$[\mathbf{k} - \omega^2 \mathbf{m}]\hat{\mathbf{v}} = \mathbf{0}$$

from which the vibration mode-shape matrix $\mathbf{\Phi}$ and frequency vector ω can be determined.

STEP 3: GENERALIZED MASS AND LOAD   With each mode-shape vector ϕ_n being used in turn, the generalized mass and generalized load for each mode can be computed [Eqs. (13-10)]:

$$M_n = \phi_n^T \mathbf{m} \phi_n \qquad P_n(t) = \phi_n^T \mathbf{p}(t)$$

STEP 4: UNCOUPLED EQUATIONS OF MOTION   The equation of motion for each mode can then be written, using the generalized mass and force for the mode together with the modal frequency ω_n and a specified value of the modal damping ratio ξ_n, as follows [Eq. (13-14b)]:

$$\ddot{Y}_n + 2\xi_n\omega_n\dot{Y}_n + \omega_n^2 Y_n = \frac{P_n(t)}{M_n} \qquad (13\text{-}38)$$

STEP 5: MODAL RESPONSE TO LOADING   The result of step 4 is a set of N independent equations of motion, one for each mode of vibration. These SDOF equations can be solved by any suitable method, depending on the type of loading. The general response expression given by the Duhamel integral [Eq. (7-14)] for each mode is

$$Y_n(t) = \frac{1}{M_n\omega_{Dn}} \int_0^t P_n(\tau)e^{-\xi_n\omega_n(t-\tau)} \sin \omega_{Dn}(t - \tau) \, d\tau \qquad (13\text{-}39)$$

STEP 6: MODAL FREE VIBRATIONS   Equation (7-14) is applicable for a system which is at rest at time $t = 0$. If the initial velocity and displacement are not zero, a free-vibration response must be added to the Duhamel integral expression for each mode. The general damped free-vibration response is given [Eq. (3-26)] for each mode by

$$Y_n(t) = e^{-\xi_n\omega_n t}\left[\frac{\dot{Y}_n(0) + Y_n(0)\xi_n\omega_n}{\omega_{Dn}} \sin \omega_{Dn}t + Y_n(0) \cos \omega_{Dn}t\right] \qquad (13\text{-}40)$$

where $Y_n(0)$ and $\dot{Y}_n(0)$ represent the initial modal displacement and velocity. These can be obtained from the specified initial displacement $\mathbf{v}(0)$ and velocity $\dot{\mathbf{v}}(0)$ expressed in the original geometric coordinates as follows for each modal component [Eq. (13-5)]:

$$Y_n(0) = \frac{\phi_n^T \mathbf{m} \mathbf{v}(0)}{M_n} \qquad (13\text{-}41)$$

$$\dot{Y}_n(0) = \frac{\phi_n^T \mathbf{m} \dot{\mathbf{v}}(0)}{M_n} \qquad (13\text{-}42)$$

STEP 7: DISPLACEMENT RESPONSE IN GEOMETRIC COORDINATES   When the response for each mode $Y_n(t)$ has been determined from Eq. (7-14) and/or Eq. (3-26), the displacements expressed in geometric coordinates are given by the normal-coordinate transformation, Eq. (13-2):

$$v(t) = \Phi Y(t)$$

Equation (13-2) may also be written

$$v(t) = \phi_1 Y_1(t) + \phi_2 Y_2(t) + \phi_3 Y_3(t) + \cdots$$

that is, it merely represents the superposition of the various modal contributions; hence the name *mode-superposition method*. It should be noted that for most types of loadings the contributions of the various modes generally are greatest for the lowest frequencies and tend to decrease for the higher frequencies. Consequently, it usually is not necessary to include all the higher modes of vibration in the superposition process [Eq. (13-2)]; the series can be truncated when the response has been obtained to any desired degree of accuracy. Moreover, it should be kept in mind that the mathematical idealization of any complex structural system also tends to be less reliable in predicting the higher modes of vibration; for this reason, too, it is well to limit the number of modes considered in a dynamic-response analysis.

STEP 8: ELASTIC FORCE RESPONSE   The displacement history of the structure may be considered to be the basic measure of its response to dynamic loading. In general, other response parameters such as stresses or forces developed in various structural components can be evaluated directly from the displacements. For example, the elastic forces f_S which resist the deformation of the structure are given directly [Eq. (10-6)] by

$$f_S(t) = kv(t) = k\Phi Y(t) \qquad (13\text{-}43)$$

An alternative expression for the elastic forces may be useful in cases where the frequencies and mode shapes have been determined from the flexibility form of the eigenvalue equation [Eq. (12-17)]. Writing Eq. (13-43) in terms of the modal contributions

$$f_S(t) = k\phi_1 Y_1(t) + k\phi_2 Y_2(t) + k\phi_3 Y_3(t) + \cdots$$

and substituting Eq. (12-39) leads to

$$f_S(t) = \omega_1{}^2 m\phi_1 Y_1(t) + \omega_2{}^2 m\phi_2 Y_2(t) + \omega_3{}^2 m\phi_3 Y_3(t) + \cdots$$

Writing the series in matrix form gives

$$f_S(t) = m\Phi[\omega_n{}^2 Y_n(t)] \qquad (13\text{-}44)$$

where $[\omega_n{}^2 Y_n(t)]$ represents a vector of modal amplitudes each multiplied by the square of its modal frequency.

In Eq. (13-44) the elastic force associated with each modal component has been replaced by an equivalent modal inertia-force expression. The equivalence of these expressions was demonstrated from the equations of free-vibration equilibrium [Eq. (13-29)]; however, it should be noted that this substitution is valid at any time, even for a static analysis.

Because each modal contribution is multiplied by the square of the modal frequency in Eq. (13-44), it is evident that the higher modes are of greater significance in defining the forces in the structure than they are in the displacements. Consequently, it will be necessary to include more modal components to define the forces to any desired degree of accuracy than to define the displacements.

EXAMPLE E13-1 Various aspects of the mode-superposition procedure will be illustrated by reference to the three-story frame structure of Example E12-1 (Fig. E12-1). For convenience, the physical and vibration properties of the structure are summarized here:

$$\mathbf{m} = \begin{bmatrix} 1.0 & 0 & 0 \\ 0 & 1.5 & 0 \\ 0 & 0 & 2.0 \end{bmatrix} \text{kip·s}^2/\text{in}$$

$$\mathbf{k} = 600 \begin{bmatrix} 1 & -1 & 0 \\ -1 & 3 & -2 \\ 0 & -2 & 5 \end{bmatrix} \text{kips/in}$$

$$\omega = \begin{bmatrix} 14.5 \\ 31.1 \\ 46.1 \end{bmatrix} \text{rad/s}$$

$$\mathbf{\Phi} = \begin{bmatrix} 1.000 & 1.000 & 1.000 \\ 0.644 & -0.601 & -2.57 \\ 0.300 & -0.676 & 2.47 \end{bmatrix}$$

Now the free vibrations which would result from the following arbitrary initial conditions will be evaluated, assuming the structure is undamped:

$$\mathbf{v}(t = 0) = \begin{bmatrix} 0.5 \\ 0.4 \\ 0.3 \end{bmatrix} \text{in} \qquad \dot{\mathbf{v}}(t = 0) = \begin{bmatrix} 0 \\ 9 \\ 0 \end{bmatrix} \text{in/s}$$

The modal coordinate amplitudes associated with the initial displacements are given by equations of the form of Eq. (13-5); writing the complete set of equations in matrix form leads to

$$\mathbf{Y}(t = 0) = \mathbf{M}^{-1}\mathbf{\Phi}^T\mathbf{m}\mathbf{v}(t = 0)$$

[which also could be derived by combining Eqs. (13-31) and (13-2)]. From the mass and mode-shape data given above, the generalized-mass matrix is

$$\mathbf{M} = \begin{bmatrix} 1.801 & 0 & 0 \\ 0 & 2.455 & 0 \\ 0 & 0 & 23.10 \end{bmatrix}$$

(where it will be noted that these terms are the same as the normalizing factors computed in Example E12-3). Multiplying the reciprocals of these terms by the mode-shape transpose and the mass matrix then gives

$$\mathbf{M}^{-1}\boldsymbol{\Phi}^T\mathbf{m} = \begin{bmatrix} 0.555 & 0.536 & 0.333 \\ 0.407 & -0.366 & -0.550 \\ 0.043 & -0.167 & 0.214 \end{bmatrix}$$

Hence the initial modal coordinate amplitudes are given as the product of this matrix and the specified initial displacements

$$\mathbf{Y}(t = 0) = \mathbf{M}^{-1}\boldsymbol{\Phi}^T\mathbf{m} \begin{bmatrix} 0.5 \\ 0.4 \\ 0.3 \end{bmatrix} = \begin{bmatrix} 0.592 \\ -0.108 \\ 0.019 \end{bmatrix} \text{ in}$$

and the modal coordinate velocities result from multiplying this by the given initial velocities

$$\dot{\mathbf{Y}}(t = 0) = \mathbf{M}^{-1}\boldsymbol{\Phi}^T\mathbf{m} \begin{bmatrix} 0 \\ 9 \\ 0 \end{bmatrix} = \begin{bmatrix} 4.83 \\ -3.30 \\ -1.50 \end{bmatrix} \text{ in/s}$$

The free-vibration response of each modal coordinate of this undamped structure is of the form

$$Y_n(t) = \frac{\dot{Y}_n(t = 0)}{\omega_n} \sin \omega_n t + Y_n(t = 0) \cos \omega_n t$$

Hence using the modal-coordinate initial conditions computed above, together with the modal frequencies, gives

$$\begin{bmatrix} Y_1(t) \\ Y_2(t) \\ Y_3(t) \end{bmatrix} = \begin{bmatrix} 0.332 \sin \omega_1 t \\ -0.106 \sin \omega_2 t \\ -0.033 \sin \omega_3 t \end{bmatrix} + \begin{bmatrix} 0.592 \cos \omega_1 t \\ -0.108 \cos \omega_2 t \\ 0.019 \cos \omega_3 t \end{bmatrix}$$

From these modal results the free-vibration motion of each story could be obtained finally from the superposition relationship $\mathbf{v}(t) = \boldsymbol{\Phi}\mathbf{Y}(t)$. It is evident that the motion of each story includes contributions at each of the natural frequencies of the structure.　　　　////

EXAMPLE E13-2 As another demonstration of mode superposition, the response of the structure of Fig. E12-1 to a sine-pulse blast-pressure load is calculated. For this purpose, the load may be expressed as

$$
\begin{bmatrix} p_1(t) \\ p_2(t) \\ p_3(t) \end{bmatrix} = \begin{bmatrix} 1 \\ 2 \\ 2 \end{bmatrix} (500 \text{ kips}) \cos \frac{\pi}{2t_1} t \quad \text{where} \quad \begin{aligned} t_1 &= 0.02 \text{ s} \\ -\frac{t_1}{2} &< t < \frac{t_1}{2} \end{aligned}
$$

With this short-duration loading, it may be assumed that the response in each mode is a free vibration with its amplitude defined by the sine-pulse spectrum of Fig. 6-6. Thus during the early response era, when the effect of damping may be neglected, the modal response may be expressed as

$$
Y_n(t) = D_n \frac{P_{0n}}{K_n} \sin \omega_n t \tag{a}
$$

in which

$$
K_n = M_n \omega_n^2 \qquad P_{0n} = \phi_n^T \begin{bmatrix} 1 \\ 2 \\ 2 \end{bmatrix} 500 \text{ kips}
$$

Using the data summarized in Example E13-1 gives

$$
\begin{bmatrix} K_1 \\ K_2 \\ K_3 \end{bmatrix} = \begin{bmatrix} 1.80 & (14.5)^2 \\ 2.455 & (31.1)^2 \\ 23.10 & (46.1)^2 \end{bmatrix} = \begin{bmatrix} 379 \\ 2,372 \\ 49,100 \end{bmatrix} \text{kips/in} \tag{b}
$$

$$
\begin{bmatrix} p_1 \\ p_2 \\ p_3 \end{bmatrix} = \phi^T \begin{bmatrix} 500 \\ 1,000 \\ 1,000 \end{bmatrix} = \begin{bmatrix} 1,444 \\ -777 \\ 400 \end{bmatrix} \text{kips} \tag{c}
$$

Also, the impulse length–period ratios for the modes of this structure are

$$
\begin{bmatrix} \dfrac{t_1}{T_1} \\[2mm] \dfrac{t_1}{T_2} \\[2mm] \dfrac{t_1}{T_3} \end{bmatrix} = \frac{0.02}{2\pi} \begin{bmatrix} \omega_1 \\ \omega_2 \\ \omega_3 \end{bmatrix} = \begin{bmatrix} 0.046 \\ 0.099 \\ 0.147 \end{bmatrix}
$$

and from Fig. 6-6, these give the following modal dynamic magnification factors:

$$
\begin{bmatrix} D_1 \\ D_2 \\ D_3 \end{bmatrix} = \begin{bmatrix} 0.18 \\ 0.39 \\ 0.57 \end{bmatrix} \tag{d}
$$

Hence, using the results given in Eqs. (*b*) to (*d*) in Eq. (*a*) leads to

$$\begin{bmatrix} Y_1(t) \\ Y_2(t) \\ Y_3(t) \end{bmatrix} = \begin{bmatrix} (0.686)\ \sin\ 14.5t \\ (-0.128)\ \sin\ 31.1t \\ (0.005)\ \sin\ 46.1t \end{bmatrix} \text{in} \qquad (e)$$

It will be noted that the motion of the top story is merely the sum of the modal expressions of Eq. (*e*), because for each mode the modal shape has a unit amplitude at the top. However, for story 2, for example, the relative modal displacement at this level must be considered, that is, the mode-superposition expression becomes

$$v_2(t) = \sum \phi_{2n} Y_n(t)$$
$$= (0.442\ \text{in})\ \sin\ 14.5t + (0.077\ \text{in})\ \sin\ 31.1t$$
$$- (0.013\ \text{in})\ \sin\ 46.1t \qquad (f)$$

Similarly, the elastic forces developed in this structure by the blast loading are given by Eq. (13-39), which for this lumped-mass system may be evaluated at story 2 as follows:

$$f_{S2}(t) = \sum \dot{m}_2 \omega_n{}^2 Y_n(t) \phi_{2n}$$
$$= (139\ \text{kips})(\sin\ 14.5t) + (112\ \text{kips})(\sin\ 31.1t)$$
$$- (41\ \text{kips})(\sin\ 46.1t) \qquad (g)$$

That the higher-mode contributions are more significant with respect to the force response than for the displacements is quite evident from a comparison of expressions (*f*) and (*g*). ////

EXAMPLE E13-3 For the structure of Example E12-1, an explicit damping matrix is to be defined such that the damping ratio in the first and third modes will be 5 percent of critical. Assuming Rayleigh damping, that is, letting *b* equal 0 and 1 in Eq. (13-17), the proportionality factors a_0 and a_1 can be evaluated from an equation of the general form of Eq. (13-24), using the frequency data listed in Example E13-1, as follows:

$$\begin{bmatrix} \xi_1 \\ \xi_3 \end{bmatrix} = \begin{bmatrix} 0.05 \\ 0.05 \end{bmatrix} = \frac{1}{2} \begin{bmatrix} \dfrac{1}{14.5} & 14.5 \\ \dfrac{1}{46.1} & 46.1 \end{bmatrix} \begin{bmatrix} a_0 \\ a_1 \end{bmatrix}$$

from which

$$\begin{bmatrix} a_0 \\ a_1 \end{bmatrix} = \begin{bmatrix} 1.10 \\ 0.00165 \end{bmatrix}$$

Hence $c = 1.10m + 0.00165k$ or, using the matrices listed in Example E13-1,

$$c = \begin{bmatrix} 2.09 & -0.99 & 0 \\ -0.99 & 4.62 & -1.98 \\ 0 & -1.98 & 7.15 \end{bmatrix} \text{kip} \cdot \text{s/in}$$

Now it is of interest to determine what damping ratio this matrix will yield in the second mode. Taking the second equation of the matrix expression of Eq. (13-24) (letting $a_{-1} = 0$) results in

$$\xi_2 = \frac{1}{2} \begin{bmatrix} \dfrac{1}{31.1} & 31.1 \end{bmatrix} \begin{bmatrix} a_0 \\ a_1 \end{bmatrix}$$

Thus, introducing the values of a_0 and a_1 found above leads to

$$\xi_2 = 0.0433 = 4.33\%$$

Hence, even though only the first and third damping ratios were specified, the resulting damping ratio for the second mode is a reasonable value.   ////

PROBLEMS

13-1 A cantilever beam supporting three equal lumped masses is shown in Fig. P13-1; also listed there are its undamped mode shapes $\boldsymbol{\Phi}$ and frequencies of vibration ω. Write an expression for the dynamic response of mass 3 of this system after an 8-kip step function load is applied at mass 2 (i.e., $8k$ is suddenly applied at time $t = 0$ and remains on the structure permanently), including all three modes and neglecting damping. Plot the history of response $v_3(t)$ for the time interval $0 < t < T_1$.

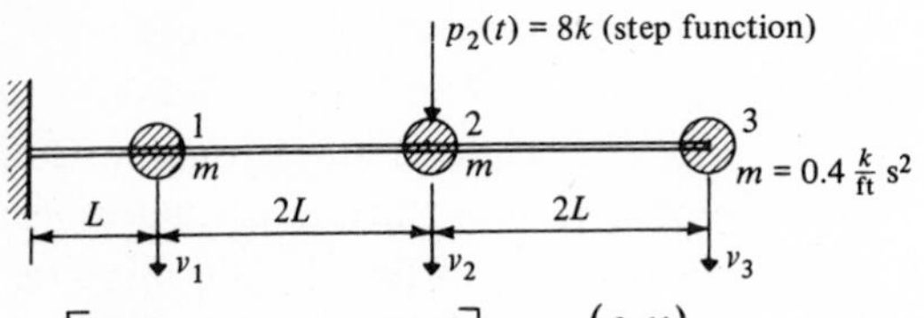

$$\Phi = \begin{bmatrix} 0.054 & 0.283 & 0.957 \\ 0.406 & 0.870 & -0.281 \\ 0.913 & -0.402 & 0.068 \end{bmatrix} ; \quad \omega = \begin{Bmatrix} 3.61 \\ 24.2 \\ 77.7 \end{Bmatrix} \text{rad/s}$$

FIGURE P13-1

13-2 Consider the beam of Prob. 13-1, but assume that a harmonic load is applied to mass 2, $p_2(t) = 3k \sin \bar{\omega} t$, where $\bar{\omega} = {}^3\!/_4 \omega_1$.
   (a) Write an expression for the steady-state response of mass 1, assuming that the structure is undamped.
   (b) Evaluate the displacements of all masses at the time of maximum response and plot the deflected shape at that time.

13-3 Repeat part (a) of Prob. 13-2, assuming that the structure has 10% critical damping in each mode.

13-4 The mass and stiffness properties of a three-story shear building, together with its undamped vibration mode shapes and frequencies, are shown in Fig. P13-2. The structure is set into free vibration by displacing the floors as follows: $v_1 = 0.3$ in, $v_2 = -0.8$ in, and $v_3 = 0.3$ in, and then releasing them suddenly at time $t = 0$. Determine the displaced shape at time $t = 2\pi/\omega_1$:

   (a) Assuming no damping.

   (b) Assuming $\xi = 10\%$ in each mode.

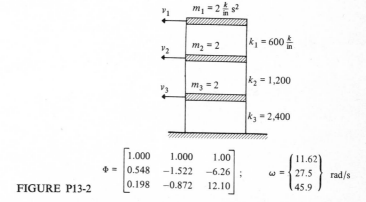

$$\Phi = \begin{bmatrix} 1.000 & 1.000 & 1.00 \\ 0.548 & -1.522 & -6.26 \\ 0.198 & -0.872 & 12.10 \end{bmatrix} ; \qquad \omega = \begin{Bmatrix} 11.62 \\ 27.5 \\ 45.9 \end{Bmatrix} \text{ rad/s}$$

FIGURE P13-2

13-5 The building of Prob. 13-4 is subjected to a harmonic loading applied at the top floor: $p_1(t) = 5k \sin \bar{\omega}t$, where $\bar{\omega} = 1.1\omega_1$. Evaluate the steady-state amplitude of motion at the three floor levels and the phase angle θ between the applied load vector and the displacement response vector at each floor.

13-6 Assuming that the building of Prob. 13-4 has Rayleigh damping, evaluate a damping matrix for the structure which will provide 5% and 15% damping ratios in the first and third modes, respectively. What damping ratio will this matrix give in the second mode?

13-7 Repeat Prob. 13-6, solving for the damping matrix by means of Eq. (13-37).

14

PRACTICAL VIBRATION ANALYSIS

14-1 PRELIMINARY COMMENTS

It is evident from the preceding discussion of the mode-superposition method that the dynamic response of any linear structure can readily be obtained after its vibration mode shapes and frequencies have been determined. Moreover, in most practical cases only a relatively small number of modes need be considered in the analysis to obtain adequate accuracy. In this regard it must be remembered that the physical properties of the structure and the loading conditions generally are known only approximately; hence the idealization and the solution procedure should be formulated to provide only a comparable level of accuracy. Nevertheless, practical problems in structural dynamics range from highly simplified mathematical models having only a few degrees of freedom and requiring only one or two modes to be considered in approximating the dynamic response to highly sophisticated finite-element models including hundreds or even thousands of degrees of freedom in which as many as 50 or 60 modes may contribute significantly to the response. To deal effectively with this wide range of analytical requirements, a variety of vibration-analysis procedures obviously is required. This chapter describes some of the analytical techniques and concepts proved efficient in practice.

First presented are two simple methods, the *Stodola method* and the *Holzer method*, which can conveniently be used with a desk calculator for the vibration analysis of small systems having up to perhaps a dozen degrees of freedom. [Since the solution of the determinantal equation (12-6) is very inefficient for systems with more than two or three degrees of freedom, it has found little use in practical analyses.] Both these methods are based on iterative improvement, but they are fundamentally different in concept. In the Stodola method, an initial assumption is made of the vibration mode shape, and it is adjusted iteratively until an adequate approximation of the true mode shape has been achieved; then the frequency of vibration is determined from the equation of motion. In the Holzer method, an initial assumption is made of the vibration frequency, and it is adjusted iteratively until the boundary conditions are satisfied; the mode shape is determined in the process of satisfying the boundary conditions.

Following the presentation of these two simple methods, the remainder of the chapter is concerned with the analysis of large-scale systems which require the use of digital computers. In principle, both the hand-solution procedures could be programmed for computer and used to evaluate vibrations in large systems, but it is better to examine the total problem in greater detail before an algorithm for a large-capacity computer program is selected. A number of the factors and concepts associated with large-scale vibration analysis are discussed here, but the presentation does not go into the details of computer programming. The first of these advanced topics is the reduction of the number of dynamic degrees of freedom which must be included in the analysis; such a reduction can be a very effective step in the treatment of large systems. Next a number of fundamental concepts concerning matrix iteration provide the basis for extending and generalizing the Stodola method of analysis. This is followed by a brief review of the problem of developing the dynamic matrix in symmetric form, so that the structural-vibration analysis can be performed by the highly efficient eigenproblem-solution programs available in many computer centers. Finally the chapter deals briefly with a special class of vibration-analysis problem involving unconstrained or only partially constrained structures which requires a modification of the typical structural-analysis procedures.

14-2 STODOLA METHOD

Fundamental-Mode Analysis

Probably the best of the hand-computation procedures is the iterative method of analysis developed by Stodola. In modern usage, the procedure is usually formulated in matrix notation; thus it is also called matrix iteration. The method is based on Eq. (12-17), which can be written

$$\frac{1}{\omega^2}\hat{\mathbf{v}} = \tilde{\mathbf{f}}\mathbf{m}\hat{\mathbf{v}} \qquad (14\text{-}1)$$

The matrix product $\tilde{\mathbf{f}}\mathbf{m}$ in this equation represents all the dynamic properties of the structure; it is called the *dynamic matrix* $\mathbf{D}$; thus

$$\mathbf{D} = \tilde{\mathbf{f}}\mathbf{m} \qquad (14\text{-}2)$$

When $\mathbf{D}$ is introduced, Eq. (14-1) becomes

$$\frac{1}{\omega^2} \hat{\mathbf{v}} = \mathbf{D}\hat{\mathbf{v}} \qquad (14\text{-}3)$$

Equation (14-3) will be satisfied, of course, only by those vectors which represent a true vibration mode shape; there are N such shapes, and the basic problem of the Stodola method is to determine them (or as many as are desired). To initiate the process, a trial shape $\mathbf{v}_1^{(0)}$ must be assumed; it should represent the best possible estimate of the first-mode *shape*; the amplitude is arbitrary. The subscript 1 used with this notation identifies the first-mode shape; the superscript (0) shows that this is the initial shape assumption. When this shape is introduced on the right side of Eq. (14-3), a new shape will be obtained by premultiplying it by $\mathbf{D}$; thus

$$\frac{1}{\omega^2} \mathbf{v}_1^{(1)} = \mathbf{D}\mathbf{v}_1^{(0)} \qquad (14\text{-}4)$$

In general, the new shape will be different from the assumed shape (unless it is a true mode shape); hence it has been identified with the superscript (1). Actually, Eq. (14-4) cannot be written directly because the vibration frequency is not known; thus the product of $\mathbf{D}$ and the assumed shape will be designated $\bar{\mathbf{v}}^{(1)}$, that is,

$$\bar{\mathbf{v}}_1^{(1)} = \mathbf{D}\mathbf{v}_1^{(0)} \qquad (14\text{-}5)$$

in which $\bar{\mathbf{v}}_1^{(1)}$ is proportional to the computed shape, $1/\omega_1^2$ being the unknown proportionality factor; thus [compare Eqs. (14-4) and (14-5)]:

$$\frac{1}{\omega_1^2} \mathbf{v}_1^{(1)} = \bar{\mathbf{v}}_1^{(1)} \qquad (14\text{-}6)$$

If it is assumed that the computed amplitude is the same as the initially assumed amplitude, an equation equivalent to Eq. (14-6) can be used to evaluate the frequency. Considering the displacement at any arbitrary point k leads to

$$\bar{v}_{k1}^{(1)} \doteq \frac{1}{\omega^2} v_{k1}^{(0)} \qquad (14\text{-}7)$$

or

$$\omega_1^2 \doteq \frac{v_{k1}^{(0)}}{\bar{v}_{k1}^{(1)}} \qquad (14\text{-}8)$$

If the assumed shape were a true mode shape, then the same frequency would be obtained by taking the ratio [Eq. (14-8)] for any coordinate of the structure. In general, the derived shape $\bar{\mathbf{v}}^{(1)}$ will differ from $\mathbf{v}^{(0)}$, and a different result will be obtained for each displacement coordinate. In this case, the true first-mode frequency lies between the maximum and minimum values obtainable from Eq. (14-8):

$$\left(\frac{v_{k1}^{(0)}}{\bar{v}_{k1}^{(1)}}\right)_{min} < \omega_1^2 < \left(\frac{v_{k1}^{(0)}}{\bar{v}_{k1}^{(1)}}\right)_{max} \qquad (14\text{-}9)$$

Because of this fact, it is evident that a better approximation of the frequency can be obtained by an averaging process. Often the best averaging procedure involves including the mass distribution as a weighting factor. Thus, writing the vector equivalent of Eq. (14-7), premultiplying both sides by $(\bar{v}_1^{(1)})^T m$, and solving for ω_1^2 gives

$$\omega_1^2 \doteq \frac{(\bar{v}_1^{(1)})^T m v_1^{(0)}}{(\bar{v}_1^{(1)})^T m \bar{v}_1^{(1)}} \qquad (14\text{-}10)$$

Equation (14-10) represents the best frequency approximation obtainable by a single iteration step, in general, from any given assumed shape $v_1^{(0)}$. [Its equivalence to the improved Rayleigh expression of Eq. (9-22) should be noted.] However, the computed shape $v_1^{(1)}$ is a better approximation of the first-mode shape than the original assumption $v_1^{(0)}$ was. Thus if $v_1^{(1)}$ and its derived shape $\bar{v}_1^{(2)}$ were used in Eq. (14-8) or (14-10), the resulting frequency approximations would be better than those computed from the initial assumption. By repeating the process sufficiently, the mode-shape approximation can be improved to any desired level of accuracy. In other words, after s cycles

$$\bar{v}_1^{(s)} = \frac{1}{\omega_1^2} v_1^{(s-1)} \qquad (14\text{-}11)$$

in which the proportionality between $\bar{v}_1^{(s)}$ and $v_1^{(s-1)}$ can be achieved to any specified number of decimal places. When the iterative process has converged to this extent, the true frequency is given by equating displacements at any desired position (preferably the position of maximum displacement):

$$\omega_1^2 = \frac{v_{k1}^{(s-1)}}{\bar{v}_{k1}^{(s)}} \qquad (14\text{-}12)$$

When the true vibrating mode shape has been determined in this way, there is no need to apply the averaging process of Eq. (14-10) to improve the result.

EXAMPLE E14-1 The Stodola method will be demonstrated by calculating the first-mode shape and frequency of the three-story building frame of Fig. E12-1 (shown again in Fig. E14-1). Although the flexibility matrix of this structure could be obtained easily by inversion of the stiffness matrix derived in that example, it will be derived here for demonstration purposes by applying

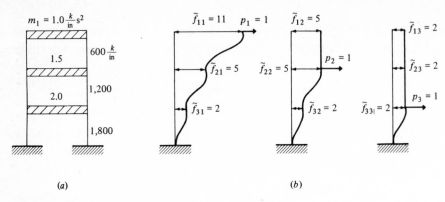

FIGURE E14-1
Frame used in example Stodola analysis: (a) structural system; (b) flexibility
influence coefficients ($\times 3,600$).

a unit load to each degree of freedom successively. By definition, the deflections
resulting from these unit loads, shown in Fig. E14-1, represent the flexibility
influence coefficients.

Thus the flexibility matrix of this structure is

$$\tilde{\mathbf{f}} = \mathbf{k}^{-1} = \frac{1}{3,600} \begin{bmatrix} 11 & 5 & 2 \\ 5 & 5 & 2 \\ 2 & 2 & 2 \end{bmatrix} \text{ in/kip}$$

Multiplying this by the mass matrix gives the dynamic matrix

$$\mathbf{D} = \tilde{\mathbf{f}}\mathbf{m} = \frac{1}{3,600} \begin{bmatrix} 11 & 7.5 & 4 \\ 5 & 7.5 & 4 \\ 2 & 3 & 4 \end{bmatrix} \text{ s}^2$$

The iteration process indicated by Eq. (14-5) can conveniently be carried
out in the tabular form shown below. A relatively poor trial vector $\mathbf{v}_1^{(0)}$ has
been used in this example to demonstrate the good convergence of the procedure.

$$\frac{1}{3,600} \overset{\mathbf{D}}{\begin{bmatrix} 11 & 7.5 & 4 \\ 5 & 7.5 & 4 \\ 2 & 3 & 4 \end{bmatrix}} \overset{\mathbf{v}_1^{(0)}}{\begin{bmatrix} 1 \\ 1 \\ 1 \end{bmatrix}}$$

$$= \overset{\bar{\mathbf{v}}_1^{(1)}}{\begin{bmatrix} 22.5 \\ 16.5 \\ 9 \end{bmatrix}} \overset{\mathbf{v}_1^{(1)}}{\begin{vmatrix} 1.00 \\ 0.73 \\ 0.40 \end{vmatrix}} \overset{\bar{\mathbf{v}}_1^{(2)}}{\begin{vmatrix} 18.1 \\ 12.1 \\ 5.8 \end{vmatrix}} \overset{\mathbf{v}_1^{(2)}}{\begin{vmatrix} 1.00 \\ 0.67 \\ 0.31 \end{vmatrix}} \overset{\bar{\mathbf{v}}_1^{(3)}}{\begin{vmatrix} 17.26 \\ 11.26 \\ 5.25 \end{vmatrix}} \overset{\mathbf{v}_1^{(3)}}{\begin{vmatrix} 1.00 \\ 0.65 \\ 0.30 \end{vmatrix}} \overset{\bar{\mathbf{v}}_1^{(4)}}{\begin{vmatrix} 17.08 \\ 11.08 \\ 5.15 \end{vmatrix}} \overset{\mathbf{v}_1^{(4)}}{\begin{vmatrix} 1.000 \\ 0.646 \\ 0.301 \end{vmatrix}} \overset{\bar{\mathbf{v}}_1^{(5)}}{\begin{vmatrix} 17.04 \\ 11.04 \\ 5.14 \end{vmatrix}}$$

Final shape

Note that the factor 1/3,600 has not been considered in this phase of the analysis because only the relative shape is important. The shapes have been normalized by dividing by the largest displacement component [see Eq. (14-114)]. After four cycles, the shape has converged to slide-rule accuracy and agrees with that obtained by the determinantal approach (Example E12-2).

From Eq. (14-12) with the largest displacement component [or Eq. (14-117)] the first-mode frequency is found to be

$$\omega_1{}^2 = \frac{v_{11}{}^{(4)}}{\bar{v}_{11}{}^{(5)}} = \frac{1.000}{(1/3,600)(17.04)} = 211 \qquad \omega_1 = 14.5 \text{ rad/s}$$

in which it will be noted that the factor 1/3,600 has now been included with the value of $\bar{v}_{11}{}^{(5)}$.

It also is of interest to determine the range of frequencies obtained after one cycle, as shown by Eq. (14-9):

$$(\omega_1{}^2)_{\text{min}} = \frac{v_{11}{}^{(0)}}{\bar{v}_{11}{}^{(1)}} = \frac{3,600}{22.5} = 160 \qquad (\omega_1{}^2)_{\text{max}} = \frac{v_{31}{}^{(0)}}{\bar{v}_{31}{}^{(1)}} = \frac{3,600}{9} = 400$$

Hence the frequency is not well established in this case after one cycle (due to the poor trial vector). However, a very good approximation can be achieved after this first cycle by applying the averaging process of Eq. (14-10):

$$\omega_1{}^2 = \frac{[22.5 \quad 24.75 \quad 18.00] \begin{bmatrix} 1 \\ 1 \\ 1 \end{bmatrix} (3,600)}{[22.5 \quad 24.75 \quad 18.00] \begin{bmatrix} 22.5 \\ 16.5 \\ 9.0 \end{bmatrix}} = \frac{65.25(3,600)}{1,077} = 218$$

This first-cycle approximation is identical to the improved Rayleigh method (R_{11}) demonstrated in Example E9-3. ////

Proof of Convergence

That the Stodola iteration process must converge to the first-mode shape, in general, can be demonstrated by recognizing that it essentially involves computing the inertia forces corresponding to any assumed shape, then computing the deflections resulting from those forces, then computing the inertia forces due to the computed deflections, etc. The concept is illustrated in Fig. 14-1 and explained mathematically in the following paragraph.

Assumed shape $v_1^{(0)}$

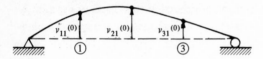

Resulting inertia forces $f_I^{(0)}$

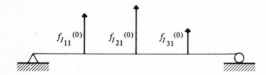

Computed shape $v_1^{(1)}$

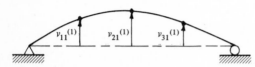

Resulting inertia forces $f_I^{(1)}$

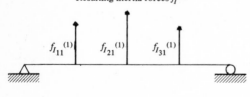

Etc.

FIGURE 14-1
Physical interpretation of Stodola iteration sequence.

The initially assumed shape is expressed in normal coordinates [see Eq. (13-2)] as

$$\mathbf{v}_1^{(0)} = \mathbf{\Phi}\mathbf{Y}^{(0)} = \boldsymbol{\phi}_1 Y_1^{(0)} + \boldsymbol{\phi}_2 Y_2^{(0)} + \boldsymbol{\phi}_3 Y_3^{(0)} + \cdots \quad (14\text{-}13)$$

in which $Y_1^{(0)}$ will be relatively large if a good guess has been made of the trial shape. The inertia forces associated with this shape vibrating at the first-mode frequency will be [see Eq. (12-33)]

$$\mathbf{f}_I^{(0)} = \omega_1^2 \mathbf{m}\mathbf{v}_1^{(0)} = \omega_1^2 \mathbf{m}\mathbf{\Phi}\mathbf{Y}^{(0)} \quad (14\text{-}14)$$

Expanding $\mathbf{v}_1^{(0)}$ as in Eq. (14-13) and writing $\omega_1^2 = \omega_n^2(\omega_1/\omega_n)^2$ gives

$$\mathbf{f}_I^{(0)} = \mathbf{m}\left[\boldsymbol{\phi}_1\omega_1^2 Y_1^{(0)} + \boldsymbol{\phi}_2\omega_2^2 Y_2^{(0)}\left(\frac{\omega_1}{\omega_2}\right)^2 + \boldsymbol{\phi}_3\omega_3^2 Y_3^{(0)}\left(\frac{\omega_1}{\omega_3}\right)^2 + \cdots \right] \quad (14\text{-}15)$$

The deflections resulting from these inertia forces are

$$\mathbf{v}_1^{(1)} = \tilde{\mathbf{f}}\mathbf{f}_I^{(0)} = \tilde{\mathbf{f}}\mathbf{m}\left[\Phi_1\omega_1^2Y_1^{(0)} + \Phi_2\omega_2^2Y_2^{(0)}\left(\frac{\omega_1}{\omega_2}\right)^2 + \cdots\right]$$

or

$$\mathbf{v}_1^{(1)} = \sum_{n=1}^{N}\tilde{\mathbf{f}}\mathbf{m}\phi_n\omega_n^2Y_n^{(1)} \qquad (14\text{-}16)$$

in which

$$Y_n^{(1)} \equiv \left(\frac{\omega_1}{\omega_n}\right)^2 Y_n^{(0)} \qquad (14\text{-}17)$$

Now noting from Eq. (12-39) (multiplying by $\tilde{\mathbf{f}}$) that

$$\phi_n = \omega_n^2\tilde{\mathbf{f}}\mathbf{m}\phi_n \qquad (14\text{-}18)$$

means that Eq. (14-16) can be written

$$\mathbf{v}_1^{(1)} = \sum_{n=1}^{N}\phi_nY_n^{(1)} = \Phi Y^{(1)} \qquad (14\text{-}19)$$

which is equivalent in form to Eq. (14-13) but refers to the derived displacements.

The same procedure for another cycle of iteration gives the deflections resulting from the second cycle

$$\mathbf{v}_1^{(2)} = \Phi Y^{(2)} = \Phi\left[\left(\frac{\omega_1}{\omega_n}\right)^2 Y_n^{(1)}\right] = \Phi\left[\left(\frac{\omega_1}{\omega_n}\right)^4 Y_n^{(0)}\right] \qquad (14\text{-}20)$$

Continuing in this fashion for s cycles leads to

$$\mathbf{v}_1^{(s)} = \Phi Y^{(s)} = \Phi\left[\left(\frac{\omega_1}{\omega_n}\right)^{2s} Y_n^{(0)}\right] \doteq \phi_1Y_1^{(0)} \qquad (14\text{-}21)$$

The last equality in this equation follows because the coefficient of the first-mode shape is much larger than all other coefficients; that is,

$$1 \gg \left(\frac{\omega_1}{\omega_2}\right)^{2s} \gg \left(\frac{\omega_1}{\omega_3}\right)^{2s} \gg \cdots \qquad (14\text{-}22)$$

It is obvious from Eq. (14-22) that the contributions of the higher modes to the shape $\mathbf{v}_1^{(s)}$ can be made as small as desired by carrying out the iteration for a sufficient number of cycles; thus the first-mode contribution represents the resulting shape. On this basis, the process must converge to the first-mode shape so long as the first-mode contribution to the initial assumed shape $\mathbf{v}_1^{(0)}$ is not zero.

Analysis of Second Mode

The proof that the Stodola iteration process converges to the first mode of vibration also provides an indication of the manner in which the procedure can be used to evaluate higher modes as well. Equation (14-21), when expanded, becomes

$$\mathbf{v}^{(s)} = \phi_1\left(\frac{\omega}{\omega_1}\right)^{2s} Y_1^{(0)} + \phi_2\left(\frac{\omega}{\omega_2}\right)^{2s} Y_2^{(0)} + \phi_3\left(\frac{\omega}{\omega_3}\right)^{2s} Y_3^{(0)} + \cdots \qquad (14\text{-}23)$$

and it is clear from this expression that if $Y_1^{(0)}$ is zero, the process must converge toward the second-mode shape; if both $Y_1^{(0)}$ and $Y_2^{(0)}$ are zero, it will converge toward the third mode, etc. Thus, to calculate the second mode by matrix iteration, it is necessary merely to assume a trial shape $\tilde{\mathbf{v}}_2^{(0)}$ which contains no first-mode component. (The tilde over the symbol indicates a shape which includes *no* first-mode contribution.)

The means of eliminating the first-mode component from any assumed second-mode shape is provided by the orthogonality condition. Consider any arbitrary assumption of the second-mode shape, expressed in terms of its modal components, as follows:

$$\mathbf{v}_2^{(0)} = \mathbf{\Phi}\mathbf{Y}^{(0)} \qquad (14\text{-}24)$$

Premultiplying both sides by $\boldsymbol{\phi}_1^T\mathbf{m}$ leads to

$$\boldsymbol{\phi}_1^T\mathbf{m}\mathbf{v}_2^{(0)} = \boldsymbol{\phi}_1^T\mathbf{m}\boldsymbol{\phi}_1 Y_1^{(0)} + \boldsymbol{\phi}_1^T\mathbf{m}\boldsymbol{\phi}_2 Y_2^{(0)} + \cdots \qquad (14\text{-}25)$$

in which the right-hand side is reduced to a first-mode term only because of the modal orthogonality properties. Hence, Eq. (14-25) can be solved for the amplitude of the first-mode component in $\mathbf{v}_2^{(0)}$:

$$Y_1^{(0)} = \frac{\boldsymbol{\phi}_1^T\mathbf{m}\mathbf{v}_2^{(0)}}{M_1} \qquad (14\text{-}26)$$

Thus, if this component is removed from the assumed shape, the vector which remains may be said to be *purified*:

$$\tilde{\mathbf{v}}_2^{(0)} = \mathbf{v}_2^{(0)} - \boldsymbol{\phi}_1 Y_1^{(0)} \qquad (14\text{-}27)$$

This purified trial vector will now converge toward the second-mode shape in the Stodola process. However, round-off errors are introduced in the numerical operations which permit first-mode components to reappear in the trial vector; therefore it is necessary to repeat this purification operation during each cycle of the iterative solution to ensure its convergence to the second mode.

A convenient means of purifying the trial vector of the first-mode component is provided by a *sweeping* matrix, which can be derived by substituting the value of $Y_1^{(0)}$ from Eq. (14-26) into Eq. (14-27), that is,

$$\tilde{\mathbf{v}}_2^{(0)} = \mathbf{v}_2^{(0)} - \frac{1}{M_1} \boldsymbol{\phi}_1\boldsymbol{\phi}_1^T\mathbf{m}\mathbf{v}_2^{(0)} \equiv \mathbf{S}_1\mathbf{v}_2^{(0)} \qquad (14\text{-}28)$$

where the first-mode sweeping matrix $\mathbf{S}_1$ is given by

$$\mathbf{S}_1 \equiv \mathbf{I} - \frac{1}{M_1} \boldsymbol{\phi}_1\boldsymbol{\phi}_1^T\mathbf{m} \qquad (14\text{-}29)$$

As is shown by Eq. (14-28), this matrix has the property of removing the first-mode component from any trial vector to which it is premultiplied, leaving only the purified shape.

The Stodola procedure can now be formulated with this sweeping matrix so that it converges toward the second mode of vibration. In this case, Eq. (14-4) can be written

$$\frac{1}{\omega_2{}^2} \, \tilde{\mathbf{v}}_2{}^{(1)} = \mathbf{D}\tilde{\mathbf{v}}_2{}^{(0)} \qquad (14\text{-}30)$$

which states that a second-mode trial shape which contains no first-mode component will converge toward the second mode. Substituting Eq. (14-28) into Eq. (14-30) gives

$$\frac{1}{\omega_2{}^2} \, \mathbf{v}_2{}^{(1)} = \mathbf{D}\mathbf{S}_1\mathbf{v}_2{}^{(0)} \equiv \mathbf{D}_2\mathbf{v}_2{}^{(0)} \qquad (14\text{-}31)$$

where

$$\mathbf{D}_2 \equiv \mathbf{D}\mathbf{S}_1 \qquad (14\text{-}32)$$

is a new dynamic matrix which eliminates the first-mode component from any trial shape $\mathbf{v}_2{}^{(0)}$ and thus automatically converges toward the second mode. When $\mathbf{D}_2$ is used, the second-mode analysis is entirely equivalent to the first-mode analysis discussed above. Thus the frequency can be approximated by the equivalent of Eq. (14-10)

$$\omega_2{}^2 \doteq \frac{(\tilde{\mathbf{v}}_2{}^{(1)})^T\mathbf{m}\mathbf{v}_2{}^{(0)}}{\tilde{\mathbf{v}}_2{}^{(1)}\mathbf{m}\tilde{\mathbf{v}}_2{}^{(1)}} \qquad (14\text{-}33)$$

in which

$$\tilde{\mathbf{v}}_2{}^{(1)} = \mathbf{D}_2\mathbf{v}_2{}^{(0)}$$

or the analysis may be carried to any desired level of convergence. It is obvious that the first mode must be evaluated before the second mode can be determined by this method. Also, the first-mode shape $\boldsymbol{\phi}_1$ must be determined with considerable accuracy in evaluating the sweeping matrix $\mathbf{S}_1$ if satisfactory results are to be obtained in the second-mode analysis. In general, the second-mode-shape ordinates will have about one less significant figure than the first-mode values.

EXAMPLE E14-2 To demonstrate the matrix-iteration analysis of a higher vibration mode, the second mode of the building of Example E14-1 will be calculated. For this analysis, the sweeping matrix will be developed in the form of Eq. (14-154) [rather than Eq. (14-29)] because it is better adapted to hand calculations, especially when only a single mode is to be swept out. The first-mode sweeping-matrix form of Eq. (14-154) is

$$\mathbf{S}_1 = \begin{bmatrix} -(\boldsymbol{\phi}_1{}^T\mathbf{m}_s)^{-1}(\boldsymbol{\phi}_1{}^T\mathbf{m}_r) \\ \mathbf{I} \end{bmatrix}$$

where for this structure (from Example E14-1) $\boldsymbol{\phi}_1{}^T = [1.000 \quad 0.646 \quad 0.301]$

and $\mathbf{m}_s$ is the first column of the mass matrix: $\mathbf{m}_s{}^T = \begin{bmatrix} 1 & 0 & 0 \end{bmatrix}$. Thus $\boldsymbol{\phi}_1{}^T\mathbf{m}_s = 1$. Also $\mathbf{m}_r$ represents the remaining columns of the mass matrix

$$\mathbf{m}_r = \begin{bmatrix} 0 & 0 \\ 1.5 & 0 \\ 0 & 2.0 \end{bmatrix}$$

which leads to $\boldsymbol{\phi}_1{}^T\mathbf{m}_r = \begin{bmatrix} 0.969 & 0.602 \end{bmatrix}$. Hence the first-mode sweeping matrix is

$$\mathbf{S}_1 = \begin{bmatrix} -0.969 & -0.602 \\ 1 & 0 \\ 0 & 1 \end{bmatrix}$$

and the second-mode dynamic matrix is

$$\mathbf{D}_2 = \mathbf{DS}_1 = \frac{1}{3,600} \begin{bmatrix} -3.16 & -2.62 \\ 2.66 & 0.99 \\ 1.06 & 2.80 \end{bmatrix}$$

The iterative solution for the second-mode shape and frequency, using this dynamic matrix, is carried out below in the same format used in Example E14-1. It is evident that only $\mathbf{v}_r{}^T = \begin{bmatrix} v_{22} & v_{32} \end{bmatrix}$ need be included in the trial vector $\mathbf{v}_2{}^{(0)}$ here because the displacement v_{12} of the top story is controlled by the orthogonality condition. This displacement need not be evaluated until the solution for $\mathbf{v}_r$ has converged.

$$\begin{array}{cc} & \mathbf{D}_2 \end{array} \quad \mathbf{v}_2{}^{(0)} \; \bar{\mathbf{v}}_2{}^{(1)} \; \mathbf{v}_2{}^{(1)} \; \bar{\mathbf{v}}_2{}^{(2)}$$

$$\frac{1}{3,600} \begin{bmatrix} -3.16 & -2.62 \\ 2.66 & 0.99 \\ 1.06 & 2.80 \end{bmatrix} \begin{vmatrix} - & - \\ 1 & 3.65 \\ 1 & 3.86 \end{vmatrix} \begin{vmatrix} - & - \\ 0.95 & 3.52 \\ 1.00 & 3.81 \end{vmatrix}$$

$$\qquad \mathbf{v}_2{}^{(2)} \; \bar{\mathbf{v}}_2{}^{(3)} \; \mathbf{v}_2{}^{(3)} \; \bar{\mathbf{v}}_2{}^{(4)} \quad \mathbf{v}_2{}^{(4)}$$

$$\times \begin{vmatrix} - & - \\ 0.92 & 3.44 \\ 1.00 & 3.77 \end{vmatrix} \begin{vmatrix} - & - \\ 0.91 & 3.41 \\ 1.00 & 3.76 \end{vmatrix} \begin{vmatrix} -1.46 \\ 0.91 \\ 1.00 \end{vmatrix} \approx \begin{bmatrix} 1.00 \\ -0.62 \\ -0.68 \end{bmatrix}$$

$$\text{Final shape}$$

In this solution, the displacements have been normalized by dividing by the larger of the two displacements considered in the iteration. Consequently, the value of v_{21} computed after $\mathbf{v}_r$ had converged was found to be greater than 1; this result has been again normalized in the final column so that the shape can be compared with that found previously in Example E12-2. The slight discrepancy is due to round-off error in these slide-rule calculations.

The second-mode frequency, computed from results obtained in the last iteration cycle, is

$$\omega_2{}^2 = \frac{v_{32}{}^{(3)}}{\bar{v}_{32}{}^{(4)}} = \frac{1.00}{(1/3,600)(3.76)} = 958$$

which compares fairly well with the value of 966 computed in Example E12-1. Of course, the method will converge to exact results if enough significant figures are carried in the calculations.   ////

Analysis of Third and Higher Modes

It should now be evident that the same sweeping process can be extended to purify a trial vector of both the first- and second-mode components, with the result that the Stodola procedure will converge toward the third mode. Expressing the purified trial third-mode shape [by analogy with Eq. (14-27)] as

$$\tilde{v}_3^{(0)} = v_3^{(0)} - \phi_1 Y_1^{(0)} - \phi_2 Y_2^{(0)} \qquad (14\text{-}34)$$

and applying the conditions that $\tilde{v}_3^{(0)}$ be orthogonal to both ϕ_1 and ϕ_2

$$\phi_1^T m \tilde{v}_3^{(0)} = 0 = \phi_1^T m v_3^{(0)} - M_1 Y_1^{(0)}$$

$$\phi_2^T m \tilde{v}_3^{(0)} = 0 = \phi_2^T m v_3^{(0)} - M_2 Y_2^{(0)}$$

leads to expressions for the first- and second-mode amplitudes in the trial vector $v_3^{(0)}$

$$Y_1^{(0)} = \frac{1}{M_1} \phi_1^T m v_3^{(0)} \qquad (14\text{-}35a)$$

$$Y_2^{(0)} = \frac{1}{M_2} \phi_2^T m v_3^{(0)} \qquad (14\text{-}35b)$$

which are equivalent to Eq. (14-26). Substituting these into Eq. (14-34) leads to

$$\tilde{v}_3^{(0)} = v_3^{(0)} - \frac{1}{M_1} \phi_1 \phi_1^T m v_3^{(0)} - \frac{1}{M_2} \phi_2 \phi_2^T m v_3^{(0)}$$

or
$$\tilde{v}_3^{(0)} = \left[I - \frac{1}{M_1} \phi_1 \phi_1^T m - \frac{1}{M_2} \phi_2 \phi_2^T m \right] v_3^{(0)} \qquad (14\text{-}36)$$

Equation (14-36) shows that the sweeping matrix S_2 which eliminates both first- and second-mode components from $v_3^{(0)}$ can be obtained by merely subtracting a second-mode term from the first-mode sweeping matrix given by Eq. (14-29), that is,

$$S_2 = S_1 - \frac{1}{M_2} \phi_2 \phi_2^T m \qquad (14\text{-}37)$$

where the sweeping-matrix operation is expressed by

$$\tilde{v}_3^{(0)} = S_2 v_3^{(0)} \qquad (14\text{-}38)$$

The Stodola relationship for analysis of the third mode can now be written by analogy with Eq. (14-31):

$$\frac{1}{\omega_3^2} v_3^{(1)} = D \tilde{v}_3^{(0)} = D S_2 v_3^{(0)} \equiv D_3 v_3^{(0)} \qquad (14\text{-}39)$$

Hence this modified dynamic matrix $\mathbf{D}_3$ performs the function of sweeping out first- and second-mode components from the trial vector $\mathbf{v}_3^{(0)}$ and thus produces convergence toward the third-mode shape.

This same process obviously can be extended successively to analysis of higher and higher modes of the system. For example, to evaluate the fourth mode, the sweeping matrix $\mathbf{S}_3$ would be calculated as follows:

$$\mathbf{S}_3 = \mathbf{S}_2 - \frac{1}{M_3} \boldsymbol{\phi}_3 \boldsymbol{\phi}_3{}^T \mathbf{m} \qquad (14\text{-}40)$$

where it would perform the function

$$\tilde{\mathbf{v}}_4^{(0)} = \mathbf{S}_3 \mathbf{v}_4^{(0)} \qquad (14\text{-}41)$$

The corresponding dynamic matrix would be

$$\mathbf{D}_4 = \mathbf{D}\mathbf{S}_3$$

The matrices suitable for calculating any mode can be obtained easily by analogy from these; that is,

$$\mathbf{S}_n = \mathbf{S}_{n-1} - \frac{1}{M_n} \boldsymbol{\phi}_n \boldsymbol{\phi}_n{}^T \mathbf{m} \qquad \mathbf{D}_{n+1} = \mathbf{D}\mathbf{S}_n \qquad (14\text{-}42)$$

Clearly the most important limitation of this procedure is that all the lower-mode shapes must be calculated before any given higher mode can be evaluated. Also, it is essential to evaluate these lower modes with great precision if the sweeping matrix for the higher modes is to perform effectively. Generally this process is used directly for the calculation of no more than four or five modes.

Analysis of Highest Mode

It is of at least academic interest to note that the Stodola method can also be applied for the analysis of the highest mode of vibration of any structure. If Eq. (14-1) is premultiplied by $\omega^2 \mathbf{m}^{-1} \mathbf{k}$, the result can be written

$$\omega^2 \hat{\mathbf{v}} = \mathbf{E} \hat{\mathbf{v}} \qquad (14\text{-}43)$$

in which the dynamic properties of the system are now contained in the matrix

$$\mathbf{E} \equiv \mathbf{m}^{-1} \mathbf{k} \equiv \mathbf{D}^{-1} \qquad (14\text{-}44)$$

If a trial shape for the highest (Nth) mode of vibration is introduced, Eq. (14-43) becomes

$$\omega_N{}^2 \mathbf{v}_N^{(1)} = \mathbf{E} \mathbf{v}_N^{(0)} \qquad (14\text{-}45)$$

which is equivalent to Eq. (14-4). By analogy with Eqs. (14-8) and (14-10), approximations of the Nth-mode frequency are given by

$$\omega_N^2 \doteq \frac{\bar{v}_{kN}^{(1)}}{v_{kN}^{(0)}} \tag{14-46a}$$

or

$$\omega_N^2 \doteq \frac{(\bar{\mathbf{v}}_N^{(1)})^T \mathbf{m} \bar{\mathbf{v}}_N^{(1)}}{(\bar{\mathbf{v}}_N^{(1)})^T \mathbf{m} \mathbf{v}_N^{(0)}} \tag{14-46b}$$

in which $\bar{\mathbf{v}}_N^{(1)} = \mathbf{E}\mathbf{v}_N^{(0)}$.

Moreover, the computed shape $\bar{\mathbf{v}}_N^{(1)}$ is a better approximation of the highest-mode shape than the original assumption was; thus if it is used as a new trial shape and the process repeated a sufficient number of times, the highest-mode shape can be determined to any desired degree of approximation.

The proof of the convergence of this process to the highest mode can be carried out exactly as for the lowest mode. The essential difference in the proof is that the term ω_N^2 is in the numerator rather than in the denominator, with the result that the equivalent of Eq. (14-22) takes the form

$$1 \gg \left(\frac{\omega_{N-1}}{\omega_N}\right)^{2s} \gg \left(\frac{\omega_{N-2}}{\omega_N}\right)^{2s} \gg \left(\frac{\omega_{N-3}}{\omega_N}\right)^{2s} \gg \cdots \tag{14-47}$$

which emphasizes the highest rather than the lowest mode.

Analysis of the next highest mode can be accomplished by developing a highest-mode-shape sweeping matrix from the orthogonality principle, and, in principle, the entire analysis could proceed from the top downward. However, since the convergence of the iteration process is much less rapid when applied with Eq. (14-45) than for the normal Stodola analysis of the lower modes, this method is seldom used except to obtain an estimate of the highest frequency of vibration which can be expected in the structure.

EXAMPLE E14-3   The analysis of the third vibration mode for the three-story structure of Example E14-1 could be carried out by evaluating the second-mode sweeping matrix and using that to obtain a dynamic matrix which would converge directly to the third mode. However, it generally is easier and more accurate to evaluate the highest mode of a structure by iterating with the stiffness form of the dynamic matrix; that approach is demonstrated here.

The stiffness matrix and the inverse of the mass matrix for the structure of Fig. E14-1 are (see Example E12-1)

$$\mathbf{k} = 600 \begin{bmatrix} 1 & -1 & -0 \\ -1 & 3 & -2 \\ 0 & -2 & 5 \end{bmatrix} \text{kips/in} \qquad \mathbf{m}^{-1} = \frac{1}{6} \begin{bmatrix} 6 & 0 & 0 \\ 0 & 4 & 0 \\ 0 & 0 & 3 \end{bmatrix} \text{in/kip·s}^2$$

Hence the stiffness form of the dynamic matrix is

$$\mathbf{E} = \mathbf{m}^{-1}\mathbf{k} = 100 \begin{bmatrix} 6 & -6 & 0 \\ -4 & 12 & -8 \\ 0 & -6 & 15 \end{bmatrix} s^{-2}$$

Using an initial shape which is a reasonable guess of the third mode, the iteration is carried out below, following the format of Example E14-1.

$$
\begin{array}{cccccc}
\mathbf{E} & \mathbf{v}_3^{(0)} & \bar{\mathbf{v}}_3^{(1)} & \mathbf{v}_3^{(1)} & \bar{\mathbf{v}}_3^{(2)} \\
100 \begin{bmatrix} 6 & -6 & 0 \\ -4 & 12 & -8 \\ 0 & -6 & 15 \end{bmatrix} & \begin{vmatrix} -1 \\ 1 \\ -1 \end{vmatrix} & \begin{vmatrix} -12 \\ 24 \\ -21 \end{vmatrix} & \begin{vmatrix} -0.50 \\ 1.00 \\ -0.88 \end{vmatrix} & \begin{vmatrix} -9.0 \\ 21.0 \\ -19.2 \end{vmatrix}
\end{array}
$$

$$
\begin{array}{ccccccc}
& \mathbf{v}_3^{(2)} & \bar{\mathbf{v}}_3^{(3)} & \mathbf{v}_3^{(3)} & \bar{\mathbf{v}}_3^{(4)} & \mathbf{v}_3^{(6)} & \bar{\mathbf{v}}_3^{(7)} \\
\times & \begin{vmatrix} -0.43 \\ 1.00 \\ -0.91 \end{vmatrix} & \begin{vmatrix} -8.58 \\ 21.00 \\ -19.65 \end{vmatrix} & \begin{vmatrix} -0.41 \\ 1.00 \\ -0.95 \end{vmatrix} & \begin{vmatrix} -8.40 \\ 21.20 \\ -20.10 \end{vmatrix} \cdots & \begin{vmatrix} -0.394 \\ 1.000 \\ -0.956 \end{vmatrix} & \begin{vmatrix} -8.36 \\ 21.24 \\ -20.34 \end{vmatrix}
\end{array}
$$

$\searrow$ Final shape

It is evident that this iteration process converges toward the highest-mode shape much more slowly than the convergence toward the lowest mode in Example E14-1; this is characteristic of the Stodola method in general. However, the final shape agrees well with that obtained from the determinantal solution (Example E12-1) and is essentially converged for slide-rule accuracy. The frequency obtained from the last iteration cycle [see Eq. (14-46a)] is

$$\omega_3^2 = \frac{\bar{v}_{23}^{(7)}}{v_{23}^{(6)}} = \frac{21.24(100)}{1} = 2,124$$

which also agrees well with the value obtained in Example E12-1. The factor of 100 in this expression is the multiplier which has been factored out of the dynamic matrix $\mathbf{E}$. ////

14-3 BUCKLING ANALYSIS BY MATRIX ITERATION

The Stodola iteration procedure for evaluating eigenvalues and eigenvectors is applicable also when axial forces act in the members of the structure if the axial forces do not vary with the vibratory motion of the structure. For any specified condition of axial loading, a Stodola equation equivalent to Eq. (14-4) may be written

$$\frac{1}{\omega_1^2}\mathbf{v}_1^{(1)} = \bar{\mathbf{D}}\mathbf{v}_1^{(0)} \qquad (14\text{-}48a)$$

in which

$$\bar{\mathbf{D}} = \bar{\mathbf{k}}^{-1}\mathbf{m} \qquad (14\text{-}48b)$$

where $\bar{\mathbf{k}} = \mathbf{k} - \mathbf{k}_G$ is the combined stiffness matrix, taking account of the geometric-stiffness effect [see Eq. (10-20)]. The vibration mode shapes and frequencies can be determined from Eq. (14-48a) by the Stodola method, just as they are without axial loads.

The effect of compressive axial forces is to reduce the stiffness of the members of the structure, thus tending to reduce the frequencies of vibration. In the limiting case, the vibration frequency goes to zero, and the static eigenvalue equation takes the form

$$(\mathbf{k} - \lambda_G \mathbf{k}_{G0})\hat{\mathbf{v}} = \mathbf{0} \qquad (12\text{-}24)$$

Premultiplying this equation by $(1/\lambda_G)\tilde{\mathbf{f}}$ gives

$$\frac{1}{\lambda_G}\hat{\mathbf{v}} = \mathbf{G}\hat{\mathbf{v}} \qquad (14\text{-}49a)$$

in which

$$\mathbf{G} = \tilde{\mathbf{f}}\mathbf{k}_{G0} \qquad (14\text{-}49b)$$

Equation (14-49a) has the same form as the Stodola equations and may be solved by the same type of iterative procedure. The eigenvalues which permit nonzero values of $\hat{\mathbf{v}}$ to be developed are the buckling loads, which are represented by the values of the load parameter λ_G. Thus, if a trial shape for the first buckling mode is designated $\mathbf{v}_1{}^{(0)}$, the iterative process is indicated by

$$\frac{1}{\lambda_{G1}}\mathbf{v}_1{}^{(1)} = \mathbf{G}\mathbf{v}_1{}^{(0)} \qquad (14\text{-}50)$$

When the iterative procedure is used to evaluate buckling modes in this way, it is called the *Vianello method*, after the man who first used it for this purpose.

The Vianello analysis of buckling is identical in principle and technique to the Stodola analysis of vibration and need not be discussed further except to mention that the orthogonality condition used in evaluating the higher buckling modes is

$$\boldsymbol{\phi}_m{}^T \mathbf{k}_{G0} \boldsymbol{\phi}_n = 0 \qquad m \neq n \qquad (14\text{-}51)$$

However, generally only the lowest mode of buckling is of interest, and there is little need to consider procedures for evaluating higher buckling modes.

EXAMPLE E14-4   The Vianello method of buckling analysis will be demonstrated by the evaluation of the initial buckling load of a uniform cantilever column loaded by its own weight (Fig. E14-2). The structure has been discretized by dividing it into three equal segments and using the lateral displacement of each node as the degrees of freedom. It is assumed that the uniformly distributed weight of the column is lumped at the ends of the segments; hence one-sixth of its total weight is concentrated at the top and one-third at the two

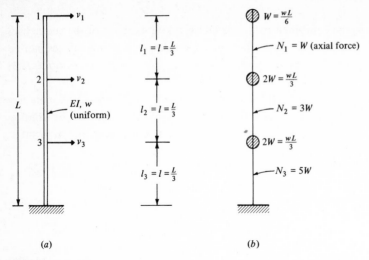

FIGURE E14-2
Analysis of column buckling due to its own weight: (*a*) uniform column; (*b*) discretized model.

interior nodes. The axial forces in the three segments of the column due to these concentrated weights are shown in the figure.

When the linear-displacement approximation [Eq. (11-36)] is used, the geometric stiffness of this column is given by

$$
\mathbf{k}_G = \begin{bmatrix} \dfrac{N_1}{l_1} & -\dfrac{N_1}{l_1} & 0 \\[2ex] -\dfrac{N_1}{l_1} & \dfrac{N_1}{l_1} + \dfrac{N_2}{l_2} & -\dfrac{N_2}{l_2} \\[2ex] 0 & -\dfrac{N_2}{l_2} & \dfrac{N_2}{l_2} + \dfrac{N_3}{l_3} \end{bmatrix} = \frac{W}{l} \begin{bmatrix} 1 & -1 & 0 \\ -1 & 4 & -3 \\ 0 & -3 & 8 \end{bmatrix}
$$

and this will be taken as the reference geometric stiffness $\mathbf{k}_{G0}$. By applying unit loads successively at the three nodes and calculating the resulting deflections by standard static-analysis procedures the flexibility matrix of the column is found to be

$$
\tilde{\mathbf{f}} = \frac{l^3}{6EI} \begin{bmatrix} 54 & 28 & 8 \\ 28 & 16 & 5 \\ 8 & 5 & 2 \end{bmatrix}
$$

Hence the stability matrix $\mathbf{G}$ is given by

$$
\mathbf{G} = \tilde{\mathbf{f}} \mathbf{k}_{G0} = \frac{Wl^2}{6EI} \begin{bmatrix} 26 & 34 & -20 \\ 12 & 21 & -8 \\ 3 & 6 & 1 \end{bmatrix}
$$

A parabola is taken as a reasonable guess for the first-mode buckled shape, and the Vianello iteration is carried out below, following the same format as the vibration examples.

$$
\frac{Wl^2}{6EI}
\begin{array}{ccc}
 & \mathbf{G} & \\
26 & 34 & -20 \\
12 & 21 & -8 \\
3 & 6 & 1
\end{array}
\begin{array}{cc}
\mathbf{v}_1^{(0)} & \bar{\mathbf{v}}_1^{(1)} \\
1.00 & 38.8 \\
0.44 & 20.3 \\
0.11 & 5.7
\end{array}
$$

$$
\times
\begin{array}{cc}
\mathbf{v}_1^{(1)} & \bar{\mathbf{v}}_1^{(2)} \\
1.00 & 40.7 \\
0.52 & 21.7 \\
0.15 & 6.3
\end{array}
\begin{array}{cc}
\mathbf{v}_1^{(2)} & \bar{\mathbf{v}}_1^{(3)} \\
1.00 & 41.1 \\
0.534 & 22.0 \\
0.155 & 6.4
\end{array}
\begin{array}{cc}
\mathbf{v}_1^{(3)} & \bar{\mathbf{v}}_1^{(4)} \\
1.000 & 41.1 \\
0.535 & 22.0 \\
0.156 & 6.4
\end{array}
$$

True shape

This process converges as quickly as the first-mode Stodola vibration analysis. The critical buckling-load factor obtained from the final iteration cycle is

$$
\lambda_{cr} = \frac{v_{11}^{(3)}}{\bar{v}_{11}^{(4)}} = \frac{1.000}{41.1(Wl^2/6EI)} = 1.315 \frac{EI}{WL^2}
$$

where the final result is expressed in terms of the total length L. From this, the critical weight per unit length is found to be

$$
w_{cr} = \frac{\lambda_{cr}W}{L/6} = 1.315(6)\frac{EI}{L^3} = 7.89 \frac{EI}{L^3}
$$

Since this compares very well with the exact result of $7.83EI/L^3$, it is evident that the geometric stiffness derived from the simple linear-displacement assumption is quite effective.

The influence of geometric stiffness on the vibration frequency of this column can also be calculated by matrix iteration. Of course, if its unit weight has the critical value calculated above, the vibration frequency will be zero. However, for any smaller value of unit weight, a corresponding frequency can be determined. Suppose, for example, that $W = {}^{27}/_{26}(EI/L^2)$, which is $({}^{27}/_{26})/1.315 = 79$ percent of the critical value. Then the geometric stiffness is given by substituting this value into the expression for k_G above.

The elastic stiffness of the column, obtained by inverting the flexibility matrix, is

$$
k = \frac{6}{26} \frac{EI}{l^3}
\begin{bmatrix}
7 & -16 & 12 \\
-16 & 44 & -46 \\
12 & -46 & 80
\end{bmatrix}
$$

Hence the combined stiffness matrix which takes account of the axial-force effects is given by [Eq. (10-20)]

$$\bar{\mathbf{k}} = \mathbf{k} - \mathbf{k}_G = \frac{6}{26}\frac{EI}{l^3}\begin{bmatrix} 7 & -16 & 12 \\ -16 & 44 & -46 \\ 12 & -46 & 80 \end{bmatrix} - \frac{27}{26}\frac{EI}{9l^3}\begin{bmatrix} 1 & -1 & 0 \\ -1 & 4 & -3 \\ 0 & -3 & 8 \end{bmatrix}$$

$$= \frac{3}{26}\frac{EI}{l^3}\begin{bmatrix} 13 & -31 & 24 \\ -31 & 84 & -89 \\ 24 & -89 & 152 \end{bmatrix}$$

Finally, the vibration analyses could be carried out by iterating with a modified dynamic matrix $\overline{\mathbf{D}} = \bar{\mathbf{k}}^{-1}\mathbf{m}$, where $\bar{\mathbf{k}}^{-1}$ is the inverse of the combined stiffness matrix shown above. The completion of this example is left to the reader.   ////

14-4   HOLZER METHOD

Basic Procedure

The basis of the Stodola method is the successive adjustment of an assumed shape until the true mode shape is achieved, followed by the evaluation of the modal frequency. The method to be discussed now, the Holzer method, proceeds in essentially the reverse sequence: the frequency is adjusted successively from an initial assumption until the true frequency is established, the mode shape being evaluated simultaneously.

The Holzer method is best adapted to analysis of structures whose components are arranged along a basic axis, a type of system frequently called a *chain structure*. Although the technique can be extended to apply to other, more complex configurations, only very elementary applications will be considered here since the purpose is to demonstrate the fundamental concept of the procedure rather than to study its applications in detail.

A practical example of the simplest type of system to which the method may be applied is the *shear building*, shown in Fig. 14-2. In this structure it is assumed that the floor slabs are rigid, so that lateral deflection results from column flexure only and no rotation takes place at the joints. In this case the story stiffness (force required to produce a unit relative displacement between floors) at level i is

$$k_i = \left(\frac{12EI}{h^3}\right)_i \qquad (14\text{-}52)$$

where I represents the total moment of inertia of all columns at level i and h is the

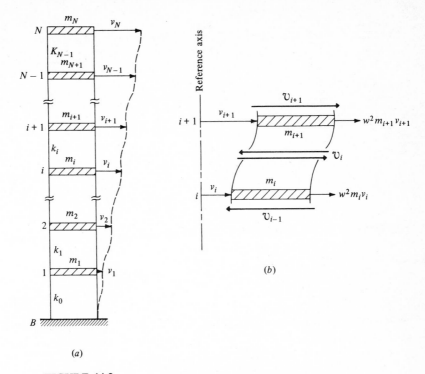

(a)

(b)

FIGURE 14-2
Holzer analysis of a shear building: (a) structural system; (b) typical story forces
and displacements.

story height. It also is assumed that all the mass is combined with the floor slabs
and that the motion takes place in the plane of the figure.

This type of system is characterized by the fact that the increment of displace-
ment in any story i depends only on the total shear force at that level $\mathcal{V}_i$, while in free
vibrations the increment of shear force at any floor level depends only on the total
deflection of that level because the inertia forces are proportional to the displacements.
This property makes it possible to compute the forces and displacements in the entire
structure by successive calculations from one end if an assumption is made of the
frequency of vibration and of the unknown boundary condition at the starting end.
For example, if the analysis were started from the top of the building of Fig. 14-2,
it would be necessary to assume a vibration displacement amplitude at this point,
but the shear-force boundary condition obviously is zero above the top slab. The
amplitude of the assumed displacement is arbitrary because free vibrations can take
place at any amplitude. The vibration frequency is definite, however, and if an incorrect frequency assumption is made, the required boundary condition at the other
end of the structure will not be satisfied. For the example of Fig. 14-2, if the analysis

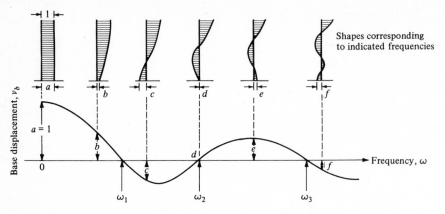

FIGURE 14-3
Variation of base displacement with applied frequency (maintaining unit displacement at top).

were started from the top with an incorrect frequency, the computed displacement at the base would not be zero. In effect, the analysis is not wrong; it is the solution to the wrong problem. The computed shape is that which would result if base of the structure moved with the assumed frequency at the computed amplitude.

The true free-vibration frequency can be found by trial and error, adjusting the frequency assumption successively until the required boundary condition is satisfied. In practice, the determination of the true frequency is greatly simplified if a plot is made of the computed boundary value as a function of the assumed frequency. Such a plot for the base-displacement amplitude computed in the building of Fig. 14-2, assuming a unit displacement at the top, is shown qualitatively in Fig. 14-3, together with a sketch of the shapes corresponding to several of the assumed frequencies. This procedure can be used to evaluate any of the vibration frequencies and the corresponding mode shape. In fact, this is one of the principal advantages of the Holzer method: it is possible to calculate any mode of vibration independently of all others. For the shear building structure the mode number for any true frequency is equal to the number of nodes (points of zero displacement amplitude) along the axis, including the base as one node.

EXAMPLE E14-5 The Holzer method of vibration analysis will be demonstrated by reevaluating the first-mode frequency and mode shape of the three-story building of Fig. E14-1. The basic properties of the building are summarized in Fig. E14-3, together with all the essential calculations of the Holzer analysis.

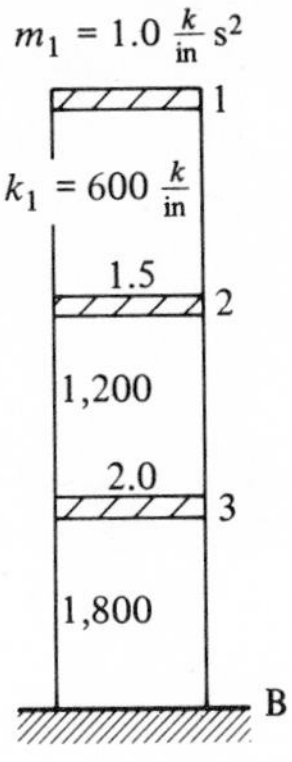

$m_1 = 1.0 \frac{k}{in} s^2$

1

$k_1 = 600 \frac{k}{in}$

1.5

2

1,200

2.0

3

1,800

B

	Trial I ($\omega^2 = 100$)				Trial II ($\omega^2 = 200$)				Trial III ($\omega^2 = 209$)		
v	Δv	$\mathcal{V}$	f_I	v	Δv	$\mathcal{V}$	f_I	v	Δv	$\mathcal{V}$	f_I
1.000			100	1.000			200	1.000			209
	0.167	100			0.333	200			0.348	209	
0.833			125	0.667			200	0.652			204
	0.187	225			0.333	400			0.344	413	
0.646			129	0.334			133	0.308			129
	0.197	354			0.296	533			0.301	542	
0.449				0.038				0.007			

FIGURE E14-3
Holzer method analysis of building vibration mode.

The first step in the procedure is to assume a trial vibration frequency; in this case it was set arbitrarily at 10 rad/s, so that $\omega^2 = 100$. Using this frequency and a top-story displacement amplitude $v_1 = 1$ in, the calculation set labeled "Trial I" was carried out, starting with the top-story inertia force

$$f_I = \omega^2 m_1 v_1 = 100(1)(1.000) = 100 \text{ kips}$$

This force then causes a top-story shear $\mathcal{V}_1 = 100$, and the corresponding top-story deformation is given by

$$\Delta v_1 = \frac{\mathcal{V}_1}{k_1} = \frac{100}{600} = 0.167 \text{ in}$$

Hence the displacement of the second story is $v_2 = v_1 - \Delta v_1 = 0.833$ in.

When this displacement is known, the inertia force at this level is given by

$$f_{12} = \omega^2 m_2 v_2 = 100(1.5)(0.833) = 125 \text{ kips}$$

Adding this to the top-story inertia gives the shear in the second story as 225 kips. Dividing this shear by the second-story stiffness gives the deformation in this story Δv_2 which is subtracted from v_2 to obtain v_3. When the procedure is continued for the bottom story, it is found that the base displacement is 0.449 in. This result indicates that a harmonic base motion of 0.449-in amplitude at a frequency of 10 rad/s will produce a 1.000-in vibration amplitude at the top. Also, it is apparent from the displacement shape (no zero displacement nodes) that the frequency $\omega = 10$ rad/s is less than the first-mode natural frequency.

Aside from the fact that 10 rad/s is too low, the first trial gives little indication of what the first-mode frequency should be; hence the second trial was arbitrarily set at $\omega^2 = 200$. Again the top displacement was taken as 1.000 in, and the deflected shape was calculated as shown by the calculation set labeled "Trial II." This time the base displacement was found to be only 0.038 in; hence the assumed $\omega^2 = 200$ is quite close to the true natural frequency.

From the results of two trials a good estimate of the true frequency can be obtained by linear extrapolation on a graph such as Fig. 14-2. Carrying out the extrapolation analytically gives

$$\left(\frac{\Delta\omega^2}{\Delta v_B}\right)_{1-2} = \left(\frac{\Delta\omega^2}{\Delta v_B}\right)_{2-3}$$

where the subscripts indicate the trial numbers for which the increments are taken. In this case the extrapolation relationship is

$$\frac{100}{0.411} = \frac{\Delta\omega^2_{2-3}}{0.038} \qquad \Delta\omega^2_{2-3} = 9$$

Hence in the third trial, the assumed frequency was $\omega^2 = 209$. The base displacement resulting from this assumption was determined in the third calculation set to be 0.007 in, which is very near to the required zero value. The fact that a zero displacement was not obtained exactly demonstrates that v_B is not a linear function of ω^2; however, if the same type of extrapolation is applied again,

$$\frac{9}{0.031} = \frac{\Delta\omega^2_{3-4}}{0.007} \qquad \Delta\omega^2_{3-4} = 2$$

the resulting $\omega^2 = 211$ is essentially exact. The displacements obtained with this frequency are $v^T = [1.000 \quad 0.648 \quad 0.301 \quad 0.000]$, which agrees to slide-rule accuracy with the shape obtained previously. ////

Transfer-Matrix Procedure

The Holzer method of vibration analysis can also be effectively carried out by means of matrix operations, using a special type of matrix called a *transfer matrix*. These matrices express the forces and displacements at one section of a chain-type structure in terms of the corresponding forces and displacements at the adjacent section; thus the complete force and displacement profile of the structure can be obtained from a sequence of transfer-matrix multiplications.

It is convenient to divide the transfer matrices into two parts in this derivation: *field* matrices, which provide for transfer across the elastic segments between the

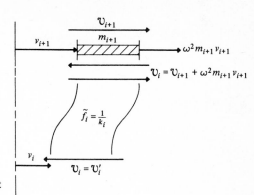

FIGURE 14-4
State vectors related by field and point
matrices in shear building.

masses; and *point* matrices, which effect the transfer across the lumped masses. The
elastic field for level i of the structure of Fig. 14-2 is shown in Fig. 14-4. The *state
vector* η_i which represents the force and displacement at level i can be expressed in
terms of the state vector at level i' (the other end of the segment) by a field-matrix
relationship as follows:

$$\begin{bmatrix} \mathcal{V}_i \\ v_i \end{bmatrix} = \begin{bmatrix} 1 & 0 \\ -\tilde{f}_i & 1 \end{bmatrix} \begin{bmatrix} \mathcal{V}_i' \\ v_i' \end{bmatrix} \quad (14\text{-}53a)$$

where $\tilde{f}_i = 1/k_i$ is the flexibility of story i, that is, the deflection produced by a unit
shear force acting in that story. Equation (14-53a) can be expressed symbolically as

$$\eta_i = \mathbf{T}_{fi}\eta_i' \quad (14\text{-}53b)$$

where $\mathbf{T}_{fi}$, the field matrix for segment i, is the square matrix in Eq. (14-53a).

The mass point for level $i + 1$ is also shown in Fig. 14-4, and the point matrix
which transfers the state vector across this level is defined by

$$\begin{bmatrix} \mathcal{V}_i' \\ v_i' \end{bmatrix} = \begin{bmatrix} 1 & \omega^2 m_{i+1} \\ 0 & 1 \end{bmatrix} \begin{bmatrix} \mathcal{V}_{i+1} \\ v_{i+1} \end{bmatrix} \quad (14\text{-}54a)$$

or symbolically

$$\eta_i' = \mathbf{T}_{p,i+1}\eta_{i+1} \quad (14\text{-}54b)$$

where $\mathbf{T}_{p,i+1}$ represents the point matrix for level $i + 1$.

The transfer matrix $\mathbf{T}_{i+1}$ for the complete segment can now be obtained by
combining the point and field matrices; thus

$$\eta_i = \mathbf{T}_{fi}\eta_i' = \mathbf{T}_{fi}\mathbf{T}_{p,i+1}\eta_{i+1} \equiv \mathbf{T}_{i+1}\eta_{i+1} \quad (14\text{-}55)$$

or with the introduction of the field and point matrices of Eq. (14-53) and (14-54) the
transfer matrix is found to be

$$\mathbf{T}_{i+1} = \begin{bmatrix} 1 & 0 \\ -\tilde{f}_i & 1 \end{bmatrix} \begin{bmatrix} 1 & \omega^2 m_{i+1} \\ 0 & 1 \end{bmatrix} = \begin{bmatrix} 1 & \omega^2 m_{i+1} \\ -\tilde{f}_i & 1 - \omega^2 \tilde{f}_i m_{i+1} \end{bmatrix} \quad (14\text{-}56)$$

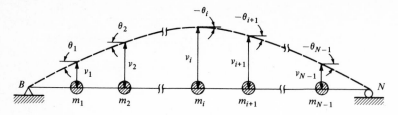

FIGURE 14-5
Degrees of freedom of flexural system.

By using a succession of transfer matrices it is possible to express the state vector at any point in terms of the state vector at any other point; e.g., the state vector at the base can be expressed in terms of the state vector at the top

$$\eta_B = T_1\eta_1 = T_1T_2\eta_2 = \cdots = T_1T_2\cdots T_{N-1}T_N\eta_N \quad (14\text{-}57)$$

in which the subscripts B and N refer to the two ends of the structure, as shown in Fig. 14-2a. When $\overline{T}_N$ is defined as

$$\overline{T}_N \equiv T_1T_2T_3T_4\cdots T_{N-1}T_N \quad (14\text{-}58)$$

Eq. (14-57) can be written

$$\eta_B = \overline{T}_N\eta_N \quad (14\text{-}59a)$$

Hence $\overline{T}_N$ is a 2×2 matrix relating the state vectors at the two ends of the structure. Writing out the matrix elements in Eq. (14-59a) leads to

$$\begin{bmatrix} \mathcal{U}_B \\ v_B \end{bmatrix} = \begin{bmatrix} t_{ss} & t_{sv} \\ t_{vs} & t_{vv} \end{bmatrix}\begin{bmatrix} \mathcal{U}_N \\ v_N \end{bmatrix} \quad (14\text{-}59b)$$

and when the given boundary conditions are introduced, the resulting Nth-degree equation in ω^2 can be solved for the natural frequencies of the system. For example, the boundary conditions in the structure of Fig. 14-2 are $v_B = 0$ and $\mathcal{U}_N = 0$. Thus, if it is assumed that the vibration amplitude is specified by $v_N = 1$, the second equation of Eq. (14-59b) is

$$v_B = t_{vv}(\omega^2) \quad (14\text{-}59c)$$

in which it is indicated that t_{vv} is a function of ω^2, and values of ω^2 which provide the required zero-displacement condition at the base are the free-vibration frequencies.

Holzer-Myklestad Method

The analysis of the shear-building type of system is particularly simple by the Holzer (or transfer-matrix) procedure because it has only one degree of freedom per joint. The extension of the method to the analysis of flexural systems, which have two degrees of freedom per joint, was first suggested by Myklestad, and the extended procedure is usually called the *Holzer-Myklestad method*. A typical idealization of a flexural system is shown in Fig. 14-5. The mass is assumed to be concentrated in a

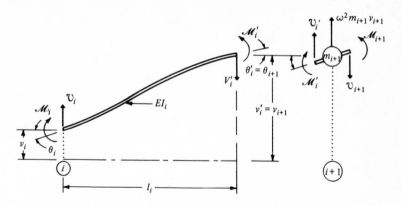

FIGURE 14-6
State vectors related by field and point matrices in beam structures.

series of points along the axis, and the degrees of freedom of the structure are the lateral translation and the rotation of these points. (The rotational inertia of the masses is usually neglected, but it can be included without difficulty.) The beam segments connecting the mass points are assumed to be weightless and of constant flexural stiffness EI in each segment.

The analysis of this type of system is also expressed most conveniently by transfer-matrix operations, and again it is necessary to consider the two types of transfer matrices: field matrices and point matrices. The forces and displacements which constitute the state vectors associated with a typical field matrix are shown in Fig. 14-6, and the field-matrix relationship is given by

$$
\begin{bmatrix} \mathcal{V}_i \\ \mathcal{M}_i \\ \theta_i \\ v_i \end{bmatrix} = \begin{bmatrix} 1 & 0 & 0 & 0 \\ -l_i & 1 & 0 & 0 \\ \dfrac{l_i^2}{2EI_i} & -\dfrac{l_i}{EI_i} & 1 & 0 \\ -\dfrac{l_i^3}{6EI_i} & \dfrac{l_i}{2EI_i} & -l_i & 1 \end{bmatrix} \begin{bmatrix} \mathcal{V}_i' \\ \mathcal{M}_i' \\ \theta_i' \\ v_i' \end{bmatrix} \qquad (14\text{-}60a)
$$

or symbolically

$$\boldsymbol{\eta}_i = \mathbf{T}_{fi}\boldsymbol{\eta}_i' \qquad (14\text{-}60b)$$

The coefficients in the field matrix $\mathbf{T}_{fi}$ are readily derived from elementary beam theory.

The state vectors associated with the point matrix for point $i + 1$ are also shown in the figure, and the relationship can be written

$$
\begin{bmatrix} \mathcal{V}_i' \\ \mathcal{M}_i' \\ \theta_i' \\ v_i' \end{bmatrix} = \begin{bmatrix} 1 & 0 & 0 & -\omega^2 m_{i+1} \\ 0 & 1 & 0 & 0 \\ 0 & 0 & 1 & 0 \\ 0 & 0 & 0 & 1 \end{bmatrix} \begin{bmatrix} \mathcal{V}_{i+1} \\ \mathcal{M}_{i+1} \\ \theta_{i+1} \\ v_{i+1} \end{bmatrix} \qquad (14\text{-}61a)
$$

or, in abbreviated notation,

$$\eta'_i = T_{p,i+1}\eta_{i+1} \qquad (14\text{-}61b)$$

When the field and point matrices from Eqs. (14-60) and (14-61) are combined, the complete transfer matrix for the segment is

$$\mathbf{T}_{i+1} = \mathbf{T}_{fi}\mathbf{T}_{p,i+1} = \begin{bmatrix} 1 & 0 & 0 & -\omega^2 m_{i+1} \\ -l_i & 1 & 0 & \omega^2 l_i m_{i+1} \\ \dfrac{l_i^2}{2EI_i} & -\dfrac{l_i}{EI_i} & 1 & -\omega^2 \dfrac{l_i^2 m_{i+1}}{2EI_i} \\ -\dfrac{l_i^3}{6EI_i} & \dfrac{l_i^2}{2EI_i} & -l_i & \omega^2 \dfrac{l_i^3 m_{i+1}}{6EI_i} + 1 \end{bmatrix} \qquad (14\text{-}62)$$

Then the state vector at one end of the structure can be expressed in terms of the state vector at the other end, as shown by Eqs. (14-57) and (14-59a), except that in this type of system the transfer matrix $\mathbf{T}_N$ for the complete structure is of dimensions 4×4. Thus expanding Eq. (14-59a) for the flexural system gives a relationship of the type

$$\begin{bmatrix} \mathcal{V}_B \\ \mathcal{M}_B \\ \theta_B \\ v_B \end{bmatrix} = \begin{bmatrix} t_{ss} & t_{sm} & t_{s\theta} & t_{sv} \\ t_{ms} & t_{mm} & t_{m\theta} & t_{mv} \\ t_{\theta s} & t_{\theta m} & t_{\theta\theta} & t_{\theta v} \\ t_{vs} & t_{vm} & t_{v\theta} & t_{vv} \end{bmatrix} \begin{bmatrix} \mathcal{V}_N \\ \mathcal{M}_N \\ \theta_N \\ v_N \end{bmatrix} \qquad (14\text{-}63)$$

If the boundary conditions are introduced into Eq. (14-63), two of the state-vector components at each end will vanish and the equation can be reduced accordingly. For example, for the simple beam supports of Fig. 14-5, the moment and displacement vanish at each end and the significant part of the relationship becomes

$$\begin{bmatrix} \mathcal{M}_B \\ v_B \end{bmatrix} = \begin{bmatrix} 0 \\ 0 \end{bmatrix} = \begin{bmatrix} t_{ms} & t_{m\theta} \\ t_{vs} & t_{v\theta} \end{bmatrix} \begin{bmatrix} \mathcal{V}_N \\ \theta_N \end{bmatrix} \qquad (14\text{-}64)$$

Introducing an arbitrary unit rotation amplitude $\theta_N = 1$ makes it possible to determine the corresponding shear value from the first equation

$$0 = t_{ms}\mathcal{V}_N + t_{m\theta}$$

as follows:

$$\mathcal{V}_N = -\frac{t_{m\theta}}{t_{ms}} \qquad (14\text{-}65)$$

Then the free-vibration condition is given by the second of Eqs. (14-64):

$$v_B = 0 = t_{vs}\mathcal{V}_N + t_{v\theta} = t_{v\theta} - \frac{t_{m\theta}t_{vs}}{t_{ms}} \qquad (14\text{-}66)$$

Each of these coefficients is a function of ω^2, of course; thus the values of the frequency which lead to a zero deflection, $v_B = 0$, are the free-vibration frequencies of the structure. [Alternatively, the free-vibration frequencies can be found by setting to zero the determinant of the square matrix in Eq. (14-64).]

Note that in this analysis procedure the boundary conditions at N should be introduced into the transfer-matrix relationships from the beginning, that is, in

$$\boldsymbol{\eta}_B = \mathbf{T}_1 \mathbf{T}_2 \cdots \mathbf{T}_{N-1} \mathbf{T}_N \boldsymbol{\eta}_N \qquad (14\text{-}67)$$

because when the matrix product is developed from right to left, it will be necessary to consider only two columns in the postmultiplying transfer matrices. The transfer-matrix procedure is also applicable to more complex chain-type structures having more than two degrees of freedom at each section.[1]

14-5 REDUCTION OF DEGREES OF FREEDOM

Background

Although modern vibration-analysis techniques are capable of solving systems with several hundred degrees of freedom, they still are not adequate for treating directly the mathematical idealizations used in static analysis of complex structures, which may include thousands of degrees of freedom. Moreover, it seldom is of interest to determine more than a few dozen vibration mode shapes even in the most complex structural systems because the mode-superposition method generally is applied to structures in which the loading excites significantly only the lowest modes. For these reasons various *eigenvalue-economizer procedures* have been developed, which serve to reduce the large number of degrees of freedom employed in defining the elastic properties of the structure to the smaller number required for a dynamic analysis.

At first glance, it might appear more consistent to use the smaller number of degrees of freedom in the original formulation of the stiffness properties of the structure as well as in the vibration analysis. In fact, however, accurate representation of the elastic properties of a structure requires a more refined mathematical model than the representation of the inertial properties does. This is because the inertial properties (or kinetic energy) depend directly on the displacements of the structure, whereas the stiffness properties (or strain energy) are functions of the derivatives of the displacements, and it is well known that the derivatives are more difficult to approximate accurately than the displacements are. Moreover, in many cases the

---

[1] For an extensive discussion of the application of this technique, see *Matrix Methods in Elastomechanics* by E. C. Pestel and F. A. Leckie, McGraw-Hill Book Company, New York, 1963.

primary objective of the analysis is the evaluation of stresses in the structure, and these require a more detailed description of the structural system than is needed to express the deflection variables involved in the dynamic analysis. These arguments suggest (and experience has borne out the fact) that a dynamic analysis can be accomplished most efficiently by establishing first the structural idealization needed for static stress-analysis purposes and then reducing the number of degrees of freedom significantly before performing the dynamic analysis.

Two general approaches have been used effectively to reduce the dynamic degrees of freedom. The simplest is based on the assumption that inertia forces are associated with only certain selected degrees of freedom of the original idealization; the remaining degrees of freedom are not explicitly involved in the dynamic analysis and can be condensed from the dynamic formulation. In the second approach, the number of dynamic degrees of freedom is limited by assuming that the displacements of the structure are combined in selected patterns, the amplitudes of which become the generalized coordinates of the dynamic analysis. Each of these approaches has been employed in a variety of specific techniques; their essential features are discussed in the following sections.

Discrete-Mass Allocations

In the analysis of framed structures, it is common practice to reduce the number of dynamic degrees of freedom by assuming the mass to be concentrated in point lumps at the structural joints. Neglecting the rotational inertia of the lumps leads to a reduction of one-third in the degrees of freedom of a general plane frame or one-half in a three-dimensional frame. If axial distortions of the members are neglected, so that relatively fewer translational degrees of freedom are involved, the elimination of the dynamic rotational degrees of freedom is much more significant; in the analysis of building frames, the dynamic degrees of freedom frequently are less than 10 percent of the number used in the static analysis. In some types of structural systems, additional reductions are made by assuming that the mass is lumped at only certain selected joints of the structure. However, this extension of the idea is much less reliable than the simple elimination of rotational degrees of freedom (which are associated with only a very small part of the kinetic energy of the structure).

The elimination of the nonessential degrees of freedom generally is accomplished by the static-condensation procedure, described earlier [Eq. (11-47)]. Consider, for example, the equation of motion in free vibrations [Eq. (12-4)] written in the form

$$\mathbf{k}\hat{\mathbf{v}} = \omega^2 \mathbf{m}\hat{\mathbf{v}} \qquad (14\text{-}68)$$

in which the vibration displacement vector $\hat{\mathbf{v}}$ represents all the degrees of freedom. If these displacements are now partitioned into a subvector $\hat{\mathbf{v}}_0$, for which it is assumed that no inertial forces are developed, and a subvector $\hat{\mathbf{v}}_t$, which is associated with the nonzero mass coefficients, and if the mass and stiffness matrices are partitioned to correspond, Eq. (14-68) can be written

$$\begin{bmatrix} \mathbf{k}_{00} & \mathbf{k}_{0t} \\ \mathbf{k}_{t0} & \mathbf{k}_{tt} \end{bmatrix} \begin{bmatrix} \hat{\mathbf{v}}_0 \\ \hat{\mathbf{v}}_t \end{bmatrix} = \omega^2 \begin{bmatrix} 0 & 0 \\ 0 & \mathbf{m}_{tt} \end{bmatrix} \begin{bmatrix} \hat{\mathbf{v}}_0 \\ \hat{\mathbf{v}}_t \end{bmatrix} \qquad (14\text{-}69)$$

in which it is assumed that the mass matrix is diagonal, that is, a lumped-mass system. Applying static condensation then leads to the reduced vibration equation

$$\mathbf{k}_t \hat{\mathbf{v}}_t = \omega^2 \mathbf{m}_{tt} \hat{\mathbf{v}}_t \qquad (14\text{-}70)$$

where $\mathbf{k}_t$ is the reduced stiffness matrix, which can be expressed in a form equivalent to Eq. (11-47).

Although the static-condensation principle is well known and is widely used for eliminating nonessential degrees of freedom in small systems, it is not always an efficient procedure for dealing with large systems. An alternative which may be more effective in many cases is the formation of a reduced flexibility matrix for the desired degrees of freedom. This can be obtained by applying unit loads successively, corresponding with each of the essential degrees of freedom, and evaluating the displacements produced in each essential degree of freedom by each load. The assembled system of flexibility coefficients is the reduced flexibility matrix $\tilde{\mathbf{f}}_t$, which is the inverse of $\mathbf{k}_t$. It can be inverted to set up a reduced free-vibration equation in the form of Eq. (14-70), or, preferably, the flexibility form of the free-vibration equation equivalent to Eq. (14-1) may be used; for the reduced system, this would be written

$$\frac{1}{\omega^2} \hat{\mathbf{v}}_t = \tilde{\mathbf{f}}_t \mathbf{m}_{tt} \hat{\mathbf{v}}_t \qquad (14\text{-}71)$$

Rayleigh Method in Discrete-Coordinate Systems

The reduction of degrees of freedom by static condensation can be very effective in certain classes of structures; however, both the applicability of the method and the extent of the possible reduction are limited. On the other hand, reduction techniques based on the use of generalized coordinates representing the amplitudes of assumed displacement patterns are applicable to any type of structural system, and they can achieve any desired degree of reduction. Consequently, a number of reduction techniques based on this concept have been used in practical structural dynamics. The essence of all of these techniques is Rayleigh's method or the generalization known as the Rayleigh-Ritz method. The application of Rayleigh's method to discrete-coordinate systems is described in this section, and the Ritz extensions are presented in the following section.

Rayleigh's method, which was applied previously in establishing an approximate

SDOF system from arbitrary structural configurations, is equally applicable if the properties of the structure are expressed in terms of discrete-coordinate matrices. To apply the method, it is necessary to express the displacement of the structure in terms of an assumed shape and a generalized-coordinate amplitude. In matrix notation, the assumed free-vibration displacements may be expressed [compare with Eq. (9-5)]

$$\mathbf{v}(t) = \boldsymbol{\psi} Z(t) = \boldsymbol{\psi} Z_0 \sin \omega t \qquad (14\text{-}72a)$$

in which $\boldsymbol{\psi}$ is the assumed shape vector and $Z(t)$ is the generalized coordinate expressing its amplitude. The velocity vector in free vibrations then is

$$\dot{\mathbf{v}}(t) = \boldsymbol{\psi} \omega Z_0 \cos \omega t \qquad (14\text{-}72b)$$

In matrix form, the maximum kinetic energy of the structure is given by

$$T_{max} = {}^{1}\!/_{2} \dot{\mathbf{v}}_{max}^{T} \mathbf{m} \dot{\mathbf{v}}_{max} \qquad (14\text{-}73a)$$

and the maximum potential energy by

$$V_{max} = {}^{1}\!/_{2} \mathbf{v}_{max}^{T} \mathbf{k} \mathbf{v}_{max} \qquad (14\text{-}73b)$$

When the maximum displacement and velocity, obtained from Eqs. (14-72), are substituted, these are written

$$T_{max} = {}^{1}\!/_{2} Z_0^{2} \omega^{2} \boldsymbol{\psi}^{T} \mathbf{m} \boldsymbol{\psi} \qquad (14\text{-}74a)$$

$$V_{max} = {}^{1}\!/_{2} Z_0^{2} \boldsymbol{\psi}^{T} \mathbf{k} \boldsymbol{\psi} \qquad (14\text{-}74b)$$

Then the frequency can be obtained by equating the maximum potential- and kinetic-energy expressions, according to the Rayleigh principle, so that

$$\omega^{2} = \frac{\boldsymbol{\psi}^{T} \mathbf{k} \boldsymbol{\psi}}{\boldsymbol{\psi}^{T} \mathbf{m} \boldsymbol{\psi}} \equiv \frac{k^{*}}{m^{*}} \qquad (14\text{-}75)$$

It should be noted that Eq. (14-75) is merely the matrix equivalent of Eq. (9-10).

The improved Rayleigh method of Eqs. (9-20) or (9-32) can also be developed in matrix form. If the initial displacement assumption is designated

$$\mathbf{v}^{(0)} = \boldsymbol{\psi} Z \qquad (14\text{-}76)$$

then the inertia forces developed in free vibrations will be [from Eq. (12-33)]

$$\mathbf{f}_I = \omega^{2} \mathbf{m} \mathbf{v}^{(0)} = \omega^{2} \mathbf{m} \boldsymbol{\psi} Z \qquad (14\text{-}77)$$

and the deflections produced by these inertia forces are

$$\mathbf{v}^{(1)} = \tilde{\mathbf{f}} \mathbf{f}_I = \omega^{2} \tilde{\mathbf{f}} \mathbf{m} \boldsymbol{\psi} Z \qquad (14\text{-}78)$$

which is a better approximation of the first-mode shape, as noted in the discussion of the Stodola method. Thus if this derived shape is used in the Rayleigh method, it will produce a better result than the initial assumption would. The result of introducing Eq. (14-78) into Eqs. (14-73) and equating them is

$$\omega^2 = \frac{\psi^T \mathbf{m}\tilde{\mathbf{f}}\mathbf{m}\psi}{\psi^T \mathbf{m}\tilde{\mathbf{f}}\mathbf{m}\tilde{\mathbf{f}}\mathbf{m}\psi} \qquad (14\text{-}79)$$

which is the improved Rayleigh method expression (method R_{11}). By comparing Eq. (14-79) with Eq. (14-10) it can be seen that the frequency obtained from the improved Rayleigh procedure is identical to that given by a single-step Stodola analysis using the mass as a weighting factor in the averaging process.

Rayleigh-Ritz Method

Although the Rayleigh method can provide a satisfactory approximation of the first mode of vibration in many structures, it frequently is necessary to include more than one mode in a dynamic analysis to give adequate accuracy in the results. The Ritz extension of the Rayleigh method is one of the most convenient procedures for evaluating the first several modes of vibration. The basic assumption of the Ritz method is that the displacement vector can be expressed in terms of a set of assumed shapes $\boldsymbol{\Psi}$ of amplitude $\mathbf{Z}$ as follows:

$$\mathbf{v} = \psi_1 Z_1 + \psi_2 Z_2 + \psi_3 Z_3 + \cdots$$

or
$$\mathbf{v} = \boldsymbol{\Psi}\mathbf{Z} \qquad (14\text{-}80)$$

in which the generalized-coordinate amplitudes $\mathbf{Z}$ are as yet unknown. To obtain the best results from the least possible number of coordinates, each of the vectors ψ_n should be taken as an approximation of the corresponding true vibration mode shape ϕ_n. However, many other schemes have been proposed for selecting the trial vectors. For example, the static-condensation process can be looked upon as a means for defining a set of Ritz shapes. The fact that a specified set of elastic forces is to be set to zero constitutes a constraint which makes it possible to express the corresponding set of displacements in terms of all the others. This type of relationship is given by Eq. (11-45) or, in the notation of Eq. (14-69), by

$$\hat{\mathbf{v}}_0 = -\mathbf{k}_{00}^{-1}\mathbf{k}_{0t}\hat{\mathbf{v}}_t$$

Hence, the complete displacement vector can also be expressed in terms of the non-zero-force degrees of freedom merely by incorporating an identity matrix of appropriate dimensions into the transformation:

$$\hat{\mathbf{v}} = \begin{bmatrix} \hat{\mathbf{v}}_0 \\ \hat{\mathbf{v}}_t \end{bmatrix} = \begin{bmatrix} -\mathbf{k}_{00}^{-1}\mathbf{k}_{0t} \\ \mathbf{I} \end{bmatrix} \hat{\mathbf{v}}_t$$

Here the second matrix in square brackets clearly is equivalent to the assumed shapes

Ψ of Eq. (14-80), and the vector $\hat{\mathbf{v}}_t$ represents the generalized coordinates $\mathbf{Z}$. As many trial vectors as desired may be used in the Ritz analysis. In general, it may be advisable to use as many as s assumed shapes Ψ if it is desired to obtain $s/2$ vibration mode shapes and frequencies with good accuracy.

Expressions for the maximum kinetic and potential energy in the system can be obtained by introducing Eq. (14-80) into Eqs. (14-73), giving

$$T_{\max} = \tfrac{1}{2}\omega^2\mathbf{Z}^T\mathbf{\Psi}^T\mathbf{m}\mathbf{\Psi}\mathbf{Z} \qquad (14\text{-}81a)$$

$$V_{\max} = \tfrac{1}{2}\mathbf{Z}^T\mathbf{\Psi}^T\mathbf{k}\mathbf{\Psi}\mathbf{Z} \qquad (14\text{-}81b)$$

Equating these then leads to the frequency expression

$$\omega^2 = \frac{\mathbf{Z}^T\mathbf{\Psi}^T\mathbf{k}\mathbf{\Psi}\mathbf{Z}}{\mathbf{Z}^T\mathbf{\Psi}^T\mathbf{m}\mathbf{\Psi}\mathbf{Z}} \equiv \frac{\tilde{\mathbf{k}}(Z)}{\tilde{\mathbf{m}}(Z)} \qquad (14\text{-}82)$$

Equation (14-82) is not an explicit expression for the frequency of vibration, of course; both the numerator and denominator are functions of the generalized-coordinate amplitudes $\mathbf{Z}$, which are not yet known. To evaluate these, the fact that the Rayleigh analysis provides an upper bound to the vibration frequency will be utilized; in other words, any assumed shape leads to a calculated frequency which is higher than the true frequency, and so the best approximation of the shape, that is, the best choice of $\mathbf{Z}$, will minimize the frequency.

Thus differentiating the frequency expression with respect to any one of the generalized coordinates Z_n and equating to zero gives

$$\frac{\partial\omega^2}{\partial Z_n} = \frac{\tilde{\mathbf{m}}(\partial\tilde{\mathbf{k}}/\partial Z_n) - \tilde{\mathbf{k}}(\partial\tilde{\mathbf{m}}/\partial Z_n)}{\tilde{\mathbf{m}}^2} = 0 \qquad (14\text{-}83)$$

But from Eq. (14-82), $\tilde{\mathbf{k}} = \omega^2\tilde{\mathbf{m}}$; thus Eq. (14-83) leads to

$$\frac{\partial\tilde{\mathbf{k}}}{\partial Z_n} - \omega^2\frac{\partial\tilde{\mathbf{m}}}{\partial Z_n} = 0 \qquad (14\text{-}84)$$

Now from the definitions given in Eq. (14-82)

$$\frac{\partial\tilde{\mathbf{k}}}{\partial Z_n} = 2\mathbf{Z}^T\psi^T\mathbf{k}\psi\frac{\partial}{\partial Z_n}(\mathbf{Z}) = 2\mathbf{Z}^T\mathbf{\Psi}^T\mathbf{k}\psi_n \qquad (14\text{-}85a)$$

and similarly

$$\frac{\partial\tilde{\mathbf{m}}}{\partial Z_n} = 2\mathbf{Z}^T\mathbf{\Psi}^T\mathbf{m}\psi_n \qquad (14\text{-}85b)$$

Substituting Eqs. (14-85) into Eq. (14-84) and transposing gives

$$\psi_n{}^T\mathbf{k}\mathbf{\Psi}\mathbf{Z} - \omega^2\psi_n{}^T\mathbf{m}\mathbf{\Psi}\mathbf{Z} = 0 \qquad (14\text{-}86)$$

Minimizing the frequency successively with respect to each of the generalized coordinates leads to an equation like Eq. (14-86) for each of the shape vectors ψ_n; thus the entire set of equations may be expressed as

$$\mathbf{\Psi}^T \mathbf{k} \mathbf{\Psi} \mathbf{Z} - \omega^2 \mathbf{\Psi}^T \mathbf{m} \mathbf{\Psi} \mathbf{Z} = 0$$

With the notation

$$\mathbf{k}^* = \mathbf{\Psi}^T \mathbf{k} \mathbf{\Psi} \qquad (14\text{-}87a)$$

$$\mathbf{m}^* = \mathbf{\Psi}^T \mathbf{m} \mathbf{\Psi} \qquad (14\text{-}87b)$$

this becomes

$$(\mathbf{k}^* - \omega^2 \mathbf{m}^*)\hat{\mathbf{Z}} = 0 \qquad (14\text{-}88)$$

where $\hat{\mathbf{Z}}$ represents each eigenvector (relative values of Z) which satisfies this eigenvalue equation.

Comparing Eq. (14-88) with Eq. (12-4) shows that the Rayleigh-Ritz analysis has the effect of reducing the system from N degrees of freedom, as represented by the geometric coordinates $\mathbf{v}$, to s degrees of freedom representing the number of generalized coordinates $\mathbf{Z}$ and the corresponding assumed shapes. Equation (14-80) is the coordinate transformation, and Eqs. (14-87) are the generalized-mass and stiffness matrices (of dimensions $s \times s$). Each element of these matrices is a generalized-mass or stiffness term; thus

$$k_{mn}^* = \psi_m{}^T \mathbf{k} \psi_n \qquad (14\text{-}89a)$$

$$m_{mn}^* = \psi_m{}^T \mathbf{m} \psi_n \qquad (14\text{-}89b)$$

In general, the assumed shapes ψ_n do not have the orthogonality properties of the true mode shapes; thus the off-diagonal terms do not vanish from these generalized-mass and stiffness matrices; however, a good choice of assumed shapes will tend to make the off-diagonal terms relatively small. In any case, it is much easier to obtain the dynamic response for the reduced number of coordinates s than for the original N equations.

Equation (14-88) can be solved by any standard eigenvalue-equation solution procedure, including the determinantal equation approach discussed earlier for systems having only a few generalized coordinates $\mathbf{Z}$. The frequency vector ω so obtained represents approximations to the true frequencies of the lower modes of vibration, the accuracy generally being excellent for the lowest modes ($1 < n < s/2$) and relatively poor in the highest modes. When the mode-shape vectors $\mathbf{Z}_n$ are normalized by dividing by some reference coordinate, they will be designated ϕ_{Zn}, where the subscript Z indicates that they represent the mode shapes expressed in generalized coordinates. The complete set of generalized-coordinate mode shapes can then be denoted $\mathbf{\Phi}_Z$, representing a square $s \times s$ matrix.

The generalized coordinates $\mathbf{Z}$ expressed in terms of the modal amplitudes [by analogy with Eq. (13-2)] are

$$\mathbf{Z} = \mathbf{\Phi}_Z \mathbf{Y} \qquad (14\text{-}90)$$

It is of interest that these mode shapes are orthogonal with respect to the generalized-mass and stiffness matrices:

$$\begin{aligned} \phi_{Zm}\mathbf{m}^*\phi_{Zn} &= 0 \\ \phi_{Zm}\mathbf{k}^*\phi_{Zn} &= 0 \end{aligned} \qquad m \neq n \qquad (14\text{-}91)$$

By introducing Eq. (14-90) into Eq. (14-80) the geometric coordinates can be expressed in terms of the normal modal coordinates

$$\mathbf{v} = \mathbf{\Psi}\mathbf{\Phi}_Z \mathbf{Y} \qquad (14\text{-}92)$$

Thus it is seen that the approximate mode shapes in geometric coordinates are given by the product of the assumed shapes and the generalized-coordinate mode shapes

$$\mathbf{\Phi} = \mathbf{\Psi}\mathbf{\Phi}_Z \qquad (14\text{-}93)$$

which is of dimensions $N \times s$. Substituting Eqs. (14-87) into Eqs. (14-91) and applying Eqs. (14-92) demonstrates that these approximate geometric mode shapes are orthogonal with respect to the mass and stiffness expressed in geometric coordinates. They can therefore be used in the standard mode-superposition dynamic-analysis procedure.

It is important to note that the same type of improvement described above for the Rayleigh method is applicable to the Rayleigh-Ritz procedure. Thus, by analogy with Eq. (14-79), the improved generalized-coordinate stiffness and mass matrices are given by

$$\mathbf{k}^* = \mathbf{\Psi}^T\mathbf{m}\tilde{\mathbf{f}}\mathbf{m}\mathbf{\Psi} \qquad (14\text{-}94a)$$

$$\mathbf{m}^* = \mathbf{\Psi}^T\mathbf{m}\tilde{\mathbf{f}}\mathbf{m}\tilde{\mathbf{f}}\mathbf{m}\mathbf{\Psi} \qquad (14\text{-}94b)$$

in place of Eqs. (14-87). The principal advantage of these equations is that the inertia-force deflections on which they are based provide reasonable assumed shapes from very crude initial assumptions. In large, complex structures it is very difficult to make detailed estimates of the shapes, and it is possible with this improved procedure merely to indicate the general character of each shape. Another major advantage in many analyses is that it avoids use of the stiffness matrix. In fact, if the initial assumed shapes are designated $\mathbf{\Psi}^{(0)}$ and the deflections resulting from inertia forces associated with those shapes are called $\mathbf{\Psi}^{(1)}$, that is,

$$\mathbf{\Psi}^{(1)} = \tilde{\mathbf{f}}\mathbf{m}\mathbf{\Psi}^{(0)} \qquad (14\text{-}95)$$

then Eqs. (14-94) may be written

$$\mathbf{k}^* = (\mathbf{\Psi}^{(1)})^T\mathbf{m}\mathbf{\Psi}^{(0)} \qquad (14\text{-}96a)$$

$$\mathbf{m}^* = (\mathbf{\Psi}^{(1)})^T\mathbf{m}\mathbf{\Psi}^{(1)} \qquad (14\text{-}96b)$$

Consequently it is not necessary to have an explicit expression for the flexibility either; it is necessary only to be able to compute the deflections resulting from a given loading (which in this case is $\mathbf{m}\Psi^{(0)}$).

This improvement process in the Rayleigh-Ritz method may be looked upon as the first cycle of an iterative solution, just as the improved Rayleigh method is equivalent to a single cycle of the Stodola method. However, the Stodola analysis results in only a single mode shape and frequency, whereas the continuation of the Ritz improvement process evaluates simultaneously the reduced set of mode shapes and frequencies. This method, called *simultaneous* or *subspace iteration*, will be described together with other aspects of matrix iteration in Sec. 14-6.

14-6 BASIC CONCEPTS IN MATRIX ITERATION

Eigenproperty Expansion of the Dynamic Matrix

The Stodola and Vianello methods are only simple applications of a general approach to the iterative solution of eigenproblems. Although a full treatment of this subject is beyond the scope of this text, discussion of a few of the basic concepts of matrix iteration in the context of vibration analysis will be helpful. One of the fundamental processes which will prove useful is the expansion of a matrix in terms of its eigenvalues and eigenvectors.

For this purpose, Eq. (14-43) will be written

$$\mathbf{E}\phi_n = \phi_n\lambda_n \qquad (14\text{-}97)$$

where $\lambda_n \equiv \omega_n^2$. It is evident from the determinantal-equation approach to evaluating eigenvalues that the eigenvalues of the transposed matrix are the same as those of the original matrix. However, the eigenvectors of the transpose of an unsymmetrical matrix like $\mathbf{E}$ are different from those of the original. Hence for the transpose $\mathbf{E}^T$, the eigenproblem can be written

$$\mathbf{E}^T\phi_{Ln} = \phi_{Ln}\lambda_n$$

where ϕ_{Ln} is the nth eigenvector of $\mathbf{E}^T$. Transposing this relationship gives

$$\phi_{Ln}{}^T\mathbf{E} = \lambda_n\phi_{Ln}{}^T \qquad (14\text{-}98)$$

Thus the eigenvectors ϕ_{Ln} are frequently called the *left-hand eigenvectors* of $\mathbf{E}$ and ϕ_n the *right-hand eigenvectors*.

The orthogonality property of the left- and right-hand eigenvectors can easily be demonstrated if Eq. (14-97) is premultiplied by the eigenvector $\phi_{Lm}{}^T$

$$\phi_{Lm}{}^T\mathbf{E}\phi_n = \phi_{Lm}{}^T\phi_n\lambda_n \qquad (14\text{-}99)$$

while Eq. (14-98) is written for mode m and postmultiplied by ϕ_n

$$\phi_{Lm}{}^T \mathbf{E} \phi_n = \lambda_m \phi_{Lm}{}^T \phi_n \qquad (14\text{-}100)$$

Subtracting Eq. (14-100) from Eq. (14-99) then gives

$$0 = (\lambda_n - \lambda_m) \phi_{Lm}{}^T \phi_n$$

which represents the orthogonality property

$$\phi_{Lm}{}^T \phi_n = 0 \qquad (\lambda_m \neq \lambda_n) \qquad (14\text{-}101)$$

If the eigenvectors are normalized to satisfy the condition $\phi_{Ln}{}^T \phi_n = 1$ (note that this does not fix the amplitude of ϕ_{Ln} or ϕ_n individually, only their product), and if the square matrices of all right- and left-hand eigenvectors are designated $\mathbf{\Phi}$ and $\mathbf{\Phi}_L$, respectively, it is evident from the normalizing and orthogonality conditions that

$$\mathbf{\Phi}_L{}^T \mathbf{\Phi} = \mathbf{I} \qquad (14\text{-}102a)$$

Hence the transpose of the left-hand eigenvectors is the inverse of the right-hand eigenvectors

$$\mathbf{\Phi}_L{}^T = \mathbf{\Phi}^{-1} \qquad (14\text{-}102b)$$

The expansion of $\mathbf{E}$ can now be demonstrated by writing the eigenproblem expression of Eq. (14-97) for the full set of eigenvectors and eigenvalues:

$$\mathbf{E}\mathbf{\Phi} = \mathbf{\Phi}\mathbf{\Lambda} \qquad (14\text{-}103)$$

in which $\mathbf{\Lambda}$ is the diagonal matrix of eigenvalues. Premultiplying Eq. (14-103) by $\mathbf{\Phi}_L{}^T$ and invoking Eq. (14-102b) leads to an expression for the eigenvalues:

$$\mathbf{\Phi}_L{}^T \mathbf{E}\mathbf{\Phi} = \mathbf{\Lambda} \qquad (14\text{-}104)$$

Alternatively, $\mathbf{E}$ can be expressed in terms of the eigenvalues and eigenvectors by premultiplying Eq. (14-104) by $\mathbf{\Phi}$, postmultiplying it by $\mathbf{\Phi}_L{}^T$, and invoking Eq. (14-102b):

$$\mathbf{E} = \mathbf{\Phi}\mathbf{\Lambda}\mathbf{\Phi}_L{}^T \qquad (14\text{-}105)$$

This result also can be expressed as the sum of the modal contributions:

$$\mathbf{E} = \sum_{n=1}^{N} \lambda_n \phi_n \phi_{Ln}{}^T \qquad (14\text{-}105a)$$

Furthermore, the square of matrix $\mathbf{E}$ is

$$\mathbf{E}^2 = \mathbf{\Phi}\mathbf{\Lambda}\mathbf{\Phi}_L{}^T \mathbf{\Phi}\mathbf{\Lambda}\mathbf{\Phi}_L{}^T = \mathbf{\Phi}\mathbf{\Lambda}^2 \mathbf{\Phi}_L{}^T \qquad (14\text{-}106)$$

and by continued multiplication, the sth power of $\mathbf{E}$ is

$$\mathbf{E}^s = \mathbf{\Phi}\mathbf{\Lambda}^s \mathbf{\Phi}_L{}^T \qquad (14\text{-}107)$$

It must be remembered that the expansion of Eq. (14-107) is based on the type of eigenvectors normalizing which has been used ($\Phi_L{}^T\Phi = I$). A specific expression for the left-hand eigenvectors can be obtained if an additional normalizing condition is introduced. For example, if Eq. (14-103) is premultiplied by $\Phi^T m$ (note that $E = m^{-1}k$), it becomes

$$\Phi^T k \Phi = \Phi^T m \Phi \Lambda \qquad (14\text{-}108)$$

Now if the right-hand eigenvectors are normalized so that

$$\Phi^T m \Phi = I \qquad (14\text{-}109)$$

it is apparent from comparison of the transpose of Eq. (14-102a) with Eqs. (14-109) and (14-108) that

$$\Phi_L = m\Phi = k\Phi\Lambda^{-1} \qquad (14\text{-}110)$$

Iterative Solution of the Eigenproblem

Equation (14-107) provides the basis for demonstrating the general matrix-iteration (or power) method for evaluating the eigenvalues and eigenvectors of the dynamic matrix E. To carry out the so-called *direct analysis*, an assumption must be made of the shape of the highest-mode, that is, highest-frequency, eigenvector, which will be designated $v_N{}^{(0)}$. This trial vector can be expressed as the sum of its true mode-shape components

$$v_N{}^{(0)} = \sum_{n=1}^{N} \phi_n Y_n = \Phi Y \qquad (14\text{-}111)$$

[see Eq. (14-13)]. The iteration procedure is initiated by calculating an improved vector $v_N{}^{(1)}$ in two steps. First an improved *shape* is obtained by the matrix multiplication

$$\bar{v}_N{}^{(1)} = E v_N{}^{(0)}$$

and then its amplitude is normalized by dividing by its largest term

$$v_N{}^{(1)} = \frac{\bar{v}_N{}^{(1)}}{\max{(\bar{v}_N{}^{(1)})}} = \frac{E v_N{}^{(0)}}{\max{(E v_N{}^{(0)})}} \qquad (14\text{-}112)$$

Similarly, the result of the next iteration cycle is given by

$$v_N{}^{(2)} = \frac{E v_N{}^{(1)}}{\max{(E v_N{}^{(1)})}} = \frac{E^2 v_N{}^{(0)}}{\max{(E^2 v_N{}^{(0)})}} \qquad (14\text{-}113)$$

and after s cycles, the result is

$$v_N{}^{(s)} = \frac{E^s v_N{}^{(0)}}{\max{(E^s v_N{}^{(0)})}} \qquad (14\text{-}114)$$

Upon substituting from Eqs. (14-107) and (14-111), Eq. (14-114) becomes

$$\mathbf{v}_N{}^{(s)} = \frac{\mathbf{\Phi\Lambda}^s\mathbf{\Phi}_L{}^T\mathbf{\Phi Y}}{\max{(\mathbf{E}^s\mathbf{v}_N{}^{(0)})}} = \frac{\mathbf{\Phi\Lambda}^s\mathbf{Y}}{\max{(\mathbf{E}^s\mathbf{v}_N{}^{(0)})}} = \frac{\sum_{n=1}^{N}\lambda_n{}^s\phi_nY_n}{\max{(\mathbf{E}^s\mathbf{v}_N{}^{(0)})}}$$

But this summation can also be written in the form

$$\mathbf{v}_N{}^{(s)} = \frac{\lambda_N{}^s}{\max{(\mathbf{E}^s\mathbf{v}_N{}^{(0)})}}\left[\phi_NY_N + \sum_{n=1}^{N-1}\left(\frac{\lambda_n}{\lambda_N}\right)^s\phi_nY_n\right] \quad (14\text{-}115)$$

and since $\lambda_N > \lambda_{N-1} > \lambda_{N-2}\cdots$, by definition, each of the terms in the remaining summation will become negligibly small when the iteration has been carried out for enough cycles. Thus the computed shape must converge finally to

$$\mathbf{v}_N{}^{(s)} = \frac{\lambda_N{}^s\phi_NY_N}{\max{(\lambda_N{}^s\phi_NY_N)}} = \frac{\phi_N}{\max{(\phi_N)}} \equiv \phi_N \quad (14\text{-}116)$$

where it will be noted that the mode shape is normalized so that its largest term is unity. Moreover, if the iteration is continued for one more step,

$$\bar{\mathbf{v}}_N^{(s+1)} = \mathbf{E}\mathbf{v}_N{}^{(s)} = \mathbf{\Phi\Lambda\Phi}_L{}^T\frac{\phi_N}{\max{(\phi_N)}} = \lambda_N\frac{\phi_N}{\max{(\phi_N)}}$$

it is evident the eigenvalue for the highest mode is given by the largest term of the eigenvector before normalization

$$\lambda_N = \max{(\bar{\mathbf{v}}_N^{(s+1)})} \quad (14\text{-}117)$$

This discussion has demonstrated the convergence of the direct-iteration procedure to the highest mode shape and frequency. In the equivalent process which converges to the lowest mode, called *inverse iteration*, the initial vector is an estimate of the lowest mode shape $\mathbf{v}_1{}^{(0)}$, and the iteration involves multiplication of the trial vector by the inverse of $\mathbf{E}$. The result of the first cycle, including normalization, is

$$\mathbf{v}_1{}^{(1)} = \frac{\mathbf{E}^{-1}\mathbf{v}_1{}^{(0)}}{\max{(\mathbf{E}^{-1}\mathbf{v}_1{}^{(0)})}} \quad (14\text{-}118)$$

and after s cycles, by analogy with the direct-iteration procedure [Eq. (14-115)], the displacement may be expressed as

$$\mathbf{v}_1{}^{(s)} = \frac{\lambda_1{}^{-s}}{\max{(\mathbf{E}^{-s}\mathbf{v}_1{}^{(0)})}}\left[\phi_1Y_1 + \sum_{n=2}^{N}\left(\frac{\lambda_1}{\lambda_n}\right)^s\phi_nY_n\right] \quad (14\text{-}119)$$

But since $\lambda_1 < \lambda_2 < \lambda_3\cdots$, by definition, each of the terms in the summation becomes negligibly small after sufficient cycles and the computed mode shape converges to

$$\mathbf{v}_1{}^{(s)} = \frac{\phi_1}{\max{(\phi_1)}} = \phi_1 \quad (14\text{-}120)$$

Also by analogy with the preceding discussion, it is evident that the corresponding eigenvalue is given by

$$\lambda_1 = \frac{1}{\max(\bar{\mathbf{v}}_1^{(s+1)})} \quad (14\text{-}121)$$

A final comment on this presentation is that *inverse* iteration with the matrix $\mathbf{E} = \mathbf{m}^{-1}\mathbf{k}$ is the same as *direct* iteration with the matrix $\mathbf{E}^{-1} = \mathbf{D} = \tilde{\mathbf{f}}\mathbf{m}$ and that this direct iteration will converge to the lowest frequency (as noted earlier with the Stodola method). However, in the analysis of complex structural systems, it frequently is more convenient to work with the stiffness formulation $\mathbf{E}$ than with the flexibility formulation $\mathbf{D}$, and it is for this situation that inverse iteration is most effective, as will be described later.

Iteration with Shifts

The iteration procedures described above are efficient means of evaluating the lowest- and highest-mode vibration properties of a structural system. In addition, they can be forced to converge to the next lowest (or next highest) mode if any contribution from the lowest (or highest) mode is swept out of the trial vector, as explained earlier. However, the formulation of the necessary sweeping matrices involves considerable computational effort, and an alternative procedure based on shifting the eigenvalues has proved effective in practice. Although shifting can be employed with either direct or inverse iteration, it is most effective with the inverse procedure and will be discussed here in that context.

The essential concept of shifting is the representation of each eigenvalue λ_n as the sum of a shift μ plus a residual δ_n; thus

$$\lambda_n = \delta_n + \mu \quad (14\text{-}122)$$

or, considering the entire diagonal matrix of eigenvalues,

$$\mathbf{\Lambda} = \hat{\boldsymbol{\delta}} + \mu\mathbf{I} \quad (14\text{-}123)$$

in which $\hat{\boldsymbol{\delta}}$ is the diagonal matrix of residuals and it will be noted that the same shift is applied to each eigenvalue.

The shift can be visualized as a displacement of the origin in a plot of the eigenvalues, as shown in Fig. 14-7. Its effect is to transform the eigenvalue problem to the analysis of the residuals rather than the actual eigenvalues, as is evident if Eq. (14-123) is substituted into Eq. (14-103):

$$\mathbf{E}\mathbf{\Phi} = \mathbf{\Phi}[\hat{\boldsymbol{\delta}} + \mu\mathbf{I}]$$

which can be rewritten

$$[\mathbf{E} - \mu\mathbf{I}]\mathbf{\Phi} = \mathbf{\Phi}\hat{\boldsymbol{\delta}} \quad (14\text{-}124)$$

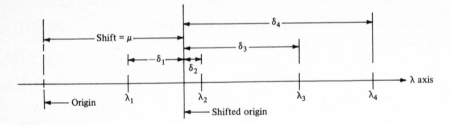

FIGURE 14-7
Demonstration of a shift on the eigenvalue axis.

Here the term in brackets represents a modified matrix to which the residual eigenvalues $\hat{\boldsymbol{\delta}}$ apply, and it will be denoted as $\hat{\mathbf{E}}$ for convenience; thus

$$\hat{\mathbf{E}}\boldsymbol{\Phi} = \boldsymbol{\Phi}\hat{\boldsymbol{\delta}} \qquad (14\text{-}125)$$

It is apparent that Eq. (14-125) is entirely equivalent to Eq. (14-103) and that the shifted matrix $\hat{\mathbf{E}}$ has the same eigenvectors as $\mathbf{E}$.

The solution of this new eigenproblem will be carried out by inverse iteration. By analogy with Eq. (14-118), the first step of the procedure can be expressed as

$$\mathbf{v}_m^{(1)} = \frac{\hat{\mathbf{E}}^{-1}\mathbf{v}_m^{(0)}}{\max \left(\hat{\mathbf{E}}^{-1}\mathbf{v}_m^{(0)}\right)}$$

where $\mathbf{v}_m^{(0)}$ is an initial approximation of the mth mode shape. After s cycles, the result is

$$\mathbf{v}_m^{(s)} = \frac{\hat{\mathbf{E}}^{-s}\mathbf{v}_m^{(0)}}{\max \left(\hat{\mathbf{E}}^{-s}\mathbf{v}_m^{(0)}\right)} = \frac{\displaystyle\sum_{n=1}^{N} \delta_n^{-s}\phi_n Y_n}{\max \left(\hat{\mathbf{E}}^{-s}\mathbf{v}_m^{(0)}\right)}$$

which may be rewritten

$$\mathbf{v}_m^{(s)} = \frac{\delta_m^{-s}}{\max \left(\hat{\mathbf{E}}^{-s}\mathbf{v}_m^{(0)}\right)} \left[\phi_m Y_m + \sum_{n=1}^{m-1} \left(\frac{\delta_m}{\delta_n}\right)^s \phi_n Y_n + \sum_{n=m+1}^{N} \left(\frac{\delta_m}{\delta_n}\right)^s \phi_n Y_n\right] \qquad (14\text{-}126)$$

where δ_m represents the smallest residual eigenvector, that is,

$$|\delta_m| < \delta_{m+1} < \delta_{m+2} \cdots \qquad \text{and} \qquad |\delta_m| < -\delta_{m-1} < -\delta_{m-2} \cdots$$

Thus it is clear that the two summations in Eq. (14-126) will become negligibly small after a sufficient number of iteration cycles, and the computed mode shape converges to

$$\mathbf{v}_m^{(s)} = \frac{\delta_m^{-s}\phi_m Y_m}{\max \left(\delta_m^{-s}\phi_m Y_m\right)} = \frac{\phi_m}{\max \left(\phi_m\right)} = \phi_m \qquad (14\text{-}127)$$

This analysis therefore shows that the process of inverse iteration with eigenvalue shift converges to the mode shape for which the eigenvalue is closest to the shift position; e.g., it would converge to the second mode for the case in Fig. 14-7. By analogy with Eq. (14-121) it may be seen that the residual eigenvalue for this mode is given by the maximum term in the computed eigenvector, before normalization:

$$\delta_m = \frac{1}{\max (\bar{\mathbf{v}}_m^{(s+1)})}$$

Hence the actual eigenvalue is obtained by adding the shift to this

$$\lambda_m = \mu + \frac{1}{\max (\bar{\mathbf{v}}_m^{(s+1)})} \qquad (14\text{-}128)$$

By appropriate selection of the shift points, this inverse-iteration process can be made to converge to any or all modes of the structural system. Moreover, since the speed of convergence can be accelerated by shifting to a point very close to the root which is sought, it is good practice to shift at intervals during iteration, as better approximations of the root are obtained. A good formula for approximating the shift point can be derived from Eq. (14-10):

$$\mu_m = \frac{\bar{\mathbf{v}}_m^{(s)}\mathbf{m}\mathbf{v}_m^{(s-1)}}{\bar{\mathbf{v}}_m^{(s)}\mathbf{m}\bar{\mathbf{v}}_m^{(s)}} \qquad (14\text{-}129)$$

It is evident that the shifting process is less effective with direct iteration, which converges to the highest root, because only the first or last residuals (δ_1 or δ_N) can be made largest by shifting. On the other hand, with inverse iteration the residual for any mode can be made smallest by a judicious selection of the shift.

A slight modification of this inverse-iteration procedure may prove to be computationally advantageous for certain types of structural systems. Consider the equation of free vibrations written in the form of Eq. (12-33):

$$\mathbf{k}\phi_n = \omega_n^2 \mathbf{m}\phi_n$$

Introducing an eigenvalue shift $\mu = \omega_n^2 - \delta_n$ leads to

$$\mathbf{k}\phi_n = (\mu + \delta_n)\mathbf{m}\phi_n$$

which may be rewritten

$$(\mathbf{k} - \mu\mathbf{m})\phi_n = \delta_n\mathbf{m}\phi_n \qquad (14\text{-}130)$$

If this were premultiplied by $\mathbf{m}^{-1}$, the result would be the same as Eq. (14-124), because $\mathbf{E} = \mathbf{m}^{-1}\mathbf{k}$. However, if both $\mathbf{m}$ and $\mathbf{k}$ are narrowly banded matrices (as is frequently the case with finite-element idealizations), less computation effort may be required if the iteration is performed directly with Eq. (14-130). (Note that the product matrix $\mathbf{E} = \mathbf{m}^{-1}\mathbf{k}$ will be quite fully populated even though $\mathbf{m}$ and $\mathbf{k}$ are banded.)

The iteration based on Eq. (14-130) is initiated by assuming a trial vector $v_n^{(0)}$ and multiplying it by the mass matrix to obtain the trial product vector $w_n^{(0)} \equiv mv_n^{(0)}$. Thus the iterative form of Eq. (14-130) becomes

$$\hat{k}\bar{v}_n^{(1)} = mv_n^{(0)} = w_n^{(0)} \quad (14\text{-}131)$$

in which $\hat{k} = k - \mu m$ is the shifted stiffness matrix. Finally, this equation is solved for the improved modal vector, with a result which may be expressed [by analogy with Eq. (14-118)] as

$$v_n^{(1)} = \frac{\hat{k}^{-1}w_n^{(0)}}{\max\left(\hat{k}^{-1}w_n^{(0)}\right)} \quad (14\text{-}132)$$

In the solution of Eq. (14-132), full advantage should be taken of the banded property of $\hat{k}$ (which will have the same bandwidth as k in normal circumstances). Instead of performing a direct inversion of $\hat{k}$, as implied by the equation, the solution can be obtained more efficiently by a Choleski decomposition. Thus the shifted stiffness matrix is decomposed to the form

$$\hat{k} = LL^T \quad (14\text{-}133)$$

where L is a lower triangular matrix. The solution for the improved but unscaled vector is then carried out in two steps:

1 Solve for $y_n^{(0)}$ in $Ly_n^{(0)} = w_n^{(0)}$ (forward reduction)
2 Solve for $\bar{v}_n^{(1)}$ in $L^T\bar{v}_n^{(1)} = y_n^{(0)}$ (back substitution)

in which each equation solution can easily be carried out because of the triangular form of L.

Since inverse iteration in the form of Eq. (14-132) requires an additional matrix multiplication for each iteration cycle [as compared with Eq. (14-118)] in obtaining $w_n = mv_n$, this version of the method will be advantageous only if the stiffness matrix $\hat{k}$ is narrowly banded, so that the extra efficiency of its solution will compensate for this extra matrix operation.

Subspace Iteration

In the discussion of matrix iteration so far, it has been assumed that only a single modal vector is considered at a time. However, the procedure can be expanded to deal with any desired number of vectors simultaneously, and in this form it may be looked upon as an extension of the improvement process for the Rayleigh-Ritz method. Thus it will be convenient to employ the notation of the Ritz procedure in this discussion.

In order to obtain accurately a system of p mode shapes and frequencies, it is desirable to start with a somewhat larger number q of trial vectors. With these trial vectors denoted by superscript (0), the displacements may be expressed as configurations of these shapes [see Eq. (14-80)] as

$$\mathbf{v}^{(0)} = \mathbf{\Psi}^{(0)}\mathbf{Z}^{(0)} = \mathbf{\Psi}^{(0)} \qquad (14\text{-}134)$$

in which the initial generalized-coordinate matrix is merely an identity matrix (indicating that the trial vectors are the assumed Ritz shapes $\mathbf{\Psi}^{(0)}$).

For the large systems to which this method is usually applied, it is important to take advantage of the banding properties of the mass and stiffness matrices; hence the free-vibration equation is written in the form of Eq. (12-33), which can be written for the set of p eigenvalues and eigenvectors as

$$\mathbf{k}\mathbf{\Phi} = \mathbf{m}\mathbf{\Phi}\mathbf{\Lambda} \qquad (14\text{-}135)$$

Introducing the q trial vectors on the right side of this equation leads to

$$\mathbf{k}\overline{\mathbf{\Psi}}^{(1)} = \mathbf{m}\mathbf{\Psi}^{(0)} \equiv \mathbf{w}^{(0)} \qquad (14\text{-}136)$$

which is equivalent to Eq. (14-133) written for multiple vectors and with no shift. The unscaled improved shapes are obtained by solving Eq. (14-136); thus

$$\overline{\mathbf{\Psi}}^{(1)} = \mathbf{k}^{-1}\mathbf{w}^{(0)} \qquad (14\text{-}137)$$

and, as explained above, it will be more efficient to use the Choleski decomposition of $\mathbf{k}$ [Eq. (14-133)] rather than its inverse in obtaining the solution.

Before the improved shapes of Eq. (14-137) can be used in a new iteration cycle, they must be modified in two ways, that is, normalized to maintain reasonable number sizes in the calculations and orthogonalized so that each vector will converge toward a different mode (rather than all toward the lowest mode). These operations can be performed in many different ways, but it is convenient to accomplish both at once by carrying out a Ritz eigenproblem analysis. Thus the first-cycle generalized-coordinate stiffness and mass matrices are computed [see Eqs. (14-87)] as follows:

$$\mathbf{k}_1^* = \overline{\mathbf{\Psi}}^{(1)T}\mathbf{k}\overline{\mathbf{\Psi}}^{(1)} \equiv \overline{\mathbf{\Psi}}^{(1)T}\mathbf{m}\mathbf{\Psi}^{(0)}$$

$$\mathbf{m}_1^* = \overline{\mathbf{\Psi}}^{(1)T}\mathbf{m}\overline{\mathbf{\Psi}}^{(1)} \qquad (14\text{-}138)$$

in which the subscripts identify the first-cycle values, and then the corresponding eigenproblem

$$\mathbf{k}_1^*\hat{\mathbf{Z}}^{(1)} = \mathbf{m}_1^*\hat{\mathbf{Z}}^{(1)}\mathbf{\Omega}_1{}^2 \qquad (14\text{-}139)$$

is solved for the generalized-coordinate mode shapes $\hat{\mathbf{Z}}^{(1)}$ and frequencies $\mathbf{\Omega}_1{}^2$. Any suitable eigenproblem-analysis procedure may be used in the solution of Eq. (14-139), but since it is a much smaller equation system than the original eigenproblem, that is, $q \ll N$, it can often be done by a standard computer-center library program. Usually it is convenient to normalize the generalized-coordinate modal vector so that

$$\hat{\mathbf{Z}}^{(1)T}\mathbf{m}_1^*\hat{\mathbf{Z}}^{(1)} = \mathbf{I}$$

When the normalized generalized-coordinate vectors are used, the improved trial vectors are given by

$$\mathbf{v}^{(1)} = \mathbf{\Psi}^{(1)} = \overline{\mathbf{\Psi}}^{(1)}\hat{\mathbf{Z}}^{(1)} \qquad (14\text{-}140)$$

The entire process can now be repeated iteratively, solving for the unscaled improved shapes $\overline{\mathbf{\Psi}}^{(2)}$, as indicated by Eq. (14-137), and then solving the corresponding Ritz eigenproblem [Eq. (14-139)] to provide for scaling and orthogonalization:

$$\mathbf{\Psi}^{(2)} = \overline{\mathbf{\Psi}}^{(2)}\hat{\mathbf{Z}}^{(2)}$$

and so on. Eventually the process will converge to the true mode shapes and frequencies, that is,

$$\begin{aligned}\mathbf{\Psi}^{(s)} &\to \mathbf{\Phi} \\ \mathbf{\Omega}_s{}^2 &\to \mathbf{\Lambda}\end{aligned} \qquad \text{as } s \to \infty \qquad (14\text{-}141)$$

In general, the lower modes converge most quickly, and the process is continued only until the desired p modes are obtained with the necessary accuracy. The additional $q - p$ trial vectors are included because they accelerate the convergence process, but obviously they require additional computational effort in each cycle, so that a reasonable balance must be maintained between the number of vectors used and the number of cycles required for convergence. By experience it has been found that a suitable choice is given by the smaller of $q = 2p$ and $q = p + 8$.

This subspace, or simultaneous-iteration, procedure has proved to be one of the most efficient methods for solving large-scale structural-vibration problems where probably no more than 40 modes are required in systems having many hundreds to a few thousand degrees of freedom. Although this may be considered as a Rayleigh-Ritz coordinate-reduction scheme, it has the great advantage that the resulting modal coordinates can be obtained to any desired degree of precision. All other coordinate-reduction procedures involve approximations which make the accuracy of the final results uncertain; hence subspace iteration is strongly recommended for practical applications.

14-7 SYMMETRIC FORM OF DYNAMIC MATRIX

In the foregoing discussion of inverse-iteration procedures, the equation of motion in free vibrations

$$\mathbf{k}\phi_n = \mathbf{m}\phi_n\lambda_n \qquad (14\text{-}142)$$

has been transformed by premultiplication by $\mathbf{m}^{-1}$ to obtain the eigenproblem [Eq. (14-97)]

$$\mathbf{E}\phi_n = \phi_n\lambda_n$$

and a number of efficient techniques for solving this eigenproblem have been discussed. However, it should be noted that the matrix $\mathbf{E} = \mathbf{m}^{-1}\mathbf{k}$ is unsymmetric, even though both $\mathbf{m}$ and $\mathbf{k}$ are symmetric, and thus this problem cannot be solved by many efficient standard solution procedures developed to take advantage of symmetry in the eigenproblem, e.g., the Householder method. For this reason it is useful to be able to transform the general vibration eigenproblem [Eq. (14-142)] to the standard symmetric form

$$\mathbf{B}\mathbf{y}_n = \mathbf{y}_n\lambda_n \quad (14\text{-}143)$$

The transformation from general to standard form can be accomplished by manipulation of the mass matrix, and the type of transformation required depends on the form of the mass matrix. Two cases will be considered here: (1) a diagonal mass matrix representing a lumped-mass system and (2) a general (nondiagonal) mass matrix which might result from a consistent finite-element formulation or as the generalized-mass matrix from a Rayleigh-Ritz analysis. In both cases, the transformation matrix which converts Eq. (14-142) into Eq. (14-143) is obtained by decomposing the mass matrix into the product of a matrix and its transpose.

Diagonal Mass Matrix

In this case the transformation matrix is obtained very simply as the square root of the mass matrix because

$$\mathbf{m} = \mathbf{m}^{1/2}\mathbf{m}^{1/2}$$

and the square-root matrix is obtained by merely taking the square root of the diagonal terms (of course the diagonal matrix is unchanged in transposition). The transformation of Eq. (14-142) is performed by expressing its eigenvectors as

$$\boldsymbol{\phi}_n = \mathbf{m}^{-1/2}\mathbf{y}_n \quad (14\text{-}144)$$

where the inverse is formed with the reciprocals of the diagonal terms in $\mathbf{m}^{1/2}$. Substituting Eq. (14-144) into Eq. (14-142) and premultiplying by $\mathbf{m}^{-1/2}$ leads to

$$\mathbf{m}^{-1/2}\mathbf{k}\mathbf{m}^{-1/2}\mathbf{y}_n = \mathbf{y}_n\lambda_n \quad (14\text{-}145)$$

which is of the form of Eq. (14-143) with $\mathbf{B} = \mathbf{m}^{-1/2}\mathbf{k}\mathbf{m}^{-1/2}$. Solving this symmetric eigenvalue problem therefore leads directly to the frequencies of the original equation (14-142); the eigenvectors $\mathbf{y}_n$ of this new eigenproblem must be transformed to obtain the desired vibration mode shapes $\boldsymbol{\phi}_n$, using Eq. (14-144).

It is evident that this transformation procedure cannot be applied if any of the diagonal mass elements is zero. Therefore, it is necessary to eliminate these degrees of freedom from the analysis by static condensation [as indicated by Eq. (14-70)] before performing the transformation to standard symmetric form.

Consistent-Mass Matrix

Two methods are available for the transformation when the mass matrix is banded rather than diagonal. The more reliable of these is based on evaluating the eigenvalues v_n and eigenvectors t_n of the mass matrix from the equation

$$\mathbf{m}t_n = t_n v_n$$

The eigenvectors of this symmetric matrix satisfy the orthogonality condition $t_m{}^T t_n = 0$ (if $m \neq n$); hence if they also are normalized, so that $t_n{}^T t_n = 1$, the complete set of eigenvectors $\mathbf{T}$ is orthonormal:

$$\mathbf{T}^T \mathbf{T} = \mathbf{I} \qquad \mathbf{T}^T = \mathbf{T}^{-1}$$

[This expression corresponds with Eq. (14-102a); note that the left-hand and right-hand eigenvectors are the same for a symmetric matrix.] Finally, the mass matrix can be expressed in terms of these eigenvectors and the set of eigenvalues $\hat{v}$ [by analogy with Eq. (14-105)]:

$$\mathbf{m} = \mathbf{T}\hat{v}\mathbf{T}^T \qquad (14\text{-}146)$$

From Eq. (14-146) it is apparent that the transformation matrix is $\mathbf{T}\hat{v}^{1/2}$, and thus the transformation of Eq. (14-142) is performed by expressing its eigenvectors as

$$\phi_n = \mathbf{T}\hat{v}^{-1/2}\mathbf{y}_n \qquad (14\text{-}147)$$

Substituting this into Eq. (14-142) and premultiplying by $\hat{v}^{-1/2}\mathbf{T}^T$ then leads to

$$(\hat{v}^{-1/2}\mathbf{T}^T \mathbf{k}\mathbf{T}\hat{v}^{-1/2})\mathbf{y}_n = \mathbf{y}_n \lambda_n \qquad (14\text{-}148)$$

where Eq. (14-146) has been used to simplify the right-hand side. Equation (14-148) is of the form of Eq. (14-143) with $\mathbf{B} = \hat{v}^{-1/2}\mathbf{T}^T \mathbf{k}\mathbf{T}\hat{v}^{-1/2}$; hence the solution of this symmetric eigenproblem gives the desired vibration frequencies directly, and the vibration mode shapes are obtained from the eigenvectors $\mathbf{y}_n$ by Eq. (14-147).

Inasmuch as this transformation requires the solution of a preliminary eigenproblem of the order of the original eigenproblem, it is apparent that the use of a vibration-analysis procedure based on the solution of the symmetric eigenproblem Eq. (14-143) will be relatively expensive if the mass matrix is not diagonal. It is possible to obtain a simpler transformation by performing a Choleski decomposition of the mass matrix, that is,

$$\mathbf{m} = \mathbf{L}\mathbf{L}^T$$

where $\mathbf{L}$ is the lower triangular component [equivalent to the decomposition of $\mathbf{k}$ in Eq. (14-133)]. The transformation can then be performed as described above with $(L^T)^{-1}$ taking the place of $\mathbf{T}\hat{v}^{-1/2}$ [in Eq. (14-147)] or of $\mathbf{m}^{-1/2}$ [in Eq. (14-144)].

However, it has been found in many practical cases that the eigenproblem resulting from the Choleski transformation may be quite sensitive and difficult to solve accurately. For this reason, the eigenvector decomposition of Eq. (14-146) is preferable for general use, even though it is more expensive.

14-8 ANALYSIS OF UNCONSTRAINED STRUCTURES

Structures which are unconstrained or only partially constrained by their external support system against rigid-body displacements present a special problem in vibration analysis because the stiffness matrix is singular and the vibration "frequencies" corresponding to the rigid-body motions are zero. Although the determinantal-equation approach (and some other formal mathematical procedures) can deal directly with a dynamic system having a singular stiffness matrix, it is evident that the Stodola method (or any other method making direct use of the stiffness inverse) cannot be applied without modification. Three of the simplest methods of avoiding difficulty with a singular stiffness matrix will be described here.

The most direct way to deal with an unconstrained structure is to modify it by adding small spring constraints to the unconstrained degrees of freedom. First a minimum set of constraints sufficient to prevent any rigid-body motions must be identified. Then if a spring is connected between the structure and the ground in each of these degrees of freedom, the singularity of the stiffness matrix will be removed. Analytically these springs are represented by adding terms to the diagonal elements of the stiffness matrix for these degrees of freedom. If the added spring stiffnesses are very small relative to the original stiffness-matrix coefficients, they will have negligible effect on the vibration mode shapes and frequencies associated with deformations of the structure, but an additional set of rigid-body modes will be defined having frequencies much smaller than the deformation modes. These constraint springs can be introduced automatically by the computer program when the eigenproblem is solved by inverse iteration. If the stiffness equations are solved by Choleski or Gauss decomposition, any singularity leads to a zero in the diagonal position and would prevent continuation of the decomposition. However, the program can be written so that each diagonal zero is replaced by a small number which physically represents the spring constraint; in this way the singularities are overcome and the decomposition process can be carried to completion.

A similar effect can be achieved mathematically by means of an eigenvalue shift. From

$$[\mathbf{k} - \mu\mathbf{m}]\phi_n = \delta_n\mathbf{m}\phi_n \quad (14\text{-}130)$$

it is evident that the shifted stiffness matrix $\hat{\mathbf{k}} = \mathbf{k} - \mu\mathbf{m}$ will be nonsingular in general even if $\mathbf{k}$ is singular. If the mass matrix is diagonal, introducing a negative shift causes a positive quantity to be added to the diagonal elements of the stiffness matrix; hence this is equivalent to connecting a spring to each degree of freedom. The essential difference between this procedure and the physical approach mentioned first is that a "spring" is added corresponding to each mass coefficient, rather than just a minimum set. The shift approach has the advantage that the mode shapes are not changed and the frequency effect is accounted for exactly by the shift.

These two procedures are well adapted to digital-computer solutions of large, complex systems. In a modern computer the analysis will be carried out with sufficient significant figures to permit the artificial-constraint springs to be several orders of magnitude smaller than the actual stiffness coefficients; in this case the artificial constraints will have no noticeable effect on the deformation-vibration properties; however, when the analysis is to be done by hand, where fewer significant figures and fewer degrees of freedom are considered in the analysis, a different approach may be more effective. This is essentially a Ritz transformation method, in which the assumed shapes are selected to satisfy the rigid-body conditions of dynamic equilibrium. In this analysis, the complete set of degrees of freedom v is partitioned into the set associated with a minimum support system v_s and the set which remains v_r; thus

$$v = \begin{bmatrix} v_s \\ v_r \end{bmatrix} \quad (14\text{-}149)$$

A system of rigid-body displacement patterns can be defined for the structure. For an unconstrained structure in three-dimensional space, these will consist of translation along, and rotation about, each of the perpendicular axes; but if the structure is partially constrained, some of these patterns will be eliminated. The set of rigid-body patterns appropriate to the given structure will be designated by ϕ_s, in which the number of columns corresponds to the number of elements in v_s.

The relationship between v_s and the remaining degrees of freedom is provided by the condition that the inertia forces acting during free vibrations must be in dynamic equilibrium. These inertia forces may be expressed

$$f_I = \omega^2 \mathbf{m} \mathbf{v}$$

Partitioning the mass matrix to correspond with Eq. (14-149) gives

$$\mathbf{f}_I = \omega^2 [\mathbf{m}_s \quad \mathbf{m}_r] \begin{bmatrix} v_s \\ v_r \end{bmatrix} \quad (14\text{-}150)$$

If these inertia forces are in equilibrium, they will do no work during any rigid-body displacement; hence considering all the rigid-body motions leads to

$$\phi_s{}^T \mathbf{f}_I = \phi_s{}^T [\mathbf{m}_s \quad \mathbf{m}_r] \begin{bmatrix} v_s \\ v_r \end{bmatrix} = 0 \quad (14\text{-}151)$$

which may be written out as

$$\phi_s{}^T m_s v_s + \phi_s{}^T m_r v_r = 0 \quad (14\text{-}152)$$

This can now be solved for the dependent "support" displacements in terms of the remaining set:

$$v_s = -[\phi_s{}^T m_s]^{-1} \phi_s{}^T m_r v_r \quad (14\text{-}153)$$

Finally, the complete displacement vector can be expressed in terms of v_r by combining this result with an identity matrix:

$$v = \begin{bmatrix} v_s \\ v_r \end{bmatrix} = \begin{bmatrix} -[\phi_s{}^T m_s]^{-1} \phi_s{}^T m_r \\ I \end{bmatrix} v_r \quad (14\text{-}154)$$

expressed symbolically as

$$v = S_r v_r \quad (14\text{-}155)$$

in which S_r is the rectangular transformation matrix in Eq. (14-154).

Now the eigenvalue problem can be expressed directly in terms of the independent degrees of freedom v_r by using S_r as a standard Ritz transformation. In other words, the generalized mass and stiffness are computed from

$$m_r^* = S_r{}^T m S_r \quad \text{and} \quad k_r^* = S_r{}^T k S_r$$

and then the resulting reduced-order eigenproblem

$$k_r^* \hat{v}_r = \omega^2 m_r^* \hat{v}_r \quad (14\text{-}156)$$

is solved for the mode shapes $\hat{v}_r$ and frequencies. Of course, the dependent degrees of freedom $\hat{v}_s$ may then be obtained by means of Eq. (14-153).

As a final note on this method of dealing with unconstrained structures, it should be mentioned that the equilibrium constraints of Eq. (14-151) may be interpreted as a condition that the deformation modes of vibration be orthogonal to the rigid-body displacements ϕ_s. Correspondingly, the transformation matrix S_r of Eq. (14-155) can be looked upon as a sweeping matrix which eliminates rigid-body components from the displacements v_r. As a matter of fact, sweeping matrices suitable for analysis of the higher modes of a fully supported stable structure can be derived with the expression of Eq. (14-154). For this purpose, ϕ_s will represent the m vibration shapes which already have been calculated for the lower modes, and v_s will represent the first m elements of the complete displacement vector v. For hand solutions, this type of sweeping matrix may be preferable to that given by Eq. (14-42) because the order of the matrices is reduced in correspondence with the number of modal components being swept out. However, the inversion of $\phi_s{}^T m_s$ becomes a major effort by hand if more than four or five modes are to be evaluated.

In general, the reduction in the order of the eigenproblem achieved by introducing constraints in the form of Eq. (14-154) will be helpful only if the total number of degrees of freedom is small (so that the reduction represents a significant part of the total). In the analysis of large systems, the reduction in the number of equations is of little importance, and the transformed system [Eq. (14-156)] often actually requires more computational effort because the banding properties of **k** and **m** are lost in the transformation. For this reason, the two methods of dealing with unconstrained structures mentioned first in this discussion generally are preferred in developing large-scale computer programs.

PROBLEMS

14-1 Evaluate the fundamental vibration-mode shape and frequency for the building of Prob. 9-3 using the Stodola (matrix iteration) method. Note that the flexibility matrix may be obtained from the given story shear stiffness either by inverting the stiffness matrix or by applying a unit load successively at each story, and evaluating the resulting displacements at each story.

14-2 Evaluate the highest mode shape and frequency for the building of Prob. 14-1 by matrix iteration.

14-3 Repeat Prob. 14-1 for the building properties of Prob. 9-4.

14-4 Repeat Prob. 14-2 for the building properties of Prob. 9-5.

14-5 Evaluate the second mode shape and frequency for the shear building of Prob. 13-4 by matrix iteration. To form the first mode sweeping matrix $\mathbf{S_1}$, use the given first mode shape $\boldsymbol{\phi}_1$ and Eq. (14-154).

14-6 Repeat Prob. 14-5 using the sweeping matrix expression of Eq. (14-29).

14-7 Repeat Prob. 14-5 using inverse iteration with shifts, as indicated by Eq. (14-125) and the discussion that follows it. For this demonstration problem, use a shift $\mu = 98\% \, (\omega_2)^2$, where ω_2 is as given in Prob. 13-4.

14-8 A beam with three lumped masses is shown in Fig. P14-1; also shown are its flexibility and stiffness matrices. By matrix iteration (Vianello method), determine the axial force N_{cr} that will cause this beam to buckle. In this analysis, use the linear approximation, Eq. (11-36), to express the geometric stiffness of the beam.

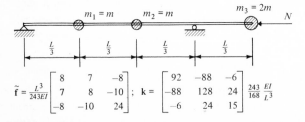

$$\tilde{\mathbf{f}} = \frac{L^3}{243EI} \begin{bmatrix} 8 & 7 & -8 \\ 7 & 8 & -10 \\ -8 & -10 & 24 \end{bmatrix}; \quad \mathbf{k} = \begin{bmatrix} 92 & -88 & -6 \\ -88 & 128 & 24 \\ -6 & 24 & 15 \end{bmatrix} \frac{243}{168} \frac{EI}{L^3}$$

FIGURE P14-1

14-9 By matrix iteration, compute the frequency of vibration of the beam of Prob. 14-8 if the axial force has the value $N = 2(EI/L^2)$.

14-10 Evaluate the mode shape and frequency of the first two modes of vibration for the shear building of Prob. 14-1, using the Holzer method.

14-11 Using the transfer matrix procedure, Eqs. (14-62) and (14-63), demonstrate that the fundamental mode frequency and shape of the cantilever beam of Prob. 13-1 are as given there.

14-12 The four-story shear frame of Fig. P14-2 has the same mass m lumped in each rigid girder and the same story-to-story stiffness k in the columns of each story. Using the indicated linear and quadratic shape functions ψ_i as generalized coordinates, obtain the approximate shape and the frequency of the first two modes of vibration by the Rayleigh-Ritz method, Eq. (14-88).

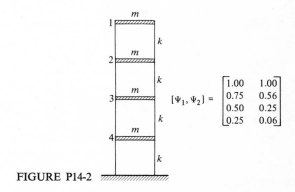

$$[\Psi_1, \Psi_2] = \begin{bmatrix} 1.00 & 1.00 \\ 0.75 & 0.56 \\ 0.50 & 0.25 \\ 0.25 & 0.06 \end{bmatrix}$$

FIGURE P14-2

14-13 Repeat Prob. 14-12 using the "improved" expressions of Eq. (14-96) to define the generalized coordinate mass and stiffness properties.

15

ANALYSIS OF NONLINEAR SYSTEMS

15-1 INTRODUCTION

In all the preceding discussion of the response of MDOF systems, it has been assumed that the structures were linear, that is, that their dynamic resisting forces could be related to the acceleration, velocity, or displacement vectors by means of linear influence coefficients. As a result of this physical characteristic, it was possible to compute vibration mode shapes and frequencies for the structures and to evaluate the response in terms of the modal coordinates. This approach has the very great advantage that an adequate estimate of the dynamic response can often be obtained by considering only a few modes of vibration, even in systems that may have dozens or hundreds of degrees of freedom; thus the computational effort may be reduced significantly.

However, as pointed out in the discussion of SDOF systems, there are many instances in which the physical properties cannot be assumed to remain constant during the dynamic response. The stiffness influence coefficients may be altered by yielding of the structural materials or by significant changes of the axial forces in the members of the structure (which will cause changes in the geometric-stiffness coefficients). Or it is possible that either the mass or damping coefficients will undergo changes during the dynamic response. Any such changes will alter the vibration characteristics of the system (actually the simple free-vibration concept is no longer applicable in a nonlinear system), and therefore the normal coordinate uncoupling of the equations of motion is not possible.

The only generally applicable method for the analysis of arbitrary nonlinear systems is the numerical step-by-step integration of the coupled equations of motion. The procedure can be carried out as the exact MDOF analog of the analysis of nonlinear SDOF systems described in Chap. 8. The response history is divided into short, equal time increments, and the response is calculated during each increment for a *linear* system having the properties determined at the beginning of the interval. At the end of the interval the properties are modified to conform to its state of deformation and stress at that time; thus the nonlinear analysis is approximated as a sequence of analyses of successively changing linear systems.

As mentioned in Chap. 8, the step-by-step integration procedure is also applicable to linear structures, in which case the computation procedure is greatly simplified because it is not necessary to modify the structural properties at each step. In some cases it may be advantageous to use this direct-integration approach rather than mode superposition because it does not require the evaluation of the vibration mode shapes and frequencies, which is a very large computational task in systems with many degrees of freedom. In general, direct step-by-step integration tends to be most useful in evaluating the response of large, complex structures to short-duration impulsive loads which tend to excite many modes of vibration but which require that only a short response history be evaluated.

One potential difficulty in the step-by-step response integration of MDOF systems is that the damping matrix c must be defined explicitly rather than in terms of modal damping ratios. It is very difficult to estimate the magnitudes of the damping influence coefficients of a complete damping matrix. In general, the most effective means for deriving a suitable damping matrix is to assume appropriate values of modal damping ratios for all the modes which are considered to be important and then to compute an orthogonal damping matrix which has those properties, as described in Chap. 13.

On the other hand, the fact that the damping matrix is defined explicitly rather than by modal damping ratios may be advantageous in that it increases the generality of the step-by-step method over mode superposition. There is no need for uncoupling the modal responses; therefore the damping matrix need not be selected to satisfy modal orthogonality conditions. Any desired set of damping-matrix coefficients can be employed in the analysis, and they may represent entirely different levels of damping in different parts of the structure. For example, in the earthquake-response analysis of a building it may be desirable to use damping coefficients representing a high damping ratio for the foundation degrees of freedom while selecting damping coefficients for the superstructure representing a much lower damping ratio.

Finally, it is worth noting that the transformation to normal coordinates may be useful even in the analysis of nonlinear systems. Of course, the undamped free-vibration mode shapes will serve to uncouple the equations of motion only so long as the stiffness matrix remains unchanged from the state for which the vibration analysis was made. As soon as the stiffness changes, due to yielding or other damage, the normal-coordinate transformation will introduce off-diagonal terms in the generalized stiffness matrix, which cause coupling of the modal response equations. However, if the nonlinear deformation mechanisms in the structure do not cause major changes in its deflection patterns, the dynamic response still may be expressed efficiently in terms of the original undamped mode shapes. Thus it often will be worthwhile to evaluate the response of a complex structure by direct step-by-step integration of a limited set of normal-coordinate equations of motion, even though the equations will become coupled as soon as any significant nonlinearity develops in the response. This treatment of a system with stiffness coupling of the normal-coordinate equations is equivalent to the approach suggested earlier for the analysis of systems in which the damping matrix is such as to introduce normal-coordinate coupling.

15-2 INCREMENTAL EQUILIBRIUM EQUATIONS

The equation expressing the equilibrium of force increments developed during a time increment Δt can be derived as the matrix equivalent of the incremental equation of motion derived for the SDOF system [Eq. (8-2)]. Thus taking the difference between the equilibrium relationships defined for times t and $t + \Delta t$ gives the incremental equilibrium equation

$$\Delta \mathbf{f}_I(t) + \Delta \mathbf{f}_D(t) + \Delta \mathbf{f}_S(t) = \Delta \mathbf{p}(t) \qquad (15\text{-}1)$$

The force increments in this equation, by analogy with the SDOF expressions [Eq. (8-3)], can be expressed as

$$\Delta \mathbf{f}_I(t) = \mathbf{f}_I(t + \Delta t) - \mathbf{f}_I(t) = \mathbf{m} \, \Delta \ddot{\mathbf{v}}(t)$$

$$\Delta \mathbf{f}_D(t) = \mathbf{f}_D(t + \Delta t) - \mathbf{f}_D(t) = \mathbf{c}(t) \, \Delta \dot{\mathbf{v}}(t)$$

$$\Delta \mathbf{f}_S(t) = \mathbf{f}_S(t + \Delta t) - \mathbf{f}_S(t) = \mathbf{k}(t) \, \Delta \mathbf{v}(t) \qquad (15\text{-}2)$$

$$\Delta \mathbf{p}(t) = \mathbf{p}(t + \Delta t) - \mathbf{p}(t)$$

where it has been assumed that the mass $\mathbf{m}$ does not change with time.

The elements of the incremental damping and stiffness matrices $\mathbf{c}(t)$ and $\mathbf{k}(t)$ are influence coefficients $c_{ij}(t)$ and $k_{ij}(t)$ defined for the time increment; typical

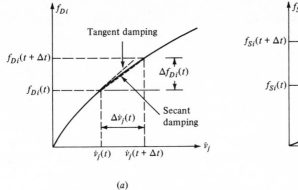
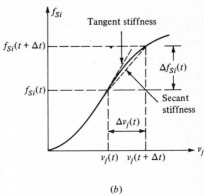

FIGURE 15-1
Definition of nonlinear influence coefficients: (a) nonlinear viscous damping c_{ij}; (b) nonlinear stiffness k_{ij}.

representations of these coefficients are shown in Fig. 15-1. As explained for the SDOF coefficients, it is convenient to use the initial tangent rather than the secant as a measure of the damping or stiffness property in order to avoid the need for iteration at each step of the solution. Hence the influence coefficients are given by

$$c_{ij}(t) = \left(\frac{df_{Di}}{d\dot{v}_j}\right)_t \qquad k_{ij}(t) = \left(\frac{df_{Si}}{dv_j}\right)_t \qquad (15\text{-}3)$$

When Eqs. (15-2) are substituted into Eq. (15-1), the incremental equation of motion becomes

$$\mathbf{m}\,\Delta\ddot{\mathbf{v}}(t) + \mathbf{c}(t)\,\Delta\dot{\mathbf{v}}(t) + \mathbf{k}(t)\,\Delta\mathbf{v}(t) = \Delta\mathbf{p}(t) \qquad (15\text{-}4)$$

The incremental force expressions on the left side of Eq. (15-4) are only approximations because of the use of the initial tangent values for $\mathbf{c}(t)$ and $\mathbf{k}(t)$. However, accumulation of errors due to this factor will be avoided if the acceleration at the beginning of each time increment is calculated from the total equilibrium of forces at that time, as was mentioned for the SDOF case.

15-3 STEP-BY-STEP INTEGRATION: LINEAR-ACCELERATION METHOD

The basic operation in the step-by-step solution of the simultaneous differential equations of motion, Eqs. (15-4), is their conversion to a set of simultaneous algebraic equations. This is accomplished by introducing a simple relationship between displacement, velocity, and acceleration which may be assumed to be valid for a short increment of time. On this basis, the incremental changes of velocity and displacement can be expressed in terms of the changes in acceleration, or, alternatively, the changes in velocity and acceleration can be expressed in terms of the incremental displacements. In either case, only one unknown vector remains in the incremental equilibrium equations, and this may be evaluated by any standard simultaneous-equation-solution procedure.

As mentioned in the discussion of SDOF systems, the relationship between displacement, velocity, and acceleration can conveniently be established by assuming the manner of variation of the acceleration vector with time. For the MDOF system, it is convenient to adopt the same linear-acceleration assumption used in Chap. 8, which then leads to a quadratic variation of the velocity vector and a cubic variation of the displacement vector. The derivation of the MDOF analysis procedure can be carried out exactly as it was in Chap. 8, and the final results, by analogy with Eqs. (8-8) and (8-9), written

$$\tilde{\mathbf{k}}(t) \, \Delta\mathbf{v}(t) = \Delta\tilde{\mathbf{p}}(t) \qquad (15\text{-}5)$$

in which

$$\tilde{\mathbf{k}}(t) = \mathbf{k}(t) + \frac{6}{\Delta t^2} \, \mathbf{m} + \frac{3}{\Delta t} \, \mathbf{c}(t) \qquad (15\text{-}6a)$$

$$\Delta\tilde{\mathbf{p}}(t) = \Delta\mathbf{p}(t) + \mathbf{m}\left[\frac{6}{\Delta t} \, \dot{\mathbf{v}}(t) + 3\ddot{\mathbf{v}}(t)\right] + \mathbf{c}(t)\left[3\dot{\mathbf{v}}(t) + \frac{\Delta t}{2} \, \ddot{\mathbf{v}}(t)\right] \qquad (15\text{-}6b)$$

Equation (15-5) has the form of a standard static-stiffness equation for the incremental displacement vector $\Delta\mathbf{v}(t)$, in which $\tilde{\mathbf{k}}(t)$ and $\Delta\tilde{\mathbf{p}}(t)$ may be interpreted as the effective dynamic-stiffness matrix and the effective load increment, respectively. The analysis is carried out by evaluating $\tilde{\mathbf{k}}(t)$ from the mass, damping, and stiffness properties defined for the conditions existing at the beginning of the time step and evaluating $\Delta\tilde{\mathbf{p}}(t)$ from the velocity and acceleration vectors at the beginning of the step as well as the load increment specified during the step. Then the displacement increment $\Delta\mathbf{v}(t)$ is computed by solving Eq. (15-5), using any standard static-equation-solution procedure. Gauss or Choleski decomposition is frequently used for this purpose in digital-computer programs; it should be noted that the changing values of $\mathbf{k}(t)$ and $\mathbf{c}(t)$ in a nonlinear problem require that the decomposition be carried out for each time step, which requires a major computational effort with large systems of equations. Many special techniques have been employed to simplify this aspect of the analysis, but such considerations are beyond the scope of this presentation.

When the displacement increment $\Delta\mathbf{v}(t)$ has been determined, the velocity increment is found from an expression analogous to Eq. (8-7b):

$$\Delta\dot{\mathbf{v}}(t) = \frac{3}{\Delta t}\,\Delta\mathbf{v}(t) - 3\dot{\mathbf{v}}(t) - \frac{\Delta t}{2}\,\ddot{\mathbf{v}}(t) \qquad (15\text{-}7)$$

The displacement and velocity vectors at the end of the increment are then given by

$$\mathbf{v}(t + \Delta t) = \mathbf{v}(t) + \Delta\mathbf{v}(t) \qquad \dot{\mathbf{v}}(t + \Delta t) = \dot{\mathbf{v}}(t) + \Delta\dot{\mathbf{v}}(t) \qquad (15\text{-}8)$$

The vectors given by Eqs. (15-8) represent the initial conditions for the next step of the analysis procedure. In addition, the acceleration vector is needed at the same time. As mentioned above, this should be calculated from the condition of dynamic equilibrium at the time $t + \Delta t$; thus

$$\ddot{\mathbf{v}}(t + \Delta t) = \mathbf{m}^{-1}[\mathbf{p}(t + \Delta t) - \mathbf{f}_D(t + \Delta t) - \mathbf{f}_S(t + \Delta t)] \qquad (15\text{-}9)$$

in which $\mathbf{f}_D(t + \Delta t)$ and $\mathbf{f}_S(t + \Delta t)$ represent the damping and stiffness force vectors, respectively, evaluated from the velocity and displacement conditions at time $t + \Delta t$ (as well as the past history if the material properties are history-dependent). The inverse $\mathbf{m}^{-1}$ of the mass matrix is used in Eq. (15-9) at each step of the analysis; thus it should be calculated and stored by the computer program.

15-4 UNCONDITIONALLY STABLE LINEAR-ACCELERATION METHOD

The linear acceleration assumption which is the basis of Eqs. (15-6) provides an efficient step-by-step integration procedure so long as a short enough time increment is used. In general, the analysis will give good accuracy for motions of the structure associated with periods of vibration at least 5 to 10 times greater than the integration interval (these periods being those of the linear systems considered during successive time increments). In many cases, the significant response of the structure is contained in the longer period components of the motion; hence the time increment required to achieve adequate accuracy of the significant response is not unreasonably short. However, as mentioned in Chap. 8, the linear-acceleration method is only conditionally stable, and it will blow up if it is applied to modal response components having periods of vibration less than about 1.8 times the integration interval. Thus the time increments must be made short relative to the least period of vibration contained in the structural system, regardless of whether the higher modes contribute significantly to the dynamic response or not.

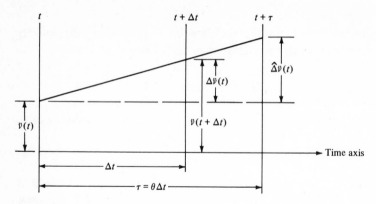

FIGURE 15-2
Linear acceleration; normal and extended time steps.

For certain types of MDOF structures, notably multistory buildings which are idealized to have only one degree of freedom per story, this limitation on the length of the integration step may be of no consequence. For the earthquake analysis of such structures, the time interval must be made rather short in order to provide an adequate description of the ground motion, and the shortest period of vibration of the mathematical model generally is considerably longer than this time increment; thus the linear-acceleration method has proved effective in both linear and nonlinear earthquake-response analysis of building frames. On the other hand, for more general classes of structures, particularly for finite-element idealizations of structures having complex geometries, the shortest period of vibration of the mathematical model may be several orders of magnitude less than the periods associated with the significant structural response. In this situation, the ordinary linear-acceleration method cannot be used because of the very short time increment required to avoid instability; instead an unconditionally stable method is required which will not blow up regardless of the time-increment : shortest-period ratio.

Several different unconditionally stable step-by-step methods have been used for the dynamic-response analysis of such systems; one of the simplest and best of these is a modification of the linear-acceleration method described above, called the *Wilson θ method.*[1] This modification is based on the assumption that the acceleration varies linearly over an extended computation interval

$$\tau = \theta \Delta t \qquad \text{where } \theta > 1.37 \qquad (15\text{-}10)$$

The parameters associated with this assumption are depicted in Fig. 15-2. The acceleration increment $\hat{\Delta}\ddot{v}(t)$ is calculated by the standard linear-acceleration procedure

[1] Devised by E. L. Wilson, University of California, Berkeley.

applied to the extended time step τ; from this the increment $\Delta\ddot{v}(t)$ for the normal time step Δt is obtained by interpolation. For a value of $\theta = 1$, the procedure reverts to the standard linear-acceleration method, but for $\theta > 1.37$ it becomes unconditionally stable.

The analysis procedure can be derived merely by rewriting the basic relationships of the linear-acceleration method for the extended time step τ. Thus, by analogy with Eqs. (8-6),

$$\hat{\Delta}\dot{v}(t) = \tau\ddot{v}(t) + \frac{\tau}{2}\hat{\Delta}\ddot{v}(t)$$

$$\hat{\Delta}v(t) = \tau\dot{v}(t) + \frac{\tau^2}{2}\ddot{v}(t) + \frac{\tau^2}{6}\hat{\Delta}\ddot{v}(t)$$

(15-11)

in which the hat identifies an increment associated with the extended time step. Solving these to express $\hat{\Delta}\ddot{v}(t)$ and $\hat{\Delta}\dot{v}(t)$ in terms of $\hat{\Delta}v(t)$ and substituting into the equation of motion leads to expressions equivalent to Eqs. (15-5) and (15-6) but written for the extended time step:

$$\hat{k}(t)\,\hat{\Delta}v(t) = \hat{\Delta}\hat{p}(t) \qquad (15\text{-}12)$$

where

$$\hat{k}(t) = k(t) + \frac{6}{\tau^2}\,m + \frac{3}{\tau}\,c(t) \qquad (15\text{-}13a)$$

$$\hat{\Delta}\hat{p}(t) = \hat{\Delta}p(t) + m\left[\frac{6}{\tau}\,\dot{v}(t) + 3\ddot{v}(t)\right] + c(t)\left[3\dot{v}(t) + \frac{\tau}{2}\ddot{v}(t)\right] \qquad (15\text{-}13b)$$

Finally the pseudostatic relationship Eq. (15-12) can be solved for $\Delta v(t)$ and substituted into the following equation [analogous to Eq. (8-7a)] to obtain the increment of acceleration during the extended time step:

$$\hat{\Delta}\ddot{v}(t) = \frac{6}{\tau^2}\,\hat{\Delta}v(t) - \frac{6}{\tau}\,\dot{v}(t) - 3\ddot{v}(t) \qquad (15\text{-}14)$$

From this, the acceleration increment for the normal time step Δt can be obtained by linear interpolation:

$$\Delta\ddot{v}(t) = \frac{1}{\theta}\,\hat{\Delta}\ddot{v}(t) \qquad (15\text{-}15)$$

and then the corresponding incremental velocity and displacement vectors are given by expressions like Eqs. (15-11) but written for the normal time step Δt. From these results, the initial conditions for the next time step are given by Eqs. (15-8) and (15-9), and the whole process can then be repeated for as many steps as desired.

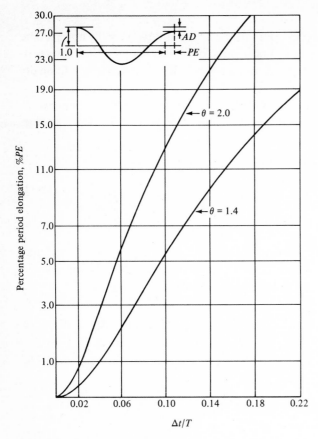

FIGURE 15-3
Period elongation, Wilson θ method.

15-5 PERFORMANCE OF THE WILSON θ METHOD

At this point it is useful to consider the general performance and efficiency of this unconditionally stable method of time integration. For this purpose, it is sufficient to discuss its performance in treating a SDOF system because it has been noted that the response of any linear system can be transformed by means of the modal coordinates into the response of a set of uncoupled SDOF systems.

The size of the errors which may be introduced by any numerical-integration scheme will depend on the characteristics of the dynamic loading and the length of the time step. However, the general nature of the computational errors is apparent in a free-vibration response analysis and may be expressed in terms of an artificial change of period and a reduction of amplitude. These period-elongation and amplitude-decay effects computed for the Wilson θ method in the free-vibration response of a simple oscillator to an initial displacement are depicted in Figs. 15-3

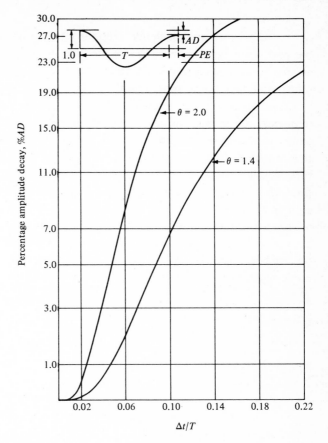

FIGURE 15-4
Amplitude decay, Wilson θ method.

and 15-4, where the magnitude of the error is shown as a function of the ratio of the time step to the period of vibration. The results shown for both $\theta = 2$ and $\theta = 1.4$ demonstrate that the accuracy is considerably better for the smaller value of θ.

Both the period-elongation and amplitude-decay effects can be important in certain cases, but generally the amplitude decay is considered to be more significant. This decay mechanism may be looked upon as a form of artificial damping which is added to any real damping that may exist, and it should be kept in mind that 6 percent amplitude decay per cycle will result from about 1 percent critical damping. Thus, it is evident in Fig. 15-4 that this artificial damping would have little significance in the analysis of typical structures having 5 percent critical damping or more, so long as the time-step ratio $\Delta t/T < {}^1/_{10}$, because in this case the added artificial damping would be within the range of uncertainty of the real damping. However, it is clear that any response components for which $\Delta t/T > {}^1/_4$ will be damped out quickly.

The dynamic analyst should keep the effect of this artificial damping in proper perspective. Clearly he must choose his time step short enough for the response of all significant modal components to be obtained without artificial reduction. On the other hand, he must be aware that the highest-mode components of his mathematical model often are not representative of behavior mechanisms of the real structure; they frequently are grossly distorted by the discretization process. Moreover, in many cases the loading is such that only the lower modes of vibration respond significantly; thus it is not necessary to integrate the higher-mode components accurately. From these considerations, it is clear that significant amplitude decay can be accepted with the higher-frequency components, and in many cases it is advantageous to have them eliminated from the response. In a sense, the amplitude-decay mechanism of the Wilson θ method may be looked upon as equivalent to the deliberate modal truncation applied with the mode-superposition method. Obviously, amplitude decay is not a serious factor with respect to any modes which would not have been included in a mode-superposition analysis.

VARIATIONAL FORMULATION OF THE EQUATIONS OF MOTION

16-1 GENERALIZED COORDINATES

The significant advantages of describing the response of dynamic systems by means of generalized coordinates, rather than by merely expressing the displacements of discrete points on the structure, have been emphasized many times in this text, and various types of generalized coordinates have been considered for this purpose. It has also been pointed out that different approaches may be used to advantage in establishing the equations of motion for a structure, depending on its geometric form and complexity as well as the type of coordinates used. In Chap. 1, the three basic techniques were outlined: (1) directly establishing the equilibrium of all dynamic forces acting in the system, (2) establishing equilibrium by application of the principle of virtual displacements, and (3) making use of Hamilton's variational principle. All these techniques were illustrated by reference to SDOF examples in Chap. 2. However, in dealing with MDOF structures up to this point, only direct equilibration and the virtual-work approach have been employed. The purpose of this chapter is to describe

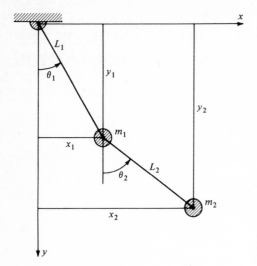

FIGURE 16-1
Double pendulum with hinge support.

and demonstrate by examples the formulation of the equations of motion for MDOF systems by the variational approach.

In formulating the variational MDOF technique, extensive use will be made of generalized coordinates, and in this development a precise definition of the concept is needed rather than the somewhat loose terminology that has sufficed until now. Generalized coordinates for a system with N degrees of freedom are defined as any set of N independent quantities which completely specify the position of every point within the system. Being completely independent, generalized coordinates must not be related in any way through geometric constraints imposed on the system.

In the classical double pendulum shown in Fig. 16-1 the position of the two masses m_1 and m_2 could be specified using the coordinates x_1, y_1, x_2, y_2; however, two geometric constraint conditions must be imposed on these coordinates, namely,

$$x_1^2 + y_1^2 - L_1^2 = 0$$
$$(x_2 - x_1)^2 + (y_2 - y_1)^2 - L_2^2 = 0 \tag{16-1}$$

Because of these constraint relations, x_1, y_1, x_2, and y_2 are not independent and therefore cannot be considered as generalized coordinates.

Suppose, on the other hand, the angles θ_1 and θ_2 were specified as the coordinates to be used in defining the positions of masses m_1 and m_2. Clearly either of these coordinates can be changed while holding the other constant; thus, they are seen to be completely independent and therefore a suitable set of generalized coordinates.

16-2 LAGRANGE'S EQUATIONS OF MOTION

The equations of motion for an N-DOF system can be derived directly from the variational statement of dynamics, namely, Hamilton's principle [Eq. (1-4), repeated here for convenience]:

$$\delta \int_{t_1}^{t_2} (T - V)\, dt + \int_{t_1}^{t_2} \delta W_{nc}\, dt = 0 \qquad (16\text{-}2)$$

by simply expressing the total kinetic energy T, the total potential energy V, and the total virtual work δW_{nc} in terms of a set of generalized coordinates, $q_1, q_2, \ldots, q_N$.

For most mechanical or structural systems, the kinetic energy can be expressed in terms of the generalized coordinates and their first time derivatives, and the potential energy can be expressed in terms of the generalized coordinates alone. In addition, the virtual work which is performed by the nonconservative forces as they act through the virtual displacements caused by an arbitrary set of variations in the generalized coordinates can be expressed as a linear function of those variations. In mathematical terms the above three statements are expressed in the form

$$T = T(q_1, q_2, \ldots, q_N, \dot{q}_1, \dot{q}_2, \ldots, \dot{q}_N) \qquad (16\text{-}3a)$$

$$V = V(q_1, q_2, \ldots, q_N) \qquad (16\text{-}3b)$$

$$\delta W_{nc} = Q_1\, \delta q_1 + Q_2\, \delta q_2 + \cdots + Q_N\, \delta q_N \qquad (16\text{-}3c)$$

where the coefficients $Q_1, Q_2, \ldots, Q_N$ are the generalized forcing functions corresponding to the coordinates $q_1, q_2, \ldots, q_N$, respectively.

Introducing Eqs. (16-3) into Eq. (16-2), and completing the variation of the first term gives

$$\int_{t_1}^{t_2} \left(\frac{\partial T}{\partial q_1}\, \delta q_1 + \frac{\partial T}{\partial q_2}\, \delta q_2 + \cdots + \frac{\partial T}{\partial q_N}\, \delta q_N + \frac{\partial T}{\partial \dot{q}_1}\, \delta \dot{q}_1 + \frac{\partial T}{\partial \dot{q}_2}\, \delta \dot{q}_2 + \cdots \right.$$

$$+ \frac{\partial T}{\partial \dot{q}_N}\, \delta \dot{q}_N - \frac{\partial V}{\partial q_1}\, \delta q_1 - \frac{\partial V}{\partial q_2}\, \delta q_2 - \cdots$$

$$\left. - \frac{\partial V}{\partial q_N}\, \delta q_N + Q_1\, \delta q_1 + Q_2\, \delta q_2 + \cdots + Q_N\, \delta q_N \right) dt = 0 \qquad (16\text{-}4)$$

Integrating the velocity dependent terms in Eq. (16-4) by parts leads to

$$\int_{t_1}^{t_2} \frac{\partial T}{\partial \dot{q}_i}\, \delta \dot{q}_i\, dt = \left[\frac{\partial T}{\partial \dot{q}_i}\, \delta q_i \right]_{t_1}^{t_2} - \int_{t_1}^{t_2} \frac{\partial}{\partial t}\left(\frac{\partial T}{\partial \dot{q}_i} \right) \delta q_i\, dt \qquad (16\text{-}5)$$

The first term on the right-hand side of Eq. (16-5) is equal to zero for each coordinate since $\delta q_i(t_1) = \delta q_i(t_2) = 0$ is the basic condition imposed upon the variations. Substituting Eq. (16-5) into Eq. (16-4) gives, after rearranging terms,

$$\int_{t_1}^{t_2} \left\{ \sum_{i=1}^{N} \left[-\frac{d}{dt}\left(\frac{\partial T}{\partial \dot{q}_i}\right) + \frac{\partial T}{\partial q_i} - \frac{\partial V}{\partial q_i} + Q_i \right] \delta q_i \right\} dt = 0 \qquad (16\text{-}6)$$

Since all variations δq_i $(i = 1, 2, \ldots, N)$ are arbitrary, Eq. (16-6) can be satisfied in general only when the term in brackets vanishes, i.e.,

$$\frac{\partial}{\partial t}\left(\frac{\partial T}{\partial \dot{q}_i}\right) - \frac{\partial T}{\partial q_i} + \frac{\partial V}{\partial q_i} = Q_i \qquad (16\text{-}7)$$

Equations (16-7) are the well-known Lagrange's equations of motion, which have found widespread application in various fields of science and engineering.

The beginning student of structural dynamics should take special note of the fact that Lagrange's equations are a direct result of applying Hamilton's variational principle, under the specific condition that the energy and work terms can be expressed in terms of the generalized coordinates, and of their time derivatives and variations, as indicated in Eqs. (16-3). Thus Lagrange's equations are applicable to all systems which satisfy these restrictions, and they may be nonlinear as well as linear. The following examples should clarify the application of Lagrange's equations in structural-dynamics analysis.

EXAMPLE E16-1   Consider the double pendulum shown in Fig. 16-1 under free-vibration conditions. The x- and y-coordinate positions along with their first time derivatives can be expressed in terms of the set of generalized coordinates $q_1 \equiv \theta_1$ and $q_2 \equiv \theta_2$ as follows:

$$
\begin{array}{ll}
x_1 = L_1 \sin q_1 & \dot{x}_1 = L_1 \dot{q}_1 \cos q_1 \\
y_1 = L_1 \cos q_1 & \dot{y}_1 = -L_1 \dot{q}_1 \sin q_1 \\
x_2 = L_1 \sin q_1 + L_2 \sin q_2 & \dot{x}_2 = L_1 \dot{q}_1 \cos q_1 + L_2 \dot{q}_2 \cos q_2 \\
y_2 = L_1 \cos q_1 + L_2 \sin q_2 & \dot{y}_2 = -L_1 \dot{q}_1 \sin q_1 - L_2 \dot{q}_2 \sin q_2
\end{array}
\qquad (a)
$$

Substituting the above velocity expressions into the basic expression for kinetic energy, namely,

$$T = \tfrac{1}{2} m_1(\dot{x}_1{}^2 + \dot{y}_1{}^2) + \tfrac{1}{2} m_2(\dot{x}_2{}^2 + \dot{y}_2{}^2) \qquad (b)$$

gives

$$T = \tfrac{1}{2} m_1 L_1{}^2 \dot{q}_1{}^2 + \tfrac{1}{2} m_2 [L_1{}^2 \dot{q}_1{}^2 + L_2{}^2 \dot{q}_2{}^2 + 2 L_1 L_2 \dot{q}_1 \dot{q}_2 \cos (q_2 - q_1)] \qquad (c)$$

The only potential energy present in the double pendulum of Fig. 16-1 is that due to gravity. If zero potential energy is assumed when $q_1 = q_2 = 0$, the potential-energy relation is

$$V = (m_1 + m_2)gL_1(1 - \cos q_1) + m_2 gL_2(1 - \cos q_2) \qquad (d)$$

where g is the acceleration of gravity. There are, of course, no nonconservative forces acting on this system; therefore, the generalized forcing functions Q_1 and Q_2 are both equal to zero.

Substituting Eqs. (c) and (d) into Lagrange's Eqs. (16-7) for $i = 1$ and $i = 2$ separately gives the two equations of motion

$$(m_1 + m_2)L_1{}^2\ddot{q}_1 + m_2 L_1 L_2 \ddot{q}_2 \cos (q_2 - q_1)$$
$$- m_2 L_1 L_2 \dot{q}_2{}^2 \sin (q_2 - q_1) + (m_1 + m_2)gL_1 \sin q_1 = 0$$
$$m_2 L_2{}^2\ddot{q}_2 + m_2 L_1 L_2 \ddot{q}_1 \cos (q_2 - q_1)$$
$$+ m_2 L_1 L_2 \dot{q}_1{}^2 \sin (q_2 - q_1) + m_2 gL_2 \sin q_2 = 0 \qquad (e)$$

These equations are highly nonlinear for large-amplitude oscillation; however, for small-amplitude oscillation Eqs. (e) can be reduced to their linear form

$$(m_1 + m_2)L_1{}^2\ddot{q}_1 + m_2 L_1 L_2 \ddot{q}_2 + (m_1 + m_2)gL_1 q_1 = 0$$
$$m_2 L_1 L_2 \ddot{q}_1 + m_2 L_2{}^2\ddot{q}_2 + m_2 gL_2 q_2 = 0 \qquad (f)$$

The small-amplitude mode shapes and frequencies can easily be obtained from the linearized equations of motion by any of the standard eigenproblem analysis methods, e.g., the determinantal-solution procedure. ////

EXAMPLE E16-2 Assume a uniform rigid bar of length L and total mass m to be supported by an elastic, massless flexure spring and subjected to a uniformly distributed time-varying external loading as shown in Fig. E16-1. If the downward vertical deflections of points 1 and 2 from their static-equilibrium positions are selected as the generalized coordinates q_1 and q_2, respectively, the governing equations of motion for small-displacement theory can be obtained from Lagrange's equations as follows.

The total kinetic energy of the rigid bar is the sum of its translational and rotational kinetic energies, that is,

$$T = {}^1\!/_2 m \left(\frac{\dot{q}_1 + \dot{q}_2}{2}\right)^2 + \frac{1}{2}\frac{mL^2}{12}\left(\frac{q_1 - q_2}{L}\right)^2$$

or

$$T = \frac{m}{6}(\dot{q}_1{}^2 + q_1 q_2 + q_2{}^2) \qquad (a)$$

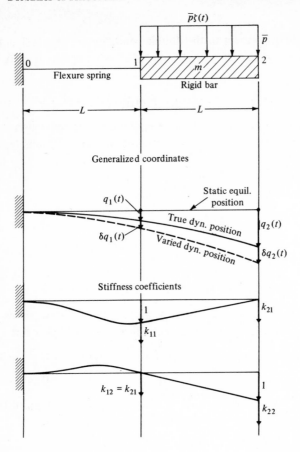

FIGURE E16-1
Rigid bar on massless flexure spring.

Since q_1 and q_2 are displacements from the static-equilibrium position, gravity forces can be ignored provided that the potential energy of the system is evaluated as only the strain energy stored in the flexure spring. Where this strain energy is expressed in terms of the stiffness influence coefficients (defined in Fig. E16-1) the potential-energy term becomes

$$V = {}^{1}\!/_{2}(k_{11}q_1{}^2 + 2k_{12}q_1q_2 + k_{22}q_2{}^2) \qquad (b)$$

The virtual work performed by the nonconservative loading $\bar{p}\zeta(t)$ as it acts through the virtual displacements produced by the arbitrary variations $\delta q_1(t)$ and $\delta q_2(t)$ is given by

$$\delta W_{nc} = \frac{\bar{p}L\zeta(t)}{2}(\delta q_1 + \delta q_2) \qquad (c)$$

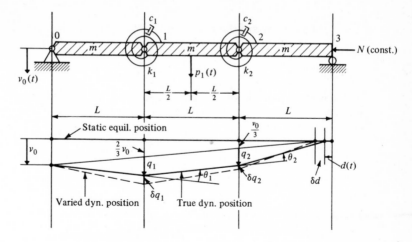

FIGURE E16-2
A 2-DOF rigid-bar assemblage with rotational springs and dashpots.

From a comparison Eq. (*c*) with Eq. (16-3*c*) it is clear of that

$$Q_1(t) = Q_2(t) = \frac{\bar{p}L}{2}\zeta(t) \qquad (d)$$

Substituting Eqs. (*a*), (*b*), and (*d*) into Lagrange's Eqs. (16-7) gives the linear equations of motion for this structure:

$$\frac{m}{6}(2\ddot{q}_1 + \ddot{q}2) + k_{11}q_1 + k_{12}q_2 = \frac{\bar{p}L}{2}\zeta(t)$$

$$\frac{m}{6}(\ddot{q}_1 + 2\ddot{q}_2) + k_{12}q_1 + k_{22}q_2 = \frac{\bar{p}L}{2}\zeta(t) \qquad (e)$$

////

EXAMPLE E16-3 Three uniform rigid bars of length L and mass m are hinged together at points 1 and 2, as shown in Fig. E16-2, and are supported by a roller at point 3 and a hinge at point 0. Concentrated moment-resisting elastic springs and viscous rotational dashpots are attached to adjoining bars at points 1 and 2, having property constants k_1, c_1, k_2, and c_2, respectively. A constant axial load N acts at point 3. If this system is excited by the applied lateral loading $p_1(t)$ and by a small vertical support motion $v_0(t)$ at end 0, the governing equations of motion based on small-deflection theory can be derived directly from Lagrange's equations as follows.

The kinetic energy of the three bars is

$$T = \frac{m}{6}(3\dot{v}_0{}^2 + 2\dot{q}_1{}^2 + 2\dot{q}_2{}^2 + 4\dot{v}_0\dot{q}_1 + 2\dot{v}_0\dot{q}_2 + \dot{q}_1\dot{q}_2) \qquad (a)$$

The movement toward the left of end 3 due to the vertical joint displacement is

$$d = \frac{1}{L}\left(\frac{v_0^2}{6} + q_1^2 + q_2^2 - q_1 q_2\right) \qquad (b)$$

The relative rotations of the bars at joints 1 and 2 and their variations are given by

$$\theta_1 = \frac{1}{L}(2q_1 - q_2) \qquad \delta\theta_1 = \frac{1}{L}(2\delta q_1 - \delta q_2) \qquad (c)$$

$$\theta_2 = \frac{1}{L}(2q_2 - q_1) \qquad \delta\theta_2 = \frac{1}{L}(2\delta q_2 - \delta q_1) \qquad (d)$$

Hence the potential energy of the springs and of the axial force N is

$$V = \left[\frac{1}{2L^2}(4k_1 + k_2) - \frac{N}{L}\right]q_1^2 + \left[\frac{1}{2L^2}(k_1 + 4k_2) - \frac{N}{L}\right]q_2^2$$

$$+ \left[\frac{1}{2L^2}(-4k_1 - 4k_2) + \frac{N}{L}\right]q_1 q_2 - \frac{Nv_0^2}{6L} \qquad (e)$$

The virtual work done by the nonconservative forces is

$$\delta W_{nc} = \frac{1}{2}p_1(t)(\delta q_1 + \delta q_2) - c_1\theta_1\,\delta\theta_1 - c_2\theta_2\,\delta\theta_2$$

or

$$\delta W_{nc} = \left[\frac{p_1}{2} - \frac{2c_1}{L^2}(2\dot{q}_1 - \dot{q}_2) + \frac{c_2}{L^2}(2\dot{q}_2 - \dot{q}_1)\right]\delta q_1$$

$$+ \left[\frac{p_1}{2} + \frac{c_1}{L^2}(2\dot{q}_1 - \dot{q}_2) - \frac{2c_2}{L^2}(2\dot{q}_2 - \dot{q}_1)\right]\delta q_2 \qquad (f)$$

from which the generalized forces are seen to be

$$Q_1 = \frac{p_1}{2} - \frac{2c_1}{L^2}(2\dot{q}_1 - \dot{q}_2) + \frac{c_2}{L^2}(2\dot{q}_2 - \dot{q}_1)$$

$$Q_2 = \frac{p_1}{2} + \frac{c_1}{L^2}(2\dot{q}_1 - \dot{q}_2) - \frac{2c_2}{L^2}(2\dot{q}_2 - \dot{q}_1) \qquad (g)$$

Substituting Eqs. (a), (e), and (g) into Eqs. (16-7) gives the following two equations of motion, from which the dynamic response can be computed:

$$\frac{2}{3}m\ddot{q}_1 + \frac{m}{6}\ddot{q}_2 + \left(\frac{4c_1}{L^2} + \frac{c_2}{L^2}\right)\dot{q}_1 + \left(-\frac{2c_1}{L^2} - \frac{2c_2}{L^2}\right)\dot{q}_2$$

$$+ \left[\frac{1}{L^2}(4k_1 + k_2) - \frac{2N}{L}\right]q_1$$

$$+ \left[\frac{1}{2L^2}(-4k_1 - 4k_2) + \frac{N}{L}\right]q_2 = \frac{p_1}{2} - \frac{m}{3}\ddot{v}_0 \qquad (h)$$

$$\frac{m}{6} \ddot{q}_1 + \frac{2m}{3} \ddot{q}_2 + \left(-\frac{2c_1}{L^2} - \frac{2c_2}{L^2} \right) \dot{q}_1 + \left(\frac{c_1}{L^2} + \frac{4c_2}{L^2} \right) \dot{q}_2$$

$$+ \left[\frac{1}{2L^2} (-4k_1 - 4k_2) + \frac{N}{L} \right] q_1$$

$$+ \left[\frac{1}{L^2} (k_1 + 4k_2) - \frac{2N}{L} \right] q_2 = \frac{p_1}{2} - \frac{m}{3} \ddot{v}_0 \qquad (i)$$

By setting the accelerations and velocities to zero and removing the sources of excitation $p_1(t)$ and $v_0(t)$ from the system, Eqs. (h) and (i) reduce to the static-equilibrium conditions

$$\left[\frac{1}{L^2} (4k_1 + k_2) - \frac{2N}{L} \right] q_1 + \left[\frac{1}{2L^2} (-4k_1 - 4k_2) + \frac{N}{L} \right] q_2 = 0$$

$$\left[\frac{1}{2L^2} (-4k_1 - 4k_2) + \frac{N}{L} \right] q_1 + \left[\frac{1}{L^2} (k_1 + 4k_2) - \frac{2N}{L} \right] q_2 = 0 \qquad (j)$$

Now a nontrivial solution of Eqs. (j) is possible only when the structure buckles under the action of the axial force N, and this is indicated when the determinant of the coefficient matrix equals zero, that is, when

$$\begin{Vmatrix} \dfrac{1}{L^2} (4k_1 + k_2) - \dfrac{2N}{L} & \dfrac{1}{2L^2} (-4k_1 - 4k_2) + \dfrac{N}{L} \\ \dfrac{1}{2L^2} (-4k_1 - 4k_2) + \dfrac{N}{L} & \dfrac{1}{L^2} (k_1 + 4k_2) - \dfrac{2N}{L} \end{Vmatrix} = 0 \qquad (k)$$

Expanding the determinant given by Eqs. (k) and solving for N gives

$$N_{cr} = -\frac{3}{2L} (k_1 - k_2) \pm \sqrt{\frac{1}{12L^2} (13k_1{}^2 - 118k_1 k_2 + 13k_2{}^2)} \qquad (l)$$

Equation (l) gives two values for N_{cr} corresponding to the first and second buckling modes. The two mode shapes are found by substituting these two critical loads separately into either of Eqs. (k) and solving for one of the generalized coordinates in terms of the other. $\qquad ////$

16-3 DERIVATION OF THE GENERAL EQUATIONS OF MOTION

As is evident in the above three examples, the kinetic and potential energies of linear engineering systems subjected to small-amplitude oscillations can be expressed in the quadratic forms

$$T = \frac{1}{2} \sum_{j=1}^{N} \sum_{i=1}^{N} m_{ij}\dot{q}_i\dot{q}_j = {}^1\!/{}_2\dot{\mathbf{q}}^T\mathbf{m}\dot{\mathbf{q}} \qquad (16\text{-}8)$$

$$V = \frac{1}{2} \sum_{j=1}^{N} \sum_{i=1}^{N} k_{ij}q_iq_j = {}^1\!/{}_2\mathbf{q}^T\mathbf{k}\mathbf{q} \qquad (16\text{-}9)$$

where N is the number of degrees of freedom in the system. For such systems, the second term of Eqs. (16-7), namely, $\partial T/\partial q_i$ ($i = 1, 2, \ldots, N$), equals zero, which reduces Lagrange's equations to the form

$$\frac{\partial}{\partial t}\left(\frac{\partial T}{\partial \dot{q}_i}\right) + \frac{\partial v}{\partial q_i} = Q_i \qquad i = 1, 2, \ldots, N \qquad (16\text{-}10)$$

When Eqs. (16-8) and (16-9) are substituted into Eqs. (16-10), Lagrange's equations of motion, when placed in matrix form, become

$$\mathbf{m}\ddot{\mathbf{q}} + \mathbf{k}\mathbf{q} = \mathbf{Q} \qquad (16\text{-}11)$$

which are similar to the discrete-coordinate equations formulated earlier by virtual work. It must be remembered, however, that *all* nonconservative forces, *including damping forces*, are contained here in the generalized forcing functions $Q_1, Q_2, \ldots, Q_N$.

Now the discretization problem will be considered, i.e., approximating infinite-DOF systems by a finite number of coordinates. For example, the lateral deflections $v(x,t)$ of a flexural member can be approximated by the relation

$$v(x,t) \doteq q_1(t)\psi_1(x) + q_2(t)\psi_2(x) + \cdots + q_N(t)\psi_N(x) \qquad (16\text{-}12)$$

where q_i ($i = 1, 2, \ldots, N$) are generalized coordinates and ψ_i ($i = 1, 2, \ldots, N$) are assumed dimensionless shape functions which satisfy the prescribed geometric boundary conditions for the member.

If $m(x)$ is the mass per unit length for the member, the kinetic energy (neglecting rotational inertia effects) can be expressed

$$T = {}^1\!/{}_2 \int m(x)\dot{v}(x,t)^2 \, dx \qquad (16\text{-}13)$$

Substituting Eq. (16-12) into Eq. (16-13) gives Eq. (16-8):

$$T = \frac{1}{2} \sum_{j=1}^{N} \sum_{i=1}^{N} m_{ij}\dot{q}_i\dot{q}_j$$

in which

$$m_{ij} = \int m(x)\psi_i(x)\psi_j(x) \, dx \qquad (16\text{-}14)$$

The flexural strain energy is given by

$$V = {}^1\!/{}_2 \int EI(x)[v''(x,t)]^2 \, dx \qquad (16\text{-}15)$$

Substituting Eq. (16-12) into Eq. (16-15) gives

$$V = \frac{1}{2} \sum_{j=1}^{N} \sum_{i=1}^{N} k_{ij} q_i q_j \qquad (16\text{-}9)$$

in which

$$k_{ij} = \int EI(x)\psi_i''(x)\psi_j''(x)\, dx \qquad (16\text{-}16)$$

To obtain the generalized forcing functions $Q_1, Q_2, \ldots, Q_N$, the virtual work δW_{nc} must be evaluated. This is the work performed by *all* nonconservative forces acting on or within the flexural member while an arbitrary set of virtual displacements $\delta q_1, \delta q_2, \ldots, \delta q_N$ is applied to the system. To illustrate the principles involved in this evaluation, it will be assumed that the material of the flexure member obeys the uniaxial stress-strain relation

$$\sigma(t) = E\varepsilon(t) + c_s \dot{\varepsilon}(t) \qquad (16\text{-}17)$$

where E is Young's modulus of elasticity and c_s is a damping modulus. Using Eq. (16-17) and the Bernoulli-Euler hypothesis that the normal strains vary linearly over the member cross section leads to the moment-displacement relation

$$m(x,t) = EI(x)v''(x,t) + c_s I(x)\dot{v}''(x,t) \qquad (16\text{-}18)$$

The first term on the right-hand side of Eq. (16-18) results from the internal conservative forces, which have already been accounted for in the potential-energy term V, while the second term results from the internal nonconservative forces. The virtual work performed by these nonconservative forces per unit length along the member equals the negative of the product of the nonconservative moment $c_s I(x)\dot{v}''(x,t)$ times the variation in the curvature $\delta v''(x,t)$. Therefore, the total virtual work performed by these internal nonconservative forces is

$$\delta W_{nc,\text{int}} = -\int c_s I(x)\dot{v}''(x,t)\, \delta v''(x,t)\, dx \qquad (16\text{-}19)$$

If the externally applied nonconservative forces are assumed in this case to be limited to a distributed transverse loading $p(x,t)$, the virtual work performed by these forces equals

$$\delta W_{nc,\text{ext}} = \int p(x,t)\, \delta v(x,t)\, dx \qquad (16\text{-}20)$$

Substituting Eq. (16-12) into Eqs. (16-19) and (16-20) and adding gives

$$\delta W_{nc,\text{total}} = \sum_{i=1}^{N} \left(p_i - \sum_{j=1}^{N} c_{ij} \dot{q}_j \right) \delta q_i \qquad (16\text{-}21)$$

where

$$p_i = \int p(x,t)\psi_i(x)\,dx \qquad (16\text{-}22)$$

$$c_{ij} = \int c_s I(x)\psi_i''(x)\psi_j''(x)\,dx \qquad (16\text{-}23)$$

When Eq. (16-21) is compared with Eq. (16-3c), it is evident that

$$Q_i = p_i - \sum_{j=1}^{N} c_{ij}\dot{q}_j \qquad (16\text{-}24)$$

Finally, substituting Eqs. (16-8), (16-9), and (16-24) into Lagrange's equations (16-7) gives the governing equations of motion in matrix form

$$\mathbf{m\ddot{q} + c\dot{q} + kq = p} \qquad (16\text{-}25)$$

Note from the definitions of m_{ij}, c_{ij}, and k_{ij} as given by Eqs. (16-14), (16-23), and (16-16), respectively, that

$$m_{ij} = m_{ji} \qquad c_{ij} = c_{ji} \qquad k_{ij} = k_{ji} \qquad (16\text{-}26)$$

Therefore, the mass, damping, and stiffness coefficient matrices of Eq. (16-25) are symmetric in form.

EXAMPLE E16-4 The formulation of the equations of motion by the general Lagrange's equation procedure described above will be illustrated for the rigid-bar assemblage shown in Fig. E16-3. The bars are interconnected by hinges, and their relative rotations are resisted by rotational springs and dashpots located at each hinge with values as indicated. The generalized coordinates of this system are taken to be the rotation angles q_i of the rigid bars, as shown in the sketch; it will be assumed that the displacements are small so that the small-deflection theory is valid.

With the kinetic energy of the rigid bars due to rotation about their individual centroids and due to translation of the controids considered separately the total kinetic energy is

$$
\begin{aligned}
T &= \frac{1}{2}\frac{WL^2}{12g}(\dot{q}_1{}^2 + \dot{q}_2{}^2 + \dot{q}_3{}^2) \\
&\quad + \frac{1}{2}\frac{W}{g}\left[\left(\frac{\dot{q}_1 L}{2}\right)^2 + \left(\dot{q}_1 L + \frac{\dot{q}_2 L}{2}\right)^2 + \left(\dot{q}_1 L + \dot{q}_2 L + \frac{\dot{q}_3 L}{2}\right)^2\right] \\
&= \frac{WL^2}{6g}(2\dot{q}_1{}^2 + 4\dot{q}_2{}^2 + \dot{q}_3{}^2 + 9\dot{q}_1\dot{q}_2 + 3\dot{q}_2\dot{q}_3 + 3\dot{q}_1\dot{q}_3) \qquad (a)
\end{aligned}
$$

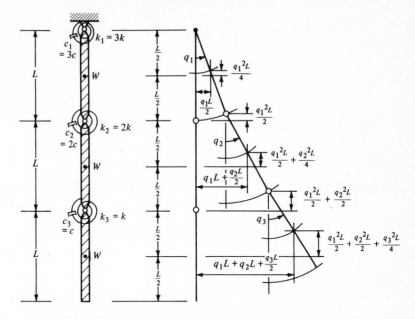

FIGURE E16-3
A 3-DOF rigid-body assemblage (including geometric-stiffness effect due to dead weight).

Also with the potential energy associated with deformation of the rotational springs and due to the raising of the bars above their vertical base position considered separately (the amounts of the vertical displacements of the centers of gravity are indicated on the sketch), the total potential energy of the system is given by

$$V = W\left[\frac{q_1^2 L}{4} + \left(\frac{q_2^2 L}{2} + \frac{q_2^2 L}{4}\right) + \left(\frac{q_1^2 L}{2} + \frac{q_2^2 L}{2} + \frac{q_3^2 L}{4}\right)\right]$$

$$+ \frac{1}{2}[k_1 q_1^2 + k_2(q_2 - q_1)^2 + k_3(q_3 - q_2)^2]$$

$$= \frac{1}{4}[(5WL + 10k)q_1^2 + (3WL + 6k)q_2^2$$

$$+ (WL + 4k)q_3^2 - 8q_1 q_2 - 4q_2 q_3] \tag{b}$$

Finally, the virtual work done by the rotational dashpots during the virtual displacements of the structure is given by

$$\delta W_{nc} = c_1 \dot{q}_1 \, \delta q_1 - c_2(\dot{q}_2 - \dot{q}_1)(\delta q_2 - \delta q_1) - c_3(\dot{q}_3 - \dot{q}_2)(\delta q_3 - \delta q_2)$$

$$= c[(-5\dot{q}_1 + 2\dot{q}_2) \, \delta q_1 + (2\dot{q}_1 - 3\dot{q}_2 + \dot{q}_3) \, \delta q_2 + (\dot{q}_2 - \dot{q}_3) \, \delta q_3]$$

from which the nonconservative forces, which are due only to damping, become

$$Q_1 = c(-5\dot{q}_1 + 2\dot{q}_2)$$

$$Q_2 = c(2\dot{q}_1 - 3\dot{q}_2 + \dot{q}_3) \qquad (c)$$

$$Q_3 = c(\dot{q}_2 - \dot{q}_3)$$

Substituting Eqs. (a) to (c) into

$$\frac{d}{dt}\left(\frac{\partial T}{\partial \dot{q}_i}\right) + \frac{\partial v}{\partial q_i} = Q_i \qquad i = 1, 2, 3 \qquad (16\text{-}10a)$$

gives the three equations of motion of the system, which, arranged in matrix form, are

$$\frac{WL^2}{6g}\begin{bmatrix} 14 & 9 & 3 \\ 9 & 8 & 3 \\ 3 & 3 & 2 \end{bmatrix}\begin{bmatrix} \ddot{q}_1 \\ \ddot{q}_2 \\ \ddot{q}_3 \end{bmatrix} + c\begin{bmatrix} 5 & -2 & 0 \\ -2 & 3 & -1 \\ 0 & -1 & 1 \end{bmatrix}\begin{bmatrix} \dot{q}_1 \\ \dot{q}_2 \\ \dot{q}_3 \end{bmatrix}$$

$$+ \frac{1}{2}\begin{bmatrix} 5WL + 10k & -4k & 0 \\ -4k & 3WL + 6k & -2k \\ 0 & -2k & WL + 4k \end{bmatrix}\begin{bmatrix} q_1 \\ q_2 \\ q_3 \end{bmatrix} = \begin{bmatrix} 0 \\ 0 \\ 0 \end{bmatrix} \qquad (d)$$

////

16-4 CONSTRAINTS AND LAGRANGE MULTIPLIERS

Usually when determining the dynamic response of an N-DOF system, the equations of motion are written in terms of a set of generalized coordinates $q_1, q_2, \ldots, q_N$; however, there are cases where in order to maintain symmetry in the equations of motion, it is preferable to select a set of coordinates $g_1, g_2, \ldots, g_c$, where $c > N$. These coordinates cannot be generalized coordinates since their number exceeds the number of degrees of freedom in the system. Therefore, one must impose m ($m = c - N$) equations of constraint to the system. For example, returning to the double pendulum shown in Fig. 16-1, it was pointed out earlier that the equations of motion could be expressed in terms of generalized coordinates θ_1 and θ_2 ($N = 2$) or in terms of coordinates x_1, y_1, x_2, y_2 ($c = 4$). If the latter coordinates are used, two equations of constraint, namely Eqs. (16-1), must be satisfied.

Suppose the m equations of constraint for a general case are expressed in the form

$$f_1(g_1, g_2, \ldots, g_c) = 0$$

$$f_2(g_1, g_2, \ldots, g_c) = 0$$

$$\cdots\cdots\cdots\cdots\cdots\cdots \qquad (16\text{-}27)$$

$$f_m(g_1, g_2, \ldots, g_c) = 0$$

Taking the variations of Eqs. (16-27) results in

$$\delta f_1 = \frac{\partial f_1}{\partial g_1} \delta g_1 + \frac{\partial f_1}{\partial g_2} \delta g_2 + \cdots + \frac{\partial f_1}{\partial g_c} \delta g_c = 0$$

$$\delta f_2 = \frac{\partial f_2}{\partial g_1} \delta g_1 + \frac{\partial f_2}{\partial g_2} \delta g_2 + \cdots + \frac{\partial f_2}{\partial g_c} \delta g_c = 0 \qquad \text{(16-28)}$$

$$\cdots\cdots\cdots\cdots\cdots\cdots\cdots\cdots\cdots\cdots\cdots\cdots\cdots$$

$$\delta f_m = \frac{\partial f_m}{\partial g_1} \delta g_1 + \frac{\partial f_m}{\partial g_2} \delta g_2 + \cdots + \frac{\partial f_m}{\partial g_c} \delta g_c = 0$$

Now if each δf_i $(i = 1, 2, \ldots, m)$ is multiplied by an unknown time function $\lambda_i(t)$ and the product is integrated over the time interval t_1 to t_2 [assuming Eqs. (16-3) to apply, when expressed in terms of coordinates $g_1, g_2, \ldots, g_c$], then if each of the above integrals is added to Hamilton's variational equation [Eq. (16.2)], the following equation is obtained after completing the variation:

$$\int_{t_1}^{t_2} \left\{ \sum_{i=1}^{c} \left[-\frac{d}{dt}\left(\frac{\partial T}{\partial \dot{g}_i}\right) + \frac{\partial T}{\partial g_i} - \frac{\partial V}{\partial g_i} + Q_i \right.\right.$$

$$\left.\left. + \lambda_1 \frac{\partial f_1}{\partial g_i} + \lambda_2 \frac{\partial f_2}{\partial g_i} + \cdots + \lambda_m \frac{\partial f_m}{\partial g_i} \right] \delta g_i \right\} dt = 0 \qquad \text{(16-29)}$$

Since the variations δg_i $(i = 1, 2, \ldots, c)$ are all arbitrary, it is necessary that each square-bracket term in Eq. (16-29) equal zero, i.e.,

$$\frac{d}{dt}\left(\frac{\partial T}{\partial \dot{g}_i}\right) - \frac{\partial T}{\partial g_i} + \frac{\partial V}{\partial g_i} = Q_i$$

$$+ \lambda_1 \frac{\partial f_1}{\partial g_i} + \lambda_2 \frac{\partial f_2}{\partial g_i} + \cdots + \lambda_m \frac{\partial f_m}{\partial g_m} = 0 \qquad i = 1, 2, \ldots, c \qquad \text{(16-30)}$$

Equation (16-30) is a modified form of Lagrange's equations which will permit the use of coordinates $g_1, g_2, \ldots, g_c$. This procedure of developing Eqs. (16-30) may seem trivial at first because a number of integrals equaling zero have been added to Hamilton's equation; however, it should be noted that while each δf_i $(i = 1, 2, \ldots, m)$ equals zero, the individual terms given on the right-hand side of Eqs. (16-28) are not equal to zero. The time-dependent functions λ_i $(i = 1, 2, \ldots, m)$ are known as *Lagrange multipliers*.

When a reduced potential-energy term $\overline{V}$ is defined as

$$\overline{V} = V(g_1, g_2, \ldots, g_c) - (\lambda_1 f_1 + \lambda_2 f_2 + \cdots + \lambda_m f_m) \qquad \text{(16-31)}$$

Eqs. (16-30) can be written

$$\frac{d}{dt}\left(\frac{\partial T}{\partial \dot{g}_i}\right) - \frac{\partial T}{\partial g_i} + \frac{\partial \overline{V}}{\partial g_i} = Q_i \qquad i = 1, 2, \ldots, c \qquad \text{(16-32)}$$

which contain the unknown time functions $g_1, g_2, \ldots, g_c, \lambda_1, \lambda_2, \ldots, \lambda_m$. Since there are $c + m$ unknown time functions, $c + m$ equations are required for their

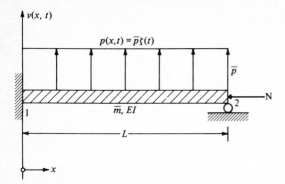

FIGURE E16-4
Uniform beam used to demonstrate
Lagrange multipliers.

solution. These equations include the c modified Lagrange's equations [Eqs. (16-32)] and the m constraint equations [Eqs. (16-27)].

EXAMPLE E16-5 The use of Lagrange multipliers in satisfying specified constraint conditions will be illustrated with reference to the end-supported cantilever beam of Fig. E16-4. This beam is subjected to a time-varying loading, uniformly distributed along its length $\bar{p}\zeta(t)$, as well as to a constant axial force N, as shown in the sketch; its stiffness is uniform along the length, and there is no damping. To obtain an approximate solution which is reasonably valid if the frequency components in the loading function are low enough, it will be assumed that the beam deflections can be expressed as

$$v(x,t) = g_1(t) \sin \frac{\pi x}{L} + g_2(t) \sin \frac{2\pi x}{L} \qquad (a)$$

Expressing the kinetic and potential energies and the virtual work performed by the external loading in terms of the coordinates g_1 and g_2 leads to

$$T = \frac{1}{2} \int_0^L \bar{m} \left(\dot{g}_1{}^2 \sin^2 \frac{\pi x}{L} + 2\dot{g}_1 \dot{g}_2 \sin \frac{\pi x}{L} \sin \frac{2\pi x}{L} + \dot{g}_2{}^2 \sin^2 \frac{2\pi x}{L} \right) dx \qquad (b)$$

$$V = \frac{1}{2} \int_0^L EI \left(g_1{}^2 \frac{\pi^4}{L^4} \sin^2 \frac{\pi x}{L} + \frac{8\pi^4}{L^4} g_1 g_2 \sin \frac{\pi x}{L} \sin \frac{2\pi x}{L} \right.$$
$$\left. + g_2{}^2 \frac{16\pi^4}{L^4} \sin^2 \frac{2\pi x}{L} \right) dx$$
$$- \frac{N}{2} \int_0^L \left(\frac{\pi^2}{L^2} g_1{}^2 \cos^2 \frac{\pi x}{L} + \frac{4\pi^2}{L^2} g_1 g_2 \cos \frac{\pi x}{L} \cos \frac{2\pi x}{L} \right.$$
$$\left. + \frac{4\pi^2}{L^2} g_2{}^2 \cos^2 \frac{2\pi x}{L} \right) dx \qquad (c)$$

$$\delta W_{nc} = \delta g_1 \int_0^L p(x,t) \sin \frac{\pi x}{L} \, dx + \delta g_2 \int_0^L p(x,t) \sin \frac{2\pi x}{L} \, dx \qquad (d)$$

Completing the integrals of Eqs. (b) to (d) gives

$$T = \frac{\bar{m}L}{4} (\dot{g}_1^{\,2} + \dot{g}_2^{\,2}) \qquad (e)$$

$$V = \frac{\pi^4 EI}{4L^3} (g_1^{\,2} + 16g_2^{\,2}) - \frac{N\pi^2}{4L} (g_1^{\,2} + 4g_2^{\,2}) \qquad (f)$$

$$\delta W_{nc} = \frac{2L}{\pi} \bar{p}\zeta(t) \, \delta g_1 \qquad (g)$$

and comparing Eq. (g) with Eq. (16-3c) gives the external loads

$$Q_1 = \frac{2L\bar{p}\zeta(t)}{\pi} \qquad Q_2 = 0 \qquad (h)$$

When the fixed-support condition at the left end of the beam is considered, it is evident that the solution must satisfy the constraint condition

$$f_1(g_1, g_2) = g_1 + 2g_2 = 0 \qquad (i)$$

Substituting Eqs. (f) and (i) into Eqs. (16-31) thus leads to the reduced potential

$$\bar{V} = \frac{\pi^4 EI}{4L^3} (g_1^{\,2} + 16g_2^{\,2}) - \frac{N\pi^2}{4L} (g_1^{\,2} + 4g_2^{\,2}) - \lambda_1(g_1 + 2g_2) \qquad (j)$$

Substituting Eqs. (e), (h), and (j) into the Lagrange reduced equations of motion [Eqs. (16-32)] finally gives

$$\frac{\bar{m}L}{2} \ddot{g}_1 + \left(\frac{\pi^4 EI}{2L^3} - \frac{\pi^2 N}{2L} \right) g_1 - \lambda_1 = \frac{2L\bar{p}\zeta(t)}{\pi}$$

$$\frac{\bar{m}L}{2} \ddot{g}_2 + \left(\frac{8\pi^4 EI}{L^3} - \frac{2\pi^2 N}{L} \right) g_2 - 2\lambda_1 = 0 \qquad (k)$$

From this point the complete solution of the problem can be obtained by solving Eq. (i) and Eqs. (k) for $g_1(t)$, $g_2(t)$, and $\lambda_1(t)$. The resulting solution shows that $\lambda_1(t)$ is proportional to the fixed end moment at $x = 0$. This moment performs zero virtual work on the member because the constraint at that location does not permit a virtual rotation of the member cross section.   ////

PROBLEMS

16-1 Applying Lagrange's equations, Eqs. (16-7), and permitting large displacements, determine the equation of motion for the system shown in Fig. E2-4. What is the linearized equation of motion for small amplitude oscillation?

16-2 Applying Lagrange's equations and permitting large displacements, determine the equations of motion for the system shown in Fig. P16-1. What are the linearized equations of motion for small-amplitude oscillations?

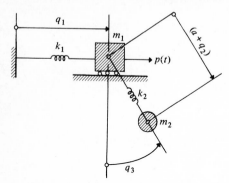

FIGURE P16-1

16-3 Repeat Prob. 16-1 for the system shown in Prob. 2-2.

16-4 Repeat Prob. 16-1 for the system shown in Prob. 2-3.

16-5 Obtain the equations of motion for the uniform cantilever beam shown in Fig. P16-2 when the deflected shape can be approximated by the relation

$$v(x, t) \doteq q_1(t)\left(\frac{x}{L}\right)^2 + q_2(t)\left(\frac{x}{L}\right)^3 + q_3(t)\left(\frac{x}{L}\right)^4$$

Assume small deflection theory.

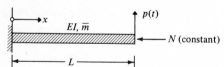

FIGURE P16-2

16-6 A ball of radius R_1 and mass m_1 is placed at rest on top of a fixed cylindrical surface of radius R_2. Assume a very slight disturbance that starts the ball rolling to the left, as shown in Fig. P16-3, under the influence of gravity. If the ball rolls without slippage and angles θ_1 and θ_2 are taken as displacement coordinates:

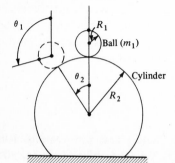

FIGURE P16-3

(a) Determine the equation of constraint between θ_1 and θ_2.

(b) Write the equation of motion in terms of one displacement coordinate by eliminating the other through the constraint equation.

(c) Write the equation of motion using both displacement coordinates and in addition using a Lagrange multiplier λ_1. (What does λ_1 represent physically in this case?)

(d) Determine the value of θ_2 when the ball leaves the surface of the cylinder.

16-7 A uniform rigid bar of total mass m_1 and length L swings as a pendulum under the influence of gravity. A concentrated mass m_2 is constrained to slide along the axis of the bar and is attached to a massless spring, as shown in Fig. P16-4. Assuming a frictionless system and large amplitude displacements, determine the equations of motion in terms of generalized coordinates q_1 and q_2.

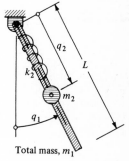

FIGURE P16-4   Total mass, m_1

16-8 Determine the linearized equations of motion for small-amplitude oscillations of the system defined in Prob. 16-7.

PART THREE

Distributed-Parameter Systems

PARTIAL DIFFERENTIAL
EQUATIONS OF MOTION

17-1 INTRODUCTION

The discrete-coordinate systems described in Part Two provide a convenient and practical approach to the dynamic-response analysis of arbitrary structures. However, the solution obtained with these coordinates can only approximate the actual dynamic behavior because the motions of the system are represented by a limited number of displacement coordinates. The precision of the results can be made as refined as desired by increasing the number of degrees of freedom considered in the analysis. But in principle an infinite number of coordinates would be required to converge to the exact results for any real structure having continuously distributed properties; hence this approach to obtaining an exact solution is manifestly impossible.

The formal mathematical procedure for considering the behavior of an infinite number of connected points is by means of differential equations in which the position coordinates are taken as independent variables. Inasmuch as time also is an independent variable in a dynamic problem, the formulation of the equations of motion in this way leads to a partial differential equation. Different classes of continuous systems can be identified in accordance with the number of independent variables required to describe the distribution of their physical properties. For example, the wave-propagation formulas used in seismology and geophysics are derived from the equations of motion expressed for general three-dimensional solids. Similarly, in studying the dynamic behavior of thin-plate or thin-shell structures, special equations of motion must be derived for these two-dimensional systems. In the present discussion, however, attention will be limited to one-dimensional structures, that is, beam- and rod-type systems in which it may be assumed that the physical properties (mass, stiffness, etc.) are expressed with reference to a single dimension, the position along the elastic axis. Thus, the partial differential equations of these systems involve only two independent variables, time and distance along the axis.

It is possible to derive the equations of motion for rather complex one-dimensional structures, including assemblages of many members in three-dimensional space. Moreover, the axes of the individual members might be arbitrarily curved in three-dimensional space, and the physical properties might vary as a complicated function of position along the axis. However, the solutions of the equations of motion for such complex systems generally can be obtained only by numerical means, and in most cases a discrete-coordinate formulation is preferable to a numerical solution of the continuum equations. For this reason, the present treatment will be limited to members having straight elastic axes and to assemblages of such members. In formulating the equations of motion, general variations of the physical properties along the axis will be permitted, although in subsequent solutions of these equations, the properties of each member will be assumed to be constant. Because of these severe limitations of the cases which may be considered, this presentation is intended mainly to demonstrate the general concepts of the partial-differential-equation formulation rather than to provide a tool for significant practical application.

17-2 BEAM FLEXURE: ELEMENTARY CASE

The first case to be considered in the formulation of continuum equations of motion is the straight, nonuniform beam shown in Fig. 17-1a. The significant physical properties of this beam are assumed to be the flexural stiffness $EI(x)$ and the mass per unit length $m(x)$, both of which may vary arbitrarily with position x along the span L. The transverse loading $p(x,t)$ is assumed to vary arbitrarily with position and time, and the transverse-displacement response $v(x,t)$ also is a function of these variables. The end-support conditions for the beam are arbitrary, although they are pictured as simple supports for illustrative purposes.

The equation of motion of this simple system can readily be derived by considering the equilibrium of forces acting on the differential segment of the beam shown in Fig. 17-1b, in much the same way that the equations were developed for a

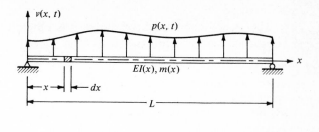

(a)

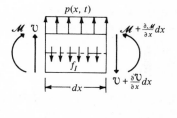

(b)

FIGURE 17-1
Basic beam subjected to dynamic loading: (a) beam properties and coordinates;
(b) forces acting on differential elements.

discrete-parameter system. Summing all forces acting vertically leads to the first dynamic-equilibrium relationship

$$\upsilon + p \, dx - \left(\upsilon + \frac{\partial \upsilon}{\partial x} \, dx\right) - f_i \, dx = 0 \qquad (17\text{-}1)$$

in which $f_i \, dx$ represents the distributed transverse inertia force and is given by the product of the differential mass and the local acceleration:

$$f_i \, dx = m \, dx \, \frac{\partial^2 v}{\partial t^2} \qquad (17\text{-}2)$$

Substituting Eq. (17-2) into Eq. (17-1) and simplifying then gives

$$\frac{\partial \upsilon}{\partial x} = p - m \, \frac{\partial^2 v}{\partial t^2} \qquad (17\text{-}3)$$

which may be recognized as the standard relationship between shear force and transverse load but with the transverse load now including the inertia force of the accelerating beam.

The second equilibrium relationship is obtained by summing moments about the elastic axis at the right-hand face of the segment, as follows:

$$\mathscr{M} + \mathscr{v}\, dx - \left(\mathscr{M} + \frac{\partial \mathscr{M}}{\partial x}\, dx\right) = 0 \qquad (17\text{-}4)$$

where it has been noted that the distributed lateral force makes only a second-order contribution to the moment. This simplifies directly to the standard static relationship between shear and moment

$$\frac{\partial \mathscr{M}}{\partial x} = \mathscr{v} \qquad (17\text{-}5)$$

No inertia forces contribute in this case to the moment equilibrium. Differentiating Eq. (17-5) with respect to x and substituting into Eq. (17-3) yields, after rearrangement,

$$\frac{\partial^2 \mathscr{M}}{\partial x^2} + m\frac{\partial^2 v}{\partial t^2} = p \qquad (17\text{-}6)$$

Finally, introducing the basic moment-curvature relationship of elementary beam theory ($\mathscr{M} = \partial^2 v/\partial x^2$) leads to the partial differential equation of motion for this elementary case of beam flexure

$$\frac{\partial^2}{\partial x^2}\left(EI\,\frac{\partial^2 v}{\partial x^2}\right) + m\frac{\partial^2 v}{\partial t^2} = p \qquad (17\text{-}7)$$

in which both EI and m are assumed to vary arbitrarily with x.

17-3 BEAM FLEXURE: INCLUDING AXIAL-FORCE EFFECTS

If the beam considered in the previous case is subjected to a force parallel to its axis in addition to the lateral loading shown in Fig. 17-1, the local equilibrium of forces is altered because the axial force interacts with the lateral displacements to produce an additional term in the moment-equilibrium expression. Consider the beam shown in Fig. 17-2, in which the axial force is assumed to be constant with respect both to time and position. (In principle, it could vary as an arbitrary function of x, but the constant case will be considered for simplicity.) It is apparent in Fig. 17-2b that transverse equilibrium is not affected by the axial force because its *direction* does not change with the beam deflection; hence Eq. (17-3) is still valid. However, the point of application of the axial force changes with the beam deflection so that the moment-equilibrium equation now becomes

$$\mathscr{M} + \mathscr{v}\, dx - N\frac{\partial v}{\partial x}\, dx - \left(\mathscr{M} + \frac{\partial \mathscr{M}}{\partial x}\, dx\right) = 0 \qquad (17\text{-}8)$$

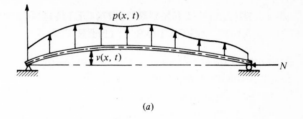

(a)

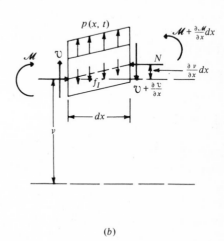

(b)

FIGURE 17-2
Beam with static axial force and dynamic lateral load: (a) beam deflected due to
loading; (b) forces acting on differential element.

from which the transverse force $\mathcal{V}$ is found to be

$$\mathcal{V} = N \frac{\partial v}{\partial x} + \frac{\partial \mathcal{M}}{\partial x} \qquad (17\text{-}9)$$

Introducing this modified expression for $\mathcal{V}$ into Eq. (17-3) and proceeding as
before gives the final equation of motion including the effects of axial force as

$$\frac{\partial^2}{\partial x^2}\left(EI \frac{\partial^2 v}{\partial x^2} \right) + N \frac{\partial^2 v}{\partial x^2} + m \frac{\partial^2 v}{\partial t^2} = p \qquad (17\text{-}10)$$

From a comparison of Eq. (17-10) with Eq. (17-7) it is evident that the product of
axial force and curvature gives rise to an additional effective transverse load acting
on the beam. Also, it should be noted that the force $\mathcal{V}$ in this derivation acts
vertically; it is *not* the section shear force because it is not acting normal to the
elastic axis.

17-4   BEAM FLEXURE: INCLUDING SHEAR DEFORMATION AND ROTATORY INERTIA

Although Eq. (17-7) or (17-10) is applicable in the vast majority of cases, these equations neglect two factors which may influence the dynamic response appreciably if the span-depth ratio of the beam is relatively small, namely the deformations due to shear forces and the inertial resistance to rotational acceleration of the beam cross section. The components of the beam deflection associated with each of these effects are shown in Fig. 17-3b. The rotatory inertia results from the rotation of the beam cross section α from its original vertical position. If there were no shear distortion, the cross sections would remain normal to the elastic axis and α would be equal to the slope of the elastic axis. However, when the distortion due to shear stress is considered, the beam deformation is much more complicated. Assuming as a first approximation that the cross section remains plane, the shear distortion is represented by a term β which reduces the slope of the elastic axis, as shown in the sketch.

Now the equilibrium of forces acting on the differential element is considered, as shown in Fig. 17-3c (axial forces are neglected for simplicity); it will be noted that the vertical equilibrium is not affected by the rotatory inertia and is still represented by Eq. (17-3). The rotational inertia per unit length $\bar{m}_I$ contributes directly to the moment-equilibrium relationship, however, as follows:

$$\mathcal{M} + \mathcal{V}\,dx + \bar{m}_I\,dx - \left(\mathcal{M} + \frac{\partial \mathcal{M}}{\partial x}\,dx\right) = 0 \qquad (17\text{-}11)$$

The rotational inertia is given by the product of the mass moment of inertia of the section and the angular acceleration

$$\bar{m}_I = \rho I \frac{\partial^2 \alpha}{\partial t^2}$$

where ρ is the mass per unit volume ($\rho = m/A$) and I is the moment of inertia of the cross-sectional area; thus

$$\bar{m}_I = m \frac{I}{A} \frac{\partial^2 \alpha}{\partial t^2} = mr^2 \frac{\partial^2 \alpha}{\partial t^2} \qquad (17\text{-}12)$$

in which $r^2 = I/A$ is the radius of gyration of the cross section. Substituting this into Eq. (17-11) and simplifying leads to

$$\frac{\partial \mathcal{M}}{\partial x} = \mathcal{V} + mr^2 \frac{\partial^2 \alpha}{\partial t^2} \qquad (17\text{-}13)$$

In order to express the equilibrium relationships [Eqs. (17-3) and (17-13)] in terms of the displacements of the beam, the shear force–deformation and moment-curvature relationships of elementary beam theory are introduced. The shear force

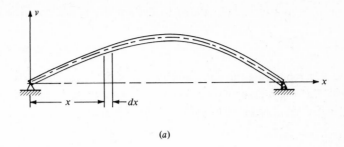

(a)

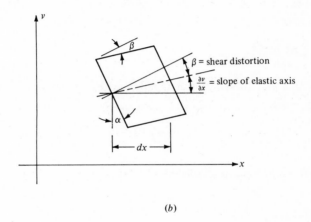

β = shear distortion

$\dfrac{\partial v}{\partial x}$ = slope of elastic axis

(b)

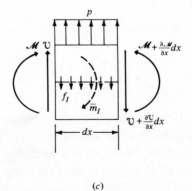

(c)

FIGURE 17-3
Effects of shear distortion and rotatory inertia: (a) dynamic beam deflection;
(b) deformations of differential element; (c) forces acting on differential element.

acting on the cross section is related to the angular rotation of the elastic axis β as follows:

$$v = k'AG\beta$$

in which $k'A$ represents the effective shear area of the section. (It is useful to recall that $k' = {}^5\!/_6$ for a rectangular section.) Differentiating this and substituting into the vertical-equilibrium relationship [Eq. (17-3)] leads to

$$\frac{\partial}{\partial x}(k'AG\beta) = p - m\frac{\partial^2 v}{\partial t^2} \qquad (17\text{-}3a)$$

The moment-curvature relationship, expressed in terms of the cross-sectional rotation angle α, is

$$\mathcal{M} = EI\frac{\partial\alpha}{\partial x}$$

Differentiating this and substituting it together with the shear-distortion expression into Eq. (17-13) leads to

$$\frac{\partial}{\partial x}\left(EI\frac{\partial\alpha}{\partial x}\right) = k'AG\beta + mr^2\frac{\partial^2\alpha}{\partial t^2} \qquad (17\text{-}13a)$$

At this point, the kinematic relationship between the various rotation angles, which is given by

$$\beta = \alpha - \frac{\partial v}{\partial x}$$

(see Fig. 17-3b) can be substituted into the two equilibrium relationships to give

$$\frac{\partial}{\partial x}\left[k'AG\left(\alpha - \frac{\partial v}{\partial x}\right)\right] = p - m\frac{\partial^2 v}{\partial t^2} \qquad (17\text{-}14a)$$

$$\frac{\partial}{\partial x}\left(EI\frac{\partial\alpha}{\partial x}\right) = k'AG\left(\alpha - \frac{\partial v}{\partial x}\right) + mr^2\frac{\partial^2\alpha}{\partial t^2} \qquad (17\text{-}14b)$$

Finally, the rotation angle α can be evaluated from the first of these equilibrium relationships and substituted into the second, leaving the transverse displacement v as the only dependent variable. In order to simplify the resulting expressions, it will be assumed now that the physical properties of the beam do not vary along its length. In this case, solving Eq. (17-14a) for $\partial\alpha/\partial x$ leads to

$$\frac{\partial\alpha}{\partial x} = \frac{\partial^2 v}{\partial x^2} + \frac{1}{k'AG}\left(p - \bar{m}\frac{\partial^2 v}{\partial t^2}\right) \qquad (17\text{-}15)$$

Differentiating Eq. (17-14b) with respect to x and substituting expressions for the appropriate derivatives of $\partial\alpha/\partial x$ from Eq. (17-15) leads finally to

$$
\underbrace{EI \frac{\partial^4 v}{\partial x^4} - \left(p - \bar{m}\frac{\partial^2 v}{\partial t^2}\right)}_{\text{Elementary case}} \underbrace{- \bar{m}r^2 \frac{\partial^4 v}{\partial x^2\,\partial t^2}}_{\text{Rotatory inertia}}
$$

$$
+ \underbrace{\frac{EI}{k'AG}\frac{\partial^2}{\partial x^2}\left(p - \bar{m}\frac{\partial^2 v}{\partial t^2}\right)}_{\text{Shear distortion}} \underbrace{- \frac{\bar{m}r^2}{k'AG}\frac{\partial}{\partial t^2}\left(p - \bar{m}\frac{\partial^2 v}{\partial t^2}\right)}_{\substack{\text{Combined shear distortion} \\ \text{and rotatory inertia}}} = 0 \qquad (17\text{-}16)
$$

As indicated in Eq. (17-16), it is possible to identify the various terms in this equation of motion which are associated with the elementary formulation and with the additional effects of shear distortion and rotatory inertia.

17-5 BEAM FLEXURE: INCLUDING VISCOUS DAMPING

In the foregoing formulations of the equations of motion of beam-type members, no consideration was given to mechanisms which absorb energy from the structure during its dynamic response. Two types of viscous (velocity-dependent) damping can be incorporated into the formulation without difficulty. These types, shown in Fig. 17-4, include a viscous resistance to transverse displacement of the beam and a viscous resistance to straining of the beam material. If the resistance to transverse velocity is represented by $c(x)$, the corresponding damping force is $f_D(x) = c(x)\,\partial v/\partial t$, and this term will contribute to the transverse-equilibrium relationship [Eq. (17-3)] with the result

$$
\frac{\partial \mathcal{V}}{\partial x} = p - m\frac{\partial^2 v}{\partial t^2} - c\frac{\partial v}{\partial t} \qquad (17\text{-}17)
$$

Similarly, if the resistance to strain velocity is represented by c_s, the damping stress is $\sigma_D = c_s\,\partial\varepsilon/\partial t$ where ε is the local normal strain. Assuming that the strains vary linearly over the section (Navier's hypothesis), it is easy to show that a damping moment results which is given by

$$
\mathcal{M}_D(x) = \int \sigma_D y\,dA = c_s I(x)\frac{\partial^3 v}{\partial x^2\,\partial t} \qquad (17\text{-}18)
$$

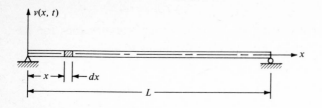

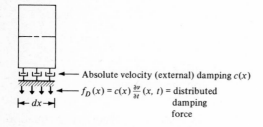

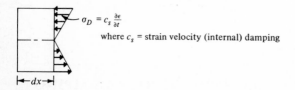

FIGURE 17-4
Viscous-damping mechanisms in a beam.

Incorporating this damping moment into the moment equilibrium relationship [Eq. (17-5)] and substituting into Eq. (17-17) leads to the differential equation of motion including damping

$$\frac{\partial}{\partial x^2}\left(EI\,\frac{\partial^2 v}{\partial t^2} + c_s I\,\frac{\partial^3 v}{\partial x^2\,\partial t}\right) + m\,\frac{\partial^2 v}{\partial t^2} + c\,\frac{\partial v}{\partial t} = p \qquad (17\text{-}19)$$

17-6 BEAM FLEXURE: GENERALIZED SUPPORT EXCITATIONS

Equation (17-19) is the partial differential equation of motion of a general beam segment having mass, stiffness, and damping properties which vary arbitrarily along its length and subjected to an external loading $p(x,t)$ which varies arbitrarily with position and time. Another general form of excitation which has great practical significance is a prescribed displacement history of the supports. A practical example of this type of excitation is the effect of an earthquake on any structure supported by the ground; however, many other examples of such excitation occur—in moving structures and due to mechanically induced vibrations.

A convenient and generally applicable method of accounting for support excitation in a distributed-parameter beam-type structure is the continuum equivalent of the formulation which is developed in the chapter on earthquake-response analysis (Sec. 27-4) for discrete-coordinate systems. The derivation for a one-dimensional continuum will be presented here for convenience and completeness. The basic step in the formulation is the expression of the total displacement of the beam v^t as the sum of the displacement which would be induced by the support motion applied statically (the so-called pseudostatic displacement v^s) plus the additional displacement due to the dynamic (inertial and viscous drag) effects v; thus

$$v^t(x,t) = v^s(x,t) + v(x,t) \qquad (17\text{-}20)$$

The various force quantities on the left-hand side of Eq. (17-19) all depend on the total displacement; thus in the context of Eq. (17-20), Eq. (17-19) becomes

$$\frac{\partial}{\partial x^2}\left(EI\,\frac{\partial^2 v^t}{\partial x^2} + c_s I\,\frac{\partial^3 v^t}{\partial x^2\,\partial t}\right) + m\,\frac{\partial^2 v^t}{\partial t^2} + c\,\frac{\partial v^t}{\partial t} = 0$$

where the loading term vanishes from the right-hand side if it is assumed that the only excitation results from support displacements. Substituting Eq. (17-20) and transferring terms associated with the pseudostatic displacement to the right-hand side leads to

$$\frac{\partial^2}{\partial x^2}\left(EI\,\frac{\partial^2 v}{\partial x^2} + c_s I\,\frac{\partial^3 v}{\partial x^2\,\partial t}\right) + m\,\frac{\partial^2 v}{\partial t^2} + c\,\frac{\partial v}{\partial t} = p_{\text{eff}} \qquad (17\text{-}21)$$

in which

$$p_{\text{eff}} = -\frac{\partial^2}{\partial x^2}\left(EI\,\frac{\partial^2 v^s}{\partial x^2} + c_s I\,\frac{\partial^3 v^s}{\partial x^2\,\partial t}\right) - m\,\frac{\partial^2 v^s}{\partial t^2} - c\,\frac{\partial v^s}{\partial t} \qquad (17\text{-}22)$$

represents the effective loading applied to the beam segment by the support excitations.

The pseudostatic displacement v^s, which is the source of the effective loading, is defined as the static displacement of the beam segment which would result from any displacements prescribed at its supports. The possible support displacements for the beam segment include the transverse displacements and rotations of the two ends, as shown in Fig. 17-5. The displaced shapes produced by applying a unit value of each of these support movements can be calculated by standard static beam-deflection analysis procedures. If these displacement patterns are designated $\varphi_r(x)$, with appropriate subscripts to identify the corresponding support movement, the pseudostatic deflection can be written

$$v^s(x,t) = \varphi_1(x)\,\delta_1(t) + \varphi_2(x)\,\delta_2(t) + \varphi_3(x)\,\delta_3(t) + \varphi_4(x)\,\delta_4(t)$$

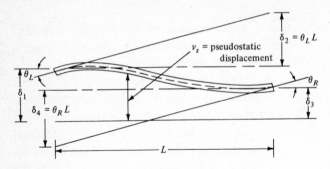

FIGURE 17-5
Pseudostatic displacement produced by beam boundary movements.

or, more concisely,

$$v^s(x,t) = \sum_{r=1}^{4} \varphi_r(x)\, \delta_r(t) \qquad (17\text{-}23)$$

Substituting Eq. (17-23) into Eq. (17-22) and noting that the first term vanishes because there can be no effective loading in an actual static situation leads to

$$p_{\text{eff}} = -\sum_{r=1}^{4} \left\{ m\varphi_r(x)\, \ddot{\delta}_r(t) + c\varphi_r(x)\, \dot{\delta}_r(t) + \frac{\partial^2}{\partial x^2}\left[c_s I(x)\, \frac{\partial^2 \varphi_r(x)}{\partial x^2}\, \dot{\delta}_r(t) \right] \right\} \qquad (17\text{-}24)$$

In most practical cases, the damping contributions to the effective loading are small compared with the inertial contribution. For this reason the last two terms of Eq. (17-24) generally are omitted, and the effective loading is represented by

$$p_{\text{eff}} = -\sum_{r=1}^{4} m\varphi_r(x)\ddot{\delta}_r(t) \qquad (17\text{-}24a)$$

Equation (17-24a) shows that there is a contribution to the effective loading from each of the specified support accelerations. For the common case of a cantilever column subjected to a simple transverse motion of the base, as in the case of an earthquake, the pseudostatic displacement pattern is $\varphi_1(x) = 1$, and no other boundary displacements are specified; thus for this important case Eq. (17-24a) becomes

$$p_{\text{eff}} = -m(x)\, \ddot{\delta}_1(t) \qquad (17\text{-}24b)$$

In some cases, however, particularly in the study of earthquake effects on piping systems, different support motions may be introduced at many points, and a general relationship such as Eq. (17-24a) must be considered.

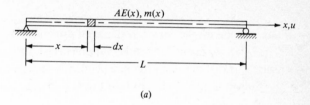

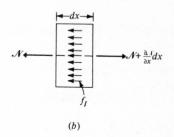

(b)

FIGURE 17-6
Bar subjected to dynamic axial deformations: (a) bar properties and coordinates;
(b) forces acting on differential element.

17-7 AXIAL DEFORMATIONS

The preceding discussions have been concerned with beam flexure, in which the dynamic displacements are in the direction transverse to the elastic axis. Although this bending mechanism is the most important type of behavior to be considered in the dynamic analysis of one-dimensional members, in some cases of dynamic loading the transverse motions are negligible and the response is associated with displacements parallel with the elastic axis. A classic example of this situation is a pile subjected to hammer blows at its end during the driving process. The equations of motion for displacements in a rod in the direction parallel to the axis can be derived by a process similar to that used in the study of transverse motions, but the analysis is much simpler because equilibrium need be considered only in one direction rather than two.

A straight bar for which the axial stiffness AE and mass per unit length vary along the length is shown in Fig. 17-6a, and the forces in the axial direction acting on a differential segment are indicated in Fig. 17-6b. Summing these forces leads to

$$\mathcal{N} + f_I\, dx - \left(\mathcal{N} + \frac{\partial \mathcal{N}}{\partial x}\, dx\right) = 0 \qquad (17\text{-}25)$$

in which f_I represents the inertial force per unit length, which may be expressed as

$$f_I = m\,\frac{\partial^2 u}{\partial t^2} \qquad (17\text{-}26)$$

where u is the displacement in the axial direction. It should be noted that the $\mathcal{N}$ in Eq. (17-25) represents the time-varying axial force associated with varying axial strains; it is distinct from the independently specified axial force N of Eq. (17-10).

Substituting Eq. (17-26) into Eq. (17-25) and simplifying leads to

$$m \frac{\partial^2 u}{\partial t^2} - \frac{\partial \mathcal{N}}{\partial x} = 0 \qquad (17\text{-}27)$$

When the axial-force–displacement relationship

$$\mathcal{N} = \sigma A = \varepsilon E A = \frac{\partial u}{\partial x} E A$$

is introduced, the partial differential equation of equilibrium for axial motions becomes finally

$$m \frac{\partial^2 u}{\partial t^2} - \frac{\partial}{\partial x} \left(EA \frac{\partial u}{\partial x} \right) = 0 \qquad (17\text{-}28)$$

in which A is the cross-sectional area of the bar and $\partial u / \partial x = \varepsilon$ is the axial strain. If any axially directed load were distributed along the length of the bar, its value (expressed in terms of force per unit length) would be introduced on the right-hand side of the equation. Generally, however, the external load is applied only at the ends of the bar, and Eq. (17-28) serves to express equilibrium within the span.

PROBLEMS

17-1 Using Hamilton's principle, Eq. (16-2), determine the differential equation of motion and boundary conditions of the uniform cantilever beam loaded as shown in Fig. P17-1. Assume small deflection theory and neglect shear and rotary inertia effects.

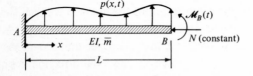

FIGURE P17-1

17-2 Using Hamilton's principle, determine the differential equation of motion and boundary conditions of the simply supported uniform pipe (shown in Fig. P17-2) through which fluid of density ρ and zero viscosity flows with constant velocity v_f relative to the pipe. Flexible moment connections are provided at each end of the pipe. Does the presence of the flowing fluid provide damping in the system? If the same pipe is supported as a cantilever member discharging the fluid at its free end, can fluid damping of the system be present

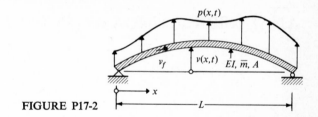

FIGURE P17-2

(neglect material damping in the pipe)? Let A equal the inside cross-sectional area of the pipe.

17-3 As shown in Fig. P17-3, a concentrated lumped mass m_1 traveling to the right with constant velocity v crosses a simply supported uniform beam during the time interval $0 < t < L/v$. Determine the governing equations of motion for this system using Lagrange's equation of motion, Eqs. (16-7), and state the required boundary and initial conditions that must be imposed to obtain the vertical forced-vibration response of the simple beam. Neglect shear and rotary inertia effects.

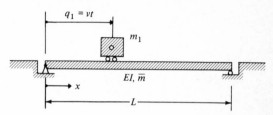

FIGURE P17-3

18

ANALYSIS OF UNDAMPED FREE VIBRATIONS

18-1 BEAM FLEXURE: ELEMENTARY CASE

Following the general approach to the dynamic-response analysis which was employed with the discrete-parameter systems, the first step in the dynamic-response analysis of a distributed-parameter system will be the evaluation of its undamped-vibration mode shapes and frequencies. To avoid unnecessary complications, this discussion will be limited to beams having uniform properties along their length or to frames assembled from such prismatic members. When we consider the elementary case, that is, we neglect shear distortions and rotatory inertia and assume that the axial-force effects are also negligible, we find that the equation of motion for the free vibrations of a prismatic beam can be written directly from Eq. (17-7) as

$$EI \frac{\partial^4 v}{\partial x^4} + \overline{m} \frac{\partial^2 v}{\partial t^2} = 0 \qquad (18\text{-}1)$$

After dividing by EI and using primes to indicate differentiation with respect to x and dots for differentiation with respect to t this becomes

$$v^{\text{iv}} + \frac{\overline{m}}{EI} \ddot{v} = 0 \qquad (18\text{-}2)$$

One form of solution of this equation can be obtained by separation of variables, assuming that the solution has the form

$$v(x,t) = \phi(x)Y(t) \qquad (18\text{-}3)$$

In other words, it is assumed that the free-vibration motions consist of a constant shape $\phi(x)$ the amplitude of which is varying with time according to $Y(t)$. Substituting Eq. (18-3) into Eq. (18-2) leads to

$$\phi^{iv}(x)Y(t) + \frac{\overline{m}}{EI}\,\phi(x)\ddot{Y}(t) = 0$$

and dividing by $\phi(x)Y(t)$ then achieves the desired separation:

$$\frac{\phi^{iv}(x)}{\phi(x)} + \frac{\overline{m}}{EI}\frac{\ddot{Y}(t)}{Y(t)} = 0 \qquad (18\text{-}4)$$

Because the first term in Eq. (18-4) is a function of x only and the second term a function of t only, the equation can be satisfied for arbitrary x and t only if each term is equal to a constant; that is,

$$\frac{\phi^{iv}(x)}{\phi(x)} = \overline{C} = -\frac{\overline{m}}{EI}\frac{\ddot{Y}(t)}{Y(t)}$$

Thus two ordinary differential equations are obtained, one involving each variable. With the constant $\overline{C} \equiv a^4$ for convenience, these two equations may be written

$$\phi^{iv}(x) - a^4\phi(x) = 0 \qquad (18\text{-}5a)$$

$$\ddot{Y}(t) + \omega^2 Y(t) = 0 \qquad (18\text{-}5b)$$

in which

$$\omega^2 = \frac{a^4 EI}{\overline{m}} \qquad \text{or} \qquad \frac{\omega^2 \overline{m}}{EI} = a^4 \qquad (18\text{-}6)$$

Equation (18-5b) is the familiar free-vibration equation for an undamped SDOF system and has the solution [see Eq. (3-10)]

$$Y(t) = A \sin \omega t + B \cos \omega t \qquad (18\text{-}7)$$

in which the constants A and B depend upon the initial velocity and displacement conditions [see Eq. (3-11); that is, $B = Y(0)$ and $A = \dot{Y}(0)/\omega$]; substituting these into Eq. (18-7) gives

$$Y(t) = \frac{\dot{Y}(0)}{\omega} \sin \omega t + Y(0) \cos \omega t \qquad (18\text{-}8)$$

Equation (18-5a) can be solved in the usual way by assuming a solution of the form

$$\phi(x) = Ce^{sx} \qquad (18\text{-}9)$$

When this is substituted, Eq. (18-5) becomes

$$(s^4 - a^4)Ce^{sx} = 0$$

from which

$$s = \pm a, \pm ia$$

Introducing these four values of s into Eq. (18-9) leads to

$$\phi(x) = C_1 e^{iax} + C_2 e^{-iax} + C_3 e^{ax} + C_4 e^{-ax}$$

Expressing these exponential functions in terms of their trigonometric and hyperbolic equivalents gives

$$\phi(x) = A_1 \sin ax + A_2 \cos ax + A_3 \sinh ax + A_4 \cosh ax \qquad (18\text{-}10)$$

The four constants A_n in Eq. (18-10) define the shape and amplitude of the beam vibration; they must be evaluated by consideration of the boundary conditions at the ends of the beam segment. Two conditions expressing the displacement, slope, moment, or shear force will be defined at each end of the beam segment. These may be used to express three of the four constants in terms of the fourth and will also provide an expression (called the *frequency equation*) from which the frequency parameter a can be evaluated. The fourth constant cannot be evaluated directly in a free-vibration analysis; it defines the amplitude of motion, which depends on the initial conditions, as shown in Eq. (18-8).

EXAMPLE E18-1 **Simple beam** To illustrate the analysis of vibration properties for a uniform beam segment, attention will be directed first to the simple beam shown in Fig. E18-1a. The four boundary conditions for this beam may be expressed as follows:

At $x = 0$:
$$\phi(0) = 0 \qquad (1)$$
$$\mathcal{M}(0) = EI\phi''(0) = 0 \qquad (2)$$

At $x = L$:
$$\phi(L) = 0 \qquad (3)$$
$$\mathcal{M}(L) = EI\phi''(L) = 0 \qquad (4)$$

Substituting the expression of Eq. (18-10) into the first two of these boundary-condition equations leads to

$$\phi(0) = A_1 \cancel{\sin} 0 + A_2 \cos 0 + A_3 \cancel{\sinh} 0 + A_4 \cosh 0 = 0 \qquad (1a)$$

$$\phi''(0) = a^2(-A_1 \cancel{\sin} 0 - A_2 \cos 0 + A_3 \cancel{\sinh} 0 + A_4 \cosh 0) = 0 \qquad (2a)$$

from which

$$A_2 + A_4 = 0 \qquad -A_2 + A_4 = 0$$

Hence
$$A_2 = A_4 = 0$$

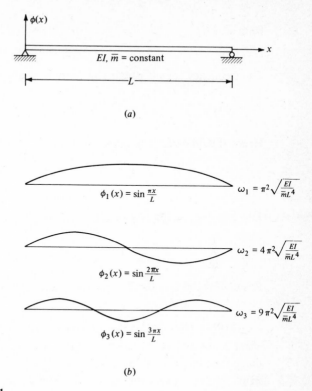

FIGURE E18-1
Simple beam-vibration analysis: (*a*) basic properties of simple beam; (*b*) first three vibration modes.

Similarly substituting into the last two boundary conditions (and setting $A_2 = A_4 = 0$) yields

$$\phi(L) = A_1 \sin aL + A_3 \sinh aL = 0 \qquad (3a)$$

$$\phi''(L) = a^2(-A_1 \sin aL + A_3 \sinh aL) = 0 \qquad (4a)$$

which when added give

$$2A_3 \sinh aL = 0$$

Hence $A_3 = 0$ inasmuch as the hyperbolic sine function cannot vanish. Thus the remaining condition may be expressed

$$\phi(L) = A_1 \sin aL = 0$$

The trivial solution ($A_1 = 0$) is excluded, and this provides the frequency equation

$$\sin aL = 0$$

from which

$$aL = n\pi \qquad n = 0, 1, 2, \ldots, \infty$$

The frequency thus is obtainable from $a = n\pi/L$, where, by definition,

$$a^4 = \frac{\omega^2 \overline{m}}{EI}$$

Hence the final result is obtained from $\omega_n{}^2 = (n\pi/L)^4 EI/\overline{m}$, that is,

$$\omega_n = n^2 \pi^2 \sqrt{\frac{EI}{\overline{m} L^4}} \qquad (a)$$

The vibration shape is given by Eq. (18-10) with $A_2 = A_3 = A_4 = 0$, that is,

$$\phi_n(x) = A_1 \sin \frac{n\pi}{L} x \qquad (b)$$

Results for the first three modes are shown in Fig. E18-1b.   ////

EXAMPLE E18-2 **Cantilever beam** The vibration analysis of the simple beam is not difficult because its mode shape is defined by only one term of the shape-function expression [Eq. (18-10)] and the coefficients of all the other terms vanish. The cantilever beam shown in Fig. E18-2a provides a more representative example of the analysis procedure. The four boundary conditions for this case are

At $x = 0$:
$$\phi(0) = 0 \qquad (1)$$
$$\phi'(0) = 0 \qquad (2)$$

At $x = L$:
$$\mathcal{M} = EI\phi''(L) = 0 \qquad (3)$$
$$\mathcal{U} = EI\phi'''(L) = 0 \qquad (4)$$

Substituting the shape-function expression [Eq. (18-10)] or its derivatives into these boundary condition equations leads to

$$\phi(0) = 0 = A_1 \cancel{\sin 0} + A_2 \cos 0 + A_3 \cancel{\sinh 0} + A_4 \cosh 0$$

or
$$A_2 = -A_4 \qquad (1a)$$

$$\phi'(0) = 0 = a(A_1 \cos 0 - A_2 \cancel{\sin 0} + A_3 \cosh 0 + A_4 \cancel{\sinh 0})$$

or
$$A_1 = -A_3 \qquad (2a)$$

$$\phi''(L) = 0 = a^2(-A_1 \sin aL - A_2 \cos aL$$
$$+ A_3 \sinh aL + A_4 \cosh aL) \qquad (3a)$$

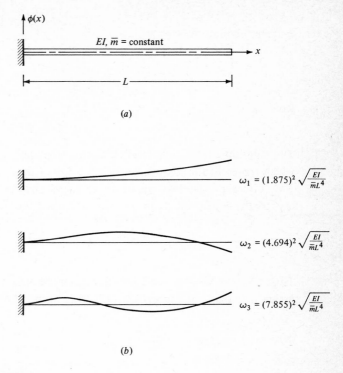

FIGURE E18-2
Cantilever-beam vibration analysis: (a) properties of cantilever beam; (b) first
three vibration modes.

$$\phi'''(L) = 0 = a^3(-A_1 \cos aL + A_2 \sin aL$$
$$+ A_3 \cosh aL + A_4 \sinh aL) \qquad (4a)$$

When the equalities defined in the first two equations are used, the last two
equations become

$$0 = A_1(\sin aL + \sinh aL) + A_2(\cos aL + \cosh aL)$$
$$0 = -A_1(\cos aL + \cosh aL) + A_2(\sin aL) - \sinh aL)$$

which may be written in matrix form as

$$\begin{bmatrix} \sin aL + \sinh aL & \cos aL + \cosh aL \\ \cos aL + \cosh aL & \sinh aL - \sin aL \end{bmatrix} \begin{bmatrix} A_1 \\ A_2 \end{bmatrix} = \begin{bmatrix} 0 \\ 0 \end{bmatrix} \qquad (a)$$

For the coefficients to be nonzero, this equation requires that the determinant
of the square matrix vanish; setting this determinant equal to zero provides the

frequency equation

$$\Delta = 0 = \sinh^2 aL - \sin^2 aL$$

$$- \cosh^2 aL - \cos^2 aL - 2\cos aL \cosh aL$$

from which

$$1 + \cos aL \cosh aL = 0 \qquad (b)$$

The solution of this transcendental equation then provides the values of aL which represent the frequencies of vibration of the cantilever beam.

Either of the two equations in matrix expression (a) may be used to express the coefficient A_2 in terms of A_1; using the first gives

$$A_2 = - \frac{\sin aL + \sinh aL}{\cos aL + \cosh aL} A_1 \qquad (c)$$

This, together with the relationships from the first two boundary conditions, makes it possible to express Eq. (18-10) in terms of only the first coefficient

$$\phi(x) = A_1 \left[\sin ax - \sinh ax + \frac{\sin aL + \sinh aL}{\cos aL + \cosh aL} (\cosh ax - \cos ax) \right] \qquad (d)$$

After the modal value of aL has been obtained from the frequency equation, it may be substituted into this shape-function expression to obtain the corresponding mode shape. Shapes and frequencies of the first three modes of vibration of a uniform cantilever beam are shown in Fig. E18-2b.   ////

EXAMPLE E18-3 **Other boundary conditions**   These examples have shown that the general procedure for analysis of mode shapes and frequencies of a uniform beam segment involves identifying the boundary conditions for the segment and then substituting the shape-function expression [Eq. (18-10)] into each equation. The end result is a set of equations equal in number to the shape-function coefficients; setting the determinant of the resulting square matrix equal to zero then provides the frequency equation. Although the analysis may be tedious, the concept is simple; it is only in identifying the boundary conditions that any real difficulty may arise. The boundary conditions for two other examples are discussed here to illustrate additional typical situations.

A cantilever beam supporting a lumped mass at its end is shown in Fig. E18-3a. The boundary conditions at the support end are the same as in the

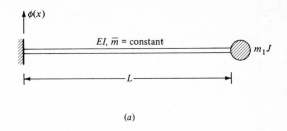

(a)

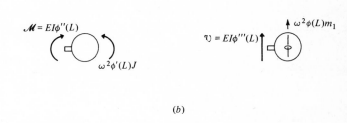

(b)

FIGURE E18-3
Beam with lumped mass at end: (a) beam properties; (b) forces acting on the
end mass.

preceding example. At the other end, however, the moment and shear no
longer vanish because of the inertia of the mass. The rotational and transverse
forces acting on the end mass, shown in Fig. E18-3b, lead to the boundary-
condition equations

$$EI\phi''(L) - \omega^2\phi'(L)J = 0 \qquad (1)$$

$$EI\phi'''(L) + \omega^2\phi(L)m_1 = 0 \qquad (2)$$

where it should be noted that J represents the mass moment of inertia at the
end of the beam and m_1 is the end mass.

A frame made up of two uniform beam segments having different
properties is shown in Fig. E18-4a. Because of the discontinuity between the
two segments, the shape-function expression [Eq. (18-10)] must be applied
separately to each segment and the problem thus requires consideration of eight
coefficients. Correspondingly, eight boundary conditions must be identified.
Two are given by the fixity condition at the origin of the column:

At $x_1 = 0$:

$$\phi_1(0) = 0 \qquad (1)$$

$$\phi_1'(0) = 0 \qquad (2)$$

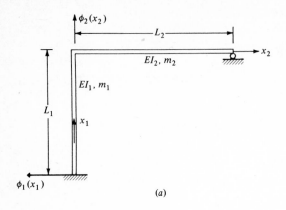

(a)

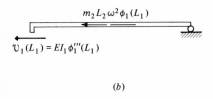

$\mathcal{V}_1(L_1) = EI_1 \phi_1'''(L_1)$

(b)

FIGURE E18-4
Simple two-member frame: (a) arrangement of frame; (b) forces acting on the beam.

Assuming that there are no axial distortions, the following three conditions are imposed by the beam-support conditions:

At $x_2 = 0$: $\qquad\qquad\qquad\qquad\qquad \phi_2(0) = 0 \qquad\qquad (3)$

At $x_2 = L_2$: $\qquad\qquad\qquad\qquad\qquad \phi_2(L_2) = 0 \qquad\qquad (4)$

$$\phi_2''(L_2) = \frac{\mathcal{M}(L_2)}{EI} = 0 \qquad\qquad (5)$$

Continuity of slope and moment equilibrium at the joint between the two members provide two additional conditions:

At $x_1 = L_1, x_2 = 0$: $\qquad\qquad\qquad \phi_1'(L_1) = \phi_2'(0) \qquad\qquad (6)$

$$EI_1 \phi_1''(L_1) = EI_2 \phi_2''(0) \qquad\qquad (7)$$

The equilibrium of the shear force at the top of the column with the inertia force developed by the sidesway motion of the beam (Fig. E18-4b) provides the eighth condition

At $x_1 = L_1$: $\qquad\qquad EI_1 \phi_1'''(L_1) + m_2 L_2 \omega^2 \phi_1(L_1) = 0 \qquad\qquad (8)$

These eight conditions can now be used to establish eight equations in terms of the eight coefficients (four for each beam segment), and setting to zero the determinant of the resulting 8 × 8 square matrix provides the frequency equation for the system. Note, however, that the solution of even this very simple two-member frame leads to a sizable computational problem if the analysis is done in this way. More practical schemes for carrying out the analysis of simple frames will be discussed later.   ////

18-2 BEAM FLEXURE: INCLUDING AXIAL-FORCE EFFECTS

Axial forces acting in a flexural element may have a very significant influence on the vibration behavior of the member, resulting generally in modifications of both the frequency and the mode shape. When free vibrations of a prismatic member are considered, the equation of motion including the effect of an axial force N which is uniform along the length of the member and not varying with time becomes [from Eq. (17-10)]

$$EI \frac{\partial^4 v}{\partial x^4} + N \frac{\partial^2 v}{\partial x^2} + \bar{m} \frac{\partial^2 v}{\partial t^2} = 0 \qquad (18\text{-}11)$$

Separating variables as before by assuming a solution in the form of Eq. (18-3) results eventually in

$$\frac{EI\phi^{iv}(x)}{\phi(x)} + \frac{N\phi''(x)}{\phi(x)} = -\bar{m} \frac{\ddot{Y}(t)}{Y(t)} = \bar{C} \qquad (18\text{-}12)$$

from which are obtained the two independent equations

$$\ddot{Y}(t) + \omega^2 Y(t) = 0 \qquad (18\text{-}13)$$

$$EI\phi^{iv}(x) + N\phi''(x) - \bar{m}\omega^2 \phi(x) = 0 \qquad (18\text{-}14)$$

in which the constant $\bar{C}$ has been replaced by $\bar{m}\omega^2$.

Equation (18-13) is the same time-variation equation obtained before, showing that a constant axial force does not affect the simple-harmonic character of the free vibrations. Equation (18-14) leads to frequency and mode-shape expressions for the freely vibrating beam, in which the axial force N is a basic parameter. When Eq. (18-14) is divided by EI, it can be written

$$\phi^{iv}(x) + g^2\phi''(x) - a^4\phi(x) = 0 \qquad (18\text{-}15)$$

in which

$$a^4 = \frac{\bar{m}\omega^2}{EI} \qquad g^2 = \frac{N}{EI} \qquad (18\text{-}16)$$

The solution of Eq. (18-15) can be obtained in the standard way by assuming a solution in the form of Eq. (18-9) and substituting, with the result

$$(s^4 + g^2 s^2 - a^4)Ce^{sx} = 0$$

from which

$$s = \pm i\delta, \pm\varepsilon$$

where
$$\delta = \sqrt{\left(a^4 + \frac{g^4}{4}\right)^{1/2} + \frac{g^2}{2}} \qquad \varepsilon = \sqrt{\left(a^4 + \frac{g^4}{4}\right)^{1/2} - \frac{g^2}{2}} \qquad (18\text{-}17)$$

Introducing these four values into Eq. (18-9) and expressing the exponential functions in terms of their trigonometric and hyperbolic equivalents leads to the final shape-function expression

$$\phi(x) = D_1 \sin \delta x + D_2 \cos \delta x + D_3 \sinh \varepsilon x + D_4 \cosh \varepsilon x \qquad (18\text{-}18)$$

Equation (18-18) defines the shape of the vibrating beam for any value of axial force which might be specified. The coefficients $D_1, D_2, \ldots$ can be evaluated by consideration of the boundary conditions, just like the system without axial force. In fact, it is evident that when the axial force vanishes, that is, $N = 0$, so that $g = 0$, then $\delta = \varepsilon = a$ and Eq. (18-18) reverts to Eq. (18-10). On the other hand, for the static (nonvibrating) case (where $\omega = 0$, so that $a = 0$) $\delta = g$ while $\varepsilon = 0$. Thus the shape function for the static buckling case can be obtained by substituting values of s corresponding to these conditions into Eq. (18-9), with the final result

$$\phi(x) = D_1 \sin gx + D_2 \cos gx + D_3 + D_4 x \qquad (18\text{-}19)$$

in which the last two terms correspond to the zero values of s. Substitution of the boundary conditions leads in this case to an expression for the critical buckling load N_{cr} in terms of the axial-force relationship of Eq. (18-16).

18-3 BEAM FLEXURE: INCLUDING SHEAR DEFORMATION AND ROTATORY INERTIA

The effect of shear deformation and rotatory inertia on the free vibration of a beam can be evaluated from the equation of motion [Eq. (17-16)] modified by omitting the external load term p:

$$EI \frac{\partial^4 v}{\partial x^4} + \bar{m} \frac{\partial^2 v}{\partial t^2} - \bar{m}r^2 \frac{\partial^4 v}{\partial x^2 \partial t^2} + \frac{\bar{m}}{k'AG}\left(\bar{m}r^2 \frac{\partial^4 v}{\partial t^4} - EI \frac{\partial^4 v}{\partial x^2 \partial t^2}\right) = 0$$

When the analysis is simplified by assuming that the displacements vary harmonically with time according to $v(x,t) = \phi(x) \sin \omega t$, this equation becomes (after dividing by $\sin \omega t$)

$$EI\phi^{iv}(x) - \bar{m}\omega^2\phi(x) + \bar{m}r^2\omega^2\phi''(x) + \frac{\bar{m}\omega^2}{k'AG}[\bar{m}r^2\omega^2\phi(x) + EI\phi''(x)] = 0$$

Letting $a^4 = \bar{m}\omega^2/EI$, as before, gives

$$\phi^{iv}(x) - a^4\phi(x) + a^4r^2\phi''(x) + \frac{\bar{m}\omega^2}{k'AG}[a^4r^2\phi(x) + \phi''(x)] = 0 \qquad (18\text{-}20)$$

The solution of Eq. (18-20) for a system with arbitrary boundary conditions is difficult. However, the case of simple beam supports is easily solved and gives considerable insight into the significance of the shear and rotatory-inertia effects. These factors have no influence on the vibration *shape* for a simple uniform beam, which is given by the expression derived previously

$$\phi(x) = A_1 \sin \frac{n\pi x}{L} \qquad (18\text{-}21)$$

Substituting Eq. (18-21) into Eq. (18-20), dividing by $\phi(x)$, and rearranging leads to

$$\left(\frac{n\pi}{L}\right)^4 - a^4 - a^4r^2\left(\frac{n\pi}{L}\right)^2\left(1 + \frac{E}{k'G}\right) + a^4r^2\left(a^4r^2\frac{E}{k'G}\right) = 0 \qquad (18\text{-}22)$$

It will be noted that the first two terms of Eq. (18-22) represent the result obtained previously for the elementary case

$$a^4 = \left(\frac{n\pi}{L}\right)^4$$

that is,

$$\omega_n = n^2\pi^2\sqrt{\frac{EI}{mL^4}}$$

The third term accounts for the principal effects of rotatory inertia and shear deformation, which are represented respectively by the terms 1 and $E/k'G$ in parentheses. For a beam of typical materials and rectangular section, the value of $E/k'G$ is about 3, so that shear distortion is about 3 times as important as rotatory inertia in this case. Temporarily omitting the last nonzero term of Eq. (18-22) leads to

$$a^4 = \left(\frac{n\pi}{L}\right)\left[\frac{1}{1 + r^2(n\pi/L)^2(1 + E/k'G)}\right]$$

in which the term in brackets provides the correction for shear and rotatory inertia. It is evident that this correction increases as the mode number increases and as the slenderness ratio L/r decreases. Where the term nr/L is small, the above expression can be approximated by

$$a^4 \doteq \left(\frac{n\pi}{L}\right)^4\left[1 - \left(\frac{nr\pi}{L}\right)^2\left(1 + \frac{E}{k'G}\right)\right]$$

from which

$$\omega_n \doteq n^2\pi^2 \sqrt{\frac{EI}{\bar{m}L^4}} \left[1 - \frac{1}{2}\left(\frac{nr\pi}{L}\right)^2 \left(1 + \frac{E}{k'G}\right)\right] \quad (18\text{-}23)$$

The last nonzero term of Eq. (18-22) is of secondary importance for practical cases, in which nr/L is small compared with 1. To establish the relative size of this term compared with the principal-shear and rotatory-inertia correction term, it may be noted that where nr/L is small, $a^4 \doteq (n\pi/L)^4$; thus the last term can be written

$$a^4 r^2 \left(a^4 r^2 \frac{E}{k'G}\right) \doteq a^4 r^2 \left(\frac{n\pi}{L}\right)^2 \left[\left(\frac{n\pi r}{L}\right)^2 \frac{E}{k'G}\right]$$

It is evident that this is small compared with the preceding term in Eq. (18-22); that is,

$$a^4 r^2 \left(\frac{n\pi}{L}\right)^2 \left[\left(\frac{n\pi r}{L}\right)^2 \frac{E}{k'G}\right] \ll a^4 r^2 \left(\frac{n\pi}{L}\right)^2 \left(1 + \frac{E}{k'G}\right)$$

because $n\pi r/L$ is small compared with 1.

EXAMPLE E18-4   As an indication of the significance of the correction for shear deformation and rotatory inertia, a rectangular beam with a cross section 6 in wide by 24 in deep ($r = 4\sqrt{3}$ in) and having a span of 40 ft will be considered. Assuming that for this material $E/k'G = 3$, the bracketed correction term in Eq. (18-23) becomes

$$1 - \frac{1}{2}n^2\pi^2 \frac{1}{4,800} (1 + 3) = 1 - \frac{n^2\pi^2}{2,400}$$

Thus the change of frequency due to shear and rotatory inertia is

Mode	1	2	3	4	5
Change, %	−0.4	−1.6	−3.7	−6.6	−10.3

of which three-fourths is due to shear distortion. The rapid increase of significance with increasing mode numbers is quite evident in this example; thus it is clear that while this correction is generally unimportant for the lowest modes of vibration, it should be kept in mind when considering the higher modes of a structural system.   ////

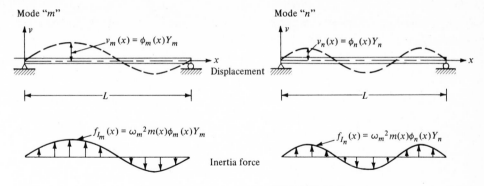

FIGURE 18-1
Two modes of vibration for the same beam.

18-4 BEAM FLEXURE: ORTHOGONALITY OF VIBRATION MODE SHAPES

The vibration mode shapes derived for beams with distributed properties have orthogonality relationships equivalent to those defined previously for the discrete-parameter systems and can be demonstrated in essentially the same way—by application of Betti's law. Consider the beam shown in Fig. 18-1. For this discussion, the beam may have arbitrarily varying stiffness and mass along its length, and it could have arbitrary support conditions although only simple supports are shown. Two different vibration modes, m and n, are shown for the beam. In each mode, the displaced shape and the inertia forces producing the displacements are indicated.

Betti's law applied to these two deflection patterns means that the work done by the inertia forces of mode n acting on the deflection of mode m is equal to the work of the forces of mode m acting on the displacement of mode n; that is,

$$\int_0^L v_m(x) f_{In}(x)\, dx = \int_0^L v_n(x) f_{Im}(x)\, dx \qquad (18\text{-}24)$$

Expressing these in terms of the modal shape functions shown in Fig. 18-1 gives

$$Y_m Y_n \omega_n^2 \int_0^L \phi_m(x) m(x) \phi_n(x)\, dx = Y_m Y_n \omega_m^2 \int_0^L \phi_n(x) m(x) \phi_m(x)\, dx$$

which may be rewritten

$$(\omega_n^2 - \omega_m^2) \int_0^L \phi_m(x) \phi_n(x) m(x)\, dx = 0 \qquad (18\text{-}25)$$

Finally if the frequencies of these two modes are different, their mode shapes must satisfy the orthogonality condition

$$\int_0^L \phi_m(x)\phi_n(x)m(x)\,dx = 0 \qquad (18\text{-}26)$$

which is clearly the distributed-parameter equivalent of the discrete-parameter orthogonality condition of Eq. (12-38a). If the two modes have the same frequency, the orthogonality condition does not apply, but this condition does not occur often in ordinary structural problems.

A second orthogonality condition, involving the stiffness property rather than the mass as a weighting parameter, can be derived for the distributed-parameter systems as it was earlier for the discrete-parameter case. For a nonuniform beam, the equation of motion in free vibrations is

$$\frac{\partial^2}{\partial x^2}\left[EI(x)\frac{\partial^2 v}{\partial x^2}\right] + m(x)\frac{\partial^2 v}{\partial t^2} = 0 \qquad (18\text{-}27)$$

When it is noted that the motion in the nth mode can be written

$$v_n(x,t) = \phi_n(x)Y_n \sin \omega_n t$$

the second term in Eq. (18-27) becomes

$$m(x)\frac{\partial^2 v}{\partial t^2} = -m(x)\omega_n^2\phi_n(x)Y_n \sin \omega_n t$$

When the time variation ($\sin \omega_n t$) is omitted, Eq. (18-27) thus leads to

$$Y_n\frac{d^2}{dx^2}\left[EI(x)\frac{d^2\phi_n}{dx^2}\right] = Y_n\omega_n^2 m(x)\phi_n \qquad (18\text{-}27a)$$

Hence it is clear that the inertia force term $m(x)\phi_n$ in the orthogonality relationship [Eq. (18-26)] can be replaced by an equivalent expression for the lateral loading in terms of the flexural rigidity:

$$\frac{1}{\omega_n^2}\frac{d^2}{dx^2}\left[EI(x)\frac{d^2\phi_n}{dx^2}\right]$$

with the final result

$$\int_0^L \phi_m(x)\frac{d^2}{dx^2}\left[EI(x)\frac{d^2\phi_n}{dx^2}\right]dx = 0 \qquad (18\text{-}28)$$

A more convenient symmetric form of this orthogonality relationship can be obtained by integrating Eq. (18-28) twice by parts, giving

$$\phi_m \mathcal{V}_n\Big|_0^L - \phi'_m \mathcal{M}_n\Big|_0^L + \int_0^L \phi''_m\phi''_n EI(x)\,dx = 0 \qquad \omega_m \neq \omega_n \qquad (18\text{-}29)$$

The first two terms in Eq. (18-29) represent the work done by the boundary shear forces of mode n acting on the end displacements of mode m and the work done by the end moments of mode n on the corresponding rotations of mode m. For the standard clamped, hinged, or free end conditions, these terms will vanish. However, they contribute to the orthogonality relationship if the beam has elastic supports or if it has a lumped mass at its end; therefore they must be retained in the expression for consideration in each particular case.

18-5 FREE VIBRATIONS IN AXIAL DEFORMATION

The analysis of free vibrations associated with axial motions of a one-dimensional number is very similar to the analysis of flexural vibrations. For a prismatic member, that is, one with properties constant along the length, the equation of axial motion, Eq. (17-28), becomes

$$\bar{m}\frac{\partial^2 u}{\partial t^2} - EA\frac{\partial^2 u}{\partial x^2} = 0 \qquad (18\text{-}30)$$

Assuming the response is of the form

$$u(x,t) = \bar{\phi}(x)Y(t) \qquad (18\text{-}31)$$

Equation (18-30) may be separated into the two ordinary differential equations

$$\bar{m}\frac{\ddot{Y}(t)}{Y(t)} = EA\frac{\bar{\phi}''(x)}{\bar{\phi}(x)} = -\bar{C}$$

or more conveniently

$$\ddot{Y}(t) + \omega^2 Y(t) = 0 \qquad (18\text{-}32)$$

$$\bar{\phi}''(x) + b^2\bar{\phi}(x) = 0 \qquad (18\text{-}33)$$

in which

$$\omega^2 = +\frac{\bar{C}}{\bar{m}} \qquad b^2 = +\frac{\bar{C}}{AE} = \frac{\omega^2\bar{m}}{AE} \qquad (18\text{-}34)$$

Equation (18-32) is the same as Eq. (18-5a); its solution shows that the free-vibration motion is a simple-harmonic form. Equation (18-33) is the same type of equation as Eq. (18-32) and has the same form of solution but is expressed in terms of position x rather than time t; thus

$$\bar{\phi}(x) = B_1 \sin bx + B_2 \cos bx \qquad (18\text{-}35)$$

in which the coefficients B_1 and B_2 determine the vibration mode shape.

By consideration of the boundary conditions, one of these can be expressed in terms of the other. The second boundary condition then leads to an expression from which the frequency parameter b can be evaluated.

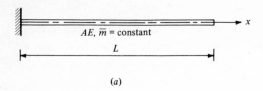

$AE, \bar{m} = \text{constant}$

L

(a)

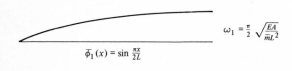

$\bar{\phi}_1(x) = \sin \frac{\pi x}{2L}$

$\omega_1 = \frac{\pi}{2} \sqrt{\frac{EA}{\bar{m}L^2}}$

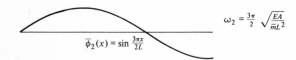

$\bar{\phi}_2(x) = \sin \frac{3\pi x}{2L}$

$\omega_2 = \frac{3\pi}{2} \sqrt{\frac{EA}{\bar{m}L^2}}$

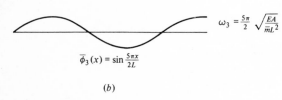

$\bar{\phi}_3(x) = \sin \frac{5\pi x}{2L}$

$\omega_3 = \frac{5\pi}{2} \sqrt{\frac{EA}{\bar{m}L^2}}$

(b)

FIGURE E18-5
Axial vibrations of bar: (a) basic properties of cantilever bar; (b) first three vibration modes.

EXAMPLE E18-5 **Cantilever bar** Consider the bar of Fig. E18-5a subjected to axial vibrations. The two boundary conditions to be considered in this case are:

At $x = 0$: $\qquad\qquad\qquad\qquad\qquad\qquad \bar{\phi}(0) = 0 \qquad\qquad\qquad$ (1)

At $x = L$: $\qquad\qquad\qquad\qquad \mathcal{N}(L) = AE\bar{\phi}'(L) = 0 \qquad\qquad$ (2)

Substituting Eq. (18-35) into the first equation leads to

$$B_1 \sin (0) + B_2 \cos (0) = 0$$

from which

$$B_2 = 0$$

Taking the first derivative of Eq. (18-35) (after setting $B_2 = 0$) and substituting into the second boundary-condition equation then yields

$$AEB_1b \cos bL = 0$$

The trivial solution $B_1 = 0$ is excluded, and the frequency equation is seen to be

$$\cos bL = 0$$

from which

$$bL = \frac{2n - 1}{2} \pi$$

The vibrating shape of the rod is thus given by

$$\bar{\phi}_n(x) = B_1 \sin \frac{2n - 1}{2} \pi \frac{x}{L} \qquad (a)$$

(where the amplitude B_1 is arbitrary) while the frequency of vibration is

$$\omega_n = \sqrt{\frac{b_n{}^2 EA}{\bar{m}}} = \frac{2n - 1}{2} \pi \sqrt{\frac{EA}{\bar{m}L^2}} \qquad (b)$$

The first three vibration mode shapes and frequencies are shown in Fig. E18-5b.

$$////$$

18-6 ORTHOGONALITY OF AXIAL VIBRATION MODES

The axial vibration mode shapes have orthogonality properties which are entirely equivalent to those demonstrated earlier for the flexural vibration modes. In fact, the orthogonality of the axial mode shapes with respect to the mass distribution can be derived by using Betti's law in the same way as for the flexural modes with the equivalent result:

$$\int_0^L \bar{\phi}_m(x)\bar{\phi}_n(x)m(x) \, dx = 0 \qquad (18\text{-}36)$$

The orthogonality relationship with respect to the axial stiffness property can be derived from the equation of motion [Eq. (17-28)] in which the harmonic time variation of free vibrations has been substituted. In other words, when the nth mode displacements are expressed as

$$u_n(x,t) = \bar{\phi}_n(x) \sin \omega_n t$$

Eq. (17-28) can be written

$$\omega_n{}^2 m(x)\bar{\phi}_n = -\frac{d}{dx}\left[EA(x) \frac{d\phi_n}{dx} \right] \qquad (18\text{-}37)$$

Thus the inertial-force term in the orthogonality relationship of Eq. (18-37) can be replaced by the equivalent axial elastic-force term, with the result

$$\int_0^L \bar{\phi}_m(x) \frac{d}{dx}\left[EA(x) \frac{d\phi_n}{dx}\right] dx = 0 \qquad (18\text{-}38)$$

Integrating Eq. (18-38) by parts leads to the more convenient symmetric form of this orthogonality relationship

$$\bar{\phi}_m \mathscr{N}_n \Big|_0^L - \int_0^L \bar{\phi}_m' \bar{\phi}_n' EA(x)\, dx = 0 \qquad \omega_m \neq \omega_n \qquad (18\text{-}39)$$

The first term in this equation represents the work done by the boundary axial forces of mode n acting on the end displacements of mode m; this term will vanish if the bar has the standard free- or fixed-end conditions but may have to be included in more complex situations.

PROBLEMS

18-1 Evaluate the fundamental frequency for the cantilever beam with a mass at the end shown in Fig. E18-3, if the end lumped mass $m_1 = 2\bar{m}L$ and if its mass moment of inertia $J = 0$. Plot the shape of this mode, evaluating at increments $L/5$ along the span.

18-2 Evaluate the fundamental frequency for the frame of Fig. E18-4 if the two members are identical, with properties L, EI, $\bar{m}$. Plot the shape of this mode, evaluating at increments $L/4$ along each span.

18-3 Evaluate the fundamental *flexural* frequency of the beam of Fig. P18-1 and plot its mode shape, evaluated at increments $L/5$ along its length. Note that the lowest frequency of this unstable structure is zero; the frequency of interest is the lowest nonzero value.

$\bar{m}$, EI = uniform

FIGURE P18-1

18-4 The uniform beam of Fig. P18-2 is continuous over two spans as shown. Evaluate the fundamental flexural frequency of this structure and plot its mode shape at increments $L/2$ along the two spans.

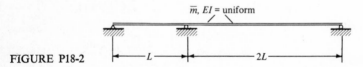

$\bar{m}$, EI = uniform

FIGURE P18-2

18-5 A reinforced concrete beam having a cross section 8 in wide by 18 in deep is simply supported with a span of 28 ft. Assuming that the modulus of the material is 3×10^6 lb/in² and that its unit weight is 150 lb/ft³, evaluate the frequency of its first five vibration modes:

(*a*) Neglecting shear distortion and rotatory inertia.

(*b*) Considering these effects, Eq. (18-23).

Assume $E/k'G = 3$.

18-6 Evaluate the fundamental frequency of *axial* vibration of the structure of Fig. E18-3 if the end lumped mass is $m_1 = 2\overline{m}L$ and if the cross-sectional area of the beam is A. Plot the shape of this mode, evaluating at increments $L/5$ along the span.

18-7 A column is assembled with two uniform bars, of the same length but having different properties, as shown in Fig. P18-3. For this structure:

(*a*) List the four boundary conditions required to evaluate the constants in deriving the axial vibration frequency equation.

(*b*) Write the transcendental axial frequency equation, and evaluate the first mode frequency and mode shape. Plot the mode shape evaluated at intervals $L/3$ along its length, normalized to unit amplitude at the free end.

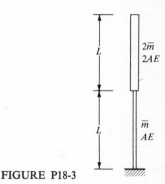

FIGURE P18-3

19

ANALYSIS OF DYNAMIC RESPONSE

19-1 NORMAL COORDINATES

The mode-superposition analysis of a distributed-parameter system is entirely equivalent to that of a discrete-coordinate system once the mode shapes and frequencies have been determined, because in both cases the amplitudes of the modal-response components are used as generalized coordinates in defining the response of the structure. In principle an infinite number of these coordinates are available for a distributed-parameter system since it has an infinite number of modes of vibration, but in practice only those modal components need be considered which provide significant contributions to the response. Thus the problem is actually converted into a discrete-parameter form in which only a limited number of modal (normal) coordinates is used to describe the response.

The essential operation of the mode-superposition analysis is the transformation from the geometric displacement coordinates to the modal-amplitude or normal coordinates. From a one-dimensional continuum, this transformation is expressed as

$$v(x,t) = \sum_{i=1}^{\infty} \phi_i(x) Y_i(t) \qquad (19\text{-}1)$$

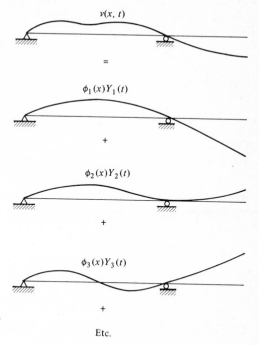

FIGURE 19-1
Arbitrary beam displacements represented by normal coordinates.

which is simply a statement that any physically permissible displacement pattern can be made up by superposing appropriate amplitudes of the vibration mode shapes for the structure. This principle is illustrated in Fig. 19-1, which shows an arbitrary displacement of a beam with an overhanging end developed as the sum of a set of modal components.

The modal components contained in any given shape, such as the top curve of Fig. 19-1, can be evaluated by applying the orthogonality conditions; usually it is most convenient to make use of the form involving the mass weighting parameter [Eq. (18-26)]. To evaluate the contribution of mode n in any arbitrary shape $v(x,t)$, Eq. (19-1) is multiplied by $\phi_n(x)m(x)$ on both sides and integrated, with the result

$$\int_0^L \phi_n(x)m(x)v(x,t)\,dx = \sum_{i=1}^{\infty} Y_i(t) \int_0^L \phi_i(x)m(x)\phi_n(x)\,dx$$

$$= Y_n(t) \int_0^L [\phi_n(x)]^2 m(x)\,dx$$

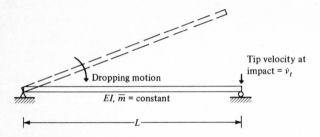

FIGURE E19-1
Example of free-vibration amplitude analysis.

where only one term remains of the infinite series on the right-hand side by virtue of the orthogonality condition. Hence the expression can be solved directly for the one remaining amplitude term

$$Y_n(t) = \frac{\displaystyle\int_0^L \phi_n(x)m(x)v(x,t)\, dx}{\displaystyle\int_0^L [\phi_n(x)]^2 m(x)\, dx} \qquad (19\text{-}2)$$

which is entirely equivalent to the discrete-parameter expression, Eq. (13-5).

EXAMPLE E19-1 Equation (19-2) can be used to evaluate the modal amplitudes expressing the initial conditions of a free-vibration problem, as indicated by Eq. (18-8). Given the beam displacement $v(x,0)$ and velocity $\dot{v}(x,0)$ at time $t = 0$, the corresponding modal amplitude $Y_n(0)$ and velocity $\dot{Y}_n(0)$ can be obtained directly from Eq. (19-2). Consider, for example, the uniform simple-beam system shown in Fig. E19-1, and assume that free vibrations of this beam are initiated by lifting the right end off the roller support and dropping it, allowing the beam to pivot about the left-end hinge support. If it is assumed that the beam rotates as a rigid body as it drops onto the right-hand support, the velocity at the time of contact ($t = 0$) varies linearly along the length of the beam, with the tip velocity represented by $\dot{v}_t$. The velocity distribution at the start of vibration thus is given by

$$\dot{v}(x,0) = \frac{x}{L}\,\dot{v}_t$$

and the displacement at the same time is $v(x,0) = 0$, corresponding to the concept that the beam has rotated as a rigid body.

The nth vibration mode shape for this simple beam is given by

$$\phi_n(x) = \sin\frac{n\pi x}{L}$$

Hence the denominator integral of Eq. (19-2) is

$$\int_0^L \phi_n^2 m(x)\, dx = \overline{m} \int_0^L \sin^2 \frac{n\pi x}{L}\, dx = \frac{\overline{m}L}{2}$$

The numerator integral of Eq. (19-2) defining the modal amplitude at $t = 0$ obviously is zero because $v(x,0)$ is zero; hence $Y_n(0) = 0$ for all modes. However, introducing the initial velocity distribution in an expression equivalent to Eq. (19-2) leads to a numerator integral

$$\int_0^L \phi_n(x) m(x) \dot{v}(x,0)\, dx = \overline{m}\dot{v}_t \int_0^L \frac{x}{L} \sin \frac{n\pi x}{L}\, dx = \pm \frac{\overline{m}L}{n\pi} \dot{v}_t$$

in which the sign is plus for $n = $ odd, minus for $n = $ even. Combining the numerator and denominator terms gives the initial normal-coordinate velocity

$$\dot{Y}_n(0) = \pm \frac{2\dot{v}_t}{n\pi}$$

and thus, from Eq. (18-8), the modal vibration is given by

$$Y_n(t) = \pm \frac{2\dot{v}_t}{n\pi\omega_n} \sin \omega_n t$$

Finally, when Eq. (19-1) is used, the free-vibration motion of the beam is

$$v(x,t) = \sum_{n=1}^\infty \phi_n(x) \left(\pm \frac{2\dot{v}_t}{n\pi\omega_n} \sin \omega_n t \right)$$

$$= \frac{2\dot{v}_t}{\pi} \left(\frac{1}{\omega_1} \sin \frac{\pi x}{L} \sin \omega_1 t - \frac{1}{2\omega_2} \sin \frac{2\pi x}{L} \sin \omega_2 t + \cdots \right)$$

Note that this analysis assumes that the right-hand end of the beam is held in contact with its support for all times after the initial impact.   ////

19-2   UNCOUPLED FLEXURAL EQUATIONS OF MOTION: UNDAMPED

The two orthogonality conditions now provide the means for decoupling the equations of motion for the distributed-parameter system in the same way that the decoupling was accomplished for the discrete-parameter analysis. Note that the orthogonality conditions which have been derived [Eqs. (18-26) and (18-28) or (18-29)] apply to a general beam with nonuniform properties; thus this decoupling analysis applies similarly to a general nonuniform beam. Introducing into the following equation of motion [Eq. (17-7)]

$$\frac{\partial^2}{\partial x^2}\left(EI\,\frac{\partial^2 v}{\partial x^2}\right) + m(x)\frac{\partial^2 v}{\partial t^2} = p(x,t)$$

the normal-coordinate expression of Eq. (19-1) leads to

$$\sum_{i=1}^{\infty} m(x)\phi_i(x)\ddot{Y}_i(t) + \sum_{i=1}^{\infty} \frac{d^2}{dx^2}\left[EI\,\frac{d^2\phi_i(x)}{\partial x^2}\right]Y_i(t) = p(x,t)$$

Multiplying each term by $\phi_n(x)$ and integrating then gives

$$\sum_{i=1}^{\infty}\ddot{Y}_i(t)\int_0^L m(x)\phi_i(x)\phi_n(x)\,dx + \sum_{i=1}^{\infty} Y_i(t)\int_0^L \phi_n(x)\frac{d^2}{dx^2}\left(EI\,\frac{d^2\phi_i}{dx^2}\right)dx$$

$$= \int_0^L \phi_n(x)p(x,t)\,dx$$

When the two orthogonality relationships [Eqs. (18-26) and (18-28)] are applied to the first two terms, it is evident that all terms of the series expansions except the nth terms vanish; thus

$$\ddot{Y}_n(t)\int_0^L m(x)\phi_n{}^2(x)\,dx + Y_n(t)\int_0^L \phi_n(x)\frac{d^2}{dx^2}\left(EI\,\frac{d^2\phi_n}{dx^2}\right)dx$$

$$= \int_0^L \phi_n(x)p(x,t)\,dx \qquad (19\text{-}3)$$

The relationship between the first two integrals of Eq. (19-3) can be established by multiplying both sides of Eq. (18-27a) by $\phi_n(x)$ and integrating, with the result

$$\int_0^L \phi_n(x)\frac{d^2}{dx^2}\left(EI\,\frac{d^2\phi_n}{dx^2}\right)dx = \omega_n{}^2\int_0^L \phi_n{}^2 m(x)\,dx \qquad (19\text{-}4)$$

It will be noted here that the right-hand integral is the generalized mass of the beam associated with the mode shape $\phi_n(x)$ [see Eq. (2-33)]. When this mass is denoted by M_n, that is,

$$M_n = \int_0^L \phi_n{}^2(x)m(x)\,dx \qquad (19\text{-}5)$$

Eq. (19-3) can be expressed in the abbreviated form

$$M_n\ddot{Y}_n(t) + \omega_n{}^2 M_n Y_n(t) = P_n(t) \qquad (19\text{-}6)$$

in which

$$P_n(t) = \int_0^L \phi_n(x)p(x,t)\,dx \qquad (19\text{-}7)$$

is the generalized load associated with the mode shape $\phi_n(x)$.

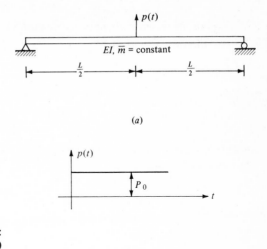

FIGURE E19-2
Example of dynamic-response analysis:
(a) arrangement of beam and load; (b)
applied step-function loading.

An equation of the type of Eq. (19-6) can be established for each vibration mode of the structure, using Eqs. (19-5) and (19-7) to evaluate the appropriate generalized mass and loading. It should be noted that these expressions are the continuum equivalents of the matrix expressions previously derived for the discrete-parameter systems. Also it should be emphasized again that they are applicable to beams of nonuniform properties if their mode shapes can be defined. When the dynamic response is excited by motion introduced at the beam supports, the effective loading acting on the structure is given by Eq. (17-24a). The normal-coordinate load term resulting from *each* support-acceleration contribution to this effective loading then becomes

$$P_{nr}(t) = \int_0^L \phi_n(x) p_{\text{eff},r}(x,t) \, dx = -\ddot{\delta}_r(t) \int_0^L m(x)\phi_n(x)\varphi_r(x) \, dx \qquad (19\text{-}7a)$$

and the total effective normal-coordinate load resulting from support excitation is the sum of the contributions from all support-point accelerations.

EXAMPLE E19-2   The mode-superposition analysis of the dynamic response of a distributed-parameter system is carried out by following exactly the same steps that were used for the analysis of a discrete-parameter system. To illustrate the process, the dynamic response of a uniform simple beam subjected to a central step-function loading will be evaluated. The structure and its loading are shown in Fig. E19-2; the analysis is carried out in the following steps.

STEP 1: COMPUTE MODE SHAPES AND FREQUENCIES   This vibration analysis is accomplished by substituting into an appropriate set of boundary-condition equations the modal-shape expression [Eq. (18-10)]. For this simple beam, the results were found in Example E18-1 to be

$$\phi_n(x) = \sin \frac{n\pi x}{L}$$

$$\omega_n = n^2\pi^2 \sqrt{\frac{EI}{\overline{m}L^4}}$$

$$n = 1, 2, \ldots, \infty$$

STEP 2: COMPUTE GENERALIZED MASS AND LOADING   From Eqs. (19-5) and (19-7), these are found to be

$$M_n = \int_0^L \phi_n^2(x)m(x)\,dx = \overline{m}\int_0^L \sin^2 \frac{n\pi x}{L}\,dx = \frac{\overline{m}L}{2}$$

$$P_n = \int_0^L \phi_n(x)p(x,t)\,dx = P_0\phi_n(x = L/2) = \alpha_n P_0$$

$$\text{where } \alpha_n = \begin{cases} 1 & n = 1, 5, 9, \ldots \\ -1 & n = 3, 7, 11, \ldots \\ 0 & n = \text{even} \end{cases}$$

STEP 3: SOLVE THE NORMAL-COORDINATE RESPONSE EQUATION   This is exactly the same equation considered previously for the discrete-parameter case:

$$M_n\ddot{Y}_n + \omega_n^2 M_n Y_n = P_n$$

The Duhamel integral expression gives

$$Y_n(t) = \frac{1}{M_n\omega_n} \int_0^t P_n(\tau) \sin \omega_n(t - \tau)\,d\tau$$

Hence $$Y_n = \frac{2\alpha_n P_0}{\overline{m}L\omega_n} \int_0^t \sin \omega_n(t - \tau)\,d\tau = \frac{2P_0\alpha_n}{\overline{m}L\omega_n^2}(1 - \cos \omega_n t)$$

STEP 4: EVALUATE DISPLACEMENT RESPONSE   Using the normal-coordinate expression [Eq. (19-1)] gives

$$v(x,t) = \sum_{n=1}^{\infty} \phi_n(x)Y_n(t) = \sum_{n=1}^{\infty} \frac{2P_0\alpha_n}{\overline{m}L\omega_n^2}(1 - \cos \omega_n t) \sin \frac{n\pi x}{L}$$

But since $\omega_n^2 = n^4\pi^4\, EI/\overline{m}L^4$, this series can be written

$$v(x,t) = \frac{2P_0 L^3}{\pi^4 EI}\left(\frac{1 - \cos \omega_1 t}{1} \sin \frac{\pi x}{L} - \frac{1 - \cos \omega_3 t}{81} \sin \frac{3\pi x}{L}\right.$$

$$\left. + \frac{1 - \cos \omega_5 t}{625} \sin \frac{5\pi x}{L} - \cdots \right)$$

which represents the complete solution of the dynamic response. It will be noted that the term n^4 in the denominator makes the contribution of the higher modes insignificant.

When the displacement of the midspan point of loading is considered as a point of interest, the time variation of this point is obtained by letting $x = L/2$ in the preceding series, with the result

$$v\left(\frac{L}{2}, t\right) = \frac{2P_0 L^3}{\pi^4 EI} \left(\frac{1 - \cos \omega_1 t}{1} + \frac{1 - \cos \omega_3 t}{81} + \frac{1 - \cos \omega_5 t}{625} + \cdots \right)$$

Considering finally the *static* deflection which remains after the vibration has died away, that is, letting $(1 - \cos \omega_n t) \to 1$, gives

$$v\left(\frac{L}{2}\right)_{\text{static}} = \frac{P_0 L^3}{48EI} \frac{1}{1.014} (1 + \tfrac{1}{81} + \tfrac{1}{625} + \cdots) = \frac{P_0 L^3}{48EI}$$

From the form of this series expression, it is apparent that the first-mode contribution provides over 98 percent of the total deflection and that the higher modes have essentially a negligible effect on the central deflection. It also is evident that the mode-superposition procedure can be used to obtain a solution to a static problem because the static loading is only a special form of dynamic loading; however, this is seldom the most advantageous way of solving a static problem.

STEP 5: EVALUATE DYNAMIC BEAM MOMENTS When the dynamic displacements of the structure have been evaluated at any time, the internal forces in the structure at that time can be found by applying the structure force-displacement relationships. For a beam element, the internal moments are proportional to the curvatures; thus taking the second derivatives of the displacement expression gives

$$\mathcal{M}(x,t) = EI \frac{\partial^2 v}{\partial x^2} = EI \sum_{n=1}^{\infty} \frac{2P_0 \alpha_n}{\bar{m} L \omega_n{}^2} (1 - \cos \omega_n t) \left(-\frac{n^2 \pi^2}{L^2} \right) \sin \frac{n\pi x}{L}$$

which can be expanded to

$$\mathcal{M}(x,t) = -\frac{2P_0 L}{\pi^2} \left(\frac{1 - \cos \omega_1 t}{1} \sin \frac{\pi x}{L} - \frac{1 - \cos \omega_3 t}{9} \sin \frac{3\pi x}{L} \right.$$
$$\left. + \frac{1 - \cos \omega_5 t}{25} \sin \frac{5\pi L}{L} - \cdots \right)$$

In this expression, the mode number is contained in the denominator only to the *second* power; thus the moment series converges much more slowly than the displacements did. This is typical of all dynamic analyses: more modes must be included in the analysis to represent the stresses to any specified degree of accuracy than will be required to represent the displacements.

If the moments at midspan are considered as a special point of interest, the preceding expression reduces to

$$\mathcal{M}\left(\frac{L}{2}, t\right) = -\frac{2P_0 L}{\pi^2} \left(\frac{1 - \cos \omega_1 t}{1} + \frac{1 - \cos \omega_3 t}{9} + \frac{1 - \cos \omega_5 t}{25} + \cdots\right)$$

When the *static* midspan moment is taken, that is, let $(1 - \cos \omega_n t) \to 1$,

$$\mathcal{M}\left(\frac{L}{2}\right)_{\text{static}} = -\frac{P_0 L}{4} \frac{1}{1.232} (1 + \frac{1}{9} + \frac{1}{25} + \frac{1}{49} + \cdots)$$

The first 11 terms of the series in parentheses in the preceding expression add up to 1.19; thus 11 modes provide only about 97 percent of the total bending moment developed at midspan, whereas only a single mode was needed to obtain 98 percent of the total displacement at the same point.   ////

19-3   UNCOUPLED FLEXURAL EQUATIONS OF MOTION: DAMPED

It now is of interest to determine the effect of the normal-coordinate transformation [Eq. (19-1)] on the damped equation of motion [Eq. (17-19)]. With this substitution, the equation of motion becomes

$$\sum_{i=1}^{\infty} m(x)\phi_i(x)\ddot{Y}_i(t) + \sum_{i=1}^{\infty} c(x)\phi_i(x)\dot{Y}_i(t) + \sum_{i=1}^{\infty} \frac{d^2}{dx^2}\left[c_s I(x)\frac{d^2\phi_i}{dx^2}\right]\dot{Y}_i(t)$$

$$+ \sum_{i=1}^{\infty} \frac{d^2}{dx^2}\left[EI(x)\frac{d^2\phi_i}{dx^2}\right]Y_i(t) = p(x,t)$$

Multiplying by $\phi_n(x)$, integrating, and applying the orthogonality relationships together with the definitions of generalized mass and generalized force leads to

$$M_n \ddot{Y}_n(t) + \sum_{i=1}^{\infty} \dot{Y}_i(t) \int_0^L \phi_n(x) \left\{c(x)\phi_i(x)\right.$$

$$\left. + \frac{d^2}{dx^2}\left[c_s I(x)\frac{d^2\phi_i}{dx^2}\right]\right\} dx$$

$$+ \omega_n^2 M_n Y_n(t) = P_n(t) \qquad (19\text{-}8)$$

Now it is evident that the equations of motion of the different modes will be coupled by the damping terms in Eq. (19-8) unless they satisfy orthogonality conditions equivalent to those associated with the mass and stiffness properties. Moreover, it is evident that such orthogonality conditions will exist if the damping effects are assumed to be proportional to the mass and stiffness properties. To investigate this case, it will be assumed that

$$c(x) = a_0 m(x) \qquad c_s = a_1 E \qquad (19\text{-}9)$$

in which a_0 and a_1 are simple proportionality factors (having dimensions of the reciprocal of time and of time, respectively). Substituting Eqs. (19-9) into Eq. (19-8) and applying the orthogonality conditions results in an uncoupled normal-coordinate equation

$$M_n \ddot{Y}_n(t) + (a_0 M_n + a_1 \omega_n^2 M_n) \dot{Y}_n(t) + \omega_n^2 M_n Y_n(t) = P_n(t) \qquad (19\text{-}10)$$

Finally, dividing through by the generalized mass and introducing the damping ratio for the nth mode, defined as

$$\xi_n = \frac{a_0}{2\omega_n} + \frac{a_1 \omega_n}{2} \qquad (19\text{-}11)$$

gives this equation the form of the standard SDOF equation

$$\ddot{Y}_n(t) + 2\xi_n \omega_n \dot{Y}_n(t) + \omega_n^2 Y_n(t) = \frac{P_n(t)}{M_n} \qquad (19\text{-}12)$$

Thus it is clear that where the damping is of the Rayleigh mass or stiffness-proportional type, the distributed-parameter equations of motion can be uncoupled in the same way as for the discrete-parameter systems. Also, Eq. (19-11) shows again that for mass-proportional damping, the damping ratio is inversely proportional to the frequency while for stiffness-proportional damping, the damping ratio is directly proportional to the frequency (as noted earlier in the discussion of damped discrete-parameter systems).

The relationship of Eq. (19-11) between the proportionality factors and the damping ratio is illustrated in Fig. 19-2. If the damping ratio corresponding to a specific mode frequency is known, it is evident that the factor a_1 or a_0 can be determined easily for a pure mass- or stiffness-proportional damping system. If both mass- and stiffness-proportional damping are present, it is necessary to establish two modal-damping ratios in order to evaluate the factors a_0 and a_1. For example, if the damping ratios ξ_m and ξ_n corresponding to modal frequencies ω_m and ω_n are given (as indicated in Fig. 19-2), Eq. (19-11) can be written in matrix form expressing the two conditions as

$$\begin{bmatrix} \xi_m \\ \xi_n \end{bmatrix} = \frac{1}{2} \begin{bmatrix} \dfrac{1}{\omega_m} & \omega_m \\ \dfrac{1}{\omega_n} & \omega_n \end{bmatrix} \begin{bmatrix} a_0 \\ a_1 \end{bmatrix}$$

from which

$$\begin{bmatrix} a_0 \\ a_1 \end{bmatrix} = 2 \frac{\omega_m \omega_n}{\omega_n^2 - \omega_m^2} \begin{bmatrix} \omega_n & -\omega_m \\ -\dfrac{1}{\omega_m} & \dfrac{1}{\omega_n} \end{bmatrix} \begin{bmatrix} \xi_m \\ \xi_n \end{bmatrix} \qquad (19\text{-}13)$$

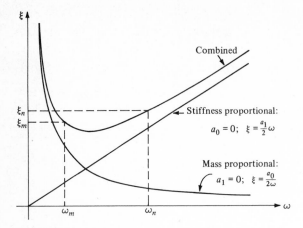

FIGURE 19-2
Relationship between damping ratio and frequency (for Rayleigh damping).

Evaluating the factors a_0 and a_1 from Eq. (19-13) leads to the desired damping ratios at the frequencies ω_m and ω_n and gives a damping-ratio–frequency relationship of the form shown by the combined curve in Fig. 19-2.

Analysis of the dynamic response to any given applied loading using the uncoupled damped equations of motion [Eq. (19-12)] is exactly equivalent to a mode-superposition analysis of a damped discrete-parameter system. The distributed-parameter characteristics of the system are contained in the computed vibration mode shapes and influence the dynamic response explicitly only in the final transformation from the normal coordinates [Eq. (19-1)].

19-4 UNCOUPLED AXIAL EQUATIONS OF MOTION: UNDAMPED

The mode-shape (normal) coordinate transformation serves to uncouple the equations of motion of any dynamic system and therefore is applicable to the axial as well as the flexural equations of motion of a one-dimensional member. Introducing Eq. (19-1) into the equation of axial motion, Eq. (17-28), leads to

$$\sum_{i=1}^{\infty} m(x)\overline{\phi}_i(x)\ddot{Y}_i(t) - \sum_{i=1}^{\infty} \frac{d}{dx}\left[AE(x)\frac{d\overline{\phi}_i}{dx}\right]Y_i(t) = p_u(x,t) \qquad (19\text{-}14)$$

in which a distributed axial-load term has been introduced on the right-hand side for generality. Multiplying each term by $\overline{\phi}_n(x)$ and applying the orthogonality relationships [Eqs. (18-36) and (18-38)] leads to

$$\ddot{Y}_n(t)\int m(x)\overline{\phi}_n{}^2(x)\,dx - Y_n(t)\int_0^L \overline{\phi}_n(x)\frac{d}{dx}\left[AE(x)\frac{d\overline{\phi}_n}{dx}\right]dx = \int_0^L \overline{\phi}_n(x)p_u(x,t)\,dx$$

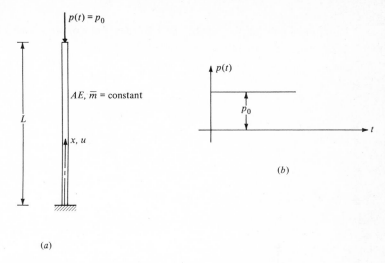

FIGURE E19-3
Pile subjected to end loading: (*a*) geometric configuration; (*b*) step-function loading.

Substituting the inertial force for the elastic-force term [from Eq. (18-37)] and introducing the standard expressions for generalized mass and load

$$M_n = \int_0^L m(x)\bar{\phi}_n{}^2(x)\, dx$$

$$P_n = \int_0^L \bar{\phi}_n(x)p_u(x,t)\, dx$$

leads to the final uncoupled axial equation of motion

$$M_n \ddot{Y}_n(t) + \omega_n{}^2 M_n Y_n(t) = P_n(t)$$

which is exactly the same as the uncoupled equation of flexural motion [Eq. (16-6)] or for any other type of vibrating system. From this discussion it is apparent that after the vibration mode shapes have been determined, the reduction to the normal-coordinate form involves exactly the same type of operations in all structures.

EXAMPLE E19-3   Because the dynamic response of a prismatic bar to axial loading has special characteristics which will be the subject of later discussion, it will be instructive to perform an example analysis of this type. Consider the pile fixed rigidly at its base and subjected to a step-function loading P_0 at the upper end, as shown in Fig. E19-3. The mode-superposition analysis of this system can be carried out by the same sequence of steps employed in the beam-response example of Fig. 19-2.

STEP 1: MODE SHAPES AND FREQUENCIES (see Example E18-5):

$$\bar{\phi}_n(x) = \sin \frac{2n-1}{2} \frac{\pi x}{L}$$

$$\omega_n = \frac{2n-1}{2} \pi \sqrt{\frac{EA}{\bar{m}L^2}}$$

$$n = 1, 2, \ldots, \infty$$

STEP 2: GENERALIZED MASS AND LOADING:

$$M_n = \int_0^L m(x)\bar{\phi}_n{}^2(x)\, dx = \bar{m} \int_0^L \sin^2\left(\frac{2n-1}{2}\frac{\pi x}{L}\right) dx = \frac{\bar{m}L}{2}$$

$$P_n = \int_0^L p_u(x,t)\bar{\phi}_n(x)\, dx = -p_0\bar{\phi}_n(L) = \pm p_0 \qquad \begin{array}{l} +n = \text{even} \\ -n = \text{odd} \end{array}$$

STEP 3: GENERALIZED-COORDINATE RESPONSE (see Example E19-2):

$$Y_n(t) = \pm \frac{2P_0}{\bar{m}L\omega_n{}^2}(1 - \cos \omega_n t)$$

STEP 4: DISPLACEMENT RESPONSE:

$$u(x,t) = \sum_{n=1}^{\infty} \bar{\phi}_n(x)Y_n(t)$$

$$= \frac{2P_0}{\bar{m}L\omega_1{}^2}\left(-\frac{1 - \cos \omega_1 t}{1}\sin\frac{\pi x}{2L} + \frac{1 - \cos \omega_2 t}{9}\sin\frac{3\pi x}{2L}\right.$$

$$\left. - \frac{1 - \cos \omega_3 t}{25}\sin\frac{5\pi x}{2L} + \cdots\right)$$

$$= \frac{8P_0}{\pi^2}\frac{L}{AE}\sum_{n=1}^{\infty}\left[(\pm)\frac{1 - \cos \omega_n t}{(2n-1)^2}\sin\frac{2n-1}{2}\frac{\pi x}{L}\right] \qquad (a)$$

STEP 5: AXIAL-FORCE RESPONSE:

$$\mathcal{N}(x,t) = EA\frac{\partial u}{\partial x}$$

$$= \frac{8P_0 L}{\pi^2}\sum_{n=1}^{\infty}\left[\pm\frac{1 - \cos \omega_n t}{(2n-1)^2}\frac{2n-1}{2}\frac{\pi}{L}\cos\frac{2n-1}{2}\frac{\pi x}{L}\right]$$

$$= \frac{4P_0}{\pi}\sum_{n=1}^{\infty}\left[(\pm)\frac{1 - \cos \omega_n t}{2n-1}\cos\frac{2n-1}{2}\frac{\pi x}{L}\right] \qquad (b)$$

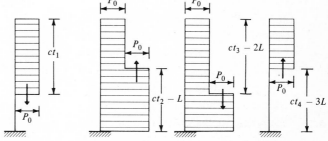

Phase 1   Phase 2   Phase 3   Phase 4
$(0 < ct_1 < L)$ $(L < ct_2 < 2L)$ $(2L < ct_3 < 3L)$ $(3L < ct_4 < 4L)$

Displacement

Axial Force

FIGURE E19-4
Response of pile to step-function loading.

The response at any time t can be obtained by summing terms in the series expressions (a) and (b) representing the displacement and force distributions. For this purpose, it is convenient to express the time-variation parameter $\omega_n t$ in the form

$$\omega_n t = \left(\frac{2n-1}{2}\pi\right)\frac{ct}{L}$$

where $c = \sqrt{EA/\overline{m}}$ has the dimensions of velocity. Thus the product ct becomes a distance, and the time parameter may be considered as the ratio of this distance to the length of the pile. The displacement and force distribution in the pile at four different values of this time parameter have been obtained by evaluating the series expressions; the results obtained by summing the series are plotted in Fig. E19-4. The simple form of the response produced by the step-function loading is evident in these sketches. For any time $t_1 < L/c$, the pile has no load ahead of the distance ct_1 but is subject to the constant force P_0

behind this distance. Thus the response may be interpreted as a force wave of amplitude P_0 propagating ahead with the velocity c. The displacement is consistent with this load distribution, of course, showing a linear variation in the section of the pile in which there is constant load and no displacement in the zone ahead of the force wave. In the time interval $L/c < t_2 < 2L/c$ the force wave is doubled in the zone from the rigid support to a point $ct_2 - L$ from this support. This response behavior may be interpreted as a reflection of the force wave, resulting in a double amplitude as it propagates back along the pile. In the time interval $2L/c < t_3 < 3L/c$ the response may be interpreted as a negative reflection from the free end of the pile, causing a reduction of the force amplitude which propagates with the velocity c. During the fourth phase, $3L/c < t_4 < 4L/c$, the negative wave of phase 3 is reflected from the rigid base and causes a reduction of the axial force to zero value. At the end of the fourth phase, at time $t = 4L/c$, the pile is completely unstressed, as it was at time $t = 0$; a negative reflection of the negative wave then initiates a positive wave propagating down the pile in a form exactly equivalent to phase 1.

From the preceding discussion it is apparent that the free vibrations of the pile subjected to the step-function loading can be interpreted as an axial-force wave propagating along the pile and being subjected to positive and negative reflections at the fixed and free ends. This wave oscillation will continue indefinitely in the absence of damping or of any change in the loading. It is important to note that this response was evaluated in this example by the superposition of the axial vibration modes of the pile, each of which involves the entire extent of the pile. For example, in phase 1 the unstressed zone in the pile ahead of the advancing wave was obtained by the superposition of an infinite number of modes, each of which included stresses ahead of the wave-front. It is evident that the mode-superposition method is a rather cumbersome way to represent the very simple wave-propagation concept, and a more direct analysis will be presented later. The fact that the mode-superposition process does account for the wave-propagation mechanism is most significant, however; this type of analysis provides the complete solution for any structure subjected to any type of dynamic loading.　　　　　////

PROBLEMS

19-1　Assume that the undamped uniform beam of Fig. E19-2 is subjected to a static central load p_0 and then set into free vibration by suddenly releasing the load at time $t = 0$. The initial deflected shape is given by

$$v(x) = \frac{p_0 x}{48EI}(3L^2 - 4x^2) \qquad 0 < x < \frac{L}{2}$$

(a) From this information, evaluate the amplitude of the midspan free-vibration displacement in each of the first three modes of vibration, expressing results as fractions of the static midspan displacement.

(b) Evaluate the amplitude of the midspan free-vibration moment in each of the first three modes of vibration, expressing the results as fractions of the static midspan moment.

19-2 Assuming that the step-function load of Fig. E19-2 is applied at the quarter span point ($x = L/4$), rather than at midspan, write expressions for the undamped displacement response and bending moment response at the load point. Plot this moment history, considering the first three modes, over the time interval $0 < t < T_1$.

19-3 Assume that the beam of Fig. E19-2 is subjected to a harmonic load applied at the quarter span point: $p(t) = p_0 \sin \bar{\omega} t$, where $\bar{\omega} = \frac{5}{4}\omega_1$. Considering the first three modes of vibration, plot the steady-state displacement response amplitude of the beam, evaluating it at increments $L/4$ along the span:

(a) Neglecting damping.

(b) Assuming the damping in each mode is 10% critical.

19-4 A uniform simple beam having flexural rigidity $EI = 78 \times 10^8$ lb-in^2 supports a total weight of 1,000 lb/ft. When immersed in a viscous fluid and set into first-mode vibration with an amplitude of 1 in, it is observed that the motion is reduced to 0.1 in amplitude in 3 cycles.

(a) Assuming that the damping resistance per unit velocity, $c(x)$, is uniform along the span, determine its numerical value.

(b) Assuming that the same beam is set into second-mode vibration with a 1 in amplitude, determine how many cycles will be required to reduce this motion to 0.1 in.

19-5 Repeat Prob. 19-3 for the uniform rod shown in Fig. P19-1. Note that the harmonic load $p(t) = p_0 \sin \bar{\omega} t$ is applied in the axial direction at midlength, and that the axial displacement response is to be plotted by both neglecting and including the influence of modal damping.

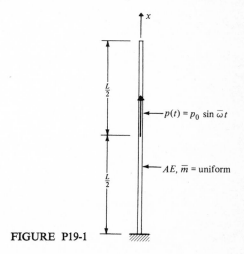

FIGURE P19-1

19-6 The uniform simple beam shown in Fig. P19-2 is subjected to a lateral loading $p(x,t) = \delta(x - a)\,\delta(t)$, where $\delta(x - a)$ and $\delta(t)$ are Dirac delta functions. (See Sec. 22-1 for a definition of the Dirac delta function.) Using elementary beam theory and the mode superposition method, determine the series expressions for lateral deflection $v(x,t)$, internal moment $\mathscr{M}(x,t)$, and internal shear $\mathscr{V}(x,t)$ caused by the loading $p(x,t)$ defined above. Discuss the relative rates of convergence of these three series expressions.

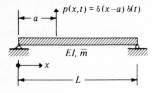

FIGURE P19-2

THE DYNAMIC DIRECT-STIFFNESS METHOD

20-1 INTRODUCTION

From the foregoing discussion it is evident that the mode-superposition analysis of any structure formed as an assemblage of beam-type elements can be carried out routinely, taking account of the actual distribution of mass, damping, and stiffness, after the vibration mode shapes and frequencies of the structure have been evaluated. In fact, it is easy to extend the analysis to include structures made up of two- and three-dimensional elements—after their vibration properties are known. The difficult part of the analysis is the evaluation of vibration mode shapes and frequencies for the given structural system.

The formulation of the vibration-analysis procedure for structures assembled from prismatic-beam segments, as described in Chap. 18, is straightforward. However, the process quickly becomes unmanageable because of the four constants which must be evaluated in each segment of the system. A simple, two-span bent involving only five uniform beam members, for example, requires the evaluation of a system of 20 shape-function constants. The analysis is much more tractable if the segment properties are expressed in terms of dynamic-stiffness coefficients rather than in terms of these arbitrary constants. After the dynamic-stiffness expressions are derived, the

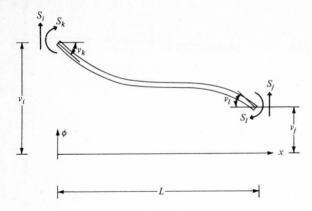

FIGURE 20-1
Boundary forces and displacements (uniform beam segment).

analysis of a system of any complexity can be carried out by exactly the same procedures used in a static direct-stiffness, that is, displacement-method, analysis: the stiffness of the structural assemblage is formed by adding appropriate contributions from the individual beam elements; then the stiffness equations of the complete system are solved for the displacements resulting from the given loading. The only new feature of the dynamic direct-stiffness analysis is that the dynamic-stiffness coefficients are functions of the frequency of the motion. For a loading applied at a given frequency, they can be computed directly and the analysis carried out in the normal fashion; in the analysis of the natural-vibration frequencies, however, the stiffness coefficients must be adjusted iteratively during the analysis procedure.

20-2   DYNAMIC FLEXURAL-STIFFNESS MATRIX

The dynamic-stiffness coefficients of a beam segment represent the end forces and moments resulting from the application of unit end displacements and rotations. The force and displacement quantities of interest are indicated in Fig. 20-1. The only difference in the definition of the dynamic-stiffness coefficients compared with the standard static-stiffness coefficients is that the end displacements vary harmonically with time, all at the same frequency and in phase; consequently, the end forces also vary at the same frequency and with the same phase relationship. (It should be noted that the dynamic stiffness is defined for an *undamped* beam segment.) The static beam stiffness therefore can be considered to represent a special case of the dynamic stiffness where the applied frequency is zero.

Inasmuch as the beam segment is assumed to be uniform and subjected to no loading within its span, its equation of motion is given by Eq. (18-2)

$$v^{\text{iv}} + \frac{\overline{m}}{EI} \ddot{v} = 0$$

Now the boundary displacements can be expressed in the form

$$v_i = v_{i0} \sin \overline{\omega} t \qquad (20\text{-}1)$$

where v_{i0} represents the *amplitude* of the boundary displacement v_i, which is varying harmonically at the applied frequency $\overline{\omega}$; therefore the displacements at any point within the span are of the form

$$v(x,t) = \phi(x) \sin \overline{\omega} t \qquad (20\text{-}2)$$

Thus Eq. (18-2) may be written

$$\phi^{\text{iv}}(x) - \overline{a}^4 \phi(x) = 0 \qquad (20\text{-}3)$$

in which

$$\overline{a}^4 \equiv \frac{\overline{m}\overline{\omega}^2}{EI} \qquad (20\text{-}4)$$

It will be noted that Eq. (20-3) is equivalent to Eq. (18-5a); thus its solution is equivalent to Eq. (18-10), that is

$$\phi(x) = A_1 \sin \overline{a}x + A_2 \cos \overline{a}x + A_3 \sinh \overline{a}x + A_4 \cosh \overline{a}x \qquad (20\text{-}5)$$

The only difference between Eq. (20-5) and Eq. (18-10) is that (20-5) contains $\overline{a}$, which is a function of the forced frequency $\overline{\omega}$, rather than a, which depends on the natural frequency ω. Inasmuch as the context in which the equation is used will indicate whether forced or natural frequencies are involved, there is no need to retain separate expressions and the form of Eq. (18-10) will be used for both situations. In other words,

$$a^4 = \frac{\overline{m}\overline{\omega}^2}{EI} \text{ or } \frac{\overline{m}\omega^2}{EI}$$

according as forced- or free-vibration conditions are involved.

Thus, expressing Eq. (18-10) and its first three derivatives in matrix form leads to

$$
\begin{bmatrix}
\phi \\
\dfrac{\phi'}{a} \\
\dfrac{\phi''}{a^2} \\
\dfrac{\phi'''}{a^3}
\end{bmatrix}
=
\begin{bmatrix}
\sin ax & \cos ax & \sinh ax & \cosh ax \\
\cos ax & -\sin ax & \cosh ax & \sinh ax \\
-\sin ax & -\cos ax & \sinh ax & \cosh ax \\
-\cos ax & \sin ax & \cosh ax & \sinh ax
\end{bmatrix}
\begin{bmatrix}
A_1 \\
A_2 \\
A_3 \\
A_4
\end{bmatrix}
\qquad (20\text{-}6)
$$

Now expressing the displacements and rotations at the two ends (nodal points) of the segment by means of the first two expressions in Eq. (20-6) leads to

$$
\begin{bmatrix} \dfrac{v_i}{L} \\[2mm] \dfrac{v_j}{L} \\[2mm] v_k \\[2mm] v_l \end{bmatrix} \equiv \begin{bmatrix} \dfrac{1}{L}\phi_{x=0} \\[2mm] \dfrac{1}{L}\phi_{x=L} \\[2mm] -\phi'_{x=0} \\[2mm] -\phi'_{x=L} \end{bmatrix} = \begin{bmatrix} 0 & \dfrac{1}{L} & 0 & \dfrac{1}{L} \\[2mm] \dfrac{s}{L} & \dfrac{c}{L} & \dfrac{S}{L} & \dfrac{C}{L} \\[2mm] -a & 0 & -a & 0 \\[2mm] -ac & as & -aC & -aS \end{bmatrix} \begin{bmatrix} A_1 \\ A_2 \\ A_3 \\ A_4 \end{bmatrix} \tag{20-7}
$$

in which

$$
\begin{aligned}
s &\equiv \sin aL & S &= \sinh aL \\
c &\equiv \cos aL & C &= \cosh aL
\end{aligned}
$$

Finally with one symbol to represent each matrix in Eq. (20-7), it is written in abbreviated form as

$$
\mathbf{v} = \mathbf{W}\boldsymbol{\eta} \tag{20-8}
$$

Equation (17-8) represents an expression of the nodal forces in terms of the constants A_i in Eq. (18-10) [or (20-5)]. Next the nodal forces will be expressed in terms of these same constants, after which the stiffness matrix can be obtained by expressing the nodal forces in terms of the nodal displacements (eliminating the constants). The nodal forces at the two ends of the beam segment can be written by means of the last two expressions in Eq. (20-6), as follows:

$$
\begin{bmatrix} S_i L \\ S_j L \\ S_k \\ S_l \end{bmatrix} = EI \begin{bmatrix} L\phi'''_{x=0} \\ -L\phi'''_{x=L} \\ \phi''_{x=0} \\ -\phi''_{x=L} \end{bmatrix}
$$

$$
= EIa^2 \begin{bmatrix} -aL & 0 & aL & 0 \\ acL & -asL & -aCL & -aSL \\ 0 & -1 & 0 & 1 \\ s & c & -S & -C \end{bmatrix} \begin{bmatrix} A_1 \\ A_2 \\ A_3 \\ A_4 \end{bmatrix} \tag{20-9}
$$

or, in abbreviated form,

$$
\mathbf{S} = \mathbf{U}\boldsymbol{\eta} \tag{20-10}
$$

It is evident that the constants $\boldsymbol{\eta}$ can be evaluated in terms of the nodal displacements by inverting Eq. (20-8):

$$
\boldsymbol{\eta} = \mathbf{W}^{-1}\mathbf{v} \tag{20-11}
$$

Hence substituting this into Eq. (20-10) leads to

$$
\mathbf{S} = \mathbf{U}\mathbf{W}^{-1}\mathbf{v} \tag{20-12}
$$

The central matrix product in Eq. (20-12) represents the dynamic stiffness of the beam segment, by definition, because it expresses the nodal forces in terms of the nodal displacements; that is,

$$\mathbf{k}(a) = \mathbf{U W}^{-1} \qquad (20\text{-}13)$$

in which it is important to note that the stiffness is a function of the frequency parameter a because both $\mathbf{U}$ and $\mathbf{W}$ depend on a.

When the inversion and multiplication are carried out, Eq. (20-12) is written

$$
\begin{bmatrix} S_i L \\ S_j L \\ S_k \\ S_l \end{bmatrix}
= \frac{EI}{L}
\begin{bmatrix}
\gamma & -\bar{\gamma} & -\beta & -\bar{\beta} \\
-\bar{\gamma} & \gamma & \bar{\beta} & \beta \\
-\beta & \bar{\beta} & \alpha & \bar{\alpha} \\
-\bar{\beta} & \beta & \bar{\alpha} & \alpha
\end{bmatrix}
\begin{bmatrix} \dfrac{v_i}{L} \\ \dfrac{v_j}{L} \\ \dfrac{v_k}{L} \\ v_l \end{bmatrix}
\qquad (20\text{-}14)
$$

in which

$$\alpha = \frac{sC - cS}{d}\lambda \qquad \bar{\alpha} = \frac{S - s}{d}\lambda$$

$$\beta = \frac{sS}{d}\lambda^2 \qquad \bar{\beta} = \frac{C - c}{d}\lambda^2 \qquad (20\text{-}15)$$

$$\gamma = \frac{sC + cS}{d}\lambda^3 \qquad \bar{\gamma} = \frac{S + s}{d}\lambda^3$$

where

$$d = 1 - cC \qquad \lambda = aL$$

Evaluation of these expressions for the static case, where $\lambda = 0$, leads to the following values for the stiffness coefficients:

$$\gamma_0 = \bar{\gamma}_0 = 12 \qquad \beta_0 = \bar{\beta}_0 = 6 \qquad \alpha_0 = \bar{\alpha}_0 = 4$$

which (when multiplied by EI/L) are the familiar static stiffness coefficients for a uniform beam. For any other value of frequency, the corresponding value of λ can be computed and the resulting value of the stiffness coefficients obtained from Eqs. (20-15). A plot of the variation of these coefficients with the frequency parameter λ is shown in Fig. 20-2a and 20-2b.[1] It is of interest that some of the stiffness coefficients become negative as the frequency is increased beyond certain critical points. These points represent natural frequencies for the beam when it has various specific support

---

[1] Tabular listings of these functions may be found in *Dynamics of Elastic Systems* by W. Nowacki, Chapman and Hall, Ltd., London, 1963 (pp. 169, 170) and in *Baudynamik der Durchlaufträger und Rahmen* by V. Kolovsek, Fachbuchverlag GMBH, Leipzig, 1953 (pp. 149-161).

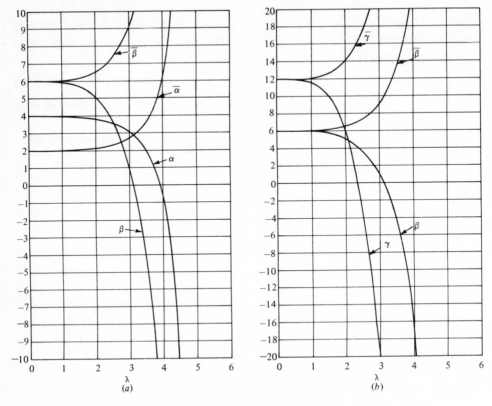

FIGURE 20-2
Dynamic-stiffness coefficients [from Eq. (20-15)].

conditions. It is important to note that the nodal forces expressed by these stiffness coefficients do *not* satisfy the normal *statics* relationships between the end moments and shears, except for the case where $\lambda = 0$. For example, it is evident that $S_i \neq -S_j$ and $(S_k + S_l) \neq -S_iL \neq S_jL$ except for this limiting case. This is because the transverse inertia of the beam must also be considered in the dynamic-equilibrium relationships.

EXAMPLE E20-1   As a preliminary example of the application of the dynamic-direct-stiffness method, consider the frame shown in Fig. E20-1, which is subjected to a harmonically varying external moment applied at the joint connecting the three uniform members (joint 1). The supports of this structure constrain it against any displacements parallel with the axes of the members if

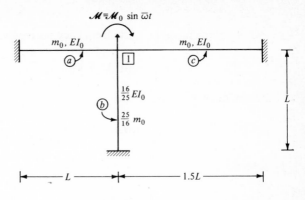

FIGURE E20-1
Frame subject to transverse displacements only.

axial distortions are neglected; thus all the member displacements are transverse, and the stiffness relationships of Eq. (20-14) are applicable. Also, it will be noted that the assemblage has only one nodal degree of freedom, rotation of the interconnection joint (joint 1). The stiffness of the assemblage at this joint is the sum of the stiffnesses of the members connecting at this joint; thus

$$k_{11} \equiv (k_{ll})_a + (k_{ll})_b + (k_{kk})_c$$

$$= \frac{EI_0}{L} \alpha_a + \frac{^{16}/_{25} EI_0}{L} \alpha_b + \frac{EI_0}{1.5L} \alpha_c \qquad (a)$$

To consider the specified frequency of the applied moment $\bar{\omega}$, the stiffness coefficient for each member can be evaluated; e.g.,

$$\alpha_a = \left(\frac{\cosh aL \sin aL - \cos aL \sinh aL}{1 - \cos aL \cosh aL} \, aL \right)_a$$

$$\text{where } (aL)_a^4 = \frac{\bar{\omega}^2}{EI_0/m_0 L^4} \qquad (b)$$

If it is assumed for the purposes of this example that the applied frequency is adjusted to have a value such that

$$\bar{\omega}^2 = (2.8)^4 \frac{EI_0}{m_0 L^4}$$

then the frequency parameter for member a is

$$(aL)_a = \left[\bar{\omega}^2 \left(\frac{\bar{m} L^4}{EI} \right)_a \right]^{1/4} = 2.8$$

Similarly for member c, the frequency parameter is

$$(aL)_c = \left[(2.8)^4 \frac{EI_0}{m_0 L^4} \frac{m_0(1.5L)^4}{EI_0} \right]^{1/4} = 4.2$$

while for member b

$$(aL)_b = \left[(2.8)^4 \frac{EI_0}{m_0 L^4} \frac{(^{25}/_{16})m_0 L^4}{(^{16}/_{25})EI_0} \right]^{1/4} = 3.5$$

From Fig. 20-2a the values of α corresponding with these frequency parameters are approximately

$$\alpha_a = 3.33 \quad \alpha_b = 2.00 \quad \alpha_c = -2.90$$

Hence the stiffness of this assemblage at joint 1 is

$$k_{11} = \frac{EI_0}{L} \left[1(3.33) + {}^{16}/_{25}(2.00) + \frac{1}{1.5}(-2.90) \right] = 2.68 \frac{EI_0}{L} \qquad (c)$$

It is important to note that the stiffness contribution of member c is negative at this applied frequency; that is, this member is tending to increase the rotation rather than reduce it. However, since the positive stiffnesses of the other two members are greater than this negative contribution, the assemblage offers a net resistance to the applied moment.

Because this assembled structure has only a single nodal displacement, the relationship between the applied moment and the resulting displacement is a simple scalar equation which may be expressed as

$$\mathcal{M} = k_{11}v_1$$

where v_1 represents the rotation of the joint. Solving for the rotation produced by the applied moment then leads to

$$v_1 = k_{11}{}^{-1}\mathcal{M} = \frac{\mathcal{M}_0 L}{2.68EI_0} \sin \bar{\omega}t \qquad (d)$$

The analysis would be entirely similar for a structure having N degrees of freedom, except that $\mathcal{M}$ and v would be vectors with N elements and k would be a square symmetric stiffness matrix of dimension N.

If the purpose of this problem were to determine the natural frequencies of the structure rather than its response to a specified harmonic loading, the analysis would have to be carried out by a trial-and-error process because the stiffnesses of the members depend on the unknown frequency. The physical basis of the frequency-analysis procedure is very simple: at the natural frequency motions can exist with no applied force; that is, in this example, a rotation v_1 could occur when the applied harmonic force is reduced to zero amplitude.

This is equivalent to stating that the total dynamic stiffness of the structure goes to zero at the natural frequency; thus in the example the natural frequency corresponds to the frequency parameter at which the negative stiffness of member c exactly equals the positive stiffness of members a and b. After some trials it is found that where $\bar{\omega}^2 = (2.89)^4(EI_0/m_0L^4)$ (that is, $aL_a = 2.89$, $aL_b = 3.66$, and $aL_c = 4.34$), the total stiffness of the joint is

$$k_{11} = \frac{EI_0}{L} \left[1(3.23) + \frac{16}{25}(1.44) - \frac{2}{3}(5.80) \right] \doteq 0$$

Thus the fundamental natural frequency is

$$\omega = (2.89)^2 \sqrt{\frac{EI_0}{m_0L^4}} \qquad (e)$$

There are an infinite number of higher-frequency values which also lead to zero rotational stiffness of the joint, corresponding to conditions in which zero-amplitude nodes are located at various positions along the three members.

   If the structure had more than one degree of freedom, its natural frequencies could be determined by essentially the same process as described above except that the criterion is the vanishing of the *determinant* of the stiffness matrix rather than setting the stiffness equal to zero. Because each element of the stiffness matrix is a function of the frequency, the expression for the determinant also depends on the frequency, and, by trial and error, the frequency can be adjusted until the expression goes to zero.

   One final comment should be made about this method of frequency analysis: the process will establish the frequencies of only those modes of vibration in which at least one of the nodal degrees of freedom participates. Any mode which does not involve displacements of the nodes will not be contained in the analysis. For example, if the three members of the frame of Fig. E20-1 had appropriate relative properties, modes of vibration corresponding to the clamped-clamped condition in each member could exist and would not be identified by setting the rotational stiffness of the central joint equal to zero.   ////

20-3   DYNAMIC STIFFNESS FOR FLEXURE AND RIGID AXIAL DISPLACEMENTS

In the development of the dynamic-stiffness matrix for a beam [Eq. (20-14)] and in its application in Example E20-1, it has been assumed implicitly that only displacements transverse to the beam axis are to be considered. If the structural system is

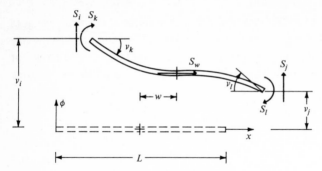

FIGURE 20-3
Beam forces and displacements including rigid axial motion.

so arranged, however, that displacement components may take place parallel to the member's axis, it is necessary to incorporate this additional degree of freedom into the dynamic-stiffness relationship. In the usual frame analysis, it is assumed that axial *distortions* of the members are very small compared with those associated with bending, and consequently the change of length of the member is neglected in establishing the structural displacements. In many structural arrangements, however, the beam may have displacement components parallel with its axis even if axial distortions are neglected, as a result of joint displacements ensuing from the flexure of supporting members. The beam displacements developed in such a situation and the corresponding member forces are shown in Fig. 20-3. Inasmuch as the axial distortions are neglected, the axial-displacement component is the same all along the length of the member and could be defined at any convenient point along the length. It is indicated at the center of the beam in the sketch to emphasize that it is not exclusively associated with either end.

If the influence of the axial force on the flexural stiffness of the beam is neglected (and in most practical situations it has a negligible effect), the axial displacement of this beam has no effect on the transverse-displacement–force relationships. Similarly, the transverse displacements have no effect on the axial-force component S_w. Thus, to account for the axial displacements in the beam-stiffness matrix, it is necessary only to include one additional term, representing the uncoupled axial-force–displacement relationship. Because the axial distortions of the beam are neglected, this axial-force term represents the inertial effect associated with its rigid-body axial accelerations, that is,

$$S_w = F_I = -\bar{m}L\bar{\omega}^2 w \qquad (20\text{-}16)$$

where $\bar{m}$ is the mass per unit length and $-\bar{\omega}^2 w$ is the axial acceleration. Incorporating this additional term into the dynamic-beam-stiffness matrix leads to the following expanded form of Eq. (20-14):

$$
\begin{bmatrix} S_i L \\ S_j L \\ S_k \\ S_l \\ S_w L \end{bmatrix} = \frac{EI}{L} \begin{bmatrix} \gamma & -\bar{\gamma} & -\beta & -\bar{\beta} & 0 \\ -\bar{\gamma} & \gamma & \bar{\beta} & \beta & 0 \\ -\beta & \bar{\beta} & \alpha & \bar{\alpha} & 0 \\ -\bar{\beta} & \beta & \bar{\alpha} & \alpha & 0 \\ 0 & 0 & 0 & 0 & -\lambda^4 \end{bmatrix} \begin{bmatrix} \dfrac{v_i}{L} \\ \dfrac{v_j}{L} \\ v_k \\ v_l \\ \dfrac{w}{L} \end{bmatrix} \tag{20-17}
$$

in which the relationship $\lambda^4 = (aL)^4 = \bar{\omega}^2(\bar{m}L^4/EI)$ has been employed. It should be evident that this term vanishes under static conditions; however, it can make a major contribution to the dynamic forces under conditions of free or forced vibration.

EXAMPLE E20-2   The stiffness expression of Eq. (20-17) makes possible the dynamic-direct-stiffness analysis of any plane-frame system, taking account of all types of joint displacements but neglecting axial deformations of the members. As an example of the formulation of a problem of this type, consider the bent with one inclined column shown in Fig. E20-2a. For simplicity, all members of the frame have been given the same length and section properties. The three degrees of freedom of the frame are sketched in Fig. E20-2b.

The stiffness of the frame is obtained by adding together the appropriate stiffnesses of the three members. The stiffness matrix serves to express the force-deflection relationship for the frame, as follows:

$$
\begin{bmatrix} p_1 \\ p_2 \\ p_3 \end{bmatrix} = \begin{bmatrix} k_{11} & k_{12} & k_{13} \\ k_{21} & k_{22} & k_{23} \\ k_{31} & k_{32} & k_{33} \end{bmatrix} \begin{bmatrix} v_1 \\ v_2 \\ v_3 \end{bmatrix}
$$

and the individual terms for the system of Fig. E20-2 are

$$ k_{11} = \frac{EI_0}{L}(\alpha_a + \alpha_b) = \frac{2EI_0}{L}\alpha $$

$$ k_{12} = \frac{EI_0}{L}\bar{\alpha}_b = \frac{EI_0}{L}\bar{\alpha} = k_{21} $$

$$ k_{13} = \frac{EI_0}{L^2}(-\,^5\!/_4\beta_a + \,^3\!/_4\beta_b) = -\frac{EI_0}{L^2}\frac{\beta}{2} = k_{31} $$

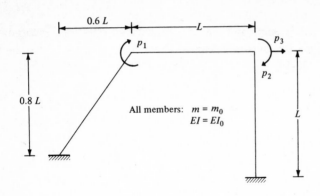

(a)

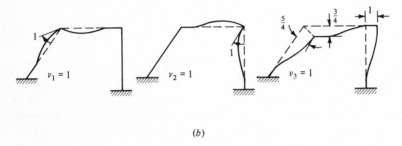

(b)

FIGURE E20-2
Frame with sidesway. (a) Geometry of frame; (b) degrees of freedom.

$$k_{22} = \frac{EI_0}{L}(\alpha_b + \alpha_c) = \frac{2EI_0}{L}\alpha$$

$$k_{23} = \frac{EI_0}{L^2}(^3/_4\bar{\beta}_b - \beta_c) = k_{32}$$

$$k_{33} = \frac{EI_0}{L^3}[^5/_4(^5/_4)\gamma_a + ^3/_4(^3/_4)\gamma_b + \gamma_c - \lambda_b{}^4] = \frac{EI_0}{L^3}[(^{25}/_8)\gamma - \lambda^4]$$

For any given frequency $\bar{\omega}^2$ of the applied loads **p**, the numerical values of these stiffness coefficients can be established and then the harmonic-displacement vector **v** can be determined by simultaneous solution of the equations of dynamic equilibrium. Alternatively the free-vibration frequencies of the frame can be formed by determining the values ω^2, that is, of aL, which cause the determinant of the stiffness matrix to vanish.   ////

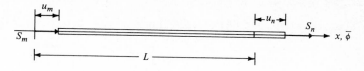

FIGURE 20-4
Nodal forces and displacements in axial deformation.

20-4 DYNAMIC AXIAL-DEFORMATION STIFFNESS MATRIX

Although the neglect of axial deformations does not lead to significant errors in the analysis of many types of structural frames and may be advantageous in hand solutions because it reduces the number of degrees of freedom to be considered, this assumption generally is not invoked in modern computer analyses of static frame behavior. In the analysis of complex frames, the advantage resulting from the reduction of degrees of freedom is outweighed by the difficulties involved in defining the kinematic relationships between the degrees of freedom imposed by the axial-bar rigidities. If the axial deformability of the members is included in the analysis, the various joint displacements can each be treated independently and the formulation of the structural-stiffness matrix can easily be automated.

Taking account of axial deformability offers the same advantages in a dynamic stiffness-method analysis as it does in the static case; therefore it is useful to develop the dynamic-axial-deformation stiffness matrix for a straight bar. The analysis procedure is entirely analogous to that used in deriving the stiffness for flexural deformations [Eq. (20-14)]. For this purpose, Eq. (18-36) and its first derivative are written in matrix form

$$\begin{bmatrix} \bar{\phi}(x) \\ \bar{\phi}'(x) \end{bmatrix} = \begin{bmatrix} \sin bx & \cos bx \\ b \cos bx & -b \sin bx \end{bmatrix} \begin{bmatrix} B_1 \\ B_2 \end{bmatrix} \qquad (20\text{-}18)$$

The displacements and forces to be considered in the axial-deformation stiffness matrix are shown in Fig. 20-4. With the first of Eqs. (20-18), the nodal displacements may be expressed as

$$\begin{bmatrix} u_m \\ u_n \end{bmatrix} = \begin{bmatrix} \bar{\phi}_{x=0} \\ \bar{\phi}_{x=L} \end{bmatrix} = \begin{bmatrix} 0 & 1 \\ s & c \end{bmatrix} \begin{bmatrix} B_1 \\ B_2 \end{bmatrix} \qquad (20\text{-}19)$$

in which

$$s = \sin bL \qquad c = \cos bL$$

Similarly, the second of Eqs. (20-18) provides an expression for the nodal axial forces

$$
\begin{bmatrix} S_m \\ S_n \end{bmatrix} = AE \begin{bmatrix} -\bar{\phi}'_{x=0} \\ \bar{\phi}'_{x=L} \end{bmatrix} = AE \begin{bmatrix} -b & 0 \\ bc & -bs \end{bmatrix} \begin{bmatrix} B_1 \\ B_2 \end{bmatrix} \qquad (20\text{-}20)
$$

Expressing the coefficients B_1 and B_2 in terms of the nodal displacements of Eq. (20-19) and substituting into Eq. (20-20) then leads to the final dynamic-stiffness relationship

$$
\begin{bmatrix} S_m \\ S_n \end{bmatrix} = \frac{AE}{L} \begin{bmatrix} \delta_1 & -\delta_2 \\ -\delta_2 & \delta_1 \end{bmatrix} \begin{bmatrix} u_m \\ u_n \end{bmatrix} \qquad (20\text{-}21)
$$

where
$$
\delta_1 = \frac{bL \cos bL}{\sin bL} \qquad \delta_2 = \frac{bL}{\sin bL}
$$

It will be noted that the terms δ_1 and δ_2 both approach unity as the frequency parameter bL approaches zero; thus this dynamic-stiffness matrix takes the form of the static axial-stiffness matrix at zero frequency.

The axial-stiffness matrix of Eq. (20-21) could be used to carry out a dynamic direct-stiffness analysis of any assemblage of bars in which only axial displacements take place, e.g., a chain of bars of different sizes arranged along a single axis. Such systems are of limited practical importance, however, and the dynamic-axial-stiffness expression alone is mainly of academic interest. It should be noted that the dynamic response of a pin-connected truss cannot be studied using only this axial-stiffness expression even though a static-truss analysis can be performed with the equivalent static stiffness. This is because a pin-connected bar offers dynamic (inertial) resistance to transverse displacement components even though this resistance vanishes at zero frequencies.

20-5 COMBINED FLEXURAL- AND AXIAL-DEFORMATION STIFFNESS

The general case of a structure in which the members are subjected to axial and transverse displacements and in which both axial and flexural deformations are considered can also be studied by the dynamic-direct-stiffness method. In this case, the nodal displacements and forces of both Figs. 20-1 and 20-4 are included in the analysis, and there are three degrees of freedom at each node. If it is assumed that the dynamic axial forces are not large enough to affect the flexural stiffness, there is no coupling between the axial and flexural stiffness mechanisms. Consequently, the combined stiffness matrix is merely a combination of the matrices of Eqs. (20-14) and (20-21), and may be written

$$
\begin{bmatrix} S_iL \\ S_jL \\ S_k \\ S_l \\ S_mL \\ S_nL \end{bmatrix} = \frac{EI}{L}
\begin{bmatrix}
\gamma & -\bar{\gamma} & -\beta & -\bar{\beta} & 0 & 0 \\
-\bar{\gamma} & \gamma & \beta & \beta & 0 & 0 \\
-\beta & \beta & \alpha & \bar{\alpha} & 0 & 0 \\
-\bar{\beta} & \beta & \bar{\alpha} & \alpha & 0 & 0 \\
0 & 0 & 0 & 0 & \delta_1\left(\dfrac{L}{r}\right)^2 & -\delta_2\left(\dfrac{L}{r}\right)^2 \\
0 & 0 & 0 & 0 & -\delta_2\left(\dfrac{L}{r}\right)^2 & \delta_1\left(\dfrac{L}{r}\right)^2
\end{bmatrix}
\begin{bmatrix} \dfrac{v_i}{L} \\ \dfrac{v_j}{L} \\ v_k \\ v_l \\ \dfrac{u_m}{L} \\ \dfrac{u_n}{L} \end{bmatrix}
\tag{20-22}
$$

where $r^2 = I/A$ and the other symbols have already been defined.

It is important to keep in mind that the axial-stiffness terms δ_1 and δ_2 in Eq. (20-22) are functions of the axial-frequency parameter b, while the remaining terms depend on the flexural-frequency parameter a. In order to establish the relative frequency ranges at which the dynamic effects become important in these two types of term, it is instructive to compare the axial natural frequencies with the flexural. For this purpose, the expression for axial vibration frequencies of a cantilever bar [Example E18-5, Eq. (b)] may be rewritten as

$$(\omega_n)_{\text{axial}} = \frac{2n-1}{2}\,\pi\,\sqrt{\frac{EI}{mL^2}\frac{A}{I}} = \frac{2n-1}{2}\,\pi\,\frac{L}{r}\,\sqrt{\frac{EI}{mL^4}}$$

When it is noted that the first flexural frequency of a simple beam is

$$(\omega_1)_{\text{flex}} = \pi^2\,\sqrt{\frac{EI}{mL^4}}$$

the ratio of the first axial to the first flexural frequency is

$$\frac{(\omega_1)_{\text{axial}}}{(\omega_1)_{\text{flex}}} = \frac{1}{2\pi}\frac{L}{r}$$

This specific result depends on the boundary conditions assumed in the axial and flexural cases; however, for any boundary conditions the frequency ratio varies with the slenderness ratio of the member. Thus, for a member of normal dimensions, the fundamental frequency of axial vibrations is significantly higher than the first frequency in flexure.

A similar conclusion can be reached by comparing directly the frequency parameters aL and bL. With the definitions of these parameters from Eqs. (18-6) and (18-34), the relationships

$$(aL)^4 = \frac{\bar{\omega}^2 \bar{m} L^4}{EI} \qquad (bL)^2 = \frac{\bar{\omega}^2 \bar{m} L^2}{AE}$$

may be written. Hence

$$(aL)^4 = (bL)^2 \frac{A}{I} L^2 = (bL)^2 \left(\frac{L}{r}\right)^2$$

from which

$$aL = \left(bL \frac{L}{r}\right)^{1/2} \qquad (20\text{-}23)$$

Equation (20-23) shows that for a member of normal slenderness ratio, the flexural-frequency parameter aL will be relatively large when the axial-frequency parameter bL is small.

A significant consequence of this conclusion is that approximate expressions can be used to evaluate the dynamic axial-stiffness terms in Eq. (17-22) for frequencies at which the flexural-stiffness terms must be evaluated exactly. By means of series expansions, it can be shown that for small values of bL, the following approximate stiffness expressions are valid:

$$\delta_1 = \frac{bL \cos bL}{\sin bL} \doteq 1 - \frac{(bL)^2}{3}$$

$$\delta_2 = \frac{bL}{\sin bL} \doteq 1 + \frac{(bL)^2}{6} \qquad (20\text{-}24)$$

The terms containing bL in Eqs. (20-24) account for the axial inertia of the bar; they make a significant contribution to the dynamic response even if the axial flexibility of the bar is of negligible importance. This fact can be demonstrated easily by substituting Eqs. (20-24) into Eq. (20-22) and then letting $u_m = u_n$, as will be the case if no axial deformations occur. In this case, the stiffness expression becomes identical to Eq. (17-17), with $S_w = S_m + S_n$. The approximations of Eq. (20-24) can be used without any significant loss of accuracy in dynamic frame analyses involving frequencies in the range of the fundamental frequencies of the complete frame. Where the frequencies are significantly higher than this range, however, the exact expressions should be used.

20-6 AXIAL-FORCE EFFECTS ON TRANSVERSE-BENDING STIFFNESS

Although axial forces have been considered in combination with the transverse-bending effects in Secs. 20-3 and 20-5, it has been assumed that the axial-force levels involved are significantly less than the critical buckling loads of the members and that they have no significant effect on the bending stiffness. However, in the earlier formulations of the equations of motion (Sec. 17-3) and free vibrations (Sec. 18-2) it was shown that a constant axial force may have a significant effect on the transverse motions of a beam. This same effect may also be considered in the derivations of the dynamic-stiffness matrix for the member. For the purpose of this discussion, it will be assumed that the prismatic member is subjected to a constant (not time-varying) axial force N which is uniform along the length. There may also be a small time-varying axial force $\mathcal{N}$, but this is assumed to be small relative to N and thus to have no important effect on the flexural stiffness.

The dynamic-flexural stiffness matrix, taking account of the constant force N, can be derived from Eq. (18-18) by exact analogy with the derivation of Eq. (20-14). The matrix relating nodal displacements to the amplitude constants D_i takes the form

$$\mathbf{W} = \begin{bmatrix} 0 & \dfrac{1}{L} & 0 & \dfrac{1}{L} \\[2mm] \dfrac{s}{L} & \dfrac{c}{L} & \dfrac{S}{L} & \dfrac{C}{L} \\[2mm] -\delta & 0 & -\varepsilon & 0 \\[2mm] -\delta c & \delta s & -\varepsilon C & -\varepsilon S \end{bmatrix} \qquad (20\text{-}25)$$

in which

$$s = \sin \delta L \qquad S = \sinh \varepsilon L$$

$$c = \cos \delta L \qquad C = \cosh \varepsilon L$$

Equation (20-25) clearly is equivalent to the corresponding matrix in Eq. (20-7). Similarly the matrix relating nodal forces to the amplitude constants is

$$\mathbf{U} = EI \begin{bmatrix} -\delta^3 L & 0 & \varepsilon^3 L & 0 \\ \delta^3 cL & -\delta^3 sL & -\varepsilon^3 CL & -\varepsilon^3 SL \\ 0 & -\delta^2 & 0 & \varepsilon^2 \\ \delta^2 s & \delta^2 c & -\varepsilon^2 S & -\varepsilon^2 C \end{bmatrix} \qquad (20\text{-}26)$$

which is the direct equivalent of the corresponding matrix in Eq. (20-9). The dynamic-stiffness matrix is then given by the matrix operations

$$\mathbf{k}(\delta,\varepsilon) = \mathbf{U}\mathbf{W}^{-1} \qquad (20\text{-}27)$$

in which it is evident that the stiffness depends on both the frequency and axial-force parameters a and g because δ and ε are expressed in terms of these quantities.

Although the derivation of an explicit expression for the dynamic-stiffness matrix is straightforward, the resulting terms of the matrix are quite lengthy and will not be listed here. Moreover, inasmuch as these terms are functions of two basic parameters, it is not practicable to depict their values in graphical form. In principle, this dynamic-stiffness matrix could be used for the analysis of complex framed structures subjected to any axial-loading and harmonic-excitation environment, but it is doubtful that it has seen much practical use to date. For such complicated loadings, the discrete-parameter finite-element methods are simpler to apply and provide adequate accuracy.

PROBLEMS

20-1 Consider a structure having member lengths and arrangement as shown in Fig. E20-1, but with all members having the same mass per unit length $\overline{m}$ and stiffness EI. Using values of the dynamic stiffness coefficients from Fig. 20-2 compute the rotation amplitude of joint 1, in terms of EI and L, if the frequency of the applied moment $\mathcal{M}(t)$ is $\overline{\omega} = 9\sqrt{EI/\overline{m}L^4}$.

20-2 Evaluate the fundamental natural frequency of the structure of Prob. 20-1.

20-3 The roller support at the right end of the beam of Fig. P20-1 constrains it so that the vertical motion of that end is the only degree of freedom of the beam. Using the appropriate stiffness coefficient value from Fig. 20-2, determine the frequency of vibration of this beam in terms of EI, $\overline{m}$, and L:

(a) Neglecting the mass of the roller system.

(b) Assuming that the moving mass of the roller system is $\overline{m}L$.

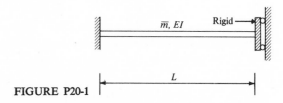

FIGURE P20-1

20-4 The two-degree of freedom structure of Fig. P20-2 is subjected to a harmonic loading $p_1(t) = p_0 \sin \overline{\omega}t$ as shown, where $\overline{\omega} = 4\sqrt{EI/\overline{m}L^4}$. Using values of the dynamic stiffness coefficients from Fig. 20-2, evaluate the amplitude of the displacement and of the moment at the base of the column in terms of p_0, EI, and L. Compare these dynamic results with the response to a static load p_0.

20-5 For the structure of Example E20-2 (Fig. E20-2), evaluate the amplitude of dynamic displacement caused by a harmonic loading $p_3(t) = p_0 \sin \overline{\omega}t$ in which $\overline{\omega} = (2.1)^2 \sqrt{EI/\overline{m}L^4}$, and where $p_1 = p_2 = 0$. Compare this result with the displacement due to a static load p_0. (*Hint:* the coordinates v_1 and v_2 can be eliminated by static condensation.)

20-6 Evaluate the frequency of axial vibration of the structure of Fig. 20-4 by the dynamic direct-stiffness method, making use of the axial stiffness coefficients of Eq. (20-21).

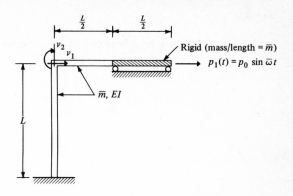

FIGURE P20-2

21
WAVE-PROPAGATION ANALYSIS

21-1 BASIC AXIAL-WAVE-PROPAGATION EQUATION

It was pointed out in Sec. 19-4 that the dynamic response of a uniform bar to a suddenly applied axial loading is of a very simple form, which can be interpreted as the propagation of a stress and deformation wave along its length. This result was obtained at that time by a mode-superposition analysis which did not take advantage of the simplicity of the mechanism involved; in fact a very large number of modal responses must be superimposed to obtain a reasonably close approximation of the wave-propagation phenomenon. It is possible to derive the equation of axial motion for a uniform bar in a different form, not based on the separation of variables, which expresses the wave-propagation concept directly.

For this derivation, the equation of motion of the uniform bar [Eq. (18-30)] is written

$$\ddot{u} - c^2 u'' = 0 \qquad (21\text{-}1)$$

in which

$$c = \sqrt{\frac{EA}{\bar{m}}} = \sqrt{\frac{E}{\rho}} \qquad (21\text{-}2)$$

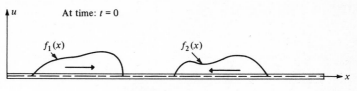

FIGURE 21-1
Axial displacement waves propagating along bar.

has the dimensions of velocity, and where ρ is the mass density. It can be shown by simple substitution that

$$u(x,t) = f_1(x - ct) + f_2(x + ct) \qquad (21\text{-}3)$$

is a solution of Eq. (21-1), f_1 and f_2 being any arbitrary functional relationships of the parameters $x - ct$ and $x + ct$. This expression represents a pair of displacement waves propagating in the positive and negative directions, respectively, along the axis of the bar, as shown in Fig. 21-1. The instant of time represented in this figure has been taken arbitrarily to be $t = 0$, so that the two waves are shown as specified functions of position only. The specific waveshapes, f_1 and f_2, shown in the sketch might be the result of specified displacement conditions applied earlier at the two ends of the bar, for example.

The nature of the wave-propagation mechanism can easily be understood by considering the forward-propagating wave at two instants of time, $t = 0$ and $t = \Delta t$, as shown in Fig. 21-2. If a new position variable $x' = x - c\,\Delta t$ is considered, then

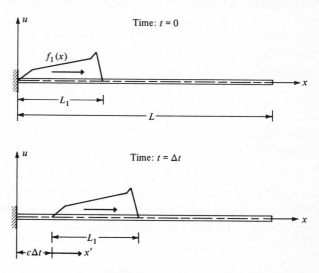

FIGURE 21-2
Propagation of wave during time interval Δt.

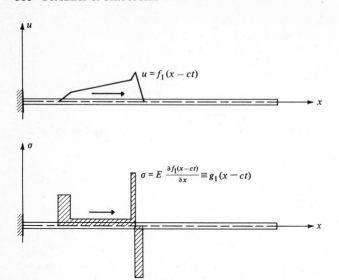

FIGURE 21-3
Relationship between displacement and stress waves.

$f_1(x - c\,\Delta t) \equiv f_1(x')$ and the shape of the wave relative to the variable x' in Fig. 21-2b is the same as the shape relative to x in Fig. 21-2a. Thus the wave has merely advanced a distance $c\,\Delta t$ during the time Δt, with no change of shape; the velocity of this wave propagation is c. By similar reasoning, it can be shown that the second term in Eq. (21-3) represents a waveform f_2 moving in the negative x direction.

The dynamic behavior of the bar can also be expressed in terms of its stress distribution rather than with respect to its displacements. With $\sigma = E\varepsilon$ and $\varepsilon = \partial u/\partial x$, the stress wave is given by

$$\sigma(x,t) = E\,\frac{\partial u}{\partial x} = E\,\frac{\partial f_1}{\partial x}\,(x - ct) + E\,\frac{\partial f_2}{\partial x}\,(x + ct) \qquad (21\text{-}4a)$$

When the stress wave functions $E\,\partial f_1/\partial x$ and $E\,\partial f_2/\partial x$ are designated by g_1 and g_2, this may be written

$$\sigma(x,t) = g_1(x - ct) + g_2(x + ct) \qquad (21\text{-}4b)$$

The relation between an arbitrary displacement waveform and the corresponding stress wave is illustrated in Fig. 21-3; obviously the stress wave also propagates with the velocity c and with unchanging shape.

EXAMPLE E21-1 The general nature of the axial wave-propagation mechanism will be demonstrated by studying the stress wave generated by the impact of a pile-driving hammer at the top of a pile. For the purpose of this example, it will be assumed that the hammer generates a force pulse $P(t) = (600 \text{ kips}) \times \sin(\pi t/0.005)$ and the stress distribution will be evaluated at the end of the pulse ($t_1 = 0.005$ s) in both the steel and the concrete piles whose properties are shown in Fig. E21-1.

To consider the steel pile first, the velocity of wave propagation given by Eq. (21-2) is

$$c_s = \sqrt{\frac{E}{\rho}} = \sqrt{\frac{(30 \times 10^6)(1,728)(386)}{490}} = 202,000 \text{ in/s} = 16,800 \text{ ft/s}$$

The stress at the origin generated by the hammer blow is

$$\sigma_0(t) = -\frac{P(t)}{A} = -(20 \text{ kips/in}^2) \sin \frac{\pi t}{0.005}$$

but from Eq. (21-4b) evaluated at the origin and considering only the forward-propagating wave,

$$\sigma_0(t) = g_1(-c_s t)$$

Hence g_1 can be evaluated by equating these expressions, giving

$$\sigma_0(t) = (-20 \text{ kips/in}^2) \sin\left(-\frac{\pi}{84}\right) c_s t$$

Thus the general expression for the forward-propagating wave is

$$\sigma(x,t) = (-20 \text{ kips/in}^2) \sin \frac{\pi}{84} (c_s t - x)$$

Evaluating this at $t_1 = 0.005$ s leads to

$$\sigma(x,0.005) = (-20 \text{ kips/in}^2) \sin \pi \left(1 - \frac{x}{84}\right)$$

which is plotted in Fig. E21-1b.

Following the same procedure for the concrete pile gives

$$c_c = \sqrt{\frac{(3 \times 10^6)(1,728)(386)}{150}} = 115,000 \text{ in/s} = 9,600 \text{ ft/s}$$

$$\sigma(x,t) = (-1.5 \text{ kips/in}^2) \sin \frac{\pi}{48} (c_c t - x)$$

$$\sigma(x,0.005) = (-1.5 \text{ kips/in}^2) \sin \pi \left(1 - \frac{x}{48}\right)$$

and the last result is plotted in Fig. E21-1c.

$/\!/\!/\!/$

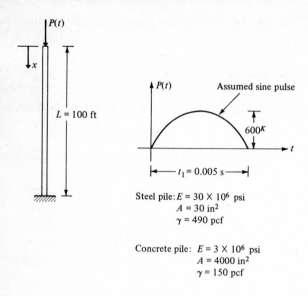

$P(t)$

x

$L = 100$ ft

$P(t)$ Assumed sine pulse

600^K

t

$t_1 = 0.005$ s

Steel pile: $E = 30 \times 10^6$ psi
$A = 30$ in^2
$\gamma = 490$ pcf

Concrete pile: $E = 3 \times 10^6$ psi
$A = 4000$ in^2
$\gamma = 150$ pcf

(a)

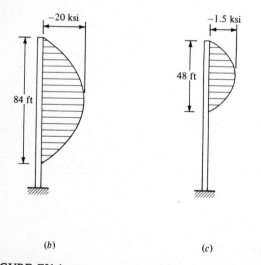

-20 ksi

84 ft

-1.5 ksi

48 ft

(b)   (c)

FIGURE E21-1
Propagation of applied stress wave: (a) properties of pile and loading; (b) stress
in steel pile at $t = 0.005$ s; (c) stress in concrete pile at $t = 0.005$ s.

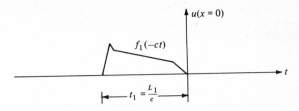

FIGURE 21-4
Displacement imposed at end ($x = 0$) of bar of Fig. 21-2.

21-2 CONSIDERATION OF BOUNDARY CONDITIONS

The function defining the shape of any wave propagating through a uniform bar is controlled by the conditions imposed at the ends of the bar; that is, the waveform within the bar is generated by the requirements of equilibrium and compatibility at the boundaries. For example, the displacement waveform shown in Fig. 21-2a could have been initiated by introducing the displacement history at $x = 0$: $u(0,t) = f_1(-ct)$, as shown in Fig. 21-4.

If the right end ($x = L$) of the bar is free, as indicated in Fig. 21-2, the condition of zero stress must be maintained at all times at that end. This condition may be satisfied by a second stress wave propagating toward the left, which, when superposed on the incident wave, cancels the end-section stresses. Expressing this concept mathematically by means of Eq. (21-4a) leads to

$$\sigma_{x=L} = 0 = E \frac{\partial f_1}{\partial x} (L - ct) + E \frac{\partial f_2}{\partial x} (L + ct)$$

from which

$$\frac{\partial f_1}{\partial x} (L - ct) = -\frac{\partial f_2}{\partial x} (L + ct) \qquad (21\text{-}5)$$

Hence it is evident that the slope $\partial u / \partial x$ of the left-propagating wave must be the negative of the slope of the forward-propagating wave as each part of the waves passes the end of the rod. The displacement waves shown in Fig. 21-5a demonstrate this condition, and the corresponding stress waves in Fig. 21-5b show clearly how the stresses at the tip are canceled.

Although the concept of a left-moving wave coming from beyond the end of the bar makes it easier to visualize the mechanism by which the boundary condition is satisfied, it should be understood that this wave actually is created at the end of the bar as the forward-propagating wave reaches that point. In other words, the incident wave is reflected at the free end; the reflected wave has the *same deflections* as the incident wave, but the *stresses* are *reversed* because the direction of travel is reversed.

Displacement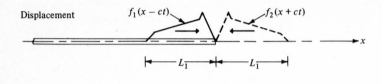

Stress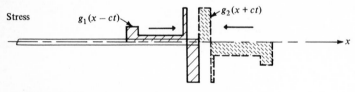

FIGURE 21-5
Reflection of displacement and stress waves at free end.

It will be noted that the total deflection at the free end is doubled by the superposition of the incident and reflected waves, while the two stress components cancel each other.

To consider now the case where the right end of the bar is fixed rather than free, it is evident that the boundary condition imposed on the two propagating waves is

$$u_{x=L} = 0 = f_1(L - ct) + f_2(L + ct)$$

from which the reflected wave may be expressed in terms of the incident wave as

$$f_2(L + ct) = -f_1(L - ct) \qquad (21\text{-}6)$$

Thus the displacement waves in this case are seen to have opposite signs, and by analogy with the preceding discussion it can be inferred that the incident and reflected stress waves have the same sign, as shown in Fig. 21-6. Hence, in satisfying the

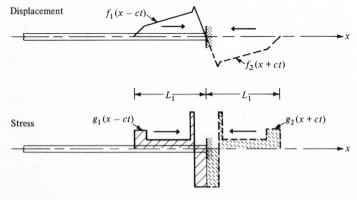

FIGURE 21-6
Reflection of displacement and stress waves at fixed end.

required zero-displacement condition, the reflected wave produces a doubling of stress at the fixed end of the bar.

EXAMPLE E21-2   To demonstrate these boundary-reflection phenomena, the stress wave produced by the driving hammer on the concrete pile in Example E21-1 will be considered further. The stress wave shown in Fig. E21-1c is traveling with a velocity of 9,600 ft/s, and so the forward end of this wave reaches the tip of the pile at a time

$$t_2 = \frac{100}{9,600} = 0.0104 \text{ s}$$

The subsequent behavior then depends on the nature of the tip support condition.

Assuming first that the pile rests on a rigid support, so that no displacement can take place at this point, the reflected stress wave must be compression, the same as the incident wave. The total stress at subsequent times is then given by the sum of the incident and reflected components. As a specific example, the distribution of stress at the time when the stress wave has traveled 128 ft,

$$t_3 = \frac{128}{9,600} = 0.0133 \text{ s}$$

is shown in Fig. E21-2a. (See p. 372.)

The other limiting case occurs if the end of the pile is resting on very soft mud, so that there is essentially no resistance to its displacement and the tip stress is required to be zero. In this case, the reflected stress wave must be tensile, and the total stress in the pile is given by the difference between the tensile and compressive components. Taking again the time t_3, when the stress wave has traveled 128 ft, the distribution of stress is as shown in Fig. E21-2b. It is significant that the net stress is tensile over the lower 28 ft of the pile and that the greatest tensile stress occurs 20 ft from the tip. This illustrates how tensile stresses will be developed in a pile during the driving process if the material through which it is driven offers little resistance, and that fracture due to these tensile stresses might occur at a significant distance from the tip. Of course, the behavior in any specific case will depend on the tensile strength of the concrete and on the duration of the hammer-force impulse.   ////

21-3   DISCONTINUITY IN BAR PROPERTIES

The wave reflections which take place at the fixed or free end of a uniform bar may be considered as special cases of the general reflection and refraction phenomena occurring at any discontinuity in the bar properties. The conditions of equilibrium

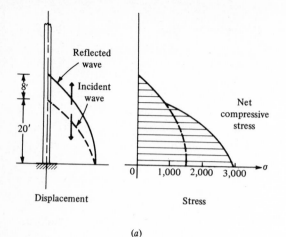

Reflected wave

Incident wave

8'

20'

Net compressive stress

0   1,000   2,000   3,000 σ

Displacement

Stress

(a)

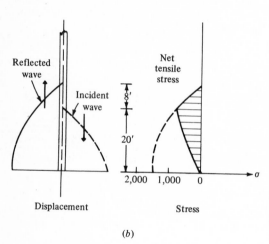

Reflected wave

Incident wave

Net tensile stress

8'

20'

2,000   1,000   0 σ

Displacement

Stress

(b)

FIGURE E21-2
Stress distribution at $t_3 = 0.0133$ s (concrete pile of Fig. E21-1): (a) fixed end at $x = L$; (b) free end at $x = L$.

and compatibility which must be satisfied at all points along the bar require that additional reflected and refracted waves be generated at the juncture between bars of different properties in response to the action of any given incident wave.

Consider, for example, the juncture between bars 1 and 2 shown in Fig. 21-7. The properties of the bars on each side of the juncture are characterized by their mass per unit length $\overline{m}$ and axial stiffness EA. Also the wave-propagation velocity on each side is given by $c = \sqrt{AE/\overline{m}} = \sqrt{E/\rho}$. The forward-propagating wave u_a which arrives at the juncture in bar 1 generates a reflection u_b which travels in the negative direction in bar 1 and at the same time creates a refracted wave u_c which propagates forward in bar 2.

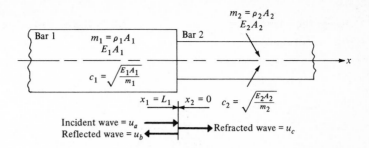

FIGURE 21-7
Wave reflection and refraction at discontinuity.

Two continuity conditions are imposed at the juncture:

Displacement: $\qquad\qquad u_1 = u_2 \qquad\qquad u_a + u_b = u_c \qquad\qquad$ (21-7a)

Force: $\qquad\qquad \mathcal{N}_1 = \mathcal{N}_2 \qquad \mathcal{N}_a + \mathcal{N}_b = \mathcal{N}_c \qquad\qquad$ (21-7b)

where the fact that both incident and reflected waves act in bar 1 has been indicated. Because these continuity conditions must be satisfied at all times, the time derivative of the displacement condition also must be satisfied, that is,

$$\frac{\partial u_a}{\partial t} + \frac{\partial u_b}{\partial t} = \frac{\partial u_c}{\partial t} \qquad (21\text{-}8)$$

But the incident wave can be expressed in the form

$$u_a = f_a(x - ct) \equiv f_a(\zeta)$$

where the variable ζ has been introduced for convenience. Now the derivatives of u_a can be expressed as

$$\frac{\partial u_a}{\partial x} = \frac{\partial f_a}{\partial \zeta}\frac{\partial \zeta}{\partial x} = \frac{\partial f_a}{\partial \zeta} \qquad \frac{\partial u_a}{\partial t} = \frac{\partial f_a}{\partial \zeta}\frac{\partial \zeta}{\partial t} = -c_1\frac{\partial f_a}{\partial \zeta}$$

from which it is evident that the time and position derivatives are related by the velocity of wave propagation

$$\frac{\partial u_a}{\partial t} = -c_1\frac{\partial u_a}{\partial x} \qquad (21\text{-}9a)$$

Similar analyses for the reflected and refracted waves result in

$$\frac{\partial u_b}{\partial t} = +c_1\frac{\partial u_b}{\partial x} \qquad (21\text{-}9b)$$

$$\frac{\partial u_c}{\partial t} = -c_2\frac{\partial u_c}{\partial x} \qquad (21\text{-}9c)$$

where the positive sign in Eq. (21-9b) is due to the negative direction of the reflected-wave propagation.

Substituting Eqs. (21-9) into (21-8) yields

$$-c_1 \frac{\partial u_a}{\partial x} + c_1 \frac{\partial u_b}{\partial x} = -c_2 \frac{\partial u_c}{\partial x} \qquad (21\text{-}10)$$

but the strains, $\partial u_a/\partial x = \varepsilon_a$, etc., can be expressed in terms of the forces acting in the bars: $\varepsilon_a = \sigma_a/E = \mathcal{N}_a/A_1 E_1$, etc.; hence, the compatibility condition of Eq. (21-10) can be expressed in terms of the force waves

$$-\frac{c_1}{A_1 E_1}\,\mathcal{N}_a + \frac{c_1}{A_1 E_1}\,\mathcal{N}_b = -\frac{c_2}{A_2 E_2}\,\mathcal{N}_c$$

or more simply

$$\mathcal{N}_c = \alpha(\mathcal{N}_a - \mathcal{N}_b) \qquad (21\text{-}11)$$

where

$$\alpha = \frac{c_1}{c_2}\frac{A_2 E_2}{A_1 E_1} = \sqrt{\frac{m_2 E_2 A_2}{m_1 E_1 A_1}} \qquad (21\text{-}12)$$

Finally, this compatibility condition [Eq. (21-11)] can be introduced into the force-equilibrium condition [Eq. (21-7b)] to express the refracted and reflected waves in terms of the incident wave

$$\mathcal{N}_a + \mathcal{N}_b = \alpha(\mathcal{N}_a - \mathcal{N}_b)$$

from which

$$\mathcal{N}_b = \mathcal{N}_a \frac{\alpha - 1}{\alpha + 1} \qquad (21\text{-}13)$$

and, from Eq. (21-11),

$$\mathcal{N}_c = \mathcal{N}_a \frac{2\alpha}{\alpha + 1} \qquad (21\text{-}14)$$

Equations (21-13) and (21-14) express the relationships between the incident, reflected, and refracted force waves at the bar discontinuity. Corresponding relationships can be obtained for the displacement waves by noting that

$$\mathcal{N} = AE\frac{\partial u}{\partial x} = \pm \frac{AE}{c}\frac{\partial u}{\partial t}$$

Substituting this into Eq. (21-13) and integrating leads to

$$\frac{A_1 E_1}{c_1}\,u_b = -\frac{A_1 E_1}{c_1}\,u_a \frac{\alpha - 1}{\alpha + 1}$$

from which

$$u_b = -u_a \frac{\alpha - 1}{\alpha + 1} \qquad (21\text{-}15)$$

Similarly, substituting into Eq. (21-14) and integrating gives

$$-\frac{A_2 E_2}{c_2} u_c = -\frac{A_1 E_1}{c_1} u_a \frac{2\alpha}{\alpha + 1}$$

from which

$$u_c = u_a \frac{2}{\alpha + 1} \qquad (21\text{-}16)$$

It is evident that the factor α defines the character of the discontinuity at the juncture between two bars and controls the relative amplitudes of the reflected and refracted waves. Where the properties of two adjoining bars are identical or related in any manner such that the value of α given by Eq. (21-12) is unity, there is no discontinuity and no reflected wave. For increasing stiffness in bar 2, the value of α increases and the reflected force wave is of the same sign as the incident wave; for decreasing stiffness in bar 2, the value of α becomes less than unity, and the reflected force wave is of opposite sign to the incident wave. In this context, the fixed and free end conditions discussed in Sec. 21-2 can be considered as limiting cases of bar discontinuity and are defined by infinite and zero values of α, respectively. The relationships between incident, reflected, and refracted waves for various cases of discontinuity are listed in Table 21-1.

EXAMPLE E21-3   To illustrate the effects caused by discontinuities on the propagation of force waves through a multiple-segment bar, the stepped bar shown in Fig. E21-3a will be considered. Since the material is the same in each

Table 21-1   WAVE RELATIONSHIPS FOR VARIOUS
             DISCONTINUITIES

Case	$\alpha = \sqrt{\dfrac{A_2 E_2 m_2}{A_1 E_1 m_1}}$	Force waves			Displacement waves		
		$\overset{\rightarrow}{\mathcal{N}_a}$ +	$\overset{\leftarrow}{\mathcal{N}_b}$ =	$\overset{\rightarrow}{\mathcal{N}_c}$	$\overset{\rightarrow}{u_a}$ +	$\overset{\leftarrow}{u_b}$ =	$\overset{\rightarrow}{u_c}$
No discontinuity	1	1	0	1	1	0	1
Fixed end	∞	1	1	2	1	-1	0
Free end	0	1	-1	0	1	1	2
$\dfrac{A_2 E_2}{A_1 E_1} = \dfrac{m_2}{m_1} = 2$	2	1	$\frac{1}{3}$	$\frac{4}{3}$	1	$-\frac{1}{3}$	$\frac{2}{3}$
$\dfrac{A_2 E_2}{A_1 E_1} = \dfrac{m_2}{m_1} = \frac{1}{2}$	$\frac{1}{2}$	1	$-\frac{1}{3}$	$\frac{2}{3}$	1	$\frac{1}{3}$	$\frac{4}{3}$

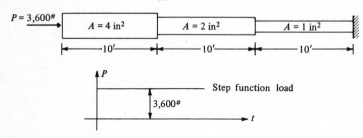

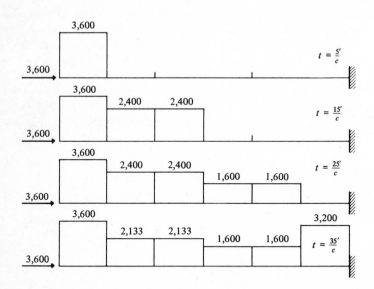

(b)

FIGURE E21-3
Force reflection and refraction at bar discontinuities: (a) definition of bar and
load; (b) force distribution at various times.

section, the discontinuities are due only to the changes of area. At each step $A_2/A_1 = {}^1\!/_2$, and so $m_2/m_1 = {}^1\!/_2$; thus $\alpha = {}^1\!/_2$. This corresponds to the last case in Table 21-1 and, as indicated there, at each step

$$\frac{\mathcal{N}_b}{\mathcal{N}_a} = -\frac{1}{3} \qquad \frac{\mathcal{N}_c}{\mathcal{N}_a} = \frac{2}{3}$$

   If the left end of the bar is subjected to a constant force of 3,600 lb, the force distributions at the times required for the stress wave to propagate 5, 15, 25, and 35 ft will be as shown in Fig. E21-3b. Stress distributions can be derived from these sketches by dividing by the appropriate area of each segment. Because the one-dimensional wave equation has been used in this analysis, it must be assumed that the segments are interconnected by rigid disks which maintain the uniaxial stress state through the discontinuities.   ////

21-4 STRESS WAVES DEVELOPED DURING PILE DRIVING

In Example E21-1 it was assumed that the force applied by the hammer in driving a pile is a sine pulse, and it can easily be shown that this will be the nature of the driving impulse if the hammer is very light in comparison with the weight of the pile because it then behaves as a SDOF system on a rigid support. However, in practical situations the hammer is heavier and generates a displacement response of the pile. The nature of the driving impulse is more complicated in this case but still can be evaluated by a relatively simple analysis.

   For the purpose of this study, the driving hammer will be assumed to be a rigid mass m, and the cushion between pile and hammer will be represented by a weightless spring m. The axial stiffness of the prismatic pile is AE and its velocity of wave propagation is c. As shown in Fig. 21-8a, the displacement of the hammer from its point of contact with the driving cushion will be denoted by Z and the displacement of the driven end of the pile by u_0. The equilibrium of forces acting on the hammer during impact (Fig. 21-8b) may be expressed as

$$\mathcal{N}_c = W + f_I$$

or, when it is noted that $W = mg$ and $f_I = -m\ddot{Z}$,

$$\mathcal{N}_c = m(g - \ddot{Z}) \qquad (21\text{-}17)$$

Also, continuity of displacements requires that the movement of the hammer be equal to the displacement of the end of the pile plus the compression of the cushion

$$Z = u_0 + \frac{\mathcal{N}_c}{k} \qquad (21\text{-}18a)$$

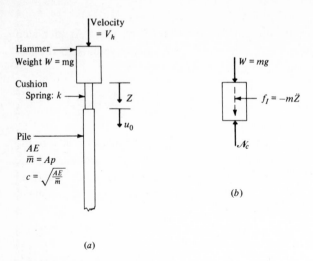

FIGURE 21-8
Analysis of pile-driving contact force: (a) definition sketch;
(b) hammer force equilibrium.

which after double differentiation gives

$$\ddot{Z} = \ddot{u}_0 + \frac{\ddot{\mathcal{N}}_c}{k} \quad (21\text{-}18b)$$

Substituting this into Eq. (21-17) then yields

$$\mathcal{N}_c = m\left(g - \ddot{u}_0 - \frac{\ddot{\mathcal{N}}_c}{k}\right) \quad (21\text{-}19)$$

Now with an expression like Eq. (21-9c), the driving force $\mathcal{N}_c$ can be expressed in terms of the motion of the pile tip, that is,

$$\mathcal{N}_c = -\sigma_c A = -AE\varepsilon_c = -AE\frac{\partial u_0}{\partial x} = \frac{AE}{c}\frac{\partial u_0}{\partial t} \quad (21\text{-}20)$$

Hence substituting this into Eq. (21-19) and rearranging leads to

$$\frac{m}{k}\frac{AE}{c}\ddot{u}_0 + m\ddot{u}_0 + \frac{AE}{c}\dot{u}_0 = mg$$

or denoting by V_0 the velocity of the end of the pile and simplifying gives

$$\ddot{V}_0 + \frac{kc}{AE}\dot{V}_0 + \frac{k}{m}V_0 = \frac{kc}{AE}g \quad (21\text{-}21)$$

which may be recognized as the equation of motion of a SDOF system subjected to a static loading.

The term on the right side of Eq. (21-21) results from the quantity mg in Eq. (21-17); that is, it represents the part of the contact force applied to the hammer weight, which is a small fraction of the inertial effect and therefore may be neglected. The remaining homogeneous equation can then be expressed in the familiar form

$$\ddot{V}_0 + 2\omega\xi \dot{V}_0 + \omega^2 V_0 = 0 \qquad (21\text{-}22)$$

where

$$\omega^2 = \frac{k}{m} \qquad 2\omega\xi = \frac{kc}{AE}$$

The solution of this SDOF damped free-vibration equation is

$$V_0 = e^{-\xi\omega t}(A \sin \omega_D t + B \cos \omega_D t) \qquad (21\text{-}23)$$

in which

$$\omega_D = \omega\sqrt{1 - \xi^2}$$

The constants A and B in Eq. (21-23) must be evaluated by consideration of the initial conditions imposed on the pile-end velocity V_0. For a stationary pile at the time of impact, this velocity must be zero, that is,

$$V_{0,t=0} = \frac{\partial u_0}{\partial t}\bigg|_{t=0} = 0 \qquad (21\text{-}24)$$

At the same time, the hammer has a prescribed velocity at the moment of impact which will be designated V_h and which depends, of course, on the height through which the hammer is dropped. Differentiating Eq. (21-18a) and substituting the specified value of hammer velocity gives:

At $t = 0$:

$$\dot{Z} \equiv V_h = \dot{u}_0 + \frac{\dot{\mathcal{N}}_c}{k}$$

or with $\dot{u}_0 = 0$ at $t = 0$

$$V_h = \frac{\dot{\mathcal{N}}_c}{k} \qquad (21\text{-}25)$$

Finally substituting $\mathcal{N}_c$ from Eq. (21-20) gives

$$V_h = \frac{AE}{c} \dot{V}_0 \frac{1}{k}$$

from which

$$\dot{V}_{0,t=0} = V_h \frac{kc}{AE} = V_h 2\omega\xi \qquad (21\text{-}26)$$

Now introducing the initial conditions of Eqs. (21-24) and (21-26) into Eq. (21-23) leads to:

At $t = 0$:

$$V_0 = 0 = [A(0) + B(1)] \qquad B = 0 \qquad (1)$$

$$\dot{V}_0 = 2\omega\xi V_h = A\omega_D(1) \qquad A = 2\xi \frac{\omega}{\omega_D} V_h \qquad (2)$$

Hence with these in Eq. (21-23), the final result for the tip velocity is

$$V_0 = e^{-\xi \omega t} \frac{2\xi \omega}{\omega_D} V_h \sin \omega_D t \qquad (21\text{-}27)$$

Of more interest than the end velocity, however, is the variation of contact force, which is given by Eq. (21-20), that is,

$$\mathcal{N}_c = \frac{AE}{c} V_0 = \frac{kV_h}{\omega_D} e^{-\xi \omega t} \sin \omega_D t \qquad (21\text{-}28)$$

This solution is valid only while the hammer is in contact with the cushion and before the stress wave reflected from the other end of the pile returns to the driven end; thus it is valid for $t \leq \pi/\omega_D$ and $t \leq 2L/c$.

EXAMPLE E21-4   To demonstrate the analysis of the force developed by a pile-driving hammer, the concrete pile considered in Example E21-1 will be considered again. The properties of the driving hammer, cushion, and the pile are shown in Fig. E21-4a. From these data, the response parameters in Eq. (21-28) can be evaluated, as follows:

$$\omega = \sqrt{\frac{kg}{W}} = 628 \text{ rad/s} \qquad \xi = \frac{ck}{2\omega AE} = 0.156$$

$$\omega_D = \omega \sqrt{1 - \xi^2} = 620 \text{ rad/s} \qquad \frac{kV_h}{\omega_D} = 606 \text{ kips}$$

These parameters completely define the contact force $\mathcal{N}_c(t)$, and, of course, this force propagates forward through the pile with the velocity c. The hammer remains in contact with the cushion until $t = \pi/\omega_D = 0.00507$ s. The distribution of force along the pile at this time is shown in Fig. E21-4b.

It may be noted in Eq. (21-28) that the parameter ξ determines the degree to which the pile response influences the contact force, and this parameter can be written

$$\xi = \frac{ck}{2\omega AE} = \frac{1}{2} \sqrt{\frac{km}{mAE}}$$

It is apparent that ξ becomes smaller as the pile becomes more massive and rigid relative to the properties of the driving system. In the limit, when the pile is a rigid immovable block, ξ becomes zero and the pile-driver–cushion system becomes a simple SDOF system. The axial force wave for this case (which is the same as that considered in Example E21-1) is plotted as a dashed curve in

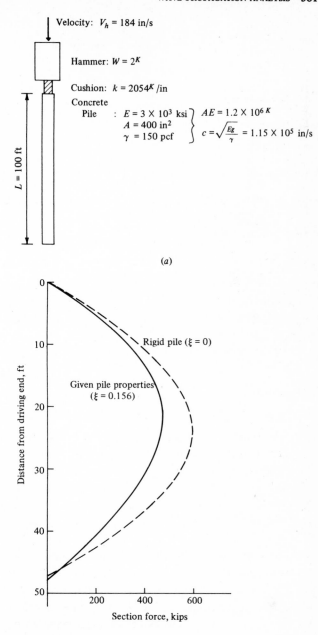

Velocity: $V_h = 184$ in/s

Hammer: $W = 2^K$

Cushion: $k = 2054^K$ /in

Concrete

Pile : $E = 3 \times 10^3$ ksi $\left.\begin{array}{l} \\ \\ \\ \end{array}\right\}$ $AE = 1.2 \times 10^{6\,K}$
$A = 400$ in^2
$\gamma = 150$ pcf $\left.\begin{array}{l} \\ \\ \\ \end{array}\right\}$ $c = \sqrt{\dfrac{Eg}{\gamma}} = 1.15 \times 10^5$ in/s

$L = 100$ ft

(*a*)

Rigid pile ($\xi = 0$)

Given pile properties
($\xi = 0.156$)

Distance from driving end, ft

Section force, kips

(*b*)

FIGURE E21-4
Analysis of pile-contact force: (*a*) properties of pile and pile driver; (*b*) force
wave propagated from driving tip ($t = 0.00507$ s).

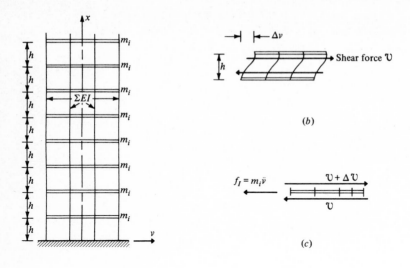

(a)

FIGURE 21-9
Shear-beam model of multistory building: (a) uniform building segment;
(b) story force-displacement relation; (c) story equilibrium relation.

Fig. E21-4b for comparison with the elastic-response case. It is evident that the
duration of the contact pressure pulse is hardly affected by the elastic-response
behavior of the pile; however, the force amplitude has been reduced by about
20 percent in this case. ////

21-5 SHEAR-WAVE PROPAGATION IN BUILDINGS

Another situation in which the dynamic response can conveniently be represented
by the simple wave-propagation concept is the case of a laterally loaded multistory
building. In any vertical segment of a building in which the column stiffnesses as
well as the story heights and weights are uniform and in which the floor systems may
be treated as *rigid*, the mechanics of shear-wave propagation through the structure
is entirely analogous to the propagation of axial force waves discussed in the preceding sections. Consider the uniform building segment shown in Fig. 21-9a, in which
the mass of each story is m_i, the *total* rigidity of all columns is EI, and the height of
each story is h.

The relation between the total story shear $\mathcal{V}$ and the story-to-story displacement Δv, as shown in Fig. 21-9b, is

$$\Delta v = \frac{h^3}{12\Sigma EI} \, \mathcal{V} \qquad (21\text{-}29)$$

taking account of the fact that the rigid floor girders prevent any rotation of the column joints. Also, the force-equilibrium relationship may be written for any floor, as shown in Fig. 21-9c

$$f_I = m_i \ddot{v} = \Delta \mathcal{V} \qquad (21\text{-}30)$$

If the building segment consists of many identical stories, it is reasonable to approximate the system as a uniform-shear beam continuum, using the approximations

$$\Delta v \doteq \frac{\partial v}{\partial x} h \qquad \Delta \mathcal{V} \doteq \frac{\partial \mathcal{V}}{\partial x} h \qquad (21\text{-}31)$$

Introducing the first of these into Eq. (21-29) leads to

$$\frac{\partial v}{\partial x} h = \frac{h^3}{12\Sigma EI} \, \mathcal{V} \qquad (21\text{-}32)$$

while substituting the second into Eq. (21-30) gives

$$m_i \ddot{v} = \frac{\partial \mathcal{V}}{\partial x} h \qquad (21\text{-}33)$$

Finally differentiating the force-deformation relationship [Eq. (21-32)] and substituting it into the equilibrium relationship [Eq. (21-33)] leads to the equation of motion, which after simplification is

$$\ddot{v} - c^2 v'' = 0 \qquad (21\text{-}34)$$

in which

$$c = \sqrt{\frac{12\Sigma EI}{m_i h}} \qquad (21\text{-}35)$$

represents the velocity of propagation of shear waves through the building segment. Because the form of Eq. (21-34) is identical to that of Eq. (21-1), it is evident that the wave-propagation solution is equally applicable here. From comparison of the wave-velocity expressions [Eqs. (21-2) and (21-35)] it may be noted that in the building system $m_i/h \approx \bar{m}$ and $12\Sigma EI/h^2 \approx EA$. With these equivalences, the discontinuity factor α of Eq. (21-12) can be rewritten for the building system as

$$\alpha = \frac{\overline{\left(\dfrac{12\Sigma EIm_i}{h^3}\right)}_2}{\left(\dfrac{12\Sigma EIm_i}{h^3}\right)_1} \qquad (21\text{-}36)$$

Hence, the influence of a discontinuity in a building frame on the shear-wave propagation phenomenon can be evaluated by using this factor in equations equivalent to Eqs. (21-13) to (21-16).

PROBLEMS

21-1   A uniform concrete pile having properties as shown in Fig. P21-1 is supported at the tip by a rigid base. Assuming the pile is subjected to a rectangular impulse of 0.010-s duration, as shown,

(*a*) Plot the distribution of axial force and of displacement along the length at time $t = 0.015$ s.

(*b*) Repeat part *a* for time $t = 0.023$ s.

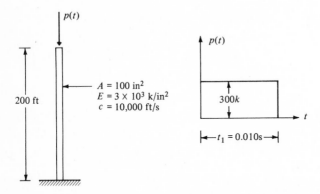

FIGURE P21-1

21-2   Repeat Prob. 21-1*b*, evaluating the response by the mode superposition method and considering only the first two modes of axial vibration.

21-3   Repeat Probs. 21-1*a* and 21-1*b*, assuming that the column is "stepped" at midheight with the area reduced to 50 in² in the lower half and the weight per foot reduced correspondingly. Note that the velocity of wave propagation is 10,000 ft/s in each section of this bar.

21-4 The forty-story shear building shown in Fig. P21-2 has uniform story heights of 12 ft, a uniform weight per story of 1,000 kips, and 25 identical concrete columns (24-in square) extending over its full height. Assume the floor slabs to be rigid; neglect damping and axial column deformations.

(*a*) Determine the frequencies of the first three modes of vibration.

(*b*) If the base of the building is subjected to a half sine wave base-displacement pulse, as shown, determine the maximum shear force developed at the base of the building and the maximum displacement at the top. Indicate the time when each of these maxima occurs.

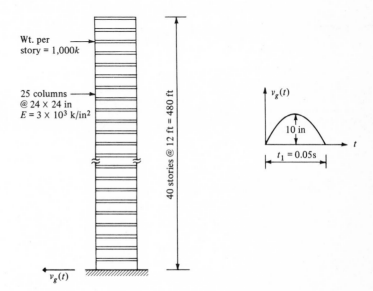

Wt. per story = 1,000k

25 columns @ 24 × 24 in
$E = 3 × 10^3$ k/in^2

40 stories @ 12 ft = 480 ft

$v_g(t)$

10 in

$t_1 = 0.05$s

$v_g(t)$

FIGURE P21-2

21-5 Repeat Prob. 21-4, assuming that the building has a "set back" at midheight, so that only 9 columns extend through the upper half of the building and the weight per story in this section is only 250 kips. Consider the response only to the time when the wave reflected from the top has traveled down to the set back, and note that the velocity of propagation is different in the upper section.

21-6 A concrete pile with the properties shown in Fig. P21-3 is being driven by a hammer weighing 2 kips and striking with a velocity at impact of 180 in/s. Assuming that the impact is cushioned by a wooden pad having a stiffness of 900 kips/in, determine the maximum *tensile* stress developed in the pile for the following support conditions at the tip:

(a) Unsupported (no constraint).
(b) Soft soil support with a reflection coefficient $\alpha = 0.05$.
(c) Firm soil support with a reflection coefficient $\alpha = 0.15$.

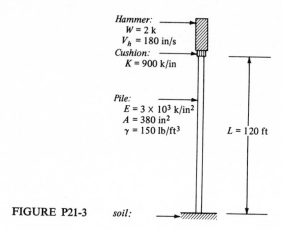

FIGURE P21-3

PART FOUR

Random Vibrations

PROBABILITY THEORY

22-1 SINGLE RANDOM VARIABLE

It is assumed that the reader has had some experience with various games of chance and an intuitive grasp of simple probability theory even though he may never have studied this subject formally. Let us begin by formalizing the basic probability concepts for a simple experiment.

Consider the familiar rotating disk shown in Fig. 22-1a, which has 10 equally spaced pegs driven into its side with the intervals between pegs representing numbers 1 through 10 as shown. When the disk is spun, it will eventually come to rest with one of these numbers at the indicator. Assuming an unbiased disk, each number has a $1/10$ probability of occurrence; that is, after sampling n times, each number will have been sampled $n/10$ times in the limit as n approaches infinity. If N represents the value of the number sampled and $p(N)$ its probability of occurrence, the probability relationship for this experiment will be the bar diagram shown in Fig. 22-2a. N is said to be a discrete random variable in this case since only discrete values can be sampled.

Consider now an unbiased rotating disk as shown in Fig. 22-1b which has no pegs but is marked off in degrees similar to a full 360° compass. In this experiment,

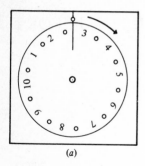 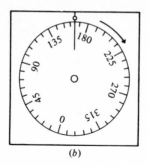

| (a) | (b) |

FIGURE 22-1
Single-random-variable experiment: (a) discrete variable N; (b) continuous variable θ.

if the disk is spun and the angle θ to which the indicator points when it comes to rest is noted, values can be sampled throughout the range $0 \leq \theta \leq 360°$ with equal chance of occurrence; that is, its probability relation will be continuous and uniform, as shown in Fig. 22-2b. Both probability relations in Fig. 22-2 are called probability density functions.

To clarify further the definition of probability density, consider a general experiment involving a single random variable x which has the probability density function shown in Fig. 22-3a. This function is defined so that $p(x_1)\,dx$ equals the chances that a sampled value of x will be in the range $x_1 < x < x_1 + dx$. When unity represents a certainty of occurrence, the above definition requires that the

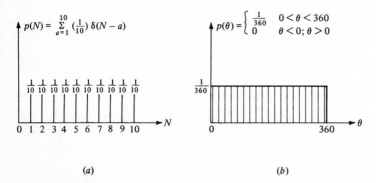

| (a) | (b) |

FIGURE 22-2
Probability density functions for single random variables N and θ: (a) discrete variable; (b) continuous variable.

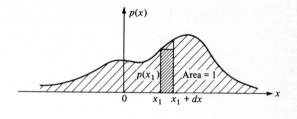

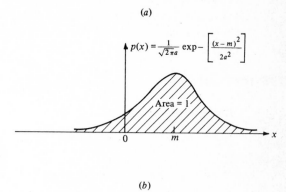

FIGURE 22-3
Probability density function for random
variable x: (a) general probability density
function; (b) normal, or gaussian, probability density function.

probability density function be normalized so that the area between the x axis and the
function itself, that is, $\int_{-\infty}^{\infty} p(x)\,dx$, equals unity.

From the above definition, it should be noted that a zero probability exists
that a sampled value of x will be exactly equal to some preselected value in the
continuous case. In other words, a finite probability can be associated only with x
falling in a certain finite range. To illustrate this point, consider again the simple
experiment shown in Fig. 22-1b, where a zero probability exists that the indicator
will point exactly to, say, 256°; however, the probability that a sampled value of θ
will be in the range $256° < \theta < 257°$ is 1/360.

Further, it should be noted that to satisfy the above definition of probability
density in the discrete case like that shown in Fig. 22-2a, the probability density
function must consist of a number of Dirac delta functions. The Dirac delta function
$\delta(x - a)$ is simply any function which satisfies the conditions

$$\delta(x - a) = \begin{cases} 0 & x \neq a \\ \infty & x = a \end{cases}$$

$$\int_{-\infty}^{\infty} \delta(x - a)\,dx = 1 \tag{22-1}$$

EXAMPLE E22-1   Show by proper selection of the constant C in the function $f(x - a)$, defined below, that this function satisfies the conditions for a Dirac delta function given by Eqs. (22-1):

$$f(x - a) = C \lim_{\varepsilon \to 0} \frac{1}{\varepsilon} \exp\left[-\frac{(x - a)^2}{2\varepsilon^2}\right] \qquad (a)$$

For $x \neq a$, it is quite apparent that the function equals zero since in the limit the exponential term approaches zero much more rapidly than ε itself. When $x = a$, the exponential term equals unity; therefore, the entire function approaches infinity in the limit at this point. Integrating as follows makes it possible to evaluate C:

$$I \equiv \int_{-\infty}^{\infty} \frac{C}{\varepsilon} \exp\left[-\frac{(x - a)^2}{2\varepsilon^2}\right] dx \qquad (b)$$

Substituting the change of variable

$$u = \frac{x - a}{\sqrt{2}\,\varepsilon} \qquad du = \frac{1}{\sqrt{2}\,\varepsilon} dx \qquad (c)$$

gives

$$I = \sqrt{2}\, C \int_{-\infty}^{\infty} e^{-u^2}\, du = \sqrt{2\pi}\, C \qquad (d)$$

Note that the value of the integral is independent of the value of ε. This is a necessary condition if the form of Eq. (a) is to serve as a Dirac delta function. Setting the integral equal to unity in accordance with the second of Eqs. (22-1) gives

$$C = \frac{1}{\sqrt{2\pi}} \qquad (e)$$

////

The most commonly used probability density function of a single random variable is the so-called *normal*, or *gaussian*, *distribution* shown in Fig. 22-3b, which is defined by the symmetric relation

$$p(x) = \frac{1}{\sqrt{2\pi}\, a} e^{-(x-m)^2/2a^2} \qquad (22\text{-}2)$$

where a and m are constants. A plot of this relation shows that a is a measure of the spread of the function in the neighborhood of $x = m$. The integral of Eq. (22-2) between the limits $x = -\infty$ and $x = +\infty$ equals unity, as it should, regardless of the numerical values of a and m.

If a random variable x is transformed into a second random variable r, which is a known single-valued function of x as defined in general form by the relation

$$r \equiv r(x) \qquad (22\text{-}3)$$

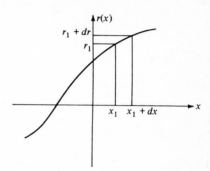

FIGURE 22-4
Relation between random variable x and
random variable r.

the probability density function for r is easily obtained from the relation

$$p(r) = p(x) \left| \frac{dx}{dr} \right| \qquad (22\text{-}4)$$

provided the inverse relation $x = x(r)$ is also a single-valued function. The validity of Eq. (22-4) is obvious since (as shown in Fig. 22-4) all sampled values of x which fall in the range $x_1 < x < x_1 + dx$ correspond to values of r in the range $r_1 < r < r_1 + dr$. The absolute value of dx/dr is necessary since for some functions $r(x)$ a positive dx corresponds to a negative dr and vice versa.

Another probability function which is useful when treating single random variables is the probability distribution function defined by

$$P(x) \equiv \int_{-\infty}^{x} p(x) \, dx \qquad (22\text{-}5)$$

In accordance with this definition, the function $P(x)$ either becomes or approaches zero and unity with increasing negative and positive values of x, respectively, as shown in Fig. 22-5. Equation (22-5) in its differential form

$$p(x) = \frac{dP(x)}{dx} \qquad (22\text{-}6)$$

is also very useful.

EXAMPLE E22-2   A random variable has the probability density function

$$p(x) = \begin{cases} \frac{1}{2} & -1 < x < +1 \\ 0 & x < 1; \ x > 1 \end{cases} \qquad (a)$$

If random variable r is related to x through the relation

$$r(x) = x|x| \qquad (b)$$

find the probability density function $p(r)$ and show that it satisfies the condition

$$\int_{-\infty}^{\infty} p(r) \, dr = 1 \qquad (c)$$

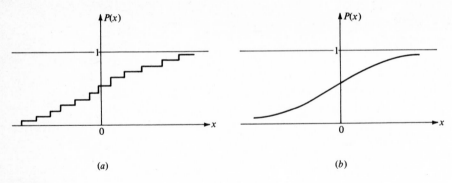

FIGURE 22-5
Probability distribution function for random variable x: (a) discrete variable;
(b) continuous variable.

First, taking the derivative of Eq. (b) gives $dr/dx = 2|x|$. Then using
Eq. (22-4) gives

$$p(r) = \begin{cases} \dfrac{1}{4|x|} & -1 < x < +1 \\ 0 & x < 1; x > 1 \end{cases} \qquad (d)$$

or
$$p(r) = \begin{cases} \dfrac{1}{4\sqrt{|r|}} & -1 < r < +1 \\ 0 & r < 1; r > 1 \end{cases} \qquad (e)$$

Substituting Eq. (e) into Eq. (c) leads to

$$I \equiv \int_{-\infty}^{\infty} p(r)\, dr = \frac{1}{4} \int_{-1}^{1} \frac{dr}{\sqrt{|r|}} = 1 \qquad (f)$$

thus showing that Eq. (c) is satisfied. ////

22-2 IMPORTANT AVERAGES OF A SINGLE RANDOM VARIABLE

If a certain random variable x is sampled n times and is each time used to evaluate a
second random variable r defined by a single-valued function $r(x)$, the average of this
second variable as n approaches infinity, that is,

$$\bar{r} \equiv \lim_{n \to \infty} \frac{1}{n} \sum_{i=1}^{n} r(x_i) \qquad (22\text{-}7)$$

where x_i is the ith sampled value of x, can be determined using the relation

$$\bar{r} = \int_{-\infty}^{\infty} r(x)p(x)\, dx \qquad (22\text{-}8)$$

A bar placed above any random variable is used to indicate average value.

Averages most commonly used in nondeterministic analyses are (1) mean value of x, (2) mean square value of x, (3) variance of x, and (4) standard deviation of x, defined as follows:

Mean value
$$\bar{x} = \int_{-\infty}^{\infty} xp(x)\, dx \qquad (22\text{-}9)$$

Mean square value
$$\overline{x^2} = \int_{-\infty}^{\infty} x^2 p(x)\, dx \qquad (22\text{-}10)$$

Variance
$$\sigma_x{}^2 = \overline{(x - \bar{x})^2} = \int_{-\infty}^{\infty} (x - \bar{x})^2 p(x)\, dx = \overline{x^2} - \bar{x}^2 \qquad (22\text{-}11)$$

Standard deviation
$$\sigma_x = \sqrt{\text{variance}} \qquad (22\text{-}12)$$

EXAMPLE E22-3   Find the mean, mean square value, and variance of a random variable x having the normal probability distribution given by Eq. (22-2).

From Eq. (22-9), the mean value can be written in the form

$$\bar{x} = \frac{1}{\sqrt{2\pi}\, a} \int_{-\infty}^{\infty} x e^{-(x-m)^2/2a^2}\, dx \qquad (a)$$

The change of variable

$$u \equiv \frac{x - m}{\sqrt{2}\, a} \qquad du = \frac{1}{\sqrt{2}\, a}\, dx \qquad (b)$$

gives
$$\bar{x} = \frac{\sqrt{2}\, a}{\sqrt{\pi}} \int_{-\infty}^{\infty} u e^{-u^2}\, du + \frac{m}{\sqrt{\pi}} \int_{-\infty}^{\infty} e^{-u^2}\, du \qquad (c)$$

The first integral in Eq. (c) equals zero while the second equals $\sqrt{\pi}$, thus showing that

$$\bar{x} = m \qquad (d)$$

The mean square value of x as given by Eq. (22-10) becomes

$$\overline{x^2} = \frac{1}{\sqrt{2\pi}\, a} \int_{-\infty}^{\infty} x^2 e^{-(x-m)^2/2a^2}\, dx \qquad (e)$$

Using the same change of variable indicated above gives

$$\overline{x^2} = \frac{2a^2}{\sqrt{\pi}} \int_{-\infty}^{\infty} u^2 e^{-u^2} \, du + \frac{2\sqrt{2} \, am}{\sqrt{\pi}} \int_{-\infty}^{\infty} u e^{-u^2} \, du + \frac{m^2}{\sqrt{\pi}} \int_{-\infty}^{\infty} e^{-u^2} \, du \qquad (f)$$

Upon integrating by parts, the first integral is shown equal to $\sqrt{\pi}/2$, the second integral equals zero, and the third equals $\sqrt{\pi}$, thus yielding

$$\overline{x^2} = a^2 + m^2 \qquad (g)$$

Substituting Eqs. (d) and (g) into Eq. (22-11) gives

$$\sigma_x^{\,2} = a^2 \qquad (h)$$

$////$

22-3   ONE-DIMENSIONAL RANDOM WALK

Assume in this experiment that n individuals are walking along a straight line without interference. If all individuals start walking from the same point ($x = 0$) and each separate step length L is controlled by the probability density function

$$p(L) = \tfrac{1}{4} \, \delta(L + \Delta L) + \tfrac{3}{4} \, \delta(L - \Delta L) \qquad (22\text{-}13)$$

that is, there exists a $^1/_4$ probability of taking a backward step of length ΔL and a $^3/_4$ probability of taking a forward step of the same length, the probability density function $p(x_i)$ for distance x_i as defined by $x_i \equiv \sum_{j=1}^{i} L_j$ will be as given in Fig. 22-6 for $i = 0, 1, 2, 3,$ and 4. (Vertical heavy arrows will be used herein to indicate Dirac delta functions.) Since all n individuals performing this experiment are at the origin before taking their first step, the probability density function $p(x_0)$ is a single Dirac delta function of unit intensity located at the origin. If n is considered to approach infinity, it follows directly from Eq. (22-13) that $3n/4$ individuals will be located at $x_1 = \Delta L$ after taking their first step and $n/4$ individuals will be located at $x_1 = -\Delta L$. Upon taking their second step three-fourths of those individuals located at $x_1 = \Delta L$, that is, $9n/16$, will move to $x_2 = 2\Delta L$ and the remaining one-fourth will step backward to the origin. Similarly, upon taking their second step, three-fourths of those individuals located at $x_1 = -\Delta L$, that is, $3n/16$, will step forward to the origin while the remaining one-fourth will step backward to $x_2 = -2\Delta L$. Such reasoning can be continued to establish each successive probability density function in the same way.

If the probability density function for length of step is given by the somewhat more general form

$$p(L) = g\delta(L + \Delta L) + h\delta(L - \Delta L) \qquad (22\text{-}14)$$

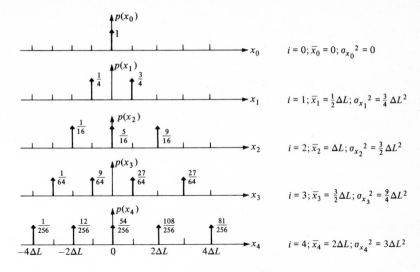

FIGURE 22-6
Example of one-dimensional random walk.

where $g + h = 1$, if the numerical values of g and h are known, it is possible to find the probability density functions $p(x_i)$ $(i = 1, 2, \ldots)$ by the same procedure used above for $g = {}^1/_4$ and $h = {}^3/_4$. While it will not be proved here, it can be easily shown that the probability density function $p(x_i)$ is given by the well-known binomial relation

$$p(x_i) = \sum_{k=-i,\,-i+2,\,\ldots}^{i} \frac{i!\ \delta(x_i - k\ \Delta L)}{[(i + k)/2]!\ [(i - k)/2]!} h^{(i+k)/2}(1 - h)^{(i-k)/2}$$

$$i = 0, 1, 2, \ldots \qquad (22\text{-}15)$$

and that the mean value and variance of x_i as defined by Eqs. (22-9) and (22-11) are, respectively,

$$\bar{x}_i = i\bar{L} = i(h - g)\ \Delta L \qquad (22\text{-}16)$$

$$\sigma_{x_i}^2 = i\sigma_L^2 = i[1 - (h - g)^2]\ \Delta L^2 \qquad (22\text{-}17)$$

The reader can easily check Eqs. (22-15) to (22-17) with the results previously obtained by straightforward means as given in Fig. 22-6.

Consider the one-dimensional random walk in its most general form, that is, one with an arbitrarily prescribed probability density function for length of step L, as shown in Fig. 22-7a. This function can be approximated by the discrete distribution shown in Fig. 22-7b obtained by simply concentrating the area $\Delta L\ p(L = q\ \Delta L)$ in the form of a Dirac delta function. Of course, in the limit as ΔL approaches zero, this discrete representation becomes exact. Likewise, the continuous probability

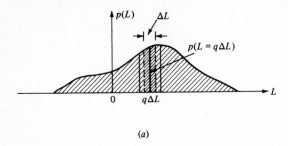

(a)

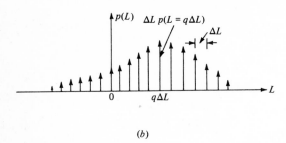

(b)

FIGURE 22-7
Arbitrary probability density function
for length of step.

density functions for distance from the origin $x_i \equiv \sum_{j=1}^{i} L_j$ (Fig. 22-8a) can be approximated by a discrete distribution as shown in Fig. 22-8b. With Δx chosen equal to ΔL, it is possible to determine the probability density function $p(x_{i+1})$ in exactly the same way as for the simpler case shown in Fig. 22-6; that is, the contribution by the Dirac delta function of intensity $\Delta x\, p(x_i = r\,\Delta x)$ to the Dirac delta function of intensity $\Delta x\, p(x_{i+1} = s\,\Delta x)$ is the product $\Delta x\, p(x_i = r\,\Delta x)\,\Delta L$ $p(L = q\,\Delta L)$ where $q \equiv s - r$. Therefore, the contribution of all delta functions in Fig. 22-8b to the intensity $p(x_{i+1} = s\,\Delta x)$ can be obtained by superposition, thus giving

$$p(x_{i+1} = s\,\Delta x) = \sum_{r=-\infty}^{\infty} p(x_i = r\,\Delta x)\, p(L = q\,\Delta L)\, \Delta L \qquad (22\text{-}18)$$

It will become apparent later in this development that it is advantageous to express the probability density functions of Fig. 22-8b and c in terms of distances X_i and X_{i+1} measured from points $x_i = iA$ and $x_{i+1} = (i + 1)A$, respectively, where A is some integer number of Δx. With this type of coordinate transformation, Eq. (22-18) becomes

$$p[X_{i+1} = s\,\Delta x - (i + 1)A] = \sum_{r=-\infty}^{\infty} p(X_i = r\,\Delta x - iA)\, p(L = q\,\Delta L)\, \Delta L \qquad (22\text{-}19)$$

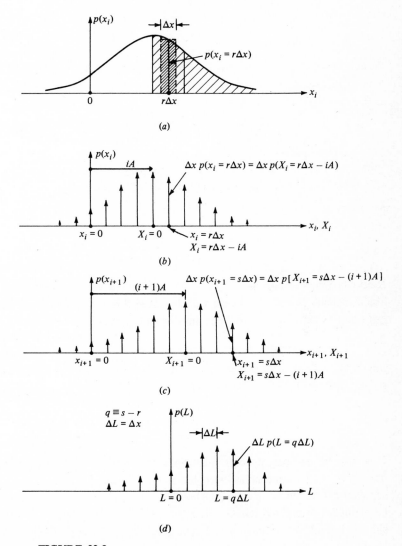

FIGURE 22-8
Probability density functions for the general one-dimensional random walk.

If in the above random walk, each individual is located at the origin $x = 0$ at time $t = 0$, and if each individual takes his ith step at the instant $t_i = i \, \Delta t$, Eq. (22-19) can be written

$$p(X; t_i + \Delta t) = \sum_{q=-\infty}^{\infty} p(X - q \, \Delta x + A; t_i) \, p(L = q \, \Delta L) \, \Delta L \qquad (22\text{-}20)$$

where

$$X_{i+1} \equiv X \qquad (22\text{-}21)$$

$$X_i = X - q \, \Delta x + A \qquad (22\text{-}22)$$

When a limiting process is now used by letting $\Delta x = \Delta L \to 0$ in such a manner that the quantity X remains finite, Eq. (22-20) converts to its continuous form with respect to distance, that is,

$$p(X; t_i + \Delta t) = \int_{-\infty}^{\infty} p(X - L + A; t_i) \, p(L) \, dL \qquad (22\text{-}23)$$

When the function $p(X - L + A; t_i)$ is expanded in a Taylor's series about point $x = 0$ and the integral is completed, Eq. (22-23) becomes

$$p(X; t_i + \Delta t) = p(X; t_i) + (A - \bar{L}) \, p'(X; t_i)$$

$$+ \frac{A^2 - 2A\bar{L} + \overline{L^2}}{2} \, p''(X; t_i) + \cdots \qquad (22\text{-}24)$$

It now becomes apparent why (as previously noted) it is helpful to express the probability density functions in terms of X rather than x, since the second term on the right-hand side of Eq. (22-24) can be eliminated by letting

$$A = \bar{L} \qquad (22\text{-}25)$$

Substituting Eq. (22-25) into Eq. (22-24) and dividing both sides of this equation by Δt and making use of Eq. (22-11) gives

$$\frac{p(X; t_i + \Delta t) - p(X; t_i)}{\Delta t} = \frac{\sigma_L^2}{2\Delta t} \, p''(X; t_i) + \cdots \qquad (22\text{-}26)$$

If during the limiting process mentioned above, the variance of the function $p(L)$, that is, σ_L^2, also approaches zero but in such a way that the ratio $\sigma_L^2/\Delta t$ equals a constant C, the terms on the right-hand side of Eq. (22-26) beyond the first term will be of higher order and can be dropped. Also in the limit as $\Delta t \to 0$, the left-hand side of Eq. (22-26) equals $\dot{p}(X,t)$; thus, Eq. (22-26) becomes the well-known one-dimensional diffusion equation

$$\frac{\partial p(X,t)}{\partial t} = \frac{C}{2} \frac{\partial^2 p(X,t)}{\partial X^2} \qquad (22\text{-}27)$$

From the known initial condition

$$p(X,0) = \delta(X) \qquad (22\text{-}28)$$

and the boundary conditions

$$\lim_{Q \to \infty} \frac{\partial p(Q,t)}{\partial X} = \lim_{Q \to \infty} \frac{\partial p(-Q,t)}{\partial X} = 0 \qquad (22\text{-}29)$$

the solution of Eq. (22-27) is

$$p(X,t) = \frac{1}{\sqrt{2\pi Ct}} e^{-X^2/2Ct} \qquad (22\text{-}30)$$

The probability density function for random variable X_i, after i steps, is given by Eq. (22-30) after substituting $C = \sigma_L^2/\Delta t$ and $i = t/\Delta t$, thus giving

$$p(X_i) = \frac{1}{\sqrt{2\pi i\sigma_L^2}} e^{-X_i^2/2i\sigma_L^2} \qquad (22\text{-}31)$$

In this case use of Eqs. (22-9) to (22-11) and (22-31) shows that

$$\overline{X}_i = 0 \qquad \sigma_{X_i}^2 = i\sigma_L^2 \qquad (22\text{-}32)$$

From Eqs. (22-25) and (22-31) and the information provided in Fig. 22-8, the relation

$$p(x_i) = \frac{1}{\sqrt{2\pi}\,\sigma_{x_i}} \exp\left[-\frac{(x_i - \bar{x}_i)^2}{2\sigma_{x_i}^2}\right] \qquad (22\text{-}33)$$

is obtained, where

$$\bar{x}_i = i\overline{L} \qquad \sigma_{x_i}^2 = i\sigma_L^2 \qquad (22\text{-}34)$$

This treatment of the general one-dimensional random walk, which follows the method used originally by Lord Rayleigh, has far-reaching significance since it shows that the probability density functions $p(x_i)$ for the algebraic sum of i random variables, namely,

$$x_i = \sum_{j=1}^{i} L_j \qquad (22\text{-}35)$$

where L_j $(j = 1\ 2, \ldots, i)$ are selected in accordance with an arbitrary probability density function $p(L)$ like that shown in Fig. 22-7, approach a gaussian distribution in the limit as $i \to \infty$. This fact is contained in the so-called central-limit theorem, which is found in most textbooks on probability theory. Fortunately, the probability density function $p(x_i)$ approaches a gaussian distribution rapidly as i increases (except for large values of x); therefore, assuming a gaussian distribution in engineering applications of Eq. (22-35) is usually justified.

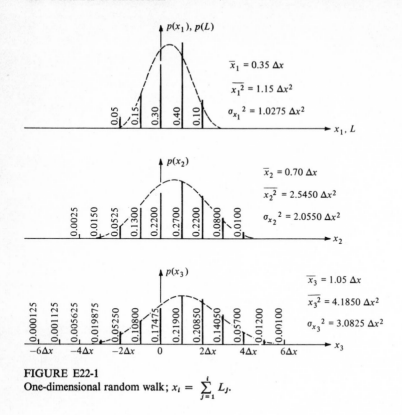

FIGURE E22-1
One-dimensional random walk; $x_i = \sum\limits_{j=1}^{i} L_j$.

EXAMPLE E22-4 Consider the one-dimensional random walk as defined by Eq. (22-35), where the probability density function for a single step length is given in the discrete form

$$p(L) = 0.05\delta(L + 2\Delta x) + 0.15\delta(L + \Delta x) + 0.30\delta(L)$$
$$+ 0.40\delta(L - \Delta x) + 0.10\delta(L - 2\Delta x) \qquad (a)$$

This function is also the probability density function for random variable x_1. By successive distributions, as used for the simpler case in Fig. 22-6, probability density functions $p(x_2)$, $p(x_3)$, etc., can be obtained, as shown in Fig. E22-1. To ensure a complete understanding of this method, it is suggested that the student check the numerical values given in the figure for distributions $p(x_2)$ and $p(x_3)$. For comparison continuous normal distributions are plotted in Fig. E22-1 by dashed lines. These distributions have the same mean values and variances as their corresponding discrete distributions. Note the very rapid rate at which the discrete distributions are approaching the normal distributions with increasing values of i.

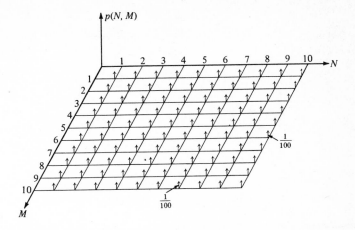

FIGURE 22-9
Joint probability density function for discrete random variables N and M.

Obviously for large values of i, the above distribution technique for obtaining $p(x_i)$ is extremely tedious and time-consuming. However, a good approximation of this function can be obtained by assuming a normal distribution having a mean value and variance as given by Eqs. (22-34). Thus for the case represented by Eq. (a), the continuous distribution is

$$p(x_i) \doteq \frac{1}{\sqrt{2\pi}\ \sigma_{x_i}} \exp\left[-\frac{(x_i - \bar{x}_i)^2}{2\sigma_{x_i}^{2}} \right] \qquad (b)$$

where
$$\bar{x}_i = 0.35i\ \Delta x \qquad \sigma_{x_i}^{2} = 1.0275i\ \Delta x^2 \qquad (c)$$

For large values of i, this distribution when discretized will give a very good approximation of the true distribution. Significant differences will appear only in the extreme "tail regions" of the distributions.   ////

22-4   TWO RANDOM VARIABLES

This section is concerned with experiments involving two random variables. Suppose, for example, a discrete random variable N is obtained by spinning the disk shown in Fig. 22-1a while a second variable M is obtained by spinning a second disk of identical design. Obtaining n ($n \rightarrow \infty$) pairs of numbers N and M in such a manner would in the limit give a discrete distribution of numbers, as shown in Fig. 22-9. This distribution $p(N,M)$, which consists of 100 two-dimensional Dirac delta functions of intensity $^1/_{100}$, is called the *joint probability density function* for random variables

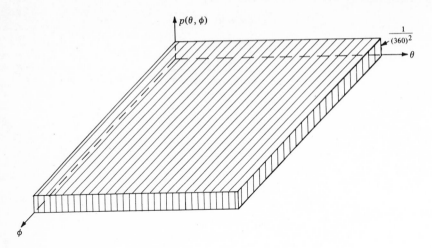

FIGURE 22-10
Joint probability density function for continuous random variables θ and ϕ.

N and M. If instead of two disks of the type shown in Fig. 22-1a two disks of the type shown in Fig. 22-1b are used to sample random variables θ and ϕ, sampling n ($n \rightarrow \infty$) pairs would give the uniform distribution shown in Fig. 22-10. Note that the volume between the plane of the two random-variable axes and the surface of the joint probability function is normalized to unity in each case.

The joint probability density function $p(x,y)$ for a general experiment involving random variables x and y is shown in Fig. 22-11. This function is defined so that the element volume $p(x_1,y_1)\,dx\,dy$ as shown in Fig. 22-10 represents the probability that a pair of sampled values will be within the region $x_1 < x < x_1 + dx$ and $y_1 < y < y_1 + dy$. This definition requires that the total volume between the xy plane and the $p(x,y)$ surface equal unity, that is, $\int_{-\infty}^{\infty} \int_{-\infty}^{\infty} p(x,y)\,dx\,dy = 1$.

The most common joint probability density function (later used extensively) is the *normal*, or *gaussian*, *distribution* given by

$$p(x,y) = \frac{1}{2\pi ab\sqrt{1 - c^2}}$$

$$\times \exp\left\{-\frac{1}{2(1 - c^2)}\left[\frac{(x - d)^2}{a^2} - \frac{2c(x - d)(x - e)}{ab}\right.\right.$$

$$\left.\left. + \frac{(y - e)^2}{b^2}\right]\right\} \qquad (22\text{-}36)$$

where a, b, c, d, and e are constants.

Suppose the joint probability density function for a new set of random variables r and s as defined by the relations

$$r = r(x,y) \qquad s = s(x,y) \qquad (22\text{-}37)$$

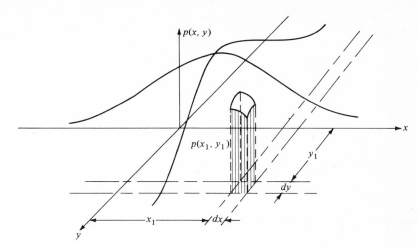

FIGURE 22-11
General joint probability density function for random variables x and y.

is desired, where Eqs. (22-37) and their inverse relations

$$x = x(r,s) \qquad y = y(r,s) \qquad (22\text{-}38)$$

are single-valued functions. Because a square infinitesimal area $dr\,ds$ in the rs plane will map as a parallelogram of area

$$\left| \frac{\partial x}{\partial r} \frac{\partial y}{\partial s} - \frac{\partial x}{\partial s} \frac{\partial y}{\partial r} \right| dr\,ds \qquad (22\text{-}39)$$

in the xy plane, it is necessary that

$$p(r,s) = \left| \frac{\partial x}{\partial r} \frac{\partial y}{\partial s} - \frac{\partial x}{\partial s} \frac{\partial y}{\partial r} \right| p(x,y) \qquad (22\text{-}40)$$

since sampled values of x and y which fall within the parallelogram correspond to values of r and s within the square. The absolute value has been indicated in Eq. (22-40), as the area of the parallelogram must always be a positive quantity. The transformation indicated by Eq. (22-40) is known as the *jacobian transformation* (Fig. 22-12).

Next certain probability functions closely associated with the joint probability density function $p(x,y)$ are defined.

Marginal probability density function $p(x)$ is defined such that $p(x_1)\,dx$ equals the chances that a sampled value of x will be in the range $x_1 < x < x_1 + dx$ regardless of the value of y sampled. Likewise, marginal probability density function $p(y)$ is defined so that $p(y_1)\,dy$ equals the chances that a sampled value of y will be in the

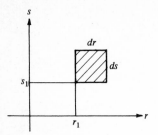
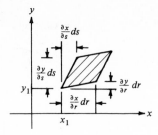

FIGURE 22-12
Jacobian transformation of two random variables.

range $y_1 < y < y_1 + dy$ regardless of the value of x sampled. In accordance with the above definitions,

$$p(x_1)\, dx = \int_{-\infty}^{\infty} \int_{x_1}^{x_1+dx} p(x,y)\, dx\, dy = dx \int_{-\infty}^{\infty} p(x_1,y)\, dy \qquad (22\text{-}41)$$

Therefore, the marginal probability density functions in unrestricted form are given by the relations

$$p(x) = \int_{-\infty}^{\infty} p(x,y)\, dy \qquad p(y) = \int_{-\infty}^{\infty} p(x,y)\, dx \qquad (22\text{-}42)$$

Probability distribution function $P(X,Y)$ is defined such that $P(X_1,Y_1)$ equals the chances that sampled values of x and y will be within the ranges $-\infty < x < X_1$ and $-\infty < y < Y_1$, respectively; thus

$$P(X,Y) = \int_{-\infty}^{Y} \int_{-\infty}^{X} p(x,y)\, dx\, dy \qquad (22\text{-}43)$$

In differential form this becomes

$$P(X,Y) = \frac{\partial^2 P(X,Y)}{\partial X\, \partial Y} \qquad (22\text{-}44)$$

Conditional probability density function $p(x \mid y)$ is defined such that $p(x_1 \mid y_1)\, dx$ equals the chances that x will be in the range $x_1 < x < x_1 + dx$ when considering *only* those sampled values of x and y which are in the ranges $-\infty < x < \infty$ and $y_1 < y < y_1 + dy$, respectively; that is, $p(x_1 \mid y_1) = p(x_1,y_1)/p(y_1)$ or, in its unrestricted form,

$$p(x \mid y) = \frac{p(x,y)}{p(y)} \qquad (22\text{-}45)$$

Likewise

$$p(y \mid x) = \frac{p(x,y)}{p(x)} \qquad (22\text{-}46)$$

It should be noted that the conditional probability density functions are ratios of marginal and joint probability density functions.

The conditional probability density functions $p(x \mid y)$ and $p(y \mid x)$ are often functions of x and y, respectively. In such cases, Eqs. (22-45) and (22-46) require that

$$p(x \mid y) = p(x) \quad \text{and} \quad p(y \mid x) = p(y) \quad (22\text{-}47)$$

and that

$$p(x,y) = p(x)p(y) \quad (22\text{-}48)$$

Random variables which satisfy Eqs. (22-47) and (22-48) are said to be *statistically independent*. Physically this means that when values of x and y are sampled, the sampled values of x are not influenced by corresponding sampled values of y and vice versa. The random variables represented in Figs. 22-9 and 22-10 are examples of statistically independent variables.

Suppose someone involved with statistically independent random variables x and y wishes to obtain the probability density function for a random variable r defined as the sum of random variables x and y. This probability density function can easily be obtained by using the jacobian transformation [Eqs. (22-40) and (22-48)] as follows. Define a new set of random variables r and s in the form

$$r \equiv x + y \quad s \equiv y \quad (22\text{-}49)$$

which, in inverse form, are

$$x = r - s \quad y = s \quad (22\text{-}50)$$

Equations (22-40), (22-48), and (22-50) give

$$p(r,s) = p(x,y) = p_x(x)p_y(y) = p_x(r - s)p_y(s) \quad (22\text{-}51)$$

Subscripts have been added here to identify the random variables involved. The marginal probability density function $p(r)$ now becomes

$$p(r) = \int_{-\infty}^{\infty} p_x(r - s)\, p_y(s)\, ds \quad (22\text{-}52)$$

EXAMPLE E22-5 Consider the one-dimensional random walk defined by Eq. (22-35), where the probability density function for a single step length is given in the continuous form

$$p(L) = \begin{cases} \dfrac{1}{\Delta x} & 0 < L < \Delta x \\ 0 & L < 0;\ L > \Delta x \end{cases} \quad (a)$$

This function is also the probability density function for random variable x_1. To obtain the probability density function for $x_2 = x_1 + L_2$, apply the convolution integral given by Eq. (22-52), writing

$$p(x_2) = \int_{-\infty}^{\infty} p_L(x_2 - s) p_{x_1}(s) \, ds \qquad (b)$$

where

$$p_{x_1}(s) = \begin{cases} \dfrac{1}{\Delta x} & 0 < s < \Delta x \\ 0 & s < 0; \, s > \Delta x \end{cases} \qquad (c)$$

$$p_L(x_2 - s) = \begin{cases} \dfrac{1}{\Delta x} & x_2 - \Delta x < s < x_2 \\ 0 & s < x_2 - \Delta x; \, s > x_2 \end{cases} \qquad (d)$$

Substituting Eqs. (c) and (d) into Eq. (b) gives

$$p(x_2) = \begin{cases} \dfrac{1}{(\Delta x)^2} \displaystyle\int_0^{x_2} ds = \dfrac{1}{(\Delta x)^2} x_2 & 0 \leq x_2 \leq \Delta x \\ \dfrac{1}{(\Delta x)^2} \displaystyle\int_{x_2-\Delta x}^{\Delta x} ds = \dfrac{1}{(\Delta x)^2} (2\Delta x - x_2) & \Delta x \leq x_2 \leq 2\Delta x \\ 0 & x_2 \leq 0; \, x_2 \geq 2\Delta x \end{cases} \qquad (e)$$

When the probability density function for x_2 is known, the same convolution integral can be used once again to find the probability density function for random variable $x_3 = x_2 + L_3$, giving

$$p(x_3) = \int_{-\infty}^{\infty} p_L(x_3 - s) p_{x_2}(s) \, ds \qquad (f)$$

With Eqs. (a) and (e), the integrand terms of this integral can be written

$$p_{x_2}(s) = \begin{cases} \dfrac{1}{(\Delta x)^2} s & 0 < s < \Delta x \\ \dfrac{1}{(\Delta x)^2} (2\Delta x - s) & \Delta x < s < 2\Delta x \\ 0 & s < 0; \, s > 2\Delta x \end{cases} \qquad (g)$$

$$p_L(x_3 - s) = \begin{cases} \dfrac{1}{\Delta x} & x_3 - \Delta x < s < x_3 \\ 0 & s < x_3 - \Delta x; \, s > x_3 \end{cases} \qquad (h)$$

Substituting Eqs. (g) and (h) into Eq. (f) gives

$$p(x_3) = \begin{cases} \dfrac{1}{(\Delta x)^3} \displaystyle\int_0^{x_3} s\, ds = \dfrac{x_3{}^2}{2(\Delta x)^3} & 0 \le x_3 \le \Delta x \\[3mm] \dfrac{1}{(\Delta x)^3} \left[\displaystyle\int_{x_3-\Delta x}^{\Delta x} s\, ds + \int_{\Delta x}^{x_3} (2\Delta x - s)\, ds \right] \\[3mm] \qquad = \dfrac{1}{(\Delta x)^3}(-x_3{}^2 + 3\Delta x\, x_3 - {}^3/_2\Delta x^2) \\[3mm] \qquad\qquad\qquad\qquad \Delta x \le x_3 \le 2\Delta x \\[3mm] \dfrac{1}{(\Delta x)^3} \displaystyle\int_{x_3-\Delta x}^{2\Delta x} (2\Delta x - s)\, ds \\[3mm] \qquad = \dfrac{1}{(\Delta x)^3}\left(\dfrac{x_3{}^2}{2} - 3\Delta x\, x_3 + {}^9/_2\Delta x^2 \right) \\[3mm] \qquad\qquad\qquad\qquad 2\Delta x \le x_3 \le 3\Delta x \\[3mm] 0 & x_3 \le 0;\ x_3 \ge 3\Delta x \end{cases}$$

$\qquad\qquad\qquad\qquad\qquad\qquad\qquad\qquad\qquad\qquad\qquad\qquad (i)$

Probability density functions $p(x_1)$, $p(x_2)$, and $p(x_3)$ as given by Eqs. (a), (e), and (i), respectively, are plotted in Fig. E22-2.

For comparison, normal distributions are plotted in Fig. E22-2 by dashed lines. These distributions have the same mean values and variances as the corresponding exact distributions shown by solid lines. Note the very rapid manner in which $p(x_i)$ approaches the normal distribution with increasing values of i. Although $p(x_4)$, $p(x_5)$, etc., could be obtained by repeated use of the convolution integral as above, this procedure would be very time-consuming. Therefore, as with the discrete case of Example 22-4, convergence toward the normal distribution as noted above allows one to assume a normal distribution having a mean value and variance given by Eq. (22-34). Thus for large values of i, one can use the normal form

$$p(x_i) \doteq \frac{1}{\sqrt{2\pi}\, \sigma_{x_i}} \exp\left[-\frac{(x_i - \bar{x}_i)^2}{2\sigma_{x_i}{}^2} \right] \qquad (j)$$

where

$$\bar{x}_i = \frac{i\,\Delta x}{2} \qquad \sigma_{x_i}{}^2 = \frac{i\,\Delta x^2}{12} \qquad (k)$$

$\qquad\qquad\qquad\qquad\qquad\qquad\qquad\qquad\qquad\qquad\qquad\qquad\qquad ////$

EXAMPLE E22-6 Given the joint probability density function

$$p(x, y) = C \exp\left[-\frac{2}{3}\left(\frac{x^2}{4} - \frac{xy}{6} + \frac{y^2}{9} \right) \right] \qquad (a)$$

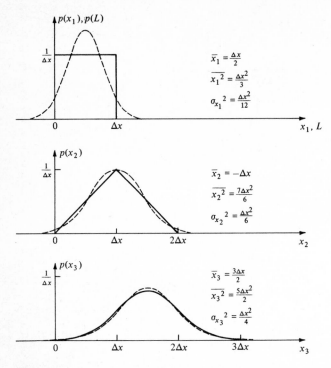

$$\bar{x}_1 = \frac{\Delta x}{2}$$

$$\overline{x_1^2} = \frac{\Delta x^2}{3}$$

$$\sigma_{x_1}^2 = \frac{\Delta x^2}{12}$$

$$\bar{x}_2 = -\Delta x$$

$$\overline{x_2^2} = \frac{7\Delta x^2}{6}$$

$$\sigma_{x_2}^2 = \frac{\Delta x^2}{6}$$

$$\bar{x}_3 = \frac{3\Delta x}{2}$$

$$\overline{x_3^2} = \frac{5\Delta x^2}{2}$$

$$\sigma_{x_3}^2 = \frac{\Delta x^2}{4}$$

FIGURE E22-2
One-dimensional random walk; $x_i \equiv \sum_{j=1}^{i} L_j$.

find (1) the numerical value of C so that the function is normalized properly, (2) the marginal probability density functions $p(x)$ and $p(y)$, and (3) the conditional probability density functions $p(x \mid y)$ and $p(y \mid x)$. Show that the random variables x and y are statistically dependent.

The function $p(x,y)$ is properly normalized when its double integral over the infinite x and y domains equals unity, that is, when

$$C \iint\limits_{-\infty}^{\infty} \exp\left[-\frac{2}{3}\left(\frac{x^2}{4} - \frac{xy}{6} + \frac{y^2}{9}\right)\right] dx\, dy = 1 \qquad (b)$$

Equation (b) can be separated and put into the equivalent form

$$C \int_{-\infty}^{\infty} \exp\left(-\frac{y^2}{18}\right)\left\{\int_{-\infty}^{\infty} \exp\left[-\frac{2}{3}\left(\frac{x}{2} - \frac{y}{6}\right)^2\right] dx\right\} dy = 1 \qquad (c)$$

By substituting the change of variable

$$u = \sqrt{\frac{2}{3}}\left(\frac{x}{2} - \frac{y}{6}\right) \qquad dx = \sqrt{6}\, du \qquad (d)$$

Eq. (c) becomes

$$\sqrt{6}\, C \int_{-\infty}^{\infty} \exp\left(-\frac{y^2}{18}\right)\left[\int_{-\infty}^{\infty} \exp\left(-u^2\right) du\right] dy = 1 \qquad (e)$$

With another change of variable $v = y/3\sqrt{2}$ and the fact that the second integral in Eq. (e) equals $\sqrt{\pi}$ the result is

$$6\sqrt{3\pi}\, C \int_{-\infty}^{\infty} \exp\left(-v^2\right) dv = 1 \qquad (f)$$

or

$$C = \frac{1}{6\sqrt{3}\,\pi} \qquad (g)$$

With the first of Eqs. (22-42), the marginal probability density function $p(x)$ can be written

$$p(x) = \frac{1}{6\sqrt{3}\,\pi} \int_{-\infty}^{\infty} \exp\left[-\frac{2}{3}\left(\frac{x^2}{4} - \frac{xy}{6} + \frac{y^2}{9}\right)\right] dy \qquad (h)$$

or

$$p(x) = \frac{1}{6\sqrt{3}\,\pi} \exp\left(-\frac{x^2}{8}\right) \int_{-\infty}^{\infty} \exp\left[-\frac{2}{3}\left(\frac{y}{3} - \frac{x}{4}\right)^2\right] dy \qquad (i)$$

With the change of variable

$$u = \sqrt{\frac{2}{3}}\left(\frac{y}{3} - \frac{x}{4}\right) \qquad dy = 3\sqrt{\frac{3}{2}}\, du \qquad (j)$$

Eq. (i) becomes

$$p(x) = \frac{1}{2\sqrt{2}\,\pi} \exp\left(-\frac{x^2}{8}\right) \int_{-\infty}^{\infty} \exp\left(-u^2\right) du \qquad (k)$$

or

$$p(x) = \frac{1}{2\sqrt{2}\,\pi} \exp\left(-\frac{x^2}{8}\right) \qquad (l)$$

Similarly the second of Eqs. (22-42) gives

$$p(y) = \frac{1}{3\sqrt{2}\,\pi} \exp\left(-\frac{y^2}{18}\right) \qquad (m)$$

Substituting Eqs. (a) and (m) into Eq. (22-45) and dividing as required gives

$$p(x \mid y) = \frac{1}{\sqrt{6\pi}} \exp\left[-\frac{2}{3}\left(\frac{x^2}{4} - \frac{xy}{6} + \frac{y^2}{36}\right)\right] \qquad (n)$$

Likewise, substituting Eqs. (a) and (l) into Eq. (22-46) gives

$$p(y \mid x) = \frac{\sqrt{2}}{3\sqrt{3\pi}} \exp\left[-\frac{2}{3}\left(\frac{x^2}{16} - \frac{xy}{6} + \frac{y^2}{9}\right)\right] \qquad (o)$$

Since the above marginal and conditional probability density functions do not satisfy Eqs. (22-47), random variables x and y are statistically dependent. ////

22-5   IMPORTANT AVERAGES OF TWO RANDOM VARIABLES

In an experiment involving random variables x and y, sampling in pairs is done n times, and each time a third random variable r is evaluated, defined by a single-valued function $r(x, y)$. The average of this third random variable as $n \rightarrow \infty$, that is,

$$\bar{r} = \lim_{n \rightarrow \infty} \frac{1}{n} \sum_{i=1}^{n} r(x_i, y_i) \qquad (22\text{-}53)$$

where x_i and y_i are the ith sampled values of x and y, respectively, can be determined using the relation

$$\bar{r} = \int\int_{-\infty}^{\infty} r(x, y) p(x, y) \, dx \, dy \qquad (22\text{-}54)$$

The validity of Eq. (22-54) can be easily rationalized since $p(x, y) \, dx \, dy$ represents the fractional number of samples falling in the infinitesimal area $dx \, dy$ located at point (x, y).

Averages most commonly used when treating two random variables are the following:

Mean values:

$$\bar{x} = \int\int_{-\infty}^{\infty} x p(x, y) \, dx \, dy = \int_{-\infty}^{\infty} x p(x) \, dx$$

$$\bar{y} = \int\int_{-\infty}^{\infty} y p(x, y) \, dx \, dy = \int_{-\infty}^{\infty} y p(y) \, dy \qquad (22\text{-}55)$$

Mean square values:

$$\overline{x^2} = \int\int_{-\infty}^{\infty} x^2 p(x, y) \, dx \, dy = \int_{-\infty}^{\infty} x^2 p(x) \, dx$$

$$\overline{y^2} = \int\int_{-\infty}^{\infty} y^2 p(x, y) \, dx \, dy = \int_{-\infty}^{\infty} y^2 p(y) \, dy \qquad (22\text{-}56)$$

Variances:

$$\sigma_x^2 = \overline{(x - \bar{x})^2} = \int\int_{-\infty}^{\infty} (x - \bar{x})^2 p(x, y) \, dx \, dy = x^2 - \bar{x}^2$$

$$\sigma_y^2 = \overline{(y - \bar{y})^2} = \int\int_{-\infty}^{\infty} (y - \bar{y})^2 p(x, y) \, dx \, dy = y^2 - \bar{y}^2 \qquad (22\text{-}57)$$

Standard deviations:

$$\sigma_x \qquad \sigma_y \qquad (22\text{-}58)$$

Covariance:

$$\mu_{xy} = \overline{(x - \bar{x})(y - \bar{y})} = \int\!\!\int_{-\infty}^{\infty} (x - \bar{x})(y - \bar{y})p(x,y)\,dx\,dy = \overline{xy} - \bar{x}\bar{y} \qquad (22\text{-}59)$$

Correlation coefficient:

$$\rho_{xy} \equiv \frac{\mu_{xy}}{\sigma_x\sigma_y} \qquad (22\text{-}60)$$

Note that when x and y are statistically independent,

$$\overline{xy} = \int\!\!\int_{-\infty}^{\infty} xy p(x)p(y)\,dx\,dy = \bar{x}\bar{y} \qquad (22\text{-}61)$$

in which case both the covariance μ_{xy} and the correlation coefficient ρ_{xy} equal zero.

Substituting the normal, or gaussian, distribution as expressed by Eq. (22-36) into the above relations gives

$$\bar{x} = d \qquad \bar{y} = e \qquad \sigma_x = a \qquad \sigma_y = b \qquad \rho_{xy} = c \qquad (22\text{-}62)$$

Therefore, the normal distribution can be expressed in the form

$$p(x,y) = \frac{1}{2\pi\sigma_x\sigma_y\sqrt{1 - \rho_{xy}{}^2}} \exp\left\{ -\frac{1}{2(1 - \rho_{xy}{}^2)} \right.$$
$$\times \left[\frac{(x - \bar{x})^2}{\sigma_x{}^2} - \frac{2\rho_{xy}(x - \bar{x})(y - \bar{y})}{\sigma_x\sigma_y} \right.$$
$$\left.\left. + \frac{(y - \bar{y})^2}{\sigma_y{}^2} \right] \right\} \qquad (22\text{-}63)$$

and usually is.

EXAMPLE E22-7   Random variables x_1 and x_2 are statistically independent and are both uniformly distributed over the range 0 to 1. Two new random variables r_1 and r_2 are defined by

$$r_1 = (-2 \ln x_1)^{1/2} \cos 2\pi x_2 \qquad r_2 = (-2 \ln x_1)^{1/2} \sin 2\pi x_2 \qquad (a)$$

Find (1) the joint probability density function $p(r_1, r_2)$, (2) the marginal probability density functions $p(r_1)$ and $p(r_2)$, (3) the mean values of r_1 and r_2, (4) the variances of r_1 and r_2, and (5) the covariance of r_1 and r_2.

Inverting Eqs. (*a*) gives

$$x_1 = \exp\left[-\frac{1}{2}(r_1{}^2 + r_2{}^2)\right]$$

$$x_2 = \frac{1}{2\pi}\cos^{-1}\frac{r_1}{\sqrt{r_1{}^2 + r_2{}^2}} = \frac{1}{2\pi}\sin^{-1}\frac{r_2}{\sqrt{r_1{}^2 + r_2{}^2}} \qquad (b)$$

Thus,

$$\frac{\partial x_1}{\partial r_1} = -r_1\exp\left[-\tfrac{1}{2}(r_1{}^2 + r_2{}^2)\right]$$

$$\frac{\partial x_1}{\partial r_2} = -r_2\exp\left[-\tfrac{1}{2}(r_1{}^2 + r_2{}^2)\right] \qquad (c)$$

$$\frac{\partial x_2}{\partial r_1} = -\frac{1}{2\pi}\frac{r_2}{r_1{}^2 + r_2{}^2}$$

$$\frac{\partial x_2}{\partial r_2} = +\frac{1}{2\pi}\frac{r_1}{r_1{}^2 + r_2{}^2}$$

With the jacobian transformation, Eq. (22-40), the joint probability density function $p(r_1, r_2)$ can be expressed as

$$p(r_1, r_2) = \left|\frac{\partial x_1}{\partial r_1}\frac{\partial x_2}{\partial r_2} - \frac{\partial x_1}{\partial r_2}\frac{\partial x_2}{\partial r_1}\right| p(x_1, x_2) \qquad (d)$$

where
$$p(x_1, x_2) = p(x_1)\,p(x_2) = \begin{cases} 1 & \begin{cases} 0 < x_1 < 1 \\ 0 < x_2 < 1 \end{cases} \\ 0 & \begin{cases} x_1 < 0;\ x_1 > 1 \\ x_2 < 0;\ x_2 > 1 \end{cases} \end{cases} \qquad (e)$$

Substituting Eqs. (*c*) and (*e*) into Eq. (*d*) gives the normal distribution

$$p(r_1, r_2) = \frac{1}{2\pi}\exp\left[-\tfrac{1}{2}(r_1{}^2 + r_2{}^2)\right] \qquad (f)$$

Making use of Eqs. (22-42) results in the relations

$$p(r_1) = \frac{1}{2\pi}\exp\left(-\frac{r_1{}^2}{2}\right)\int_{-\infty}^{\infty}\exp\left(-\frac{r_2{}^2}{2}\right)dr_2 = \frac{1}{\sqrt{2\pi}}\exp\left(-\frac{r_1{}^2}{2}\right) \qquad (g)$$

$$p(r_2) = \frac{1}{2\pi}\exp\left(-\frac{r_2{}^2}{2}\right)\int_{-\infty}^{\infty}\exp\left(-\frac{r_1{}^2}{2}\right)dr_1 = \frac{1}{\sqrt{2\pi}}\exp\left(-\frac{r_2{}^2}{2}\right)$$

Integrating in accordance with Eqs. (22-55) and (22-56) gives

$$\bar{r}_1 = \frac{1}{\sqrt{2\pi}} \int_{-\infty}^{\infty} r_1 \exp\left(-\frac{r_1^2}{2}\right) dr_1 = 0$$

$$\bar{r}_2 = \frac{1}{\sqrt{2\pi}} \int_{-\infty}^{\infty} r_2 \exp\left(-\frac{r_2^2}{2}\right) dr_2 = 0$$

$$\overline{r_1^2} = \frac{1}{\sqrt{2\pi}} \int_{-\infty}^{\infty} r_1^2 \exp\left(-\frac{r_1^2}{2}\right) dr_1 = 1 \qquad (h)$$

$$\overline{r_2^2} = \frac{1}{\sqrt{2\pi}} \int_{-\infty}^{\infty} r_2^2 \exp\left(-\frac{r_2^2}{2}\right) dr_2 = 1$$

Thus,

$$\sigma_{r_1}^2 = \overline{r_1^2} - \bar{r}_1^2 = 1 \qquad \sigma_{r_2}^2 = \overline{r_2^2} - \bar{r}_2^2 = 1 \qquad (i)$$

Since r_1 and r_2 appear in an uncoupled form in Eq. (f), the mean value of $r_1 r_2$ is of the form given by Eq. (22-61), that is,

$$\overline{r_1 r_2} = \overline{r}_1 \, \overline{r}_2 \qquad (j)$$

Therefore the covariance becomes

$$\mu_{r_1 r_2} = \overline{r_1 r_2} - \overline{r}_1 \, \overline{r}_2 = 0 \qquad (k)$$

which shows that the random variables r_1 and r_2 are statistically independent.

////

EXAMPLE E22-8 Given the joint probability density function used in Example E22-6, namely,

$$p(x, y) = \frac{1}{6\sqrt{3}\,\pi} \exp\left[-\frac{2}{3}\left(\frac{x^2}{4} - \frac{xy}{6} + \frac{y^2}{9}\right)\right] \qquad (a)$$

find (1) the mean values, (2) the mean square values, (3) the variances, and (4) the covariance of random variables x and y.

These quantities could be obtained from the general relations given by Eqs. (22-55) to (22-57) and (22-59). However, comparison of this equation with the general form of the normal distribution given by Eq. (22-63) shows that it is obviously of similar form. Therefore these quantities can be obtained directly by setting the coefficients of terms in Eq. (a) equal to their corresponding coefficients in Eq. (22-63), giving

$$\frac{1}{6} = \frac{1}{2\sigma_x^2(1 - \rho_{xy}^2)} \qquad \frac{1}{9} = \frac{\rho_{xy}}{\sigma_x \sigma_y(1 - \rho_{xy}^2)} \qquad \frac{2}{27} = \frac{1}{2\sigma_y^2(1 - \rho_{xy}^2)} \qquad (b)$$

Solving Eqs. (b) for the three unknowns gives

$$\sigma_x = 2 \qquad \sigma_y = 3 \qquad \rho_{xy} = \frac{1}{2} \qquad (c)$$

The mean values $\bar{x}$ and $\bar{y}$ are obviously zero from the form of the equation; therefore,

$$\overline{x^2} = 4 \qquad \overline{y^2} = 9 \qquad (d)$$

The covariance is easily obtained since

$$\mu_{xy} = \sigma_x \sigma_y \rho_{xy} = 3 \qquad (e)$$

$////$

22-6 SCATTER DIAGRAM AND CORRELATION OF TWO RANDOM VARIABLES

The so-called scatter diagram can be helpful to the beginner in understanding the basic concepts and definitions of probability related to two random variables x and y. This diagram is obtained by sampling pairs of random variables and each time plotting them as a point on the xy plane, as shown in Fig. 22-13. Suppose n pairs are sampled and that $(x_1, y_1), (x_2, y_2), \ldots, (x_n, y_n)$ represent their coordinates on the scatter diagram. If n_1, n_2, and n_3 represent the numbers of sampled pairs falling in regions $X < x < X + \Delta x$ and $Y < y < Y + \Delta y$, $X < x < X + \Delta x$ and $-\infty < y < +\infty$, and $-\infty < x < \infty$ and $Y < y < Y + \Delta y$, respectively, the joint, marginal, and conditional probability density functions as previously defined will be given by

$$p(x, y) = \lim_{\substack{\Delta x \to 0 \\ \Delta y \to 0 \\ n \to 0}} \frac{n_1}{n \, \Delta x \, \Delta y} \qquad (22\text{-}64)$$

$$p(x) = \lim_{\substack{\Delta x \to 0 \\ n \to \infty}} \frac{n_2}{n \, \Delta x} \qquad p(y) = \lim_{\substack{\Delta y \to 0 \\ n \to \infty}} \frac{n_3}{n \, \Delta y} \qquad (22\text{-}65)$$

$$p(x \mid y) = \lim_{\substack{\Delta x \to 0 \\ \Delta y \to 0 \\ n \to \infty}} \frac{n_1}{n_3 \, \Delta x} \qquad p(y \mid x) = \lim_{\substack{\Delta x \to 0 \\ \Delta y \to 0 \\ n \to \infty}} \frac{n_1}{n_2 \, \Delta y} \qquad (22\text{-}66)$$

Further, it is quite apparent that

$$\bar{x} = \lim_{n \to \infty} \frac{1}{n} \sum_{i=1}^{n} x_i \qquad \bar{y} = \lim_{n \to \infty} \frac{1}{n} \sum_{i=1}^{n} y_i \qquad (22\text{-}67)$$

$$\overline{x^2} = \lim_{n \to \infty} \frac{1}{n} \sum_{i=1}^{n} x_i^2 \qquad \overline{y^2} = \lim_{n \to \infty} \frac{1}{n} \sum_{i=1}^{n} y_i^2 \qquad (22\text{-}68)$$

$$\sigma_x^2 = \lim_{n \to \infty} \frac{1}{n} \sum_{i=1}^{n} (x_i - \bar{x})^2 \qquad \sigma_y^2 = \lim_{n \to \infty} \frac{1}{n} \sum_{i=1}^{n} (y_i - \bar{y})^2 \qquad (22\text{-}69)$$

$$\mu_{xy} = \lim_{n \to \infty} \frac{1}{n} \sum_{i=1}^{n} (x_i - \bar{x})(y_i - \bar{y}) \qquad (22\text{-}70)$$

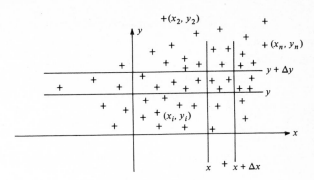

FIGURE 22-13
Scatter diagram for random variables x and y.

The correlation coefficient ρ_{xy} as defined by Eq. (22-60) should be fully understood, as it represents the degree of statistical dependence present between random variables x and y. First, to establish the range of possible numerical values which it may possess, consider two new random variables r and s as defined by the relations

$$r \equiv \frac{x - \bar{x}}{\sigma_x} \qquad s \equiv \frac{y - \bar{y}}{\sigma_y} \qquad (22\text{-}71)$$

This transformation represents a translation of the coordinate axes and a scale-factor change along each axis, so that

$$\bar{r} = \bar{s} = 0 \qquad \overline{r^2} = \sigma_r^2 = \overline{s^2} = \sigma_s^2 = 1 \qquad \overline{rs} = \rho_{rs} = \rho_{xy} \qquad (22\text{-}72)$$

Consider now the mean square value of $r \pm s$. Use of Eqs. (22-72) leads to

$$\overline{(r \pm s)^2} = 2(1 \pm \rho_{rs}) \qquad (22\text{-}73)$$

Since the mean square values given above must always be positive, the correlation coefficient must always be in the range

$$-1 < \rho_{rs} < +1 \qquad (22\text{-}74)$$

From the normal distribution as given by Eq. (22-63), the joint probability density function for variables r and s as defined by Eq. (22-71) is easily obtained by using the jacobian transformation, Eq. (22-40), thus yielding the relation

$$p(r,s) = \frac{1}{2\pi\sqrt{1 - \rho_{rs}^2}} \exp\left[-\frac{1}{2(1 - \rho_{rs}^2)}(r^2 - 2\rho_{rs}rs + s^2)\right] \qquad (22\text{-}75)$$

Contour lines representing equal values of $p(r,s)$ are shown in Fig. 22-14 for one particular value of ρ_{rs}. To obtain the analytical expression for such contour lines, the natural logarithm of both sides of Eq. (22-75) is taken, giving

$$r^2 - 2\rho_{rs}rs + s^2 = C^2 \qquad (22\text{-}76)$$

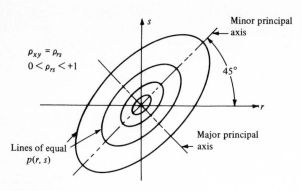

FIGURE 22-14
Contour lines of equal $p(r,s)$.

where C^2 is a constant which can be varied to correspond to a particular value of $p(r,s)$. When the correlation coefficient is positive, that is, in the range $0 < \rho_{rs} < 1$, Eq. (22-76) is the equation of an ellipse with its major and minor axes oriented as shown in Fig. 22-14. On the other hand, when the correlation coefficient is in the range $-1 < \rho_{rs} < 0$, this same equation represents an ellipse but with the directions of major and minor principal axes reversed from those shown in Fig. 22-14. When the correlation coefficient equals zero, Eq. (22-76) is the equation of a circle. As the correlation coefficient approaches $+1$, profiles of $p(r,s)$ normal to the minor principal axis approach Dirac delta functions centered on the minor axis. Likewise, as the correlation coefficient approaches -1, profiles of $p(r,s)$ normal to the major principal axis approach Dirac delta functions centered on the major axis.

Contour lines of equal probability $p(r,s)$ as given by Eq. (22-75) along with a limited number of scatter points are shown in Fig. 22-15 for each of the above five cases. It is clear from the diagrams of Fig. 22-15 that random variables r and s (or

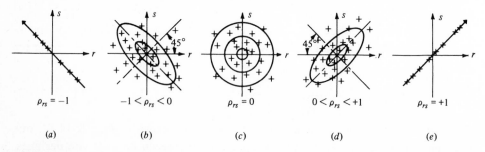

FIGURE 22-15
Contour lines of equal probability $p(r,s)$ with limited number of scatter points ($\rho_{rs} = \rho_{xy}$).

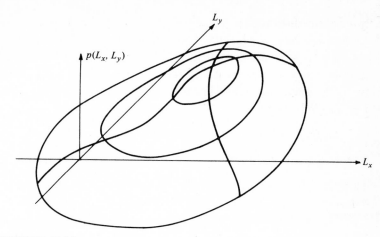

FIGURE 22-16
Arbitrary joint probability density function for x and y components of a single random step.

x and y) are completely dependent upon each other when the correlation coefficient is either $+1$ or -1. In other words, only one random variable really exists in these cases, as one of the random variables can be determined directly from the other. However, when the correlation coefficient equals zero, as in Fig. 22-15c, the random variables are completely independent of each other. The cases in Fig. 22-15b and d are intermediate examples, representing partial statistical dependence of one random variable upon the other.

22-7 TWO-DIMENSIONAL RANDOM WALK

Assume in this experiment that n individuals are walking on an xy plane without interference but in accordance with an arbitrarily prescribed joint probability density function $p(L_x,L_y)$, as shown in Fig. 22-16, where L_x and L_y represent the x and y components, respectively, of a single random step of length $L \equiv \sqrt{L_x{}^2 + L_y{}^2}$.

When x_i and y_i are defined as the coordinate positions of an individual after taking i steps, namely,

$$x_i = \sum_{j=1}^{i} (L_x)_j \qquad y_i = \sum_{j=1}^{i} (L_y)_j \qquad (22\text{-}77)$$

the joint probability density function $p(x_i,y_i)$ can be established by procedures similar to those used previously for the one-dimensional random walk. First, the probability density functions $p(x_i,y_i)$, $p(x_{i+1},y_{i+1})$, and $p(L_x,L_y)$ are approximated by discrete distributions as shown in Fig. 22-17a, b, and c, respectively. With $\Delta x = \Delta y = \Delta L_x = \Delta L_y$, the joint probability density function $p(x_{i+1}, y_{i+1})$ is obtained by superposition; that is, the contribution by the two-dimensional Dirac delta function of intensity $\Delta x\, \Delta y\, p(x_i = r\, \Delta x,\, y_i = m\, \Delta y)$ shown in Fig. 22-17a to the Dirac delta function of intensity $\Delta x\, \Delta y\, p(x_{i+1} = s\, \Delta x,\, y_{i+1} = n\, \Delta y)$ shown in Fig. 22-17b is the product

$$\Delta x\, \Delta y\, p(x_i = r\, \Delta x,\, y_i = m\, \Delta y)\, \Delta L_x\, \Delta L_y\, p(L_x = q\, \Delta x,\, L_y = g\, \Delta y) \qquad (22\text{-}78)$$

where
$$q = s - r \qquad g = n - m \qquad (22\text{-}79)$$

Therefore the contribution of all delta functions in Fig. 22-17a to the intensity $p(x_{i+1} = s\, \Delta x,\, y_{i+1} = n\, \Delta y)$ can be obtained by superposition, thus giving

$$p(x_{i+1} = s\, \Delta x,\, y_{i+1} = n\, \Delta y) = \sum_{m=-\infty}^{\infty} \sum_{r=-\infty}^{\infty} p(x_i = r\, \Delta x,\, y = m\, \Delta y)$$
$$p(L_x = q\, \Delta x,\, L_y = g\, \Delta y)\, \Delta L_x\, \Delta L_y \qquad (22\text{-}80)$$

As for the one-dimensional random walk, it is advantageous to express the joint probability density functions shown in Fig. 22-17a and b in terms of coordinates (X_i, Y_i) and (X_{i+1}, Y_{i+1}), respectively, where

$$X_i \equiv x_i - iA \qquad Y_i \equiv y_i - iB \qquad (22\text{-}81)$$

When this coordinate transformation is used, Eq. (22-80) becomes

$$p[X_{i+1} = s\, \Delta x - (i + 1)A,\, Y_{i+1} = n\, \Delta y - (i + 1)B]$$
$$= \sum_{m=-\infty}^{\infty} \sum_{r=-\infty}^{\infty} p(X_i = r\, \Delta x - iA,\, Y_i = m\, \Delta y - iB)$$
$$p(L_x = q\, \Delta x,\, L_y = g\, \Delta y)\, \Delta L_x\, \Delta L_y \qquad (22\text{-}82)$$

If in the above random walk each individual is located at the origin ($x = 0$, $y = 0$) at time $t = 0$, and if each individual takes his ith step at the instant $t_i = i\, \Delta t$, Eq. (22-82) can be written

$$p(X, Y; t_i + \Delta t) = \sum_{g=-\infty}^{\infty} \sum_{q=-\infty}^{\infty} p(X - q\, \Delta x + A,\, Y - g\, \Delta y + B; t_i)$$
$$p(L_x = q\, \Delta x,\, L_y = g\, \Delta y)\, \Delta L_x\, \Delta L_y \qquad (22\text{-}83)$$

where
$$X_{i+1} \equiv X \qquad Y_{i+1} \equiv Y \qquad (22\text{-}84)$$
$$X_i = X - q\, \Delta x + A \qquad Y_i = Y - g\, \Delta y + B \qquad (22\text{-}85)$$

When a limiting process is used by letting $\Delta x = \Delta y = \Delta L_x = \Delta L_y \to 0$, in such a manner that the quantities X and Y remain finite, Eq. (22-83) converts to the continuous form with respect to the space coordinates:

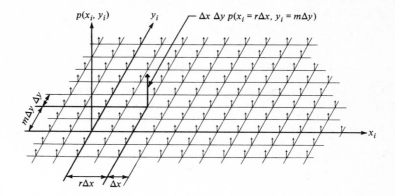

(a)

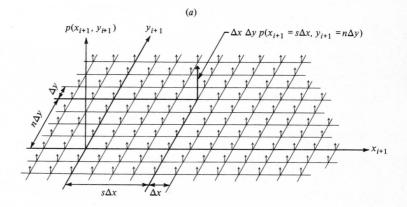

(b)

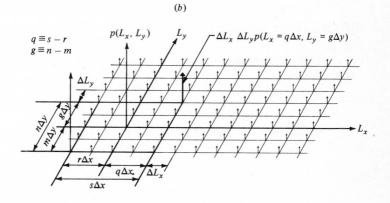

(c)

FIGURE 22-17
Joint probability density functions for the general two-dimensional random walk.

$p(X,Y; t_i + \Delta t)$

$$= \int\limits_{-\infty}^{\infty} \int p(X - L_x + A, Y - L_y + B; t_i) \, p(L_x, L_y) \, dL_x \, dL_y \quad (22\text{-}86)$$

Expanding the function $p(X - L_x + A, Y - L_y + B; t_i)$ in a Taylor's series about point $X = Y = 0$ and completing the double integral changes Eq. (22-86) into

$p(X,Y; t_i + \Delta t)$

$$= p(X,Y; t_i) + (A - L_x) \frac{\partial p(X,Y; t_i)}{\partial X} + (B - L_y) \frac{\partial p(X,Y; t_i)}{\partial Y}$$

$$+ \tfrac{1}{2}(A^2 - 2A\bar{L}_x + \overline{L_x^2}) \frac{\partial^2 p(X,Y; t_i)}{\partial X^2}$$

$$+ \tfrac{1}{2}(B^2 - 2B\bar{L}_y + \overline{L_y^2}) \frac{\partial^2 p(X,Y; t_i)}{\partial Y^2}$$

$$+ (AB - A\bar{L}_y - B\bar{L}_x + \overline{L_x L_y}) \frac{\partial^2 p(X,Y; t_i)}{\partial X \, \partial Y} + \cdots \quad (22\text{-}87)$$

Note that if A and B are selected equal to $\bar{L}_x$ and $\bar{L}_y$, respectively, the second and third terms on the right-hand side of Eq. (22-87) are eliminated. Further, when use is made of Eqs. (22-57) and (22-59), Eq. (22-87) reduces to

$$p(X,Y; t_i + \Delta t) = p(X,Y; t_i) + \frac{\sigma_{L_x}^2}{2} \frac{\partial^2 p(X,Y; t_i)}{\partial X^2}$$

$$+ \frac{\sigma_{L_y}^2}{2} \frac{\partial^2 p(X,Y; t_i)}{\partial Y^2} + \mu_{L_x L_y} \frac{\partial^2 p(X,Y; t_i)}{\partial X \, \partial Y} + \cdots \quad (22\text{-}88)$$

If L_x and L_y are statistically independent, the covariance $\mu_{L_x L_y}$ will of course equal zero.

Let us now transform Eq. (22-88) using the linear transformation

$$u \equiv Y \sin \theta + X \cos \theta \quad (22\text{-}89)$$

$$v \equiv Y \cos \theta - X \sin \theta$$

which corresponds to selecting a new set of orthogonal axes u and v, as shown in Fig. 22-18. Solving for the inverse relations of Eq. (22-89) gives

$$X = u \cos \theta - v \sin \theta \quad (22\text{-}90)$$

$$Y = u \sin \theta + v \cos \theta$$

When the jacobian transformation [Eq. (22-40)] is used, the joint probability density function $p(u,v)$ can be written

$$p(u,v; t_i) = \left| \frac{\partial X}{\partial u} \frac{\partial Y}{\partial v} - \frac{\partial X}{\partial v} \frac{\partial Y}{\partial u} \right| p(X,Y; t_i) \quad (22\text{-}91)$$

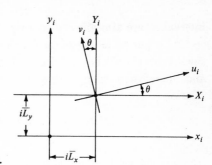

FIGURE 22-18
Coordinate transformations.

Substituting Eqs. (22-90) into Eq. (22-91) gives

$$p(u, v; t_i) = p(X, Y; t_i) \qquad (22\text{-}92)$$

When Eqs. (22-89) and (22-92) are used, Eq. (22-88) transforms into the relation

$$p(u,v; t_i + \Delta t) = p(u,v; t_i) + \frac{A_1^2}{2} \frac{\partial^2 p(u,v; t_i)}{\partial u^2}$$
$$+ \frac{A_2^2}{2} \frac{\partial^2 p(u,v; t_i)}{\partial v^2} + A_3 \frac{\partial^2 p(u,v; t_i)}{\partial u \, \partial v} + \cdots \qquad (22\text{-}93)$$

where

$$A_1^2 \equiv \sigma_{L_x}^2 \cos^2 \theta + \sigma_{L_y}^2 \sin^2 \theta + 2\mu_{L_x L_y} \sin \theta \cos \theta$$
$$A_2^2 \equiv \sigma_{L_x}^2 \sin^2 \theta + \sigma_{L_y}^2 \cos^2 \theta - 2\mu_{L_x L_y} \sin \theta \cos \theta \qquad (22\text{-}94)$$
$$A_3 \equiv (\sigma_{L_y}^2 - \sigma_{L_x}^2) \sin \theta \cos \theta + \mu_{L_x L_y}(\cos^2 \theta - \sin^2 \theta)$$

The cross-derivative term on the right-hand side of Eq. (22-93) can be eliminated by selecting a transformation angle which will make A_3 equal to zero, that is, one which satisfies the condition

$$(\sigma_{L_y}^2 - \sigma_{L_x}^2) \sin \theta \cos \theta + \mu_{L_x L_y}(\cos^2 \theta - \sin^2 \theta) = 0 \qquad (22\text{-}95)$$

Solving Eq. (22-95) for θ gives the desired transformation angle

$$\theta = \frac{1}{2} \tan^{-1} \frac{2\mu_{L_x L_y}}{\sigma_{L_x}^2 - \sigma_{L_y}^2} \qquad (22\text{-}96)$$

Substituting Eq. (22-96) into Eq. (22-93) and dividing both sides by Δt gives

$$\frac{p(u,v; t_i + \Delta t) - p(u,v; t_i)}{\Delta t} = \frac{A_1^2}{2\Delta t} \frac{\partial^2 p(u,v; t_i)}{\partial u^2} + \frac{A_2^2}{2\Delta t} \frac{\partial^2 p(u,v; t_i)}{\partial v^2} + \cdots \qquad (22\text{-}97)$$

If during the previously mentioned limiting process, where $\Delta x = \Delta y = \Delta L_x = \Delta L_y \to 0$ and $i \to \infty$, the variances $\sigma_{L_x}^2$ and $\sigma_{L_y}^2$, the covariance $\mu_{L_x L_y}$, and the time interval Δt are also allowed to approach zero but in such a manner that $A_1^2/\Delta t$ and $A_2^2/\Delta t$ equal finite constants B_1 and B_2, respectively, the terms on the right-hand side of Eq. (22-97) beyond the second terms are of higher order and can be dropped and the left-hand side of this same equation becomes $\dot{p}(u,v,t)$. This establishes the relation

$$\frac{\partial p(u,v,t)}{\partial t} = \frac{B_1}{2}\frac{\partial^2 p(u,v,t)}{\partial u^2} + \frac{B_2}{2}\frac{\partial^2 p(u,v,t)}{\partial v^2} \qquad (22\text{-}98)$$

which is the well-known two-dimensional diffusion equation.

When initial and boundary conditions for this two-dimensional case are used which are comparable to those used in the one-dimensional case, that is, Eqs. (22-28) and (22-29), the solution of Eq. (22-98) is

$$p(u,v,t) = \frac{1}{2\pi\sqrt{B_1 B_2 t^2}} \exp\left[-\frac{1}{2}\left(\frac{u^2}{B_1 t} + \frac{v^2}{B_2 t}\right)\right] \qquad (22\text{-}99)$$

Equation (22-95) can now be changed back to its discrete form with respect to time by making use of the previously defined relations

$$t_i \equiv i\,\Delta t \qquad B_1 \equiv \frac{A_1^2}{\Delta t} \qquad B_2 \equiv \frac{A_2^2}{\Delta t} \qquad (22\text{-}100)$$

thus giving

$$p(u_i,v_i) = \frac{1}{2\pi i A_1 A_2} \exp\left[-\frac{1}{2}\left(\frac{u_i^2}{iA_1^2} + \frac{v_i^2}{iA_2^2}\right)\right] \qquad (22\text{-}101)$$

When Eqs. (22-55), (22-60), (22-94), and (22-101) are used, the result is

$$\overline{u}_i = 0 \qquad \overline{v}_i = 0 \qquad (22\text{-}102)$$

$$\overline{u_i^2} = \sigma_{u_i}^2 = iA_1^2 = i\sigma_{L_x}^2\cos^2\theta + i\sigma_{L_y}^2\sin^2\theta + 2i\mu_{L_xL_y}\sin\theta\cos\theta \qquad (22\text{-}103)$$

$$\overline{v_i^2} = \sigma_{v_i}^2 = iA_2^2 = i\sigma_{L_x}^2\sin^2\theta + i\sigma_{L_y}^2\cos^2\theta - 2i\mu_{L_xL_y}\sin\theta\cos\theta \qquad (22\text{-}104)$$

$$\mu_{u_iv_i} = 0 \qquad \rho_{u_iv_i} = 0 \qquad (22\text{-}105)$$

Substituting Eqs. (22-103) and (22-104) into Eq. (22-101) gives the standard form of the normal, or gaussian, distribution, namely,

$$p(u_i,v_i) = \frac{1}{2\pi\sigma_{u_i}\sigma_{v_i}} \exp\left[-\frac{1}{2}\left(\frac{u_i^2}{\sigma_{u_i}^2} + \frac{v_i^2}{\sigma_{v_i}^2}\right)\right] \qquad (22\text{-}106)$$

Note that random variables u_i and v_i are statistically independent.

When the joint probability density function for random variables u_i and v_i

has been obtained, it is possible at this point to work backward through the transformations previously defined to obtain

$$p(X_i, Y_i) = \frac{1}{2\pi\sigma_{X_i}\sigma_{Y_i}\sqrt{1 - \rho_{X_iY_i}^2}}$$

$$\times \exp\left[-\frac{1}{2(1 - \rho_{X_iY_i}^2)}\left(\frac{X_i^2}{\sigma_{X_i}^2} - \frac{2\rho_{X_iY_i}X_iY_i}{\sigma_{X_i}\sigma_{Y_i}} + \frac{Y_i^2}{\sigma_{Y_i}^2}\right)\right] \quad (22\text{-}107)$$

where

$$\sigma_{X_i}^2 = i\sigma_{L_x}^2 \qquad \sigma_{Y_i}^2 = i\sigma_{L_y}^2$$

$$\rho_{X_iY_i} = \rho_{L_xL_y} = \frac{\mu_{L_xL_y}}{\sigma_{L_x}\sigma_{L_y}} \quad (22\text{-}108)$$

and

$$p(x_i, y_i) = \frac{1}{2\pi\sigma_{x_i}\sigma_{y_i}\sqrt{1 - \rho_{x_iy_i}^2}}$$

$$\times \exp\left\{-\frac{1}{2(1 - \rho_{x_iy_i}^2)}\left[\frac{(x_i - \bar{x}_i)^2}{\sigma_{x_i}^2}\right.\right.$$

$$- \frac{2\rho_{x_iy_i}(x_i - \bar{x}_i)(y_i - \bar{y}_i)}{\sigma_{x_i}\sigma_{y_i}}$$

$$\left.\left. + \frac{(y_i - \bar{y}_i)^2}{\sigma_{y_i}^2}\right]\right\} \quad (22\text{-}109)$$

where

$$\bar{x}_i = i\bar{L}_x \qquad \bar{y}_i = i\bar{L}_y \quad (22\text{-}110)$$

$$\sigma_{x_i}^2 = i\sigma_{L_x}^2 \qquad \sigma_{y_i}^2 = i\sigma_{L_y}^2 \quad (22\text{-}111)$$

$$\mu_{x_iy_i} = i\mu_{L_xL_y} \quad (22\text{-}112)$$

$$\rho_{x_iy_i} = \rho_{L_xL_y} = \frac{\mu_{L_xL_y}}{\sigma_{L_x}\sigma_{L_y}} \quad (22\text{-}113)$$

If the reader is familiar with Mohr's circle construction for transformation of stress, strain, or moments of inertia of an area, he can easily see by a one-to-one comparison of the controlling transformation equations with those presented above for transformation of joint probability that the Mohr's circle for joint probability is constructed as shown in Fig. 22-19. From this circle, the principal variances are seen to be given by the relations

$$\sigma_{u_i}^2 = \frac{\sigma_{x_i}^2 + \sigma_{y_i}^2}{2} + \sqrt{\left(\frac{\sigma_{x_i}^2 - \sigma_{y_i}^2}{2}\right)^2 + \mu_{x_iy_i}^2} \quad (22\text{-}114)$$

$$\sigma_{v_i}^2 = \frac{\sigma_{x_i}^2 + \sigma_{y_i}^2}{2} - \sqrt{\left(\frac{\sigma_{x_i}^2 - \sigma_{y_i}^2}{2}\right)^2 + \mu_{x_iy_i}^2} \quad (22\text{-}115)$$

and the covariance is

$$\mu_{u_iv_i} = 0 \quad (22\text{-}116)$$

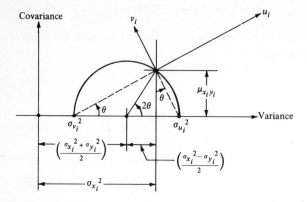

FIGURE 22-19
Mohr's circle for locating principal axes of joint probability.

EXAMPLE E22-9 Again consider random variables x and y as defined in Examples E22-6 and E22-8. Defining two new random variables u and v through the linear transformations

$$u = y \sin \theta + x \cos \theta \qquad (a)$$

$$v = y \cos \theta - x \sin \theta$$

find the angle θ which will uncouple u and v statistically and for this particular angle find the variance of u and v.

From Example E22-8 the variances of x and y are 4 and 9, respectively, and the covariance of these same variables is 3. These numerical values can be used to construct the Mohr's circle for locating principal axes of joint probability, as shown in Fig. 22-19. The resulting circle is shown in Fig. E22-3. From this circle it is readily seen that

$$\theta = {}^1\!/_2 \tan^{-1} \frac{2\mu_{xy}}{\sigma_x^{\,2} - \sigma_y^{\,2}} = 64°54' \qquad (b)$$

and that

$$\sigma_v^{\,2} = 2.60 \qquad \sigma_u^{\,2} = 10.40 \qquad (c)$$

////

Note from the above treatment of the general two-dimensional random walk that the joint probability density function $p(x_i, y_i)$ approaches a gaussian distribution as $i \to \infty$. Further, note that, in the limit, the only characteristics of the joint probability density function $p(L_x, L_y)$ which influence $p(x_i, y_i)$ are $\bar{L}_x$, $\bar{L}_y$, $\sigma_{L_x}^{\,2}$, $\sigma_{L_y}^{\,2}$, and $\mu_{L_x L_y}$. These observations, which are comparable to those noted previously for the one-dimensional random walk, are contained in the so-called central-limit theorem.

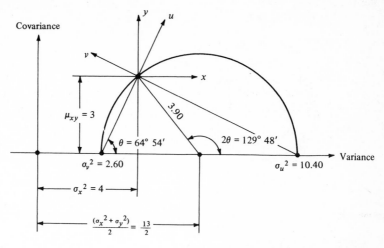

FIGURE E22-3
Mohr's circle for joint probability.

Fortunately the joint probability density function $p(x_i, y_i)$ approaches a gaussian distribution rapidly as i increases; therefore, the often assumed gaussian distribution used in engineering practice is usually justified.

With the completion of the development of the general two-dimensional random walk, one special case most often used in engineering studies will be considered briefly, namely, one for which

$$\bar{L}_x = \bar{L}_y = 0 \qquad \sigma_{L_x}^2 = \sigma_{L_y}^2 = {}^1/_2\sigma_L^2 \qquad \mu_{L_x L_y} = 0 \qquad (22\text{-}117)$$

When Eqs. (22-110) to (22-113) are used, the joint probability density function for random variables x_i and y_i reduces to the simplified form

$$p(x_i, y_i) = \frac{1}{\pi i \sigma_L^2} \exp\left(-\frac{x_i^2 + y_i^2}{i \sigma_L^2}\right) \qquad (22\text{-}118)$$

Because of complete symmetry of Eq. (22-118) about the origin, as shown by the equal-probability contours in Fig. 22-20, the probability density function for distance from the origin after i steps (regardless of direction) is of interest and is easily found, since

$$p(r_i)\, dr_i = 2\pi r_i p(x_i, y_i)\, dr_i \qquad (22\text{-}119)$$

Substituting Eq. (22-118) into Eq. (22-119) and making use of the relation $x_i^2 + y_i^2 = r_i^2$ gives

$$p(r_i) = \frac{r_i}{a^2} \exp\left(-\frac{r_i^2}{2a^2}\right) \qquad 0 \leq r_i < \infty \qquad (22\text{-}120)$$

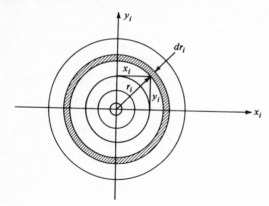

FIGURE 22-20
Equal probability contours as defined by
Eq. (22-116).

where a^2 is defined by

$$a^2 = \frac{i\sigma_L{}^2}{2} \quad (22\text{-}121)$$

Equation (22-120), known as the *Rayleigh distribution*, is plotted in Fig. 22-21.

EXAMPLE E22-10 Consider a random variable x having the Rayleigh distribution

$$p(x) = \frac{x}{a^2} \exp\left(-\frac{x^2}{2a^2}\right) \quad x \geq 0 \quad (a)$$

Find its (1) most probable value, (2) mean value, (3) mean square value, (4) variance, and (5) probability distribution function $P(X)$. What is the probability that x will exceed $1\sigma_x$, $2\sigma_x$, $3\sigma_x$, $4\sigma_x$, and $5\sigma_x$?

The most probable value of x is that value which maximizes Eq. (a); therefore it is found by differentiating Eq. (a) with respect to x, setting the resulting equation equal to zero, and solving for x as follows:

$$\frac{dp(x)}{dx} = -\frac{1}{a^4}(x^2 - a^2) \exp\left(-\frac{x^2}{2a^2}\right) = 0 \quad x = a \quad (b)$$

The mean value is found using Eq. (22-9), which gives

$$\bar{x} = \int_0^\infty \frac{x^2}{a^2} \exp\left(-\frac{x^2}{2a^2}\right) dx$$

Substituting the change of variable $u = x/a$ changes the equation to the form

$$\bar{x} = a \int_0^\infty u^2 \exp\left(-\frac{u^2}{2}\right) du$$

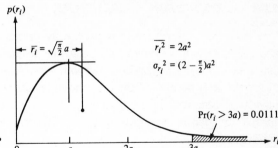

FIGURE 22-21
Rayleigh probability density function,
Eq. (22-120).

which can be integrated to give

$$\bar{x} = a\sqrt{\frac{\pi}{2}} \qquad (c)$$

With Eq. (22-10), the mean square value is expressed as

$$\overline{x^2} = \int_0^\infty \frac{x^3}{a^2} \exp\left(-\frac{x^2}{2a^2}\right) dx$$

Again using the above change of variable and integrating by parts gives

$$\overline{x^2} = 2a^2 \qquad (d)$$

The variance is given by Eq. (22-11) as

$$\sigma_x^2 = \overline{x^2} - \bar{x}^2 = \left(2 - \frac{\pi}{2}\right)a^2 = 0.429a^2 \qquad (e)$$

From Eq. (22-5) the probability distribution function is found to be of the form

$$P(X) = 1 - \exp\left(-\frac{X^2}{2a^2}\right) \qquad (f)$$

If $Q(X)$ is the probability that x will exceed the value X, then

$$Q(X) = 1 - P(X) = \exp\left(-\frac{X^2}{2a^2}\right) \qquad (g)$$

This equation then gives

$$Q(1\sigma_x) = 0.808 \qquad Q(2\sigma_x) = 0.424 \qquad Q(3\sigma_x) = 0.145$$

$$Q(4\sigma_x) = 0.0324 \qquad Q(5\sigma_x) = 0.0049 \qquad (h)$$
$$////$$

22-8 *m* RANDOM VARIABLES

The basic probability concepts previously presented for experiments involving one and two random variables can easily be extended to m random variables. For example, suppose a single set of random variables $x_1, x_2, \ldots, x_m$ is obtained by spinning separately m disks of the type shown in either Fig. 22-1a or b. When n such sets have been obtained, in the limit as $n \to \infty$ it is possible to establish a multivariate probability density function $p(x_1, x_2, \ldots, x_m)$ which is defined to satisfy the condition

$$p(X_1, X_2, \ldots, X_m)\, dx_1\, dx_2 \cdots dx_m = \Pr\,(X_1 < x_1 < X_1 + dx_1,$$
$$X_2 < x_2 < X_2 + dx_2, \ldots, X_m < x_m < X_m + dx_m) \quad (22\text{-}122)$$

This probability density function will be of discrete form with disks of the type shown in Fig. 22-1a and of continuous form with disks of the type shown in Fig. 22-1b.

For a general experiment involving m random variables, the probability density function defined by Eq. (22-122) may be discrete, continuous, or some combination of these forms. However, in engineering, the normal, or gaussian, distribution

$$p(x_1, x_2, \ldots, x_m) = \frac{1}{(2\pi)^{m/2}|\boldsymbol{\mu}|^{1/2}} \exp\,\{-\tfrac{1}{2}[\mathbf{x} - \bar{\mathbf{x}}]^T \boldsymbol{\mu}^{-1}[\mathbf{x} - \bar{\mathbf{x}}]\} \quad (22\text{-}123)$$

is usually assumed, where $\mathbf{x}$ and $\bar{\mathbf{x}}$ are the vectors

$$\mathbf{x} = \begin{bmatrix} x_1 \\ x_2 \\ \vdots \\ x_m \end{bmatrix} \qquad \bar{\mathbf{x}} = \begin{bmatrix} \bar{x}_1 \\ \bar{x}_2 \\ \vdots \\ \bar{x}_m \end{bmatrix} \quad (22\text{-}124)$$

and where matrix $\boldsymbol{\mu}$ is the covariance matrix

$$\boldsymbol{\mu} \equiv \begin{bmatrix} \mu_{11} & \mu_{12} & \cdots & \mu_{1m} \\ \mu_{21} & \mu_{22} & \cdots & \mu_{2m} \\ \cdots\cdots\cdots\cdots\cdots\cdots \\ \mu_{m1} & \mu_{m2} & \cdots & \mu_{mm} \end{bmatrix} \quad (22\text{-}125)$$

which is composed of the individual covariance terms

$$\mu_{ij} \equiv \overline{(x_i - \bar{x}_i)(x_j - \bar{x}_j)} \qquad i,j = 1, 2, \ldots, m \quad (22\text{-}126)$$

The statistical dependence of random variables x_i and x_j upon each other is given by the correlation coefficient

$$\rho_{ij} \equiv \frac{\mu_{ij}}{\sqrt{\mu_{ii}\mu_{jj}}} = \frac{\mu_{ij}}{\sigma_{ii}\sigma_{jj}} \quad (22\text{-}127)$$

If all m random variables are statistically independent, the covariance matrix Eq. (22-125) will be a diagonal matrix and all m random variables will appear in an uncoupled form in Eq. (22-123). The reader can easily verify that Eq. (22-123) reduces to the form of Eq. (22-63) when $m = 2$.

The multivariate probability density function for a new set of random variables $y_1, y_2, \ldots, y_m$ as defined by the general relations

$$
\begin{aligned}
y_1 &= y_1(x_1, x_2, \ldots, x_m) \\
y_2 &= y_2(x_1, x_2, \ldots, x_m) \\
&\cdots\cdots\cdots\cdots\cdots\cdots \\
y_m &= y_m(x_1, x_2, \ldots, x_m)
\end{aligned}
\qquad (22\text{-}128)
$$

can be obtained using the jacobian transformation

$$
p(y_1, y_2, \ldots, y_m) = \begin{vmatrix}
\dfrac{\partial x_1}{\partial y_1} & \dfrac{\partial x_2}{\partial y_1} & \cdots & \dfrac{\partial x_m}{\partial y_1} \\[2ex]
\dfrac{\partial x_1}{\partial y_2} & \dfrac{\partial x_2}{\partial y_2} & \cdots & \dfrac{\partial x_m}{\partial y_2} \\[1ex]
\cdots & \cdots & \cdots & \cdots \\[1ex]
\dfrac{\partial x_1}{\partial y_m} & \dfrac{\partial x_2}{\partial y_m} & \cdots & \dfrac{\partial x_m}{\partial y_m}
\end{vmatrix} p(x_1, x_2, \ldots, x_m) \qquad (22\text{-}129)
$$

provided that Eqs. (22-128) and their inverse relations

$$
\begin{aligned}
x_1 &= x_1(y_1, y_2, \ldots, y_m) \\
x_2 &= x_2(y_1, y_2, \ldots, y_m) \\
&\cdots\cdots\cdots\cdots\cdots\cdots \\
x_m &= x_m(y_1, y_2, \ldots, y_m)
\end{aligned}
\qquad (22\text{-}130)
$$

are all single-valued functions. This procedure is a straightforward extension of the two-dimensional case.

The statistical average of a new random variable $r = r(x_1, x_2, \ldots, x_m)$ can be obtained using the relation

$$
\bar{r} = \int\!\!\int_{-\infty}^{\infty}\!\!\cdots\int_{-\infty}^{\infty} r(x_1, x_2, \ldots, x_m)\, p(x_1, x_2, \ldots, x_m)\, dx_1\, dx_2 \cdots dx_m \qquad (22\text{-}131)
$$

which is simply a generalization of Eq. (22-54).

Finally, suppose the multivariate probability density function $p(x_{1n}, x_{2n}, \ldots, x_{mn})$ is to be found, where the set of random variables $x_{1n}, x_{2n}, \ldots, x_{mn}$ is defined in terms of a second set of random variables $L_1, L_2, \ldots, L_m$ by the relation

$$
x_{in} = \sum_{k=1}^{n} (L_i)_k \qquad i = 1, 2, \ldots, m \qquad (22\text{-}132)
$$

and where $(L_i)_k$ is the kth sampled value of random variable L_i, which is assumed to be statistically independent with respect to k. Since this problem is an extension of the one- and two-dimensional random walks, it is immediately apparent that the probability density function $p(x_{1n}, x_{2n}, \ldots, x_{mn})$ approaches the gaussian distribution given by Eq. (22-123) with increasing values of n regardless of the form of the probability density function $p(L_1, L_2, \ldots, L_m)$ and that

$$\bar{x}_{in} = n\bar{L}_i$$

$$\sigma_{x_{in}}^2 = n\sigma_{L_i}^2 \qquad i,k = 1, 2, \ldots, m \qquad (22\text{-}133)$$

$$\mu_{x_{in}x_{kn}} = n\mu_{L_iL_k}$$

This statement is contained in the central-limit theorem.

22-9   LINEAR TRANSFORMATIONS OF NORMALLY DISTRIBUTED RANDOM VARIABLES

If random variables $y_1, y_2, \ldots, y_m$ are defined in terms of random variables $x_1, x_2, \ldots, x_m$ by the linear relations

$$y_1 \equiv a_{11}x_1 + a_{12}x_2 + \cdots + a_{1m}x_m$$
$$y_2 \equiv a_{21}x_1 + a_{22}x_2 + \cdots + a_{2m}x_m$$
$$\cdots\cdots\cdots\cdots\cdots\cdots\cdots\cdots\cdots\cdots\cdots\cdots\cdots \qquad (22\text{-}134)$$
$$y_m \equiv a_{m1}x_1 + a_{m2}x_2 + \cdots + a_{mm}x_m$$

variables $y_1, y_2, \ldots, y_m$ will always have a gaussian distribution when variables $x_1, x_2, \cdots, x_m$ are normally distributed. To prove this very important characteristic of linear transformations, substitute the matrix form of Eq. (22-134), namely,

$$\mathbf{y} = \mathbf{a}\mathbf{x} \qquad (22\text{-}135)$$

into the right-hand side of Eq. (22-123) and apply the jacobian transformation given by Eq. (22-129) to obtain

$$p(y_1, y_2, \ldots, y_m) = \frac{|\mathbf{a}^{-1}|}{(2\pi)^{m/2}|\boldsymbol{\mu}|^{1/2}} \exp\left\{-\tfrac{1}{2}[\mathbf{y} - \bar{\mathbf{y}}]^T[\mathbf{a}^T]^{-1}\boldsymbol{\mu}^{-1}\mathbf{a}^{-1}[\mathbf{y} - \bar{\mathbf{y}}]\right\} \qquad (22\text{-}136)$$

or

$$p(y_1, y_2, \ldots, y_m) = \frac{1}{(2\pi)^{m/2}|\mathbf{a}\boldsymbol{\mu}\mathbf{a}^T|^{1/2}} \exp\left\{-\tfrac{1}{2}[\mathbf{y} - \bar{\mathbf{y}}]^T[\mathbf{a}\boldsymbol{\mu}\mathbf{a}^T]^{-1}[\mathbf{y} - \bar{\mathbf{y}}]\right\} \qquad (22\text{-}137)$$

Evaluating the individual covariance terms

$$v_{ij} \equiv \overline{(y_i - \bar{y}_i)(y_j - \bar{y}_j)} \qquad i,j = 1, 2, \ldots, m \qquad (22\text{-}138)$$

directly from Eqs. (22-134) shows the covariance matrix for random variables $y_1, y_2, \ldots, y_m$ to be

$$v = a\mu a^T \quad (22\text{-}139)$$

When Eq. (22-139) is substituted into Eq. (22-137), the desired probability density function becomes

$$p(y_1, y_2, \ldots, y_m) = \frac{1}{(2\pi)^{m/2}|v|^{1/2}} \exp\{-\tfrac{1}{2}[\mathbf{y} - \overline{\mathbf{y}}]^T v^{-1}[\mathbf{y} - \overline{\mathbf{y}}]\} \quad (22\text{-}140)$$

which is obviously gaussian when compared with Eq. (22-124).

PROBLEMS

22-1 A random variable x has the probability density function

$$p(x) = \begin{cases} 1 - |x| & 0 \le |x| \le 1 \\ 0 & |x| \ge 1 \end{cases}$$

If a new random variable y is defined by the relation $y = ax^2$, find and plot the probability density function $p(y)$.

22-2 The probability density function for random variable x has the exponential form

$$p(x) = a \exp(-b|x|)$$

where a and b are constants. Determine the required relation between constants a and b and, for $a = 1$, find the probability distribution function $P(X)$.

22-3 Consider the one-dimensional random walk when the probability density function for a single step length is

$$p(L) = 0.6\delta(L - \Delta L) + 0.4\delta(L + \Delta L)$$

Find the probability density function for random variable x_4 as defined by

$$x_4 = \sum_{j=1}^{4} L_j$$

which represents distance from the origin after four steps.

22-4 Consider the one-dimensional random walk when the probability density function for a single step length is

$$p(L) = 0.1\delta(L + \Delta L) + 0.3\delta(L) + 0.5\delta(L - \Delta L) + 0.1\delta(L - 2\Delta L)$$

Approximately what is the probability of being at location $6\Delta L$ after 10 steps?

22-5 Let x and y represent two stochastically independent random variables and define a third random variable z as the product of x and y; that is, $z = xy$. Derive an expression for the probability density function $p(z)$ in terms of the probability density functions $p(x)$ and $p(y)$.

22-6   Two statistically independent random variables x and y have identical probability density functions:

$$p(x) = \begin{cases} {}^1/_4 & -1 < x < 1 \\ 0 & x < -1; x > 1 \end{cases} \qquad p(y) = \begin{cases} {}^1/_4 & -1 < y < 1 \\ 0 & y < -1; y > 1 \end{cases}$$

What is the probability density function for random variable z in the range $0 < z < 1$ when z is defined by the relation $z = yx^{-2}$?

22-7   The joint probability density function for two random variables x and y is

$$p(x,y) = \begin{cases} \dfrac{y}{\pi\sqrt{1 - x^2}} \exp\left(-\dfrac{y^2}{2}\right) & y \geq 0; |x| < 1 \\ 0 & \text{otherwise} \end{cases}$$

What are the expressions for the marginal probability density function $p(y)$ and the conditional probability density function $p(x \mid y)$, and what is the mean value of x? Are random variables x and y statistically independent?

22-8   Prove the validity of Eqs. (22-62).

22-9   The probability density function for the random variables x and y is

$$p(x,y) = \begin{cases} a \exp(-x - y) & x > 0; y > 0 \\ 0 & x < 0; y < 0 \end{cases}$$

Find the numerical value of a so that this function is properly normalized. What is the probability that x will be in the range $0 < x < 1$ when $y = 1$? Are random variables x and y statistically independent? What is the probability that x and y will fall outside the square $OABC$ of unit area as shown in Fig. P22-1? Find the probability distribution function $P(X, Y)$.

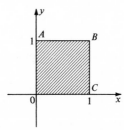

FIGURE P22-1
Region $OABC$ in the xy plane of Prob. 9.

22-10   Random variables x and y are statistically independent and can be sampled in accordance with the marginal probability density functions

$$p(x) = \begin{cases} 2(1 - x) & 0 < x < 1 \\ 0 & x < 0; x > 1 \end{cases} \qquad p(y) = \begin{cases} 2(1 - y) & 0 < y < 1 \\ 0 & y < 0; y > 1 \end{cases}$$

Sketch the joint probability density function $p(x,y)$ and find mean values $\bar{x}$ and $\bar{y}$, mean square values $\overline{x^2}$ and $\overline{y^2}$, covariance μ_{xy}, and the mean value $\overline{x + y}$.

22-11 The joint probability density function for two random variables x and y equals a constant C over the region shown in Fig. P22-2 and equals zero outside that region.

(*a*) Find the numerical value of C so that $p(x,y)$ is properly normalized.

(*b*) Plot the marginal probability density functions $p(x)$ and $p(y)$.

(*c*) Plot the conditional probability density functions $p(x \mid y = 0.5)$ and $p(y \mid x = 1.5)$.

(*d*) Are random variables x and y statistically independent?

(*e*) Find mean values $\bar{x}$ and $\bar{y}$, variances $\sigma_x{}^2$ and $\sigma_y{}^2$, and the covariance μ_{xy}.

(*f*) Consider sampling values of x and y, say $x_1, x_2, x_3, \ldots$ and $y_1, y_2, y_3, \ldots$, respectively. If two new random variables r and s are defined as

$$r_n = x_1 + x_2 + x_3 + \cdots + x_n$$

$$s_n = y_1 + y_2 + y_3 + \cdots + y_n$$

find an appropriate expression for the joint probability density function $p(r_n, s_n)$ when $n = 20$.

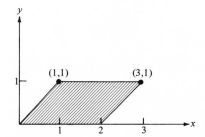

FIGURE P22-2
Region of nonzero joint probability in the xy plane of Prob. 11.

22-12 Consider again random variables x and y as defined in Prob. 22-11. Defining two new random variables u and v through the transformation

$$u = (y - A) \sin \theta + (x - B) \cos \theta$$

$$v = (y - A) \cos \theta + (x - B) \sin \theta$$

find the values of A and B which will give zero-mean values for u and v and find the angle θ which will uncouple u and v statistically. For this particular angle find the variances of u and v.

23
RANDOM PROCESSES

23-1 DEFINITION

A random process is a family, or ensemble, of n dependent random variables related to a similar phenomenon which may be functions of one or more independent variables. For example, suppose n accelerometers are mounted on the frames of n automobiles for the purpose of measuring vertical accelerations as these automobiles travel over a rough country road. The recorded accelerometer signals $x_i(t)$ $(i = 1, 2, \ldots, n)$, which are functions of one independent variable, namely, time t, might look something like the waveforms shown in Fig. 23-1. Each waveform in such a process differs from all other waveforms; that is, $x_r(t) \neq x_s(t)$ for $r \neq s$. To characterize this process completely in a probabilistic sense, it is necessary to establish the multivariate probability density function $p(x_1, x_2, \ldots, x_m)$ as defined by the relation

$$p(X_1, X_2, \ldots, X_m) \, dx_1 \, dx_2 \cdots dx_m$$
$$= \Pr \, (X_1 < x_1 < X_1 + dx_1, X_2 < x_2 < X_2 + dx_2, \ldots,$$
$$X_m < x_m < X_m + dx_m) \qquad (23\text{-}1)$$

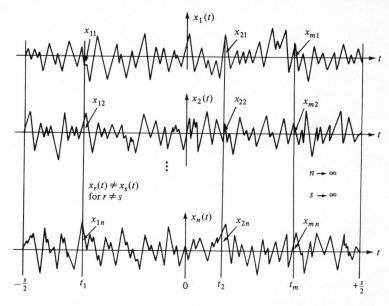

FIGURE 23-1
Random process (one independent variable).

for $m = 1, 2, \ldots$, where $x_i = x(t_i)$ and t_i is a discrete value of time. Usually in engineering fields, it is sufficient to establish only the first two of these functions, that is, $p(x_1)$ and $p(x_1, x_2)$.

The number of members n in the ensemble required to characterize a random process depends upon the type of process and the accuracy desired. Should it be necessary to establish the probability density functions statistically by sampling values of the random variables across the ensemble, exact results are obtained only in the limit as n approaches infinity. In practice, however, sufficient accuracy can be obtained using a finite number of members.

For some random processes, the desired probability density functions can be determined from an analysis of a single member of each process, in which case their exact characterizations are obtained only in the limit as the durations s of such processes approach infinity. In practice these processes are always limited in duration; therefore, the characterizations obtained can only be approximate; however, engineering accuracy can usually be obtained with relatively short-duration sample waveforms.

In the above example, time t happens to be the independent variable, but it should be recognized that in general the independent variable can be any quantity.

As a second example of a random process, consider the wind drag force per unit height $p(x,t)$ acting on a tall industrial smokestack during a strong windstorm.

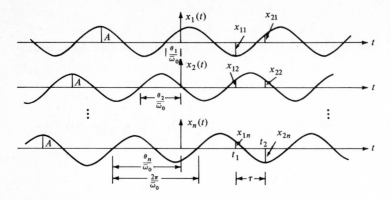

FIGURE 23-2
Random process of harmonic waveform.

This forcing function will contain a large steady-state or static component but will in addition contain a significant random component due to air turbulence. Clearly such turbulence produces drag forces which are not only random with respect to time t but are random with respect to the vertical space coordinate x as well. This process therefore involves two independent variables.

The pressure fluctuations over the surface of an aircraft during flight are an example of a random process involving three independent variables, namely, time and two surface coordinates.

Obviously, the larger the number of independent variables involved in a random process the more difficult it is to characterize the process.

23-2   STATIONARY AND ERGODIC PROCESSES

A specific random process will now be described in detail to help the reader develop a better understanding of random processes involving one independent variable. Consider the random process $x(t)$ shown in Fig. 23-2, which is defined by the relation

$$x_r(t) = A \sin (\overline{\omega}_0 t + \theta_r) \qquad r = 1, 2, \ldots, \infty \qquad (23-2)$$

where $x_r(t) = r$th member of the ensemble

$\quad\quad A$ = fixed amplitude for each harmonic waveform

$\quad\quad \overline{\omega}_0$ = fixed circular frequency

$\quad\quad \theta_r$ = rth sampled value of a random phase angle θ with uniform probability density function in range $0 < \theta < 2\pi$ of intensity $1/2\pi$

This process shows that waveforms need not be irregular, that is, contain many frequency components, to be classified as random. Harmonic, periodic, or aperiodic waveforms may or may not be random, depending upon whether they are fully prescribed or not. If known in a probabilistic sense only, they are defined as random. From this definition it is clear that once a random signal has been sampled, that particular waveform immediately becomes fully known and can no longer by itself be considered random; however, it still is considered part of the random process from which it was sampled. By statistically studying a sufficient number of sampled waveforms, the probability density functions for the process can be estimated, in which case any unsampled waveform becomes known in a probabilistic sense.

To establish the probability density function for random variable $x_1 \equiv x(t_1)$, a transformation relation similar to that given by Eq. (22-4) is used, namely,

$$p(x_1) = 2p(\theta) \left| \frac{d\theta}{dx_1} \right| \qquad (23\text{-}3)$$

This equation differs slightly from Eq. (22-4) since the latter is valid only when $x_1 = x_1(\theta)$ and $\theta = \theta(x_1)$, its inverse relation, are single-valued functions. In this example, however, as random variable θ is allowed to change over its full range $0 < \theta < 2\pi$, random variable x_1 changes not once but twice over the range $-A < x_1 < +A$, which explains why the factor of 2 appears in Eq. (23-3). When Eq. (23-2) is substituted into Eq. (23-3) and the known information

$$p(\theta) = \begin{cases} \dfrac{1}{2\pi} & 0 < \theta < 2\pi \\ 0 & \theta < 0; \theta > 2\pi \end{cases} \qquad (23\text{-}4)$$

is used, the probability density function $p(x_1)$ becomes

$$p(x_1) = \begin{cases} \dfrac{1}{\pi |\sqrt{A^2 - x_1{}^2}|} & -A < x_1 < A \\ 0 & x_1 < -A; x_1 > A \end{cases} \qquad (23\text{-}5)$$

Equations (23-4) and (23-5) are plotted in Fig. 23-3.

The joint probability density function $p(x_1,x_2)$, where $x_1 \equiv x(t_1)$ and $x_2 \equiv x(t_2)$, can be obtained for the above process in the following manner. First, by using the appropriate trigonometric identity, x_2 can be expressed in the form

$$x_2 \equiv x(t_2) = x_1 \cos \bar{\omega}_0 \tau \pm |\sqrt{A^2 - x_1{}^2}| \sin \bar{\omega}_0 \tau \qquad -A \leq x_1 \leq A \qquad (23\text{-}6)$$

Clearly this relation shows that for any sampled value of x_1 random variable x_2 has only two possible values with equal chances of occurring. In other words, for a

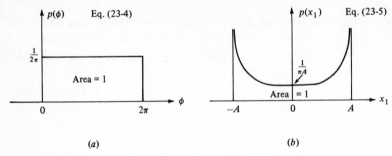

FIGURE 23-3
Probability density functions for θ and x_1, where $x_1 = A \sin(\bar{\omega}_0 t_1 + \theta)$.

given time interval $\tau = t_2 - t_1$, the conditional probability density function $p(x_2 \mid x_1)$ consists of two Dirac delta functions, namely,

$$p(x_2 \mid x_1) = {}^1\!/_2[\delta(x_2 - x_1 \cos \bar{\omega}_0 \tau + |\sqrt{A^2 - x_1{}^2}| \sin \bar{\omega}_0 \tau)$$
$$+ \delta(x_2 - x_1 \cos \bar{\omega}_0 \tau - |\sqrt{A^2 - x_1{}^2}| \sin \bar{\omega}_0 \tau)] \qquad (23\text{-}7)$$

Substituting Eqs. (23-5) and (23-7) into the following form of Eq. (22-46)

$$p(x_1, x_2) = p(x_1)p(x_2 \mid x_1) \qquad (23\text{-}8)$$

leads to

$$p(x_1, x_2)$$
$$= \frac{1}{2\pi|\sqrt{A^2 - x_1{}^2}|} [\delta(x_2 - x_1 \cos \bar{\omega}_0 \tau + |\sqrt{A^2 - x_1{}^2}| \sin \bar{\omega}_0 \tau)$$
$$+ \delta(x_2 - x_1 \cos \bar{\omega}_0 \tau - |\sqrt{A^2 - x_1{}^2}| \sin \bar{\omega}_0 \tau)] \qquad (23\text{-}9)$$

which is valid in the range $-A < x_2 < +A$ and $-A < x_1 < +A$. Outside this range $p(x_1, x_2)$ equals zero.

EXAMPLE E23-1　Consider the single harmonic random process defined by Eq. (23-2), namely,

$$x_r(t) = A \sin(\bar{\omega}_0 t + \theta_r) \qquad r = 1, 2, \ldots, \infty \qquad (a)$$

where A is a fixed amplitude, $\bar{\omega}_0$ is a fixed circular frequency, and θ_r is the rth sampled value of a random phase angle θ having a uniform probability density function over the range $0 < \theta < 2\pi$. Defining the random variables x_1 and x_2 as

$$x_1 \equiv x(t) \qquad x_2 \equiv x(t + \tau) \qquad (b)$$

characterize the form of the scatter diagram for variables x_1 and x_2 and plot the diagram for $\bar{\omega}_0\tau = 0$, $\pi/4$, $\pi/2$, $3\pi/4$, and π.

The form of the scatter diagram can easily be obtained from Eq. (23-9) by noting that sample pairs of random variables x_1 and x_2 must satisfy the condition

$$x_2 - x_1 \cos \bar{\omega}_0\tau = \pm\sqrt{A^2 - x_1{}^2} \sin \bar{\omega}_0\tau \qquad (c)$$

Squaring both sides of Eq. (c) gives

$$x_2{}^2 - 2 \cos \bar{\omega}_0\tau\, x_1 x_2 + x_1{}^2 = A^2 \sin^2 \bar{\omega}_0\tau \qquad (d)$$

This equation represents an ellipse with its major and minor axes at 45° from the x_1 and x_2 axes. To determine the dimensions of the major and minor axes, transform Eq. (d) to a new set of orthogonal axes u and v located on the principal axes of the ellipse; that is, use the linear transformation

$$u = \frac{1}{\sqrt{2}}(x_1 + x_2) \qquad v = \frac{1}{\sqrt{2}}(x_2 - x_1) \qquad (e)$$

to obtain

$$\frac{u^2}{a^2} + \frac{v^2}{b^2} = 1 \qquad (f)$$

where

$$a^2 = \frac{\sin^2 \bar{\omega}_0\tau}{1 - \cos \bar{\omega}_0\tau} A^2 \qquad b^2 = \frac{\sin^2 \bar{\omega}_0\tau}{1 + \cos \bar{\omega}_0\tau} A^2 \qquad (g)$$

Thus it is shown that the scatter diagram is in the form of an ellipse with its principal axes at 45° from the x_1 and x_2 axes and with the dimensions of its principal axes

$$2a = \frac{2 \sin \bar{\omega}_0\tau}{\sqrt{1 - \cos \bar{\omega}_0\tau}} A \qquad 2b = \frac{2 \sin \bar{\omega}_0\tau}{\sqrt{1 + \cos \bar{\omega}_0\tau}} A \qquad (h)$$

as shown in Fig. E23-1. Substituting the values 0, $\pi/4$, $\pi/2$, $3\pi/4$, and π, separately, into Eqs. (h) for $\bar{\omega}_0\tau$ gives the corresponding values $\sqrt{2}\,A$, 1.31A, 1.00A, 0.54A, and 0 for a and 0, 0.54A, 1.00A, 1.31A and $\sqrt{2}\,A$ for b. Plots of the scatter diagrams for each of these five cases are shown in Fig. E23-2. Note from the figure that the ellipse degenerates into a straight line for $\bar{\omega}_0\tau = \pi/2$. From the above it is clear that a straight line with positive slope of 1 will occur for $\bar{\omega}_0\tau = 0$, 2π, 4π, 6π, $\ldots$, a straight line with negative slope of 1 will occur for $\bar{\omega}_0\tau = \pi$, 3π, 5π, $\ldots$, a circle will occur for $\bar{\omega}_0\tau = \pi/2$, $3\pi/2$, $5\pi/2$, $\ldots$, and an ellipse will occur for all other values of $\bar{\omega}_0\tau$.   ////

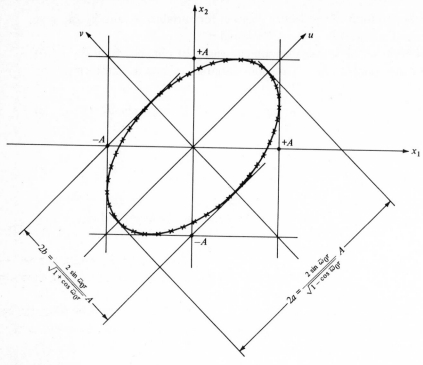

FIGURE E23-1
Scatter diagram for random variables x_1 and x_2 derived from single harmonic process, Eq. (23-2).

Usually of main interest are the mean values, mean square values, variances, the covariance, and the correlation coefficient for random variables x_1 and x_2. Using Eqs. (22-55) to (22-60) and (23-9) gives the following ensemble averages for the process:

Mean values:
$$E(x_1) = E(x_2) = 0$$

Mean square values:
$$E(x_1{}^2) = E(x_2{}^2) = \frac{A^2}{2}$$

Variances:
$$\sigma_{x_1}{}^2 = \sigma_{x_2}{}^2 = \frac{A^2}{2} \qquad (23\text{-}10)$$

Covariance:
$$\mu_{x_1 x_2} = \frac{A^2}{2} \cos \bar{\omega}_0 \tau$$

Correlation coefficient:
$$\rho_{x_1 x_2} = \cos \bar{\omega}_0 \tau$$

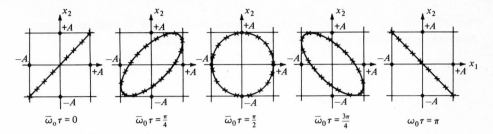

FIGURE E23-2
Scatter diagrams for five cases of the more general diagram in Fig. E23-1.

The letter E has been introduced as a substitute for the bar previously placed above the random variable. It indicates that the variable has been averaged across the ensemble.

It is significant to note that all ensemble averages for this example process are independent of time t. Processes having this characteristic are defined as *stationary processes*.

It is also significant that for this process, any average obtained with respect to time t along any member r of the ensemble is exactly equal to the corresponding average across the ensemble at an arbitrary time t. Mathematically, this statement can be expressed in the form

$$\langle f(x_r) \rangle \equiv \lim_{s \to \infty} \frac{1}{S} \int_{-s/2}^{s/2} f(x_r)\, dt = E[f(x_i)] \qquad (23\text{-}11)$$

where $f(x_r)$ is any function of the variable $x_r(t)$, $x_i = x(t_i)$, and where the angle brackets indicate time average. Processes having this characteristic are defined as *ergodic processes*.

It is suggested that the reader check the results given by Eq. (23-10) using Eq. (23-11) to show that the example process being considered, Eq. (23-2), is indeed ergodic; that is, show

$$\langle x_r \rangle = \lim_{s \to \infty} \frac{1}{S} \int_{-s/2}^{s/2} x_r(t)\, dt = 0$$

$$\langle x_r^2 \rangle = \lim_{s \to \infty} \frac{1}{S} \int_{-s/2}^{s/2} x_r(t)^2\, dt = \frac{A^2}{2}$$

$$\sigma_{x_r}^2 = \frac{A^2}{2} \qquad r = 1, 2, \ldots \qquad (23\text{-}12)$$

$$\mu_{x_r}(\tau) = \frac{A^2}{2} \cos \bar{\omega}_0 \tau$$

$$\rho_{x_r}(\tau) = \cos \bar{\omega}_0 \tau$$

According to the above definitions, an ergodic process must always be stationary; however, a stationary process may or may not be ergodic.

23-3 AUTOCORRELATION FUNCTION FOR STATIONARY PROCESSES

Consider again the general random process $x(t)$ shown in Fig. 23-1, which involves one independent variable. Assume for this discussion that this process is stationary (but not necessarily ergodic) and that it has a zero ensemble mean value, that is, $E(x) = 0$.

The covariance function $E[x(t)x(t + \tau)]$ in this case, like all ensemble averages, will be independent of time t and therefore will be a function of τ only. This function of τ will be referred to subsequently as the *autocorrelation function* and will be expressed in the form

$$R_x(\tau) = E[x(t)x(t + \tau)] \qquad (23\text{-}13)$$

Certain important properties of the autocorrelation function should be noted, namely,

$$R_x(0) = \sigma_x^2 \qquad R_x(\tau) = R_x(-\tau) \qquad |R_x(\tau)| \leq R_x(0) \qquad (23\text{-}14)$$

The first of Eq. (23-14) is obvious since $R_x(0) = E[x(t)x(t)]$ is the variance when $E(x) = 0$. The second equation is a direct result of the assumed stationarity of the process, and the third equation can readily be proved using the fact that the following mean square average must always be greater than or equal to zero:

$$E\{[x(t) \pm x(t + \tau)]^2\} = R_x(0) \pm 2R_x(\tau) + R_x(0) \geq 0 \qquad (23\text{-}15)$$

or

$$|R_x(\tau)| \leq R_x(0) \qquad (23\text{-}16)$$

From the definition given by Eq. (23-13) it is clear that the autocorrelation function $R_x(\tau)$ gives a direct measure of the statistical dependence of random variables $x(t + \tau)$ and $x(t)$ upon each other. For most stationary processes, the autocorrelation function decays rapidly with increasing values of τ, thus showing a similar rapid loss of statistical dependency of the two random variables as they are separated with respect to time. One notable exception, however, is the random process consisting of discrete harmonic waveforms, as shown in Fig. 23-2. This process has the autocorrelation function

$$R_x(\tau) = E(x_1 x_2) = \frac{A^2}{2} \cos \bar{\omega}_0 \tau \qquad (23\text{-}17)$$

Clearly, regardless of the process, the two random variables $x(t)$ and $x(t + \tau)$ approach each other numerically as the time separation τ approaches zero. Therefore, these variables become completely dependent upon each other in the limit as reflected by the correlation coefficient

$$\rho_x(0) = \frac{R_x(0)}{\sigma_x{}^2} = 1 \qquad (23\text{-}18)$$

It is very significant to note that if the general process $x(t)$ being considered is stationary, has a zero mean value $E[x(t)] = 0$, and has the gaussian distribution given by Eq. (22-123), the autocorrelation function $R_x(\tau)$ completely characterizes the process. This fact is evident since all variance and covariance functions given by Eq. (22-126) are directly related to the autocorrelation function as follows:

$$\mu_{ik} = \begin{cases} R_x(0) & i = k \\ R_x(\tau) & i \neq k \end{cases} \qquad \tau = t_k - t_i \qquad (23\text{-}19)$$

For an ergodic process, the ensemble average given by Eq. (23-13) can be obtained by averaging along any single member (x_r) of the ensemble, in which case the autocorrelation function is more easily obtained using the relation

$$R_x(\tau) = \lim_{s \to \infty} \frac{1}{s} \int_{-s/2}^{s/2} x_r(t) x_r(t + \tau) \, dt \qquad (23\text{-}20)$$

It should now be obvious to the reader why a gaussian ergodic process is so easily characterized in a probabilistic sense.

EXAMPLE E23-2   A sample function $x_r(t)$ of random process $x(t)$ is established by assigning statistically independent sampled values of a random variable x to successive ordinates spaced at equal intervals along the time abscissa and by assuming a linear variation of the ordinates over each interval as shown in Fig. E23-3. A complete ensemble of such sample functions ($r = 1, 2, \ldots$) can be obtained in a similar manner.

If the probability density function for x is prescribed arbitrarily, except that its mean value $\bar{x}$ is held equal to zero, and if the ordinate x_{1r} occurs at time $t = \alpha_r$, where α_r is a sampled value of a random variable α uniformly distributed over the range $0 < \alpha < \Delta\varepsilon$, determine the mean value, mean square value, and variance of $x(t)$ and the covariance of $x(t)$ and $x(t + \tau)$. What kind of random process is $x(t)$?

First, consider the above process but with all values of α_r ($r = 1, 2, \ldots$) set equal to zero, thus forcing all ordinates x_{ir} ($i, r = 1, 2, \ldots$) to occur at

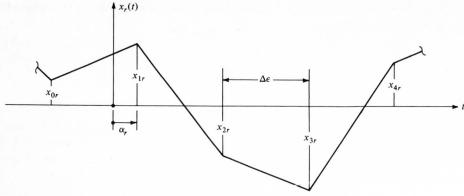

FIGURE E23-3
Sample function $x_r(t)$ from random process $x(t)$.

time $t = (i - 1) \Delta\varepsilon$. The linear variation of ordinates shown in Fig. E23-3 leads to

$$x_r(t) = \left(1 - \frac{t}{\Delta\varepsilon}\right) x_{1r} + \frac{t}{\Delta\varepsilon} x_{2r} \qquad 0 < t < \Delta\varepsilon$$

$$x_r(t + \tau) = \begin{cases} \left(1 - \frac{t + \tau + \Delta\varepsilon}{\Delta\varepsilon}\right) x_{0r} + \frac{t + \tau + \Delta\varepsilon}{\Delta\varepsilon} x_{1r} \\ \qquad\qquad -\Delta\varepsilon < t + \tau < 0 \\ \left(1 - \frac{t + \tau}{\Delta t}\right) x_{1r} + \frac{t + \tau}{\Delta\varepsilon} x_{2r} \qquad 0 < t + \tau < \Delta\varepsilon \\ \left(1 - \frac{t + \tau - \Delta\varepsilon}{\Delta\varepsilon}\right) x_{2r} + \frac{t + \tau - \Delta\varepsilon}{\Delta\varepsilon} x_{3r} \\ \qquad\qquad \Delta\varepsilon < t + \tau < 2\Delta\varepsilon \end{cases} \qquad (a)$$

Taking the ensemble average of the first of Eqs. (a) gives

$$E[x(t)] = \left(1 - \frac{t}{\Delta\varepsilon}\right) E(x_1) + \frac{t}{\Delta\varepsilon} E(x_2)$$

However, when it is noted that

$$E(x_i) = \bar{x} = \int_{-\infty}^{\infty} x p(x)\, dx \qquad i = 1, 2, \ldots \qquad (b)$$

the result is

$$E[x(t)] = \bar{x} = 0 \qquad (c)$$

Squaring the first of Eqs. (a) and taking the ensemble average gives

$$E[x(t)^2] = \left(1 - \frac{t}{\Delta\varepsilon}\right)^2 E(x_1^2) + 2\left(1 - \frac{t}{\Delta\varepsilon}\right)\frac{t}{\Delta\varepsilon} E(x_1 x_2) + \left(\frac{t}{\Delta\varepsilon}\right)^2 E(x_2^2)$$

Making use of the relations

$$E(x_i{}^2) = \overline{x^2} = \int_{-\infty}^{\infty} x^2 p(x)\, dx \qquad i,j = 1, 2, \ldots \tag{d}$$

$$E(x_i x_j) = 0 \quad i \neq j$$

results in

$$E[x(t)^2] = \overline{x^2}\left(1 - \frac{2t}{\Delta\varepsilon} + \frac{2t^2}{\Delta\varepsilon^2}\right) \tag{e}$$

Therefore,

$$\sigma^2_{x(t)} = \overline{x^2}\left(1 - \frac{2t}{\Delta\varepsilon} + \frac{2t^2}{\Delta\varepsilon^2}\right) \tag{f}$$

From Eqs. (a) and (d)

$$E[x(t)x(t+\tau)] = \begin{cases} \left[\left(-\dfrac{1}{\Delta\varepsilon^2}\right)t^2 + \left(-\dfrac{\tau}{\Delta\varepsilon^2}\right)t + \left(\dfrac{\tau}{\Delta\varepsilon} + 1\right)\right]\overline{x^2} \\[1em] \qquad 0 \leq t \leq \Delta\varepsilon \qquad -\Delta\varepsilon \leq t + \tau \leq 0 \\[1em] \left[\dfrac{2}{\Delta\varepsilon^2}t^2 + \left(\dfrac{2\tau}{\Delta\varepsilon^2} - \dfrac{2}{\Delta\varepsilon}\right)t + \left(1 - \dfrac{\tau}{\Delta\varepsilon}\right)\right]\overline{x^2} \\[1em] \qquad 0 \leq t \leq \Delta\varepsilon \qquad 0 \leq t + \tau \leq \Delta\varepsilon \\[1em] \left[\left(-\dfrac{1}{\Delta\varepsilon^2}\right)t^2 + \left(\dfrac{2}{\Delta\varepsilon} - \dfrac{\tau}{\Delta\varepsilon^2}\right)t\right]\overline{x^2} \\[1em] \qquad 0 \leq t \leq \Delta\varepsilon \qquad \Delta\varepsilon \leq t + \tau \leq 2\Delta\varepsilon \end{cases} \tag{g}$$

Note that the covariance of $x(t)$ and $x(t + \tau)$ as given by Eq. (g) is time-dependent; therefore, the random process treated above is nonstationary. Further, note that this covariance equals zero for values of τ outside the ranges indicated for Eqs. (g). The ranges indicated for the first, second, and third of Eqs. (g) are shown by the shaded regions 1, 2, and 3, respectively, in Fig. E23-4. If the origin of time $t = 0$ had been selected coincident with x_{ir} ($r = 1, 2, \ldots$) rather than x_{1r}, as above, Eqs. (a) would obviously be of exactly the same form except that x_{0r}, x_{1r}, x_{2r}, and x_{3r} would be replaced by $x_{i-1,r}$, x_{ir}, $x_{i+1,r}$, and $x_{i+2,r}$, respectively. Thus, the covariance function $E[x(t)x(t + \tau)]$ must be periodic in time with period $\Delta\varepsilon$. This periodic behavior is also indicated in Fig. E23-4 by a repetition of the shaded regions in each interval along the time t axis.

If the probability density function $p(x)$ used in sampling values of x were gaussian in form, then the entire process $x(t)$ would be gaussian, in which case Eqs. (a) would completely characterize the process in a probabilistic sense even though it is nonstationary.

The restriction placed on α_r ($r = 1, 2, \ldots$) above is now removed, and it is sampled for a uniform distribution over the range $0 < \alpha < \Delta\varepsilon$ as originally stated. Since any arbitrary time t will now occur uniformly over the intervals

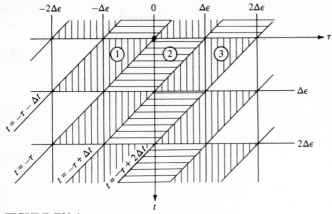

FIGURE E23-4
Regions of nonzero covariance for random variables $x(t)$ and $x(t + \tau)$.

$\Delta\varepsilon$ looking across the ensemble, the process must be stationary and the covariance function $E[x(t)x(t + \tau)]$ is obtained by simply averaging that function as given by Eqs. (g) over time. Since the resulting function is independent of time and depends only upon the time difference τ, it becomes the autocorrelation function $R_x(\tau)$ for the process. Carrying out this averaging procedure gives

$$
R_x(\tau) = \begin{cases}
\dfrac{\overline{x^2}}{\Delta\varepsilon} \displaystyle\int_{-\tau-\Delta\varepsilon}^{\Delta\varepsilon} \left[\left(-\dfrac{1}{\Delta\varepsilon^2}\right)t^2 + \left(-\dfrac{\tau}{\Delta\varepsilon^2}\right)t + \left(\dfrac{\tau}{\Delta\varepsilon} + 1\right)\right]dt \\
\qquad\qquad\qquad\qquad\qquad\qquad\qquad -2\Delta\varepsilon < \tau < -\Delta\varepsilon \\[2mm]
\dfrac{\overline{x^2}}{\Delta\varepsilon}\left\{\displaystyle\int_{0}^{-\tau}\left[\left(-\dfrac{1}{\Delta\varepsilon^2}\right)t^2 + \left(-\dfrac{\tau}{\Delta\varepsilon^2}\right)t + \left(\dfrac{\tau}{\Delta\varepsilon} + 1\right)\right]dt \right. \\
\qquad\left. + \displaystyle\int_{-\tau}^{\Delta\varepsilon}\left[\dfrac{2}{\Delta\varepsilon^2}t^2 + \left(\dfrac{2\tau}{\Delta\varepsilon^2} - \dfrac{2}{\Delta\varepsilon}\right)t + \left(1 - \dfrac{\tau}{\Delta\varepsilon}\right)\right]dt\right\} \\
\qquad\qquad\qquad\qquad\qquad\qquad\qquad -\Delta\varepsilon < \tau < 0 \\[2mm]
\dfrac{\overline{x^2}}{\varepsilon\Delta}\left\{\displaystyle\int_{0}^{-\tau+\Delta\varepsilon}\left[\dfrac{2}{\Delta\varepsilon^2}t^2 + \left(\dfrac{2\tau}{\Delta\varepsilon^2} - \dfrac{2}{\Delta\varepsilon}\right)t + \left(1 - \dfrac{\tau}{\Delta\varepsilon}\right)\right]dt \right. \\
\qquad\left. + \displaystyle\int_{-\tau+\Delta\varepsilon}^{\Delta\varepsilon}\left[\left(-\dfrac{1}{\Delta\varepsilon^2}\right)t^2 + \left(\dfrac{2}{\Delta\varepsilon} - \dfrac{\tau}{\Delta\varepsilon^2}\right)t\right]dt\right\} \\
\qquad\qquad\qquad\qquad\qquad\qquad\qquad 0 < \tau < \Delta\varepsilon \\[2mm]
\dfrac{\overline{x^2}}{\varepsilon\Delta} \displaystyle\int_{0}^{-\tau+2\Delta\varepsilon}\left[\left(-\dfrac{1}{\Delta\varepsilon^2}\right)t^2 + \left(\dfrac{2}{\Delta\varepsilon} - \dfrac{\tau}{\Delta\varepsilon^2}\right)t\right]dt \\
\qquad\qquad\qquad\qquad\qquad\qquad\qquad \Delta\varepsilon < \tau < 2\Delta\varepsilon
\end{cases}
\tag{h}
$$

When the above integrals are completed and terms are collected, the result is

$$R_x(\tau) = \begin{cases} \left(\dfrac{4}{3} + \dfrac{2\tau}{\Delta\varepsilon} + \dfrac{\tau^2}{\Delta\varepsilon^2} + \dfrac{\tau^3}{6\Delta\varepsilon^3}\right)\overline{x^2} & -2\Delta\varepsilon \le \tau \le -\Delta\varepsilon \\[2ex] \left(\dfrac{2}{3} - \dfrac{\tau^2}{\Delta\varepsilon^2} - \dfrac{\tau^3}{2\Delta\varepsilon^3}\right)\overline{x^2} & -\Delta\varepsilon \le \tau \le 0 \\[2ex] \left(\dfrac{2}{3} - \dfrac{\tau^2}{\Delta\varepsilon^2} + \dfrac{\tau^3}{2\Delta\varepsilon^3}\right)\overline{x^2} & 0 \le \tau \le \Delta\varepsilon \\[2ex] \left(\dfrac{4}{3} - \dfrac{2\tau}{\Delta\varepsilon} + \dfrac{\tau^2}{\Delta\varepsilon^2} - \dfrac{\tau^3}{6\Delta\varepsilon^3}\right)\overline{x^2} & \Delta\varepsilon \le \tau \le 2\Delta\varepsilon \end{cases} \qquad (i)$$

Because of the second of Eqs. (d), $R_x(\tau) = 0$ for $\tau \le -2\Delta\varepsilon$ and $\tau \ge 2\Delta\varepsilon$.

If a random variable x has a normal distribution, the process is gaussian, in which case Eqs. (i) completely characterize the process. ////

23-4 POWER SPECTRAL DENSITY FUNCTION FOR STATIONARY PROCESSES

As demonstrated in Chap. 5, any sample waveform $x_r(t)$ taken from a real stationary random process having a zero mean value, that is, $E[x(t)] = 0$, can be separated into its frequency components using a standard Fourier analysis. If this waveform is represented only over the finite interval $-s/2 < t < +s/2$, the Fourier series representation can be used, namely,

$$x_r(t) = \sum_{n=-\infty}^{\infty} C_{nr} \exp(in\bar{\omega}_0 t)$$

$$\tag{23-21}$$

where

$$C_{nr} = \frac{1}{s} \int_{-s/2}^{s/2} x_r(t) \exp(-in\bar{\omega}_0 t)\, dt$$

and where $\bar{\omega}_0 \equiv 2\pi/s$. If $x_r(t)$ is periodic, Eqs. (23-21) give an exact representation of the entire waveform provided the integration interval s is taken as one full period. Such periodic waveforms consist of discrete harmonics having circular frequencies $\bar{\omega}_0, 2\bar{\omega}_0, 3\bar{\omega}_0, \ldots,$ with corresponding finite amplitudes $A_{1r} = 2|C_{1r}|, A_{2r} = 2|C_{2r}|, A_{3r} = 2|C_{3r}|, \ldots,$ provided, of course, corresponding negative and positive frequency components are combined.

Usually the quantity of most interest when analyzing stationary random processes is the mean square value of $x_r(t)$ over the interval $-s/2 < t < +s/2$, which can be obtained by substituting the first of Eqs. (23-21) into the relation

$$\langle x_r(t)^2 \rangle = \frac{1}{s} \int_{-s/2}^{s/2} x_r(t)^2\, dt \qquad (23-22)$$

to obtain

$$\langle x_r(t)^2 \rangle = \sum_{n=-\infty}^{\infty} |C_{nr}|^2 = \sum_{n=1}^{\infty} \frac{A_{nr}^2}{2} \qquad (23\text{-}23)$$

When $\Delta\bar{\omega}$ represents the frequency spacing of the discrete harmonics, that is,

$$\Delta\bar{\omega} = \bar{\omega}_0 = \frac{2\pi}{s} \qquad (23\text{-}24)$$

and the second of Eqs. (23-21) is used, Eq. (23-23) becomes

$$\langle x_r(t)^2 \rangle = \sum_{n=-\infty}^{\infty} \frac{\left| \int_{-s/2}^{s/2} x_r(t) \exp\left(-in\bar{\omega}_0 t\right) dt \right|^2}{2\pi s} \Delta\bar{\omega} \qquad (23\text{-}25)$$

If s is now allowed to approach infinity, $\Delta\bar{\omega} \to d\bar{\omega}$, $n\bar{\omega}_0 \to \bar{\omega}$ and the summation becomes an integral; thus, Eq. (23-25) is converted into the form

$$\langle x_r(t)^2 \rangle = \int_{-\infty}^{\infty} S_{x_r}(\bar{\omega}) \, d\bar{\omega} \qquad (23\text{-}26)$$

where the function

$$S_{x_r}(\bar{\omega}) \equiv \lim_{s \to \infty} \frac{\left| \int_{-s/2}^{s/2} x_r(t) \exp\left(-i\bar{\omega}t\right) dt \right|^2}{2\pi s} \qquad (23\text{-}27)$$

is defined as the *power spectral density function* for waveform $x_r(t)$ provided a limit actually exists. According to this definition, the power-spectral-density function is an even function when $x_r(t)$ is a real function, is positive and finite for all values of $\bar{\omega}$, and yields the mean square value of $x_r(t)$ when integrated over the entire range $-\infty < \bar{\omega} < +\infty$.

The power spectral density function for the entire stationary process $x(t)$ is obtained by simply averaging the power spectral density functions for individual members across the ensemble as follows:

$$S_x(\bar{\omega}) = \lim_{n \to \infty} \frac{1}{n} \sum_{r=1}^{n} S_{x_r}(\bar{\omega}) \qquad (23\text{-}28)$$

The ensemble average of the mean square value of $x(t)$ can now be obtained by integrating $S_x(\bar{\omega})$ over the entire range $-\infty < \bar{\omega} < +\infty$.

If the random process is ergodic, each member of the ensemble will yield the same power spectral density function, in which case it is unnecessary to average across the ensemble. It is sufficient simply to generate the power spectral density function using one member. For most ergodic processes encountered in engineering, the power spectral density function given by Eq. (23-27) approaches its limit rapidly with increasing values of s, so that sufficient accuracy can usually be obtained with a relatively short sample of the waveform.

23-5 RELATIONSHIP BETWEEN POWER SPECTRAL DENSITY AND AUTOCORRELATION FUNCTIONS

Let a function $F_{x_r}(\bar{\omega})$ be defined as the Fourier transform of the time average $\langle x_r(t)x_r(t + \tau)\rangle$; that is, let

$$F_{x_r}(\bar{\omega}) \equiv \int_{-\infty}^{\infty} \left[\lim_{s\to\infty} \frac{1}{s} \int_{-s/2}^{s/2} x_r(t)x_r(t + \tau)\, dt \right] \exp\left(-i\bar{\omega}\tau\right)\, d\tau \qquad (23\text{-}29)$$

Assuming that the function $F_{x_r}(\bar{\omega})$ does indeed exist, Fourier transform theory requires that the quantity in square brackets in Eq. (23-29), which is a function of τ only, decay with increasing values of $|\tau|$ so that the integral

$$I \equiv \int_{-\infty}^{\infty} \left| \lim_{s\to\infty} \frac{1}{s} \int_{-s/2}^{s/2} x_r(t)x_r(t + \tau)\, dt \right| d\tau \qquad (23\text{-}30)$$

exists. When Eq. (23-29) is expressed in its equivalent form

$$\frac{1}{2\pi} F_{x_r}(\bar{\omega}) = \lim_{s\to\infty} \frac{1}{2\pi s} \iint_{-s/2}^{s/2} x_r(t)x_r(t + \tau) \exp\left(-i\bar{\omega}\tau\right)\, d\tau\, dt \qquad (23\text{-}31)$$

and a change of variable as defined by

$$\theta \equiv t + \tau \qquad (23\text{-}32)$$

is substituted, Eq. (23-31) becomes

$$\frac{1}{2\pi} F_{x_r}(\bar{\omega}) = \lim_{s\to\infty} \frac{1}{2\pi s} \int_{-s/2}^{s/2} x_r(t) \exp\left(i\bar{\omega}t\right)\, dt \int_{t-s/2}^{t+s/2} x_r(\theta) \exp\left(-i\bar{\omega}\theta\right)\, d\theta \qquad (23\text{-}33)$$

The expanding domain of integration given by Eq. (23-33) is shown in Fig. 23-4a. Since the function $F_{x_r}(\bar{\omega})$ can exist only when the total integrand of this equation decays rapidly with increasing values of $|\tau|$, it is valid to change the limits of the second integral as shown by the relation

$$\frac{1}{2\pi} F_{x_r}(\bar{\omega}) = \lim_{s\to\infty} \frac{1}{2\pi s} \int_{-s/2}^{s/2} x_r(t) \exp\left(i\bar{\omega}t\right)\, dt \int_{-s/2}^{s/2} x_r(\theta) \exp\left(-i\bar{\omega}\theta\right)\, d\theta \qquad (23\text{-}34)$$

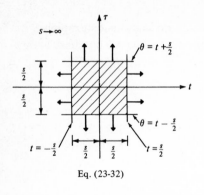

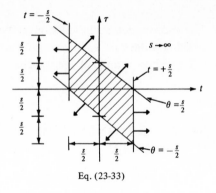

Eq. (23-32)   Eq. (23-33)

(a)   (b)

FIGURE 23-4
Expanding domains of integration.

which simply changes the expanding domain of integration to that shown in Fig. 23-4b. At this point θ can be changed to t since it is serving only as a dummy time variable. Equation (23-34) then can be expressed in the form

$$\frac{1}{2\pi} F_{x_r}(\bar{\omega}) = \lim_{s \to \infty} \frac{\left| \int_{-s/2}^{s/2} x_r(t) \exp(-i\bar{\omega}t)\, dt \right|^2}{2\pi s} \qquad (23\text{-}35)$$

When Eq. (23-35) is compared with Eq. (23-27), it is clear that

$$\frac{1}{2\pi} F_{x_r}(\bar{\omega}) = S_{x_r}(\bar{\omega}) \qquad (23\text{-}36)$$

If the stationary process being considered is ergodic, $F_{x_r}(\bar{\omega})$ is simply the Fourier transform of the autocorrelation function $R_x(\tau)$, and $S_{x_r}(\bar{\omega})$ equals the power spectral density for the process $S_x(\bar{\omega})$. Thus, it has been shown that for an ergodic process, the autocorrelation and power spectral density functions for the process are related through the Fourier integrals given by

$$S_x(\bar{\omega}) = \frac{1}{2\pi} \int_{-\infty}^{\infty} R_x(\tau) \exp(-i\bar{\omega}\tau)\, d\tau$$

$$R_x(\tau) = \int_{-\infty}^{\infty} S_x(\bar{\omega}) \exp(i\bar{\omega}\tau)\, d\bar{\omega} \qquad (23\text{-}37)$$

If the stationary process being considered is nonergodic, an additional step must be taken by averaging Eq. (23-36) across the ensemble as expressed by the relation

$$\frac{1}{2\pi} \lim_{n \to \infty} \frac{1}{n} \sum_{r=1}^{n} F_{x_r}(\overline{\omega}) = \lim_{n \to \infty} \frac{1}{n} \sum_{r=1}^{n} S_{x_r}(\overline{\omega}) \qquad (23\text{-}38)$$

When Eq. (23-31) is used, it is observed that the left-hand side of Eq. (23-38) is equal to $1/2\pi$ times the Fourier transform of $R_x(\tau)$. Since the right side of this same equation is $S_x(\overline{\omega})$, Eqs. (23-37) must also be valid for a nonergodic stationary process.

It was previously demonstrated that if a stationary process having zero mean values is gaussian, it is completely characterized by the autocorrelation function. Now that it has been shown that the power spectral density function can be obtained by a Fourier transformation of the autocorrelation function, that function must also completely characterize such a process.

EXAMPLE E23-3   Derive the power spectral density function for random process $x(t)$ as given in stationary form by Example E23-2.

Substituting Eqs. (*i*) of Example E23-2 into the first of Eqs. (23-37), namely,

$$S_x(\overline{\omega}) = \frac{1}{2\pi} \int_{-\infty}^{\infty} R_x(\tau) \exp(-i\overline{\omega}\tau)\, d\tau \qquad (a)$$

gives

$$S_x(\overline{\omega}) = \frac{\overline{x^2}}{2\pi} \left[\int_{-2\Delta\varepsilon}^{-\Delta\varepsilon} \left(\frac{4}{3} + \frac{2\tau}{\Delta\varepsilon} + \frac{\tau^2}{\Delta\varepsilon^2} + \frac{\tau^3}{6\Delta\varepsilon^3} \right) \exp(-i\overline{\omega}\tau)\, d\tau \right.$$

$$+ \int_{-\Delta\varepsilon}^{0} \left(\frac{2}{3} - \frac{\tau^2}{\Delta\varepsilon^2} - \frac{\tau^3}{2\Delta\varepsilon^3} \right) \exp(-i\overline{\omega}\tau)$$

$$+ \int_{0}^{\Delta\varepsilon} \left(\frac{2}{3} - \frac{\tau^2}{\Delta\varepsilon^2} + \frac{\tau^3}{2\Delta\varepsilon^3} \right) \exp(-i\overline{\omega}\tau)\, d\tau$$

$$\left. + \int_{\Delta\varepsilon}^{2\Delta\varepsilon} \left(\frac{4}{3} - \frac{2\tau}{\Delta\varepsilon} + \frac{\tau^2}{\Delta\varepsilon^2} - \frac{\tau^3}{6\Delta\varepsilon^3} \right) \exp(-i\overline{\omega}\tau)\, d\tau \right]$$

After integrating and collecting all terms, the result is

$$S_x(\overline{\omega}) = \frac{\overline{x^2}}{2\pi} \left\{ \frac{1}{\overline{\omega}^4\, \Delta\varepsilon^3} \left[6 - 4\exp(-i\overline{\omega}\tau) - 4\exp(i\overline{\omega}\tau) \right. \right.$$

$$\left. \left. + \exp(-2i\overline{\omega}\tau) + \exp(2i\overline{\omega}\tau) \right] \right\}$$

which can be converted to the trigonometric form

$$S_x(\overline{\omega}) = \frac{\overline{x^2}}{2\pi} \frac{6 - 8 \cos \overline{\omega} \, \Delta\varepsilon + 2 \cos 2\overline{\omega} \, \Delta\varepsilon}{\overline{\omega}^4 \, \Delta\varepsilon^3} \qquad -\infty < \overline{\omega} < \infty \qquad (b)$$

$$////$$

23-6 POWER SPECTRAL DENSITY AND AUTOCORRELATION FUNCTIONS FOR DERIVATIVES OF PROCESSES

When the power spectral density and autocorrelation functions for the random variable $x(t)$ are known, these same functions can easily be obtained for time derivatives of this variable such as $\dot{x}(t)$ and $\ddot{x}(t)$. To illustrate the method, consider the autocorrelation function for $x(t)$ in its most basic form, that is,

$$R_x(\tau) \equiv E[x(t)x(t + \tau)] \qquad (23\text{-}39)$$

Differentiating with respect to τ gives

$$R_x'(\tau) = \frac{dR_x(\tau)}{d\tau} = E[x(t)\dot{x}(t + \tau)] \qquad (23\text{-}40)$$

Since the process $x(t)$ is stationary, Eq. (23-40) can also be expressed in the form

$$R_x'(\tau) = E[x(t - \tau)\dot{x}(t)] \qquad (23\text{-}41)$$

Differentiating once more with respect to τ gives

$$R_x''(\tau) = -E[\dot{x}(t - \tau)\dot{x}(t)] = -E[\dot{x}(t)\dot{x}(t + \tau)] \qquad (23\text{-}42)$$

Since the ensemble average in Eq. (23-42) is by definition the autocorrelation function for $\dot{x}(t)$, it becomes apparent that

$$R_{\dot{x}}(\tau) = -R_x''(\tau) \qquad (23\text{-}43)$$

Differentiating in the same manner two more times shows that

$$R_{\ddot{x}}(\tau) = -R_{\dot{x}}''(\tau) = R_x^{\text{iv}}(\tau) \qquad (23\text{-}44)$$

The above autocorrelation functions can be expressed in the form of the second of Eqs. (23-37), namely,

$$R_x(\tau) = \int_{-\infty}^{\infty} S_x(\overline{\omega}) \exp(i\overline{\omega}\tau) \, d\overline{\omega}$$

$$R_{\dot{x}}(\tau) = \int_{-\infty}^{\infty} S_{\dot{x}}(\overline{\omega}) \exp(i\overline{\omega}\tau) \, d\overline{\omega} \qquad (23\text{-}45)$$

$$R_{\ddot{x}}(\tau) = \int_{-\infty}^{\infty} S_{\ddot{x}}(\overline{\omega}) \exp(i\overline{\omega}\tau) \, d\overline{\omega}$$

Substituting the first of Eq. (23-45) into Eqs. (23-43) and (23-44) gives

$$R_{\dot{x}}(\tau) = \int_{-\infty}^{\infty} \bar{\omega}^2 S_x(\bar{\omega}) \exp(i\bar{\omega}\tau) \, d\bar{\omega}$$

$$R_{\ddot{x}}(\tau) = \int_{-\infty}^{\infty} \bar{\omega}^4 S_x(\bar{\omega}) \exp(i\bar{\omega}\tau) \, d\bar{\omega}$$

(23-46)

Comparing Eqs. (23-46) with the second and third of Eqs. (23-45) shows that

$$S_{\dot{x}}(\bar{\omega}) = \bar{\omega}^2 S_x(\bar{\omega}) \qquad S_{\ddot{x}}(\bar{\omega}) = \bar{\omega}^4 S_x(\bar{\omega}) \qquad (23\text{-}47)$$

EXAMPLE E23-4 If random process $x(t)$ has the autocorrelation function

$$R_x(\tau) = (1 - \tau^2)e^{-\tau^2} \qquad (a)$$

find the corresponding autocorrelation functions for random processes $\dot{x}(t)$ and $\ddot{x}(t)$.

Taking derivatives of Eq. (a) gives

$$R_x'(\tau) = (2\tau^3 - 4\tau)e^{-\tau^2}$$
$$R_x''(\tau) = (-4\tau^4 + 14\tau^2 - 4)e^{-\tau^2}$$
$$R_x'''(\tau) = (8\tau^5 - 44\tau^3 + 36\tau)e^{-\tau^2}$$
$$R_x^{\text{iv}}(\tau) = (-16\tau^6 + 128\tau^4 - 204\tau^2 + 36)e^{-\tau^2}$$

Thus from Eqs. (23-43) and (23-44)

$$R_{\dot{x}}(\tau) = (4\tau^4 - 14\tau^2 + 4)e^{-\tau^2} \qquad (b)$$
$$R_{\ddot{x}}(\tau) = (-16\tau^6 + 128\tau^4 - 204\tau^2 + 36)e^{-\tau^2} \qquad (c)$$

23-7 SUPERPOSITION OF STATIONARY PROCESSES

Consider a stationary process $q(t)$ which is defined as the sum of three separate stationary processes $x(t)$, $y(t)$, and $z(t)$ all of which have zero mean values. To find the autocorrelation function for this process, namely,

$$R_q(\tau) \equiv E[q(t)q(t + \tau)] \qquad (23\text{-}48)$$

substitute the relation

$$q(t) = x(t) + y(t) + z(t) \qquad (23\text{-}49)$$

into Eq. (23-48) to obtain

$$R_q(\tau) = E[x(t)x(t + \tau)] + E[y(t)y(t + \tau)] + E[z(t)z(t + \tau)]$$
$$+ E[x(t)y(t + \tau)] + E[y(t)z(t + \tau)] + E[x(t)z(t + \tau)]$$
$$+ E[y(t)x(t + \tau)] + E[z(t)y(t + \tau)] + E[z(t)x(t + \tau)] \qquad (23\text{-}50)$$

The first three ensemble averages on the right-hand side of this equation are the autocorrelation functions for processes $x(t)$, $y(t)$, and $z(t)$, respectively, and the last six ensemble averages are *cross-correlation functions* (or covariance functions), which will be designated as

$$R_{xy}(\tau) = E[x(t)y(t + \tau)] \qquad R_{yx}(\tau) = E[y(t)x(t + \tau)]$$
$$R_{yz}(\tau) = E[y(t)z(t + \tau)] \qquad R_{zy}(\tau) = E[z(t)y(t + \tau)] \qquad (23\text{-}51)$$
$$R_{xz}(\tau) = E[x(t)z(t + \tau)] \qquad R_{zx}(\tau) = E[z(t)x(t + \tau)]$$

Thus, the autocorrelation function for process $q(t)$ can be expressed in terms of the autocorrelation and cross-correlation functions for $x(t)$, $y(t)$, and $z(t)$ as follows:

$$R_q(\tau) = R_x(\tau) + R_y(\tau) + R_z(\tau) + R_{xy}(\tau) + R_{yz}(\tau)$$
$$+ R_{xz}(\tau) + R_{yx}(\tau) + R_{zy}(\tau) + R_{zx}(\tau) \qquad (23\text{-}52)$$

If random processes $x(t)$, $y(t)$, and $z(t)$ are statistically independent, their cross-correlation functions will equal zero, in which case

$$R_q(\tau) = R_x(\tau) + R_y(\tau) + R_z(\tau) \qquad (23\text{-}53)$$

It should be noted that for real stationary processes

$$R_{xy}(\tau) = R_{yx}(-\tau) \qquad R_{yz}(\tau) = R_{zy}(-\tau) \qquad R_{xz}(\tau) = R_{zx}(-\tau) \qquad (23\text{-}54)$$

The power spectral density function for process $q(t)$ is obtained using the first of Eqs. (23-37), that is,

$$S_q(\bar{\omega}) = \frac{1}{2\pi} \int_{-\infty}^{\infty} R_q(\tau) \exp(-i\bar{\omega}\tau) \, d\tau \qquad (23\text{-}55)$$

Substituting Eq. (23-52) into Eq. (23-55) gives

$$S_q(\bar{\omega}) = S_x(\bar{\omega}) + S_y(\bar{\omega}) + S_z(\bar{\omega}) + S_{xy}(\bar{\omega}) + S_{yz}(\bar{\omega})$$
$$+ S_{xz}(\bar{\omega}) + S_{yx}(\bar{\omega}) + S_{zy}(\bar{\omega}) + S_{zx}(\bar{\omega}) \qquad (23\text{-}56)$$

where $S_{xy}(\bar{\omega})$, $S_{yz}(\bar{\omega})$, ... are *cross-spectral density functions* which are related to their respective cross-correlation functions through the Fourier transform relation

$$S_{xy}(\bar{\omega}) = \frac{1}{2\pi} \int_{-\infty}^{\infty} R_{xy}(\tau) \exp(-i\bar{\omega}\tau) \, d\tau \qquad (23\text{-}57)$$

Note that $S_{yx}(\bar{\omega})$ is the complex conjugate of $S_{xy}(\bar{\omega})$. The inverse of Eq. (23-57) is, of course,

$$R_{xy}(\tau) = \int_{-\infty}^{\infty} S_{xy}(\bar{\omega}) \exp(i\bar{\omega}\tau) \, d\bar{\omega} \qquad (23\text{-}58)$$

When the procedure of Sec. 23-4 is followed, the time average of the product $x_r(t)\,y_r(t)$ becomes

$$\langle x_r(t)\,y_r(t)\rangle = \int_{-\infty}^{\infty} S_{x_r y_r}(\bar{\omega})\,d\bar{\omega} \qquad (23\text{-}59)$$

where

$$S_{x_r y_r}(\bar{\omega}) \equiv \lim_{s \to \infty} \frac{\left[\int_{-s/2}^{s/2} x_r(t)\exp(i\bar{\omega}t)\,dt\right]\left[\int_{-s/2}^{s/2} y_r(t)\exp(-i\bar{\omega}t)\,dt\right]}{2\pi s} \qquad (23\text{-}60)$$

Note that $S_{x_r y_r}(-\bar{\omega})$ is the complex conjugate of $S_{x_r y_r}(\bar{\omega})$. Therefore, only the real part of $S_{x_r y_r}(\bar{\omega})$ contributes to the integral in Eq. (23-59). If processes $x(t)$ and $y(t)$ are ergodic, $S_{x_r y_r}(\bar{\omega})$ as given by Eq. (23-60) represents the cross-spectral density for these processes. However, if processes $x(t)$ and $y(t)$ are nonergodic, the cross-spectral density function for these processes must be obtained by averaging across the ensemble, that is,

$$S_{xy}(\bar{\omega}) = \lim_{n \to \infty} \frac{1}{n} \sum_{r=1}^{n} S_{x_r y_r}(\bar{\omega}) \qquad (23\text{-}61)$$

23-8 STATIONARY GAUSSIAN PROCESSES: ONE INDEPENDENT VARIABLE

In engineering it is common practice to assume a gaussian, or normal, distribution for random processes. To help in establishing a rational basis for this assumption, consider a stationary random process $x(t)$ having zero mean values of the form

$$x_{jr}(t) = \sum_{n=-j}^{j} C_{nr}\exp(in\bar{\omega}_0 t) \qquad r = 1, 2, \ldots \qquad (23\text{-}62)$$

where $x_{jr}(t)$ is the rth member of the ensemble which contains j discrete harmonics having frequencies $\bar{\omega}_0, 2\bar{\omega}_0, \ldots, n\bar{\omega}_0$, and where C_{nr} represents random complex constants. For the process to have zero mean values it is necessary, of course, that coefficients C_{0r} equal zero; and since it is assumed that the process contains real functions only, it is necessary that complex coefficients C_{nr} and C_{mr} be conjugate pairs when $n = -m$.

To define the randomness of coefficients C_{nr}, assume first that $|C_{nr}| = C$ (a constant) for all permissible values of n and r but that their corresponding phase angles α_{nr} are sampled values of a random variable α which has a uniform probability density function of intensity $1/2\pi$ in the range $0 < \alpha < 2\pi$. Under these conditions Eq. (23-62) can be written in the form

$$x_{jr}(t) = \sum_{n=-j}^{j} |C_{nr}| \exp\left[i(n\bar{\omega}_0 t + \alpha_{nr})\right] \tag{23-63}$$

or

$$x_{jr} = 2C \sum_{n=1}^{j} \sin(n\bar{\omega}_0 t + \theta_{nr}) \qquad r = 1, 2, \ldots \tag{23-64}$$

where $\theta_{nr} = +\pi/2 - \alpha_{nr}$. Since this process contains discrete harmonics at frequency intervals of $\bar{\omega}_0$, each ensemble member will be periodic with a period $s = 2\pi/\bar{\omega}_0$. When a new random variable $L(t)$ is defined so that

$$L_{nr}(t) = 2C \sin(n\bar{\omega}_0 t + \theta_{nr}) \tag{23-65}$$

Eq. (23-64) can be written in the form of the one-dimensional random walk:

$$x_{jr}(t) = \sum_{n=1}^{j} L_{nr}(t) \tag{23-66}$$

From Eqs. (23-5) and (23-10), it is clear that

$$p[L(t)] = \begin{cases} \dfrac{1}{\pi|\sqrt{4C^2 - L^2}|} & -2C < L < 2C \\ 0 & L < -2C; L > 2C \end{cases} \tag{23-67}$$

and that

$$\overline{L(t)} = 0 \qquad \sigma_{L(t)}^2 = 2C^2 \tag{23-68}$$

When the one-dimensional random-walk relations given by Eq. (22-34) are used, it follows that

$$\overline{x_j(t)} = 0 \qquad \sigma_{x(t)}^2 = 2jC^2 \tag{23-69}$$

At this point, apply the same limiting procedure previously used in the one-dimensional random-walk development, that is, let $\bar{\omega}_0 \to 0$, $j \to \infty$, and $C^2 \to 0$, but in such a manner that

$$n\bar{\omega}_0 \to \bar{\omega} \text{ (a variable)} \quad j\bar{\omega}_0 \to \bar{\omega}_1 \text{ (a constant)} \quad C^2/\bar{\omega}_0 \to S_0 \text{ (a constant)} \tag{23-70}$$

Since $\bar{\omega}_0 = 2\pi/s$, period $s \to \infty$ by this limiting procedure.

When the above relations are used, from the fact that all coefficients $|C_{nr}|$ must satisfy the condition, Eq. (23-21),

$$|C_{nr}| = C = \frac{1}{s}\left|\int_{-s/2}^{s/2} x_{jr}(t) \exp(-in\bar{\omega}_0 t)\, dt\right| \tag{23-71}$$

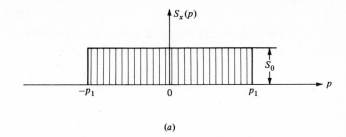

(a)

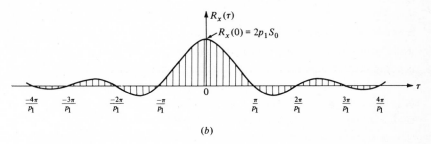

(b)

FIGURE 23-5
Power spectral density and autocorrelation functions for random process $x(t)$.

it is evident that in the limit

$$S_0 = \lim_{s \to \infty} \frac{\left| \int_{-s/2}^{s/2} x_r(t) \exp(-i\overline{\omega}t)\, dt \right|^2}{2\pi s} \qquad r = 1, 2, \ldots \qquad (23\text{-}72)$$

where

$$x_r(t) = \lim_{\substack{j \to \infty \\ \overline{\omega}_0 \to 0}} x_{jr}(t) \qquad (23\text{-}73)$$

A comparison of Eq. (23-72) with Eq. (23-27) and recognition of the limiting conditions given by the first and second of Eqs. (23-70) lead to the conclusion that ensemble member $x_r(t)$ has a uniform power spectral density function $S_{x_r}(\overline{\omega})$ of intensity S_0 over the frequency range $-\overline{\omega}_1 < \overline{\omega} < \overline{\omega}_1$ and of intensity zero outside this range and that since this power spectral density function is invariant with r, the process is ergodic; thus, the power spectral density for the entire process $x(t)$ is that function shown in Fig. 23-5a. Further, the earlier one-dimensional random-walk development leads to the conclusion that this random process is gaussian and that its variance [see the second of Eqs. (23-69)] is given by

$$\sigma_{x(t)}{}^2 = 2\overline{\omega}_1 S_0 \qquad (23\text{-}74)$$

With the power spectral density function shown in Fig. 23-5a and the second of Eqs. (23-37), the autocorrelation function for random process $x(t)$ is

$$R_x(\tau) = \frac{2S_0}{\tau} \sin \bar{\omega}_1 \tau \qquad -\infty < \tau < \infty \qquad (23\text{-}75)$$

This relation is plotted in Fig. 23-5b.

Note that when the power spectral density function for this process becomes uniform over the entire frequency range, that is, when $\bar{\omega}_1 \to \infty$, the variance $\sigma_x{}^2 \to \infty$ and the autocorrelation function $R_x(\tau) \to 2\pi S_0 \delta(\tau)$, where $\delta(\tau)$ is a Dirac delta function located at the origin. This process, which is commonly referred to as a *white* process or simply *white noise*, can be considered as totally random since $x(t)$ is completely independent of $x(t + \tau)$ for all values of $\tau \neq 0$.

Consider again random process $x_r(t)$ but this time assume that coefficients $|C_{nr}|$ equal zero for all values of n in the range $-k < n < +k$, where $k < j$, and that they equal a constant C for all values of n in the ranges $-j \leq n \leq -k$ and $+k \leq n \leq +j$. The same procedures as before are followed, but this time $\bar{\omega}_0 \to 0$, $k \to \infty$, $j \to \infty$, and $C^2 \to 0$ in such a manner that $n\bar{\omega}_0 \to \bar{\omega}$, $k\bar{\omega}_0 \to \bar{\omega}_1$, $j\bar{\omega}_0 \to \bar{\omega}_2$, and $C^2/\bar{\omega}_0 \to S_0$; again the process becomes gaussian in the limit and its power spectral density and autocorrelation functions are of the form

$$S_x(\bar{\omega}) = \begin{cases} S_0 & -\bar{\omega}_2 < \bar{\omega} < -\bar{\omega}_1; \bar{\omega}_1 < \bar{\omega} < \bar{\omega}_2 \\ 0 & \bar{\omega} < -\bar{\omega}_2; -\bar{\omega}_1 < \bar{\omega} < \bar{\omega}_1; \bar{\omega} > \bar{\omega}_2 \end{cases}$$

$$R_x(\tau) = \frac{2S_0}{\tau}(\sin \bar{\omega}_2 \tau - \sin \bar{\omega}_1 \tau) \qquad -\infty < \tau < \infty \qquad (23\text{-}76)$$

To generalize one step further, consider a random process $z(t)$ defined as the sum of the statistically independent gaussian ergodic processes $x(t)$ and $y(t)$, both of which are developed separately from Eq. (23-62) using the same limiting procedure as before. From the proof given in Sec. 22-9, process $z(t)$ will also have a gaussian distribution.

If the power spectral density functions for processes $x(t)$ and $y(t)$ are as shown in Fig. 23-6a and b, respectively, it follows from the discussion in Sec. 23-7 [see Eq. (23-56)] that the power spectral density function for $z(t)$ is obtained by direct superposition, giving the result shown in Fig. 23-6c. Since this example makes it apparent that a random process having any arbitrary power spectral density function can be built up by superposition of statistically independent gaussian processes of the type represented in Fig. 23-6a and b, such a process must also be gaussian.

Finally, once more use the process given by Eq. (23-63) as expressed in the equivalent form

$$x_{jr}(t) = \sum_{n=1}^{j} 2|C_{nr}| \sin (n\bar{\omega}_0 t + \theta_{nr}) \qquad r = 1, 2, \ldots \qquad (23\text{-}77)$$

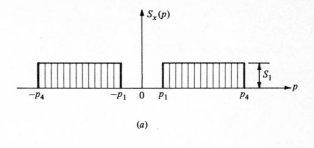

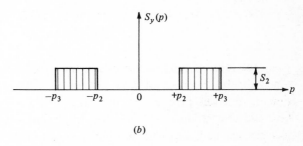

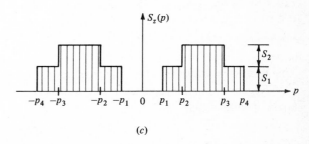

FIGURE 23-6
Power spectral density functions for processes $x(t)$, $y(t)$, and $z(t)$.

For this process assume that phase angles θ_{nr} are sampled values of random variable θ which has the uniform probability density function shown in Fig. 23-7a and that coefficients $|C_{nr}|$ are sampled values of a second random variable C which has an arbitrary, but prescribed, probability density as shown in Fig. 23-7b. When $L_{nr}(t)$ is defined by the relation

$$L_{nr}(t) \equiv 2|C_{nr}| \sin (n\bar{\omega}_0 t + \theta_{nr}) \qquad (23\text{-}78)$$

$x_r(t)$ can again be expressed in the form

$$x_{jr}(t) \equiv \sum_{n=1}^{j} L_{nr}(t) \qquad (23\text{-}79)$$

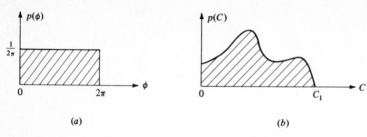

FIGURE 23-7
Probability density functions for random variables ϕ and C.

When the form of probability density functions $p(\theta)$ and $p(C)$ is known, the probability density function for $L(t)$, as defined by Eq. (23-78), could be established if desired. For this process, however, this step is unnecessary since the mean value $\bar{L}(t)$ and the variance $\sigma^2_{L(t)}$ are the only quantities required in the random-walk development and they can be obtained without establishing the function $p[L(t)]$. From the form of Eq. (23-78), it can be reasoned that

$$\overline{L(t)} = 0 \qquad \sigma_{L(t)}{}^2 = 2\overline{C^2} = 2 \int_{-\infty}^{\infty} C^2 p(C)\, dC \qquad (23\text{-}80)$$

This process is stationary since the variance for $L(t)$ is independent of time t.

When the one-dimensional random-walk relations given by Eq. (22-34) are used again, it follows that

$$x_j(t) = 0 \qquad \sigma^2_{x_j(t)} = 2j\overline{C^2} \qquad (23\text{-}81)$$

When Eq. (23-81) is compared with Eq. (23-69), it is clear that the same limiting procedures used previously can once again be used provided C^2 is replaced by $\overline{C^2}$. In this case

$$S_0 \equiv \frac{\overline{C^2}}{\bar{\omega}_0} = \lim_{n \to \infty} \frac{1}{n} \sum_{r=1}^{n} S_{x_r}(\bar{\omega}) \qquad (23\text{-}82)$$

where

$$S_{x_r}(\bar{\omega}) = \lim_{s \to \infty} \frac{\left| \int_{-s/2}^{s/2} x_r(t) \exp(-i\bar{\omega}t)\, dt \right|^2}{2\pi s} \qquad (23\text{-}83)$$

Because coefficients $|C_{nr}|$ are random for this process, the power spectral density function for member r as defined by Eq. (23-83) will vary with frequency $\bar{\omega}$ and will be different for each value of r. Therefore, this process is nonergodic, and the power spectral density function for the entire process must be obtained by averaging across the ensemble, as indicated by Eq. (23-82). While the power spectral density function for each member of the ensemble varies with frequency $\bar{\omega}$, the power spectral density function for the entire process as given by Eq. (23-82) will be constant and of intensity S_0 over the range $-\bar{\omega}_1 < \bar{\omega} < \bar{\omega}_1$.

The earlier development which restricted the nonzero-frequency components to the range $\bar{\omega}_1 < \bar{\omega} < \bar{\omega}_2$ and the development which presented the principle of superposition obviously both apply equally well to the present process involving two random variables. Therefore, it may be concluded that any stationary process $x(t)$ (whether ergodic or not) will be gaussian when its power spectral density function $S_x(\bar{\omega})$ truly exists and when all the phase angles between frequency components which are randomly distributed in a uniform manner over 360° are statistically independent of each other.

When the phase angles between frequency components are not uniformly distributed over the full 360°, gaussian processes will still result in the limit; however, stationarity will no longer be maintained. For example, if the random phase angle θ for the process defined by Eq. (23-2) has a uniform probability density function of intensity $1/\theta_1$ over the range $0 < \theta < \theta_1$, where $\theta_1 < 2\pi$, the ensemble mean square value $E[x(t)^2]$ (or variance in this case) will be time-dependent. To prove this statement, substitute Eqs. (23-2) and (22-4) into Eq. (22-10) to obtain

$$E[x(t)^2] = \int_{A \sin \bar{\omega}_0 t}^{A \sin (\bar{\omega}_0 t + \theta_1)} \frac{x^2}{\theta_1 |\sqrt{A^2 - x^2}|} \, dx \qquad (23\text{-}84)$$

After the integration is completed, this equation becomes

$$E[x(t)^2] = \frac{A^2}{2} \left\{ 1 - \frac{1}{2\theta_1} \left[\sin 2\theta_1 \cos 2\bar{\omega}_0 t - (1 - \cos 2\theta_1) \sin 2\bar{\omega}_0 t \right] \right\} \qquad (23\text{-}85)$$

which clearly shows the time dependency. Note that as $\theta_1 \to 2\pi$, the time dependency is gradually removed; that is, $E[x(t)^2] \to A^2/2$, and as $\theta_1 \to 0$, the random character of the process is gradually lost, so that $E[x(t)^2] \to A^2 \sin^2 \bar{\omega}_0 t$.

It is important to recognize that gaussian processes result only when the random variables involved are statistically independent.

EXAMPLE E23-5 Assume random variables r_1 and r_2 as defined in Example E22-7 are used as successive discrete ordinates for all members of random process $x(t)$ given in Example E23-2. What is the joint probability density function for random variables $x(t)$ and $x(t + \tau)$?

First it should be recognized that random variables $x(t)$ and $x(t + \tau)$ are linearly related to random variables r_1 and r_2 in accordance with the first and second of Eqs. (a) in Example E23-2. Since random variables r_1 and r_2 have the normal distribution given by Eq. (f) of Example E22-7, random variables $x(t)$ and $x(t + \tau)$ must also have a normal distribution in accordance with the principle of linear transformation treated in Sec. 22-9. Thus the probability density functions must be of the form

$$
p(x_1) = \frac{1}{\sqrt{2\pi}\ \sigma_{x_1}} \exp\left[-\frac{(x_1 - \bar{x}_1)^2}{2\sigma_{x_1}^2} \right]
$$

$$
p(x_1,x_2) = \frac{1}{2\pi\sigma_{x_1}\sigma_{x_2}\sqrt{1 - \rho_{x_1 x_2}^2}}
$$

$$
\times \exp\left\{ -\frac{1}{2(1 - \rho_{x_1 x_2}^2)} \left[\frac{x_1 - \bar{x}_1^{\ 2}}{\sigma_{x_1}^2} \right.\right.
$$

$$
\left.\left. -\frac{2\rho_{x_1 x_2}(x_1 - \bar{x}_1)(x_2 - \bar{x}_2)}{\sigma_{x_1}\sigma_{x_2}} \right.\right.
$$

$$
\left.\left. +\frac{(x_2 - \bar{x}_2)^2}{\sigma_{x_2}^2} \right] \right\} \tag{a}
$$

where $x_1 \equiv x(t)$ and $x_2 \equiv x(t + \tau)$. With the results in Examples E22-7 and E23-2, it is shown that

$$
\overline{x_1} = \overline{x_2} = 0
$$

$$
\sigma_{x_1}^2 = \sigma_{x_2}^2 = R_x(0) = {}^2\!/\!_3 \overline{x_2^2} = {}^2\!/\!_3 \tag{b}
$$

$$
\rho_{x_1 x_2}(\tau) = \frac{R_x(\tau)}{R_x(0)}
$$

$$
= \begin{cases}
1 - \dfrac{3\tau^2}{2\Delta\varepsilon^2} + \dfrac{3|\tau|^3}{4\Delta\varepsilon^3} & -\Delta\varepsilon \le \tau \le \Delta\varepsilon \\[2ex]
2 - \dfrac{3|\tau|}{\Delta\varepsilon} + \dfrac{3\tau^2}{2\Delta\varepsilon^2} - \dfrac{|\tau|^3}{4\Delta\varepsilon^3} & \begin{array}{l} -2\Delta\varepsilon \le \tau \le -\Delta\varepsilon \\ \Delta\varepsilon \le \tau \le 2\Delta\varepsilon \end{array} \\[2ex]
0 & \tau \le -2\Delta\varepsilon;\ \tau \ge 2\Delta\varepsilon
\end{cases}
$$

Substituting Eqs. (b) into Eqs. (a) gives the desired probability density function.

////

23-9 STATIONARY GAUSSIAN WHITE NOISE

In the previous discussion on stationary gaussian processes, white noise was defined as a process having a uniform power spectral density function of intensity S_0 over the entire frequency range $-\infty < \bar{\omega} < \infty$, which corresponds to a Dirac delta function of intensity $2\pi S_0$ at the origin for the autocorrelation function. By this definition it is clear that such processes contain frequency components of equal intensity (based on squared amplitude as a measure of intensity) over the entire frequency

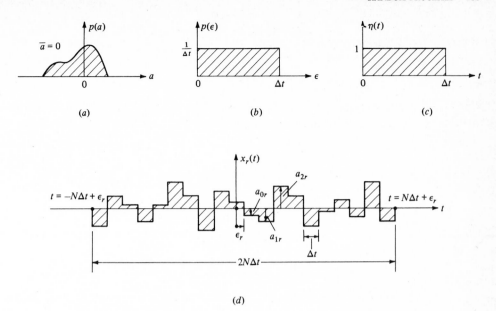

FIGURE 23-8
White-noise process, Eq. (23-86).

range, thus providing statistical independence between the random variables at time t and $t + \tau$ for all $\tau \neq 0$.

In subsequent developments, it will be found desirable to express white-noise processes in an equivalent but quite different manner. To develop this new type of representation, consider the random process

$$x_r(t) = \lim_{N \to \infty} \sum_{k=-N}^{N-1} a_{kr}\eta(t - k\,\Delta t - \varepsilon_r) \qquad r = 1, 2, \ldots \qquad (23\text{-}86)$$

where coefficients a_{kr} are statistically independent random variables having a zero mean value and sampled in accordance with the arbitrary but prescribed probability density function $p(a)$ shown in Fig. 23-8a, Δt is a constant time interval, variables ε_r are statistically independent random phase parameters having the uniform probability density function shown in Fig. 23-8b, and $\eta(t)$ is the function defined in Fig. 23-8c. The rth member of this ensemble is shown in Fig. 23-8d. The uniformly random phase shift ε over a full interval Δt is a necessary condition for the process to be stationary.

The power spectral density function for member $x_r(t)$ can be derived by using Eq. (23-27) in its equivalent form

$$S_{x_r}(\bar{\omega}) = \lim_{N \to \infty} \frac{|Q_{x_r}(i\bar{\omega})|^2}{4\pi N\,\Delta t} \qquad (23\text{-}87)$$

where

$$Q_{x_r}(i\bar{\omega}) = \int_{-N\Delta t + \varepsilon_r}^{N\Delta t + \varepsilon_r} \left[\sum_{k=-N}^{N-1} a_{kr}\eta(t - k\,\Delta t - \varepsilon_r) \right] \exp{(-i\bar{\omega}t)}\,dt \qquad (23\text{-}88)$$

When the change of variable $\theta \equiv t - k\,\Delta t - \varepsilon_r$ is substituted into this equation and the order of summation and integration is changed, it becomes

$$Q_{x_r}(i\bar{\omega}) = \sum_{k=-N}^{N-1} a_{kr} \exp{[-i\bar{\omega}(k\,\Delta t + \varepsilon_r)]} \int_{-(N+k)\Delta t}^{(N-k)\Delta t} \eta(\theta)\exp{(-i\bar{\omega}\theta)}\,d\theta \qquad (23\text{-}89)$$

or

$$Q_{x_r}(i\bar{\omega}) = \frac{i}{\bar{\omega}} \left[\exp{(-i\bar{\omega}\,\Delta t)} - 1 \right] \sum_{k=-N}^{N-1} a_{kr} \exp{[-i\bar{\omega}(k\,\Delta t + \varepsilon_r)]} \qquad (23\text{-}90)$$

Substituting this equation into Eq. (23-87) gives

$$S_{x_r}(\bar{\omega}) = \lim_{N\to\infty} \frac{1}{4\pi N\,\Delta t\,\bar{\omega}^2} \left[\exp{(-i\bar{\omega}\,\Delta t)} - 1 \right] \left[\exp{(i\bar{\omega}\,\Delta t)} - 1 \right]$$

$$\times \sum_{k=-N}^{N-1} \sum_{j=-N}^{N-1} a_{kr}a_{jr} \exp{[-i\bar{\omega}(k - j)\,\Delta t]} \qquad (23\text{-}91)$$

Since the process as defined is stationary but nonergodic, the power spectral density function for the process must be obtained by averaging Eq. (23-91) across the ensemble.

Since random variables a_{kr} and a_{jr} ($r = 1, 2, \ldots, \infty$) are statistically independent, their covariances, that is, $E(a_{kr}a_{jr})$ for $j \neq k$, must all equal zero. Therefore, the double summation in Eq. (23-91) reduces to a single summation, which obviously equals $2N\sigma_a^2$ when averaged with respect to r across the ensemble. Thus, the power spectral density function for the process becomes

$$S_x(\bar{\omega}) = \frac{\sigma_a^2}{2\pi\,\Delta t\,\bar{\omega}^2} \left[\exp{(-i\bar{\omega}\,\Delta t)} - 1 \right] \left[\exp{(i\bar{\omega}\,\Delta t)} - 1 \right] \qquad (23\text{-}92)$$

or

$$S_x(\bar{\omega}) = \frac{\sigma_a^2\,\Delta t}{2\pi} \frac{\sin^2{[(\bar{\omega}\,\Delta t)/2]}}{[(\bar{\omega}\,\Delta t)/2]^2} \qquad (23\text{-}93)$$

When $\sigma_a^2 \to \infty$ and $\Delta t \to 0$ in such a way that $\sigma_a^2\,\Delta t = C$ (a constant), this equation becomes

$$S_x(\bar{\omega}) = \frac{C}{2\pi} = S_0 \qquad (23\text{-}94)$$

showing that the process becomes white noise in the limit.

As a special case of the above process, let the probability density function $p(a)$ consist of two Dirac delta functions of intensity $1/2$ located at $a = \pm A$. This process becomes white noise having a uniform power spectral density function of intensity $S_0 = C/2\pi$ when $A^2 \to \infty$, $\Delta t \to 0$, and $A^2\,\Delta t \to C$.

EXAMPLE E23-6 For the stationary random process defined in Example E22-5, (1) show that this process approaches white noise in the limit as $\Delta\varepsilon \to 0$ and (2) find the normalization factor C which would force this limiting process to have a constant power spectral density equal to S_0.

From the form of the autocorrelation function given in Example E23-2

$$\lim_{\Delta\varepsilon \to 0} R_x(\tau) \begin{cases} = 0 & \tau \neq 0 \\ \neq 0 & \tau = 0 \end{cases} \tag{a}$$

which suggests the form of a Dirac delta function. Integrating $R_x(\tau)$ over the infinite τ domain gives

$$\int_{-\infty}^{\infty} R_x(\tau)\, d\tau = 2\overline{x^2} \left\{ \int_0^{\Delta\varepsilon} \left(\frac{2}{3} - \frac{\tau^2}{\Delta\varepsilon^2} + \frac{\tau^3}{2\Delta\varepsilon^3} \right) d\tau \right.$$

$$\left. + \int_{\Delta\varepsilon}^{2\Delta\varepsilon} \left(\frac{4}{3} - \frac{2\tau}{\Delta\varepsilon} + \frac{\tau^2}{\Delta\varepsilon^2} - \frac{\tau^3}{6\Delta\varepsilon^3} \right) d\tau \right.$$

or

$$\int_{-\infty}^{\infty} R_x(\tau)\, d\tau = \overline{x^2}\, \Delta\varepsilon = \Delta\varepsilon \tag{b}$$

If all discrete ordinates of process $x(t)$ were multiplied by the constant $(2\pi S_0/\Delta\varepsilon)^{1/2}$ giving a new process $a(t)$, following the above procedures would demonstrate that

$$\lim_{\Delta\varepsilon \to 0} R_a(\tau) \begin{cases} = 0 & \tau \neq 0 \\ \neq 0 & \tau = 0 \end{cases} \tag{c}$$

$$\int_{-\infty}^{\infty} R_a(\tau)\, d\tau = 2\pi S_0$$

thus showing that

$$R_a(\tau) \to 2\pi S_0 \delta(\tau) \tag{d}$$

which means that process $a(t)$ approaches white noise of intensity S_0. Therefore the normalization factor C is given by

$$C = \left(\frac{2\pi S_0}{\Delta\varepsilon} \right)^{1/2} \tag{e}$$

The solution to this example can be obtained more easily by noting that the power spectral density function for the process $x(t)$, as given by Eq. (b) in Example E23-3, becomes in the limit

$$\lim_{\Delta\varepsilon \to 0} S_x(\overline{\omega}) = \frac{\overline{x^2}\, \Delta\varepsilon}{2\pi} = \frac{\Delta\varepsilon}{2\pi} \tag{f}$$

Likewise the limiting power spectral density function for process $a(t)$ would be

$$\lim_{\Delta\varepsilon \to 0} S_a(\overline{\omega}) = S_0 \tag{g}$$

again showing that the normalization factor is given by Eq. (e).   ////

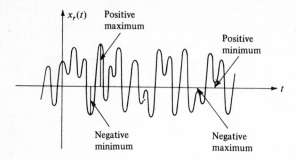

FIGURE 23-9
Sample function of process $x(t)$.

23-10 PROBABILITY DISTRIBUTION FOR MAXIMA[1]

Consider a zero-mean stationary gaussian process $x(t)$ having an arbitrary power spectral density function $S_x(p)$. A sample function taken from this process (Fig. 23-9) shows positive and negative maxima and positive and negative minima. From Fig. 23-10, it is clear that for a maximum ($+$ or $-$) to occur in the time interval $(t, t + dt)$, it is necessary that $\dot{x}_r(t)$ be positive and $\ddot{x}_r(t)$ be negative and that

$$0 < \dot{x}_r(t) < |\ddot{x}_r(t)|\ dt \qquad (23\text{-}95)$$

With the definition of three new random variables $\zeta_1 \equiv x(t)$, $\zeta_2 \equiv \dot{x}(t)$, and $\zeta_3 \equiv \ddot{x}(t)$, the probability density function $p(\zeta_1, \zeta_2, \zeta_3)$ can be written in its normal form

$$p(\zeta_1, \zeta_2, \zeta_3) = \frac{1}{(2\pi)^{3/2}|\mu|^{1/2}} \exp\left\{-\tfrac{1}{2}[\zeta - \bar{\zeta}]^T \mu^{-1} [\zeta - \bar{\zeta}]\right\} \qquad (23\text{-}96)$$

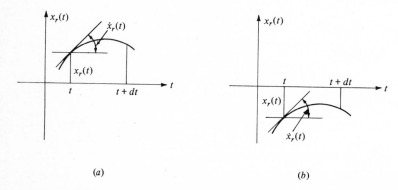

(a)

(b)

FIGURE 23-10
Maxima occurring in the time interval $(t, t + dt)$: (a) positive maxima; (b) negative maxima.

[1] D. E. Cartwright and M. S. Longuet-Higgins, The Statistical Distributions of the Maxima of a Random Function, *Proc. R. Soc.*, ser. A, vol. 237, pp. 212–232, 1956; A. G. Davenport, Note on the Distribution of the Largest Value of a Random Function with Application to Gust Loading, *Proc. Inst. Civ. Eng.*, vol. 28, pp. 187–196, 1964.

where ζ^T is the vector $[\zeta_1 \quad \zeta_2 \quad \zeta_3]$, $\bar{\zeta}^T$ is the vector $[\bar{\zeta}_1 \quad \bar{\zeta}_2 \quad \bar{\zeta}_3] = 0$, and μ is the covariance matrix

$$\mu = \begin{bmatrix} \mu_{11} & \mu_{12} & \mu_{13} \\ \mu_{21} & \mu_{22} & \mu_{23} \\ \mu_{31} & \mu_{32} & \mu_{33} \end{bmatrix} \quad (23\text{-}97)$$

where

$$\mu_{ik} = E(\zeta_i \zeta_k) \quad (23\text{-}98)$$

When

$$m_n = \int_{-\infty}^{\infty} \bar{\omega}^n S_x(\bar{\omega}) \, d\bar{\omega} \quad (23\text{-}99)$$

it is easily shown, using the techniques of derivation in Sec. 23-6, that

$$\mu = \begin{bmatrix} m_0 & 0 & -m_2 \\ 0 & m_2 & 0 \\ -m_2 & 0 & m_4 \end{bmatrix} \quad (23\text{-}100)$$

Thus, Eq. (23-96) becomes

$$p(\zeta_1,\zeta_2,\zeta_3) = \frac{1}{(2\pi)^{3/2}(m_2\Delta)^{1/2}}$$

$$\times \exp\left[-\frac{1}{2}\left(\frac{\zeta_2{}^2}{m_2} + \frac{m_4\zeta_1{}^2 + 2m_2\zeta_1\zeta_3 + m_0\zeta_3{}^2}{\Delta} \right) \right] \quad (23\text{-}101)$$

where

$$\Delta \equiv m_0 m_4 - m_2{}^2 \quad (23\text{-}102)$$

From Fig. 23-11 it becomes apparent that the probability of a maximum (+ or −) occurring in the range $(\zeta_1, \zeta_1 + d\zeta_1)$ during the time interval $(t, t + dt)$ is expressed by the relation

$$F(\zeta_1) \, d\zeta_1 \, dt = \left[\int_{-\infty}^{0} p(\zeta_1,0,\zeta_3)|\zeta_3| \, d\zeta_3 \right] d\zeta_1 \, dt \quad (23\text{-}103)$$

Thus, it follows that the mean frequency of occurrence of maxima (+ and −) over the complete range $-\infty < \zeta < \infty$ is given by

$$N_1 \equiv \int_{-\infty}^{\infty} \left[\int_{-\infty}^{0} p(\zeta_1,0,\zeta_3)|\zeta_3| \, d\zeta_3 \right] d\zeta_1 = \frac{1}{2\pi}\left(\frac{m_4}{m_2} \right)^{1/2} \quad (23\text{-}104)$$

The probability density function for maxima is now obtained as the ratio $F(\zeta_1)/N_1$.

Substituting Eq. (23-101) into Eqs. (23-103) and (23-104), carrying out the integrals, and letting maxima be expressed in the nondimensional form

$$\eta \equiv \frac{\zeta_1}{m_0{}^{1/2}} \quad (23\text{-}105)$$

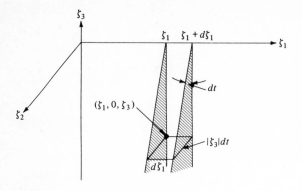

FIGURE 23-11
Shaded region satisfies the conditions for a maximum ($+$ or $-$) occurring in the range $(\zeta_1, \zeta_2 + d\zeta_1)$ and in the time interval $(t, t + dt)$.

yields the probability density function for maxima in the form

$$p(\eta) = \frac{1}{(2\pi)^{1/2}} \left[\varepsilon e^{-\eta^2/2\varepsilon^2} + (1 - \varepsilon^2)^{1/2} \eta e^{-\eta^2/2} \int_{-\infty}^{[\eta(1-\varepsilon^2)^{1/2}]/\varepsilon} e^{-x^2/2} \, dx \right] \quad (23\text{-}106)$$

where

$$\varepsilon^2 \equiv \frac{m_0 m_4 - m_2^2}{m_0 m_4} = \frac{\Delta}{m_0 m_4} \quad (23\text{-}107)$$

From Eq. (23-99) it can easily be shown that Δ is always positive; therefore, ε, as defined by Eq. (23-107), must always be in the range

$$0 < \varepsilon < 1 \quad (23\text{-}108)$$

Equation (23-106) is plotted in Fig. 23-12 for different values of ε throughout this range. Note that for a narrowband process approaching the single harmonic process given by Eq. (23-2), $\varepsilon \to 0$, in which case Eq. (23-106) reduces to the form of a Rayleigh distribution, Eq. (22-120). When the process is white noise or band-limited white noise, as given by Eq. (23-75), $\varepsilon = 2/3$. The limiting case $\varepsilon = 1$ can be approached by superposition of a single harmonic process $y(t)$ at frequency $\bar{\omega}_2$ and a band-limited process $z(t)$ within the frequency range $-\bar{\omega}_1 < \bar{\omega} < \bar{\omega}_1$, provided that $\bar{\omega}_2/\bar{\omega}_1 \to \infty$ and $\sigma_y^2/\sigma_z^2 \to 0$. This is equivalent to placing a very high-frequency, low-amplitude "dither" signal on top of a low-frequency band-limited signal. The resulting distribution of maxima as given by Eq. (23-106) approaches the form of a gaussian distribution. If the value of ε is to be estimated for a given sample process $x(t)$, this can easily be accomplished by first counting the total number of maxima ($+$ and $-$) N and the number of negative maxima N^- occurring in a sample function of reasonable duration. Dividing N^- by N gives the proportion r of negative maxima present in the total, which must be equal to the area under the probability density

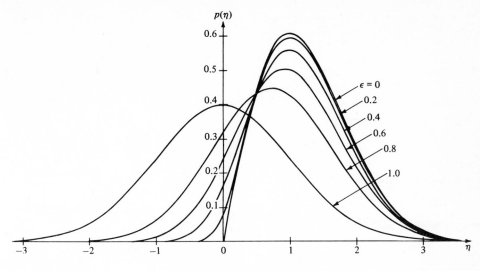

FIGURE 23-12
Probability density function for maxima for different values of ε.

function $p(\eta)$ to the left of the origin in Fig. 23-12. It can be shown that ε is approximately related to this area by

$$\varepsilon^2 = 4r(1 - r) \qquad (23\text{-}109)$$

Thus after $r = N^-/N$ has been determined, ε can immediately be estimated by this relation.

EXAMPLE E23-7   Compute the numerical value of ε for stationary process $x(t)$ which has a uniform power spectral density function of intensity S_0 over the ranges $-\bar{\omega}_2 < \bar{\omega} < -\bar{\omega}_1$ and $\bar{\omega}_1 < \bar{\omega} < \bar{\omega}_2$ as given by Eqs. (23-76).

Substituting the first of Eqs. (23-76) into Eq. (23-99) and completing the integral for $n = 0, 2,$ and 4 gives, respectively,

$$m_0 = 2S_0(\bar{\omega}_2 - \bar{\omega}_1)$$

$$m_2 = \frac{2S_0}{3} (\bar{\omega}_2{}^3 - \bar{\omega}_1{}^3) \qquad (a)$$

$$m_4 = \frac{2S_0}{5} (\bar{\omega}_2{}^5 - \bar{\omega}_1{}^5)$$

Substituting these relations into Eq. (23-107) yields

$$\varepsilon^2 = 1 - \frac{5}{9} \frac{(1 - \gamma^3)^2}{(1 - \gamma)(1 - \gamma^5)} \qquad (b)$$

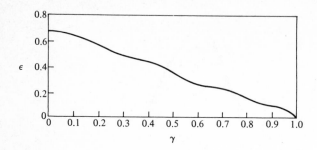

FIGURE E23-5
Parameter ε versus frequency ratio $\bar{\omega}_1/\bar{\omega}_2$.

where γ is the dimensionless frequency parameter

$$\gamma = \frac{\bar{\omega}_1}{\bar{\omega}_2} \qquad (c)$$

Equation (b) is plotted in Fig. E23-5, showing that ε varies in an approximately linear fashion from a value of $^2/_3$ at $\gamma = 0$ to a value of zero at $\gamma = 1$, thus (from Fig. 23-12) showing how the probability density function for maxima approaches the Rayleigh distribution as the frequency bandwidth narrows.

////

23-11 PROBABILITY DISTRIBUTION FOR EXTREME-VALUES[1]

Consider N independently observed maxima having the probability density function $p(\eta)$ given by Eq. (23-106). The probability that all N maxima will be less than η is given by

$$\text{Pr (all } N \text{ maxima} < \eta) = P(\eta)^N \qquad (23\text{-}110)$$

where $P(\eta)$ is the probability distribution function for maxima as defined by

$$P(\eta) \equiv \int_{-\infty}^{\eta} p(\eta)\, d\eta \qquad (23\text{-}111)$$

Obviously, the probability distribution function for the largest maxima $P_e(\eta)$ (extreme-value) must also be given by Eq. (23-110), that is,

$$P_e(\eta) = P(\eta)^N \qquad (23\text{-}112)$$

Taking the derivative of Eq. (23-112) gives the probability density function for the extreme-value in the form

$$p_e(\eta) = NP(\eta)^{N-1}p(\eta) \qquad (23\text{-}113)$$

[1] D. E. Cartwright and M. S. Longuet-Higgins, The Statistical Distribution of the Maxima of a Random Function, loc. cit.; A. G. Davenport, Note on the Distribution of the Largest Value of a Random Function with Application to Gust Loading, loc. cit.

For large values of N, it is quite apparent that relatively large values of η are of interest; therefore the accuracy with which the extreme-value distribution $p_e(\eta)$ can be defined depends very much on the accuracy of the function $P(\eta)$ as it approaches unity asymptotically with increasing values of η.

When Eq. (23-101) is substituted into Eq. (23-104) and the double integral is carried out along with multiplication by T, the number of maxima N occurring in a sample function $x_r(t)$ over a time duration T is given by Rice's equation

$$N = \frac{1}{2\pi} \left(\frac{m_4}{m_2} \right)^{1/2} T \quad (23\text{-}114)$$

Using Eqs. (23-106), (23-111), and (23-112), Davenport has shown, relying in part on earlier work by Cartwright and Lonquet-Higgins, that

$$P_e(\eta) = \exp\left[-vT \exp\left(-\frac{\eta^2}{2} \right) \right] \quad (23\text{-}115)$$

where

$$v \equiv \frac{1}{2\pi} \left(\frac{m_2}{m_0} \right)^{1/2} \quad (23\text{-}116)$$

The probability density function $p_e(\eta)$ can easily be obtained by differentiating Eq. (23-115) with respect to η.

Using the extreme-value probability distribution function given by Eq. (23-115), it has been shown by Davenport that the mean extreme-value is given by the approximate relation

$$\bar{\eta}_e \doteq (2 \ln vT)^{1/2} + \frac{\gamma}{(2 \ln vT)^{1/2}} \quad (23\text{-}117)$$

where γ is Euler's constant, equal to 0.5772, and that the standard deviation of the extreme-values is given by

$$\sigma_{\eta_e} = \frac{\pi}{\sqrt{6}} \frac{1}{(2 \ln vT)^{1/2}} \quad (23\text{-}118)$$

Figure 23-13 shows a plot of the probability density function for process $x(t)$, a plot of the probability density function for maxima $\eta(\varepsilon = 2/3)$, and plots of the extreme-value probability density function for four different values of vT (10^2, 10^3, 10^4, 10^5). It should be noted that the probability density functions for extreme-values are sharply peaked and that the degree of peaking increases with increasing values of vT. Because of this characteristic, engineering designs can often be based on the mean extreme-value $\bar{\eta}_e$ as expressed by Eq. (23-117), which is plotted in Fig. 23-14. It is clear from this figure that arbitrarily assuming $\bar{\eta}_e$ equal to 3, as is often done in practice, can be considerably on the unsafe side, especially for large values of vT.

Since the general form of the probability distribution function for largest

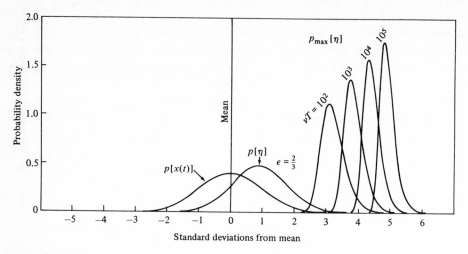

FIGURE 23-13
Probability density functions for $x(t)$, η, and η_e.

maxima $P_e(\eta)$ closely depends on the accuracy of the probability distribution function for maxima $P(\eta)$ as it nears unity with increasing values of η, other forms of $P_e(\eta)$ have been derived by making various assumptions regarding the manner in which $P(\eta)$ approaches unity. One such assumption is that $P(\eta)$ approaches unity in the manner

$$P(\eta) = 1 - e^{-\eta} \quad (23\text{-}119)$$

With this asymptotic form, the extreme-value distribution (Gumbel Type I)[1] can be expressed as

$$P_e(\eta) = \exp\{-\exp[-\alpha(\eta - u)]\} \quad (23\text{-}120)$$

where α and u are constants. Since the second derivative of Eq. (23-120) vanishes for $\eta = u$, constant u must equal the most probable value of η. Equation (23-120) gives the mean and standard deviation for the extreme-value in the form

$$\bar{\eta}_e = u + \frac{\gamma}{\alpha} \quad (23\text{-}121)$$

$$\sigma_{\eta e} = \frac{\pi}{\sqrt{6}\,\alpha} \quad (23\text{-}122)$$

where γ is Euler's constant (0.5772). From Eq. (23-122) it is clear that constant α is a measure of the dispersion of the extreme values.

[1] E. J. Gumbel and P. G. Carlson, Extreme Values in Aeronautics, *Jour. of Aero. Sci.*, pp. 389–398, June, 1954. E. J. Gumbel, Probability Tables for the Analysis of Extreme-Value Data, *Natl. Bur. Stds. Appl. Math. Ser.* 22, July, 1953.

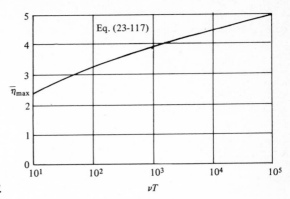

FIGURE 23-14
Normalized mean extreme-value vs. νT.

If a very large number of experimental extreme-values are known, the mean and standard deviation can be calculated fairly accurately, whereupon Eqs. (23-121) and (23-122) can be used to solve for α and u. However, if the number of extreme-values is relatively small, a correction should be made to this procedure, as reported by Gumbel.

EXAMPLE E23-8 The extreme-values of 50 sample members of random process $x(t)$ have been measured giving the following numerical values rearranged in order of rank:

0.82	1.14	−1.54	1.97	2.67
−0.90	1.16	1.60	1.99	−2.74
0.98	−1.20	−1.64	−2.02	−2.98
−1.03	1.29	−1.67	−2.09	3.33
−1.06	−1.39	1.70	2.11	3.50
1.08	−1.44	1.75	2.13	−3.63
1.10	1.46	−1.77	−2.23	3.85
−1.11	1.48	1.84	2.37	−4.07
−1.12	−1.50	−1.90	−2.51	−4.18
−1.13	1.51	−1.93	2.60	4.33

Assuming a Gumbel Type I distribution, write the corresponding probability distribution function for random variable x_e.

An approximation of the distribution function can be obtained using Eqs. (23-121) and (23-122). When the signs of the measured extreme values are ignored, the result is

$$\bar{x}_e = \frac{1}{50} \sum_{i=1}^{50} x_{e_i} = 1.97 \qquad \sigma_{x_e}^2 = \frac{1}{50} \sum_{i=1}^{50} (x_{e_i} - \bar{x}_e)^2 = 0.839 \qquad (a)$$

Using Eqs. (23-121) and (23-122) gives

$$\alpha = \frac{\pi}{\sqrt{6}\,\sigma_{x_e}} = 1.40 \qquad u = \bar{x}_e - \frac{0.577}{\alpha} = 1.56 \qquad (b)$$

Substituting Eqs. (b) into Eq. (23-120) results in

$$P(x_e) \doteq \exp\{-\exp[-1.40(x_e - 1.56)]\} \qquad (c)$$

Using the correction as given by Gumbel for the case of 50 sample values gives the more accurate expression

$$P(x_e) = \exp\{-\exp[-1.27(x_e - 1.54)]\} \qquad (d)$$

$////$

23-12 NONSTATIONARY GAUSSIAN PROCESSES

A stationary process has previously been defined as one for which all ensemble averages are independent of time; therefore, a nonstationary process is one for which these same ensemble averages are time-dependent. Thus the ensemble average $E[x(t)x(t + \tau)]$, which completely characterizes a nonstationary gaussian process $x(t)$, will be dependent upon time t as well as the time interval τ.

In engineering, a nonstationary process $x(t)$ can often be represented fairly well using the form

$$x(t) = \zeta(t)z(t) \qquad (23\text{-}123)$$

where $\zeta(t)$ is a fully prescribed function of time and $z(t)$ is a stationary process. If $x(t)$ is a gaussian process, $z(t)$ will also be gaussian, in which case the function

$$E[x(t)x(t + \tau)] = \zeta(t)\zeta(t + \tau)R_z(\tau) \qquad (23\text{-}124)$$

completely characterizes the process.

The above characterization of nonstationary gaussian processes involving one independent variable can be extended directly to processes involving more than one independent variable.

23-13 STATIONARY GAUSSIAN PROCESS: TWO OR MORE INDEPENDENT VARIABLES

All the stationary gaussian processes characterized previously involved one independent variable which was considered to be time t. The basic concepts developed for these processes will now be extended to stationary gaussian processes involving two or more independent variables. To illustrate this extension, suppose the variable of interest is random not only with respect to time but with respect to certain space coordinates as well. For example, consider the wind drag force per unit height acting on a tall industrial smokestack during a strong windstorm, as described in Sec. 23-1. This loading involves two independent variables, x and t.

To characterize the random component of drag $p(x,t)$ in a probabilistic sense, it is necessary to establish probability density functions involving random variables $p(x,t)$ and $p(\alpha, t + \tau)$, where α and τ are dummy space and time variables, respectively. If the process is gaussian, these probability density functions will be completely known provided the covariance function as given by the ensemble average $E[p(x,t)p(\alpha, t + \tau)]$ can be defined. If the process is stationary, this ensemble average will be independent of time but will depend upon the time difference τ, in which case the covariance function defined by the relation

$$R_p(x,\alpha,\tau) \equiv E[p(x,t)p(\alpha,t + \tau)] \quad (23\text{-}125)$$

completely characterizes the process.

Assuming the above process is ergodic, that is, the mean wind velocity remains constant for all members of the ensemble, the cross-spectral density function for the rth member, that is,

$$S_{p_r}(x,\alpha,\bar{\omega})$$

$$\equiv \lim_{s \to \infty} \frac{\left[\int_{-s/2}^{s/2} p_r(x,t) \exp(i\bar{\omega}t) \, dt \right] \left[\int_{-s/2}^{s/2} p_r(\alpha,t) \exp(-i\bar{\omega}t) \, dt \right]}{2\pi s} \quad (23\text{-}126)$$

will also characterize the process. This cross-spectral density function is related to the covariance function through the Fourier relations

$$S_p(x,\alpha,\bar{\omega}) = \frac{1}{2\pi} \int_{-\infty}^{\infty} R_p(x,\alpha,\tau) \exp(-i\bar{\omega}\tau) \, d\tau$$

$$R_p(x,\alpha,\tau) = \int_{-\infty}^{\infty} S_p(x,\alpha,\bar{\omega}) \exp(i\bar{\omega}\tau) \, d\bar{\omega} \quad (23\text{-}127)$$

Extending the above characterizations to stationary gaussian processes involving more than two independent variables is straightforward. For example, to characterize a field potential $\Phi(x,y,z,t)$ which is random with respect to time and each space coordinate, one must establish either the covariance function

$$R_\Phi(x,y,z,\alpha,\beta,\gamma,\tau) \equiv E[\Phi(x,y,z,t)\Phi(\alpha, \beta, \gamma, t + \tau)] \quad (23\text{-}128)$$

or the corresponding cross-spectral density function $S_\Phi(x,y,z,\alpha,\beta,\gamma,\bar{\omega})$. Terms α, β, and γ are dummy variables for x, y, and z, respectively.

If the field potential $\Phi(x,y,z,t)$ is homogeneous, the covariance and cross-spectral density functions depend only on the differences in coordinates, that is, on

$$X \equiv x - \alpha \qquad Y \equiv y - \beta \qquad Z \equiv z - \gamma \quad (23\text{-}129)$$

The process is then characterized either by the function $R_\Phi(X,Y,Z,\tau)$ or by the function $S_\Phi(X,Y,Z,\overline{\omega})$.

If the potential function $\Phi(x,y,z,t)$ happens to be isotropic as well as homogeneous, the covariance and cross-spectral density functions will depend only upon the distance between points, that is, the distance

$$\rho \equiv [(x - \alpha)^2 + (y - \beta)^2 + (z - \gamma)^2]^{1/2} \quad (23\text{-}130)$$

in which case the process will be characterized either by $R_\Phi(\rho,\tau)$ or $S_\Phi(\rho,\overline{\omega})$.

PROBLEMS

23-1 Show that the Fourier transform of an even function and of an odd function are real and imaginary, respectively.

23-2 Find the Fourier transform of each function $x(t)$ shown in Fig. P23-1.

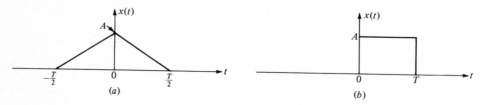

FIGURE P23-1
Functions $x(t)$ referred to in Prob. 2.

23-3 Consider the function $x(t) = A \cos at$ in the range $-T/2 < t < T/2$ and $x(t) = 0$ outside this range. Find and sketch the Fourier transform $X(\overline{\omega})$ when (a) $T = \pi/a$, (b) $T = 3\pi/a$, (c) $T = 5\pi/a$, and (d) $T \to \infty$.

23-4 Evaluate the integral

$$I = \int_1^\infty \left[\int_1^\infty \frac{x^2 - y^2}{(x^2 + y^2)^2} \, dy \right] dx$$

by integrating first with respect to y and then with respect to x. Then reverse the order of integration and reevaluate the integral. Finally evaluate the limit L of integral I by integrating over the finite domain and then taking the limit as follows:

$$L = \lim_{T=\to} \left\{ \int_1^T \left[\int_1^T \frac{x^2 - y^2}{(x^2 + y^2)^2} \, dy \right] dx \right\}$$

Noting that the integrand in integral I is antisymmetric about the line $x = y$, which form of integration would you recommend for engineering applications?

23-5 Evaluate the integral

$$I = \int_{-\infty}^{\infty} \frac{\sin^2 x}{x^2} \, dx$$

23-6 Consider the stationary random process $x(t)$ defined by

$$x_r(t) = \sum_{n=1}^{10} A_{nr} \cos (n\overline{\omega}_0 t + \theta_{nr}) \qquad r = 1, 2, \ldots$$

where $x_r(t) = r$th member of ensemble

A_{nr} = sample values of random variable A

$\overline{\omega}_0$ = fixed circular frequency

θ_{nr} = sample values of random phase angle θ having a uniform probability density function in range $0 < \theta < 2\pi$ of intensity $1/2\pi$

If random variable A is gaussian having a known mean value $\overline{A}$ and a known variance σ_A^2, find the ensemble mean value of $x(t)$ and the ensemble variance of $x(t)$. Is process $x(t)$ a gaussian process?

23-7 Derive the autocorrelation function for the stationary random process $x(t)$ defined in Prob. 23-6.

23-8 Derive the power spectral density function for the stationary random process $x(t)$ defined in Prob. 23-6, assuming that Dirac delta functions are permitted in the answer.

23-9 A stationary random process $x(t)$ has the autocorrelation function

$$R_x(\tau) = A \exp (-a|\tau|)$$

where A and a are real constants. Find the power spectral density function for this process.

23-10 Consider a random process $x(t)$ which takes the value $+A$ or $-A$, with equal probability, throughout each interval $n \, \Delta\varepsilon < t < (n + 1) \, \Delta\varepsilon$ of each member of the process, where n is an integer running from $-\infty$ to $+\infty$. Find and plot the ensemble covariance function $E[x(t)x(t + \tau)]$. Is this process stationary or nonstationary?

23-11 If the origin of time, that is, $t = 0$, for each member of process $x(t)$ defined in Prob. 23-10 is selected randomly over an interval $\Delta\varepsilon$ with uniform probability of occurrence, what is the covariance function $E[x(t)x(t + \tau)]$? Is this process stationary or nonstationary?

23-12 Assuming that you find the process in Prob. 23-11 stationary, what are the autocorrelation and power spectral density functions for this process? Use Eqs. (23-35) and (23-38) in finding the power spectral density function.

23-13 Show that the autocorrelation and power spectral density functions obtained in Prob. 23-12 are Fourier transform pairs in accordance with Eqs. (23-37).

23-14 Each member of a stationary random process $x(t)$ consists of a periodic infinite train of triangular pulses, as shown in Fig. P23-2. All members of the process are identical except for phase, which is a random variable uniformly distributed over the interval $(0,T)$. Assuming that the period T is not less than $2a$, where a is the duration of a single pulse, find the autocorrelation function for this process.

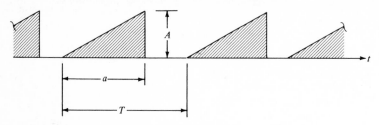

FIGURE P23-2
One sample member of process $x(t)$ referred to in Prob. 14.

23-15 Each member of a random process $x(t)$ consists of the superposition of rectangular pulses of duration $\Delta\varepsilon$ and of constant intensity A which are located in a random fashion with respect to time as shown in Fig. P23-3a. Each value of ε_n is sampled in accordance with the uniform probability density function $p(\varepsilon)$ given in Fig. P23-3b. What is the ensemble value $\overline{x(t)}$ for this process? What are the autocorrelation and power spectral density functions for the process $x(t) - \overline{x(t)}$?

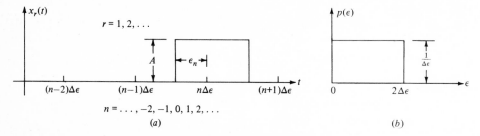

(a)   *(b)*

FIGURE P23-3
One sample member of process $x(t)$ and probability density function for random variable ε referred to in Prob. 15.

23-16 Assume the autocorrelation and power spectral density functions $R_{xx}(\tau)$ and $S_{xx}(\overline{\omega})$ for a stationary random process $x(t)$ are known. Derive the expressions for $S_{x\dot{x}}(\overline{\omega})$, $S_{\dot{x}x}(\overline{\omega})$, $S_{x\ddot{x}}(\overline{\omega})$, $S_{\ddot{x}x}(\overline{\omega})$, $S_{\dot{x}\dot{x}}(\overline{\omega})$, $S_{\ddot{x}\ddot{x}}(\overline{\omega})$ and $R_{x\dot{x}}(\tau)$, $R_{\dot{x}x}(\tau)$, $R_{x\ddot{x}}(\tau)$, $R_{\ddot{x}x}(\tau)$, $R_{\dot{x}\dot{x}}(\tau)$, $R_{\ddot{x}\ddot{x}}(\tau)$ in terms of $S_{xx}(\overline{\omega})$ and $R_{xx}(\tau)$, respectively.

23-17 Considering two stationary random processes $x(t)$ and $y(t)$, show that $S_{yx}(\overline{\omega})$ is the complex conjugate of $S_{xy}(\overline{\omega})$.

23-18 Two stationary random processes $x(t)$ and $y(t)$ have the joint probability density function

$$p[x(t)y(t + \tau)] = \frac{1}{2ab\sqrt{1 - c^2}} \exp\left[-\frac{1}{2(1 - c^2)}\left(\frac{x^2}{a^2} - \frac{2cxy}{ab} + \frac{y^2}{b^2}\right)\right]$$

Define a, b, and c in terms of the appropriate autocorrelation and/or cross-correlation functions for processes $x(t)$ and $y(t)$. What is the corresponding joint probability density function $p[\dot{x}(t)\dot{y}(t + \tau)]$? Define the coefficients in this function in terms of the appropriate autocorrelation and/or cross-correlation functions for processes $x(t)$ and $y(t)$.

24

STOCHASTIC RESPONSE OF LINEAR SDOF SYSTEMS

24-1 TRANSFER FUNCTIONS

This chapter develops the appropriate input-output relationships for stable linear SDOF systems having constant coefficients and characterizes the stationary output processes of such systems in terms of their corresponding stationary input processes and their transfer relationships.

Suppose that a stationary gaussian process $p(t)$ is the input to a linear SDOF system and that $v(t)$ is the desired output process, as shown in Fig. 24-1, where $TF_1, TF_2, \ldots, TF_n$ represent the transfer functions of systems $1, 2, \ldots, n$, respectively. Since uncontrollable random variables are always present during construction of real systems (even though of identical design), these transfer functions will also have random characteristics. Usually, however, in vibration analysis, the randomness of these characteristics is small in comparison with the randomness of the input $p(t)$ and therefore can be neglected, in which case $TF_1 = TF_2 = \cdots = TF_n = TF$. Thus, in the subsequent treatment of linear systems, the coefficients appearing in all mathematical representations will be considered as fixed constants; that is, the transfer

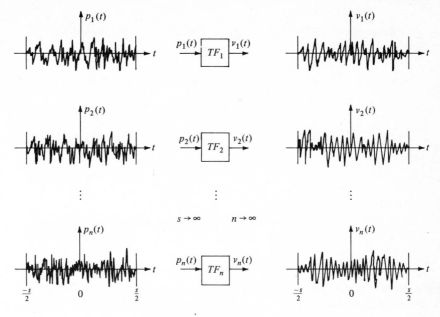

FIGURE 24-1
Input and output processes of a stable linear SDOF system.

function (or functions) TF_r will in each case be treated as independent of r. When the transfer function TF and either the autocorrelation function $R_p(\tau)$ or the power spectral density function $S_p(\bar{\omega})$ are known, the output process $v(t)$ can be completely characterized.

In Chap. 7 it was shown that the response of a SDOF system as governed by the equation

$$m\ddot{v} + c\dot{v} + kv = p(t) \qquad (24\text{-}1)$$

can be obtained through the time domain using the convolution, or Duhamel, integral

$$v(t) = \int_{-\infty}^{t} p(\tau)h(t - \tau)\, d\tau \qquad (24\text{-}2)$$

where τ is a dummy time variable and $h(t - \tau)$ is the free-vibration response $v(t)$ due to a unit impulse at time τ, that is, due to $p(t) = \delta(t - \tau)$. When this approach is used, the forcing function $p(t)$ is separated into infinitesimal impulses $p(\tau)\, d\tau$ and the total response is obtained by superposition of the individual free-vibration responses.

Response of a SDOF system can also be obtained through the frequency domain using the relation

$$v(t) = \frac{1}{2\pi} \int_{-\infty}^{\infty} H(i\bar{\omega})c(i\bar{\omega}) \exp{(i\bar{\omega}t)} \, d\bar{\omega} \qquad (24\text{-}3)$$

where $\bar{\omega}$ is a variable frequency, $H(i\bar{\omega})$ is the complex frequency-response function, and $c(i\bar{\omega})$ is the Fourier transform of the forcing function $p(t)$, that is,

$$c(i\bar{\omega}) = \int_{-\infty}^{\infty} p(t) \exp{(-i\bar{\omega}t)} \, dt \qquad (24\text{-}4)$$

which has the inverse relation

$$p(t) = \frac{1}{2\pi} \int_{-\infty}^{\infty} c(i\bar{\omega}) \exp{(i\bar{\omega}t)} \, d\bar{\omega} \qquad (24\text{-}5)$$

In this case, the forcing function $p(t)$ is separated into a complete spectrum of simple harmonics which cover the entire time range $-\infty < t < \infty$. For this reason, and because the right-hand sides of Eqs. (24-3) and (24-5) differ only by the presence of the complex-frequency-response function $H(i\bar{\omega})$ in one equation, this function must be defined as the ratio of the *steady-state* output response $v(t)$ to the harmonic excitation

$$p(t) = c(i\bar{\omega}) \exp{(i\bar{\omega}t)} \qquad (24\text{-}6)$$

which produces it, that is,

$$v(t) = H(i\bar{\omega})c(i\bar{\omega}) \exp{(i\bar{\omega}t)} \qquad (24\text{-}7)$$

Substituting Eqs. (24-6) and (24-7) into Eq. (24-1) and solving for the complex frequency function gives

$$H(i\bar{\omega}) = \frac{1}{k[1 + 2i\xi(\bar{\omega}/\omega) - (\bar{\omega}/\omega)^2]} \qquad \xi \geq 0 \qquad (24\text{-}8)$$

which is equivalent to Eq. (5-11).

To obtain the unit-impulse-response function $h(t)$, substitute $p(t) = \delta(t)$ into Eq. (24-1) and solve for $v(t)$, which equals $h(t)$ in this case, obtaining

$$h(t) = \begin{cases} \dfrac{1}{\omega_D m} \exp{(-\omega\xi t)} \sin{\omega_D t} & 0 \leq \xi \leq 1 \\[2em] \dfrac{1}{2\omega m \sqrt{\xi^2 - 1}} \exp{(-\omega\xi t)} & \\[1em] \quad \times \left[\exp{(\omega\sqrt{\xi^2 - 1}\, t)} \exp{(-\omega\sqrt{\xi^2 - 1}\, t)}\right] & \xi \geq 1 \end{cases} \qquad (24\text{-}9)$$

where ω_D is the damped frequency $\omega\sqrt{1 - \xi^2}$ in the undercritically damped case. The function $h(t)$ is, of course, zero for $t < 0$.

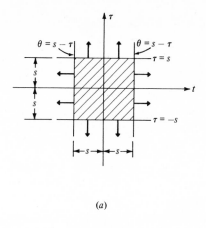

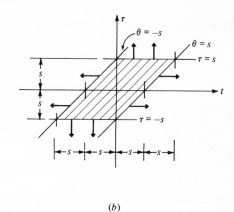

(a)

(b)

FIGURE 24-2
Expanding domains of integration.

24-2   RELATIONSHIP BETWEEN UNIT-IMPULSE AND COMPLEX-FREQUENCY-RESPONSE FUNCTIONS

When a complex function $V(i\bar{\omega})$ is defined as the Fourier transform of the response function $v(t)$, Eq. (24-2) in its transformed state becomes

$$V(i\bar{\omega}) = \int_{-\infty}^{\infty} \left[\int_{-\infty}^{t} p(\tau)h(t - \tau)\, d\tau \right] \exp\left(-i\bar{\omega}t\right) dt \qquad (24\text{-}10)$$

It will be assumed here that $v(t)$ damps out rapidly with respect to t so that its Fourier transform does indeed exist.

Since the function $h(t - \tau)$ equals zero for $\tau > t$, the upper limit of the second integral in Eq. (24-10) can be changed from t to ∞ without influencing the final result. Therefore, Eq. (24-10) can be expressed in the equivalent form

$$V(i\bar{\omega}) = \lim_{s \to \infty} \int\int_{-s}^{s} p(\tau)h(t - \tau) \exp\left(-i\bar{\omega}t\right) dt\, d\tau \qquad (24\text{-}11)$$

When a new variable $\theta \equiv t - \tau$ is introduced, Eq. (24-11) becomes

$$V(i\bar{\omega}) = \lim_{s \to \infty} \int_{-s}^{s} p(\tau) \exp\left(-i\bar{\omega}\tau\right) d\tau \int_{-s-\tau}^{s-\tau} h(\theta) \exp\left(-i\bar{\omega}\theta\right) d\theta \qquad (24\text{-}12)$$

The expanding domain of integration given by Eq. (24-12) is shown in Fig. 24-2a. Since the function $V(i\bar{\omega})$ exists only when the response $v(t)$ damps out rapidly with time, $p(t)$ and $h(t)$ must also damp out rapidly. Therefore, it is valid to drop τ from the limits of the second integral in Eq. (24-12), obtaining the relation

$$V(i\bar{\omega}) = \lim_{s \to \infty} \int_{-s}^{s} p(\tau) \exp(-i\bar{\omega}\tau) \, d\tau \int_{-s}^{s} h(\theta) \exp(-i\bar{\omega}\theta) \, d\theta \qquad (24\text{-}13)$$

which changes the expanding domain of integration to that shown in Fig. 24-2b. Variable θ can now be changed to t since it is serving only as a dummy time variable. Equation (24-13) then becomes

$$V(i\bar{\omega}) = c(i\bar{\omega}) \int_{-\infty}^{\infty} h(t) \exp(-i\bar{\omega}t) \, dt \qquad (24\text{-}14)$$

When it is noted that Eq. (24-3) in its inverse form gives

$$V(i\bar{\omega}) = c(i\bar{\omega})H(i\bar{\omega}) \qquad (24\text{-}15)$$

comparison of Eqs. (24-14) and (24-15) makes it apparent that

$$H(i\bar{\omega}) = \int_{-\infty}^{\infty} h(t) \exp(-i\bar{\omega}t) \, dt \qquad h(t) = \frac{1}{2\pi} \int_{-\infty}^{\infty} H(i\bar{\omega}) \exp(i\bar{\omega}t) \, d\bar{\omega} \qquad (24\text{-}16)$$

This derivation has shown that the unit-impulse-response function $h(t)$ and the complex-frequency-response function $H(i\bar{\omega})$ are Fourier transform pairs provided the unit-impulse-response function decays with respect to time as a result of damping present in the system. Systems which satisfy this condition are said to be stable.

EXAMPLE E24-1 Show that the complex-frequency-response function given by Eq. (24-8) and the unit-impulse-response functions given by Eqs. (24-9) are Fourier transform pairs in accordance with Eqs. (24-16) for $\xi > 0$.

Substituting Eq. (24-8) into the second of Eqs. (24-16), namely,

$$h(t) = \frac{1}{2\pi} \int_{-\infty}^{\infty} H(i\bar{\omega}) \exp(i\bar{\omega}t) \, d\bar{\omega} \qquad (a)$$

gives

$$h(t) = \frac{-1}{2\pi m\omega} \int_{-\infty}^{\infty} \frac{\exp(i\omega\beta t)}{(\beta - r_1)(\beta - r_2)} \, d\beta \qquad (b)$$

after introducing

$$\beta = \frac{\bar{\omega}}{\omega} \qquad\qquad k = m\omega^2 \qquad (c)$$

$$r_1 = i\xi + \sqrt{1 - \xi^2} \qquad r_2 = i\xi - \sqrt{1 - \xi^2} \qquad (d)$$

The integration of Eq. (b) is best carried out using the complex β plane and contour integration. The integrand in this equation is an analytic function everywhere in the β plane except at $\beta = r_1$ and $\beta = r_2$. At these two points poles of order 1 exist, as shown in Fig. E24-1 for damping in the ranges $0 < \xi < 1$ and $\xi < 1$. Note that for $\xi = 1$, points $\beta = r_1$ and $\beta = r_2$ coincide at location $(0,i)$, thus forming a single pole of order 2 in this case. The arrows along the closed paths in Fig. E24-1 indicate the directions of contour integration for the ranges of time shown. The poles mentioned above have residues as follows:

$$\text{Res} \,(\beta = r_1) = \frac{\exp\left[i\omega(i\xi + \sqrt{1 - \xi^2})\,t\right]}{2\sqrt{1 - \xi^2}} \qquad 0 < \xi < 1; \xi > 1$$

$$\text{Res} \,(\beta = r_2) = \frac{\exp\left[i\omega(i\xi - \sqrt{1 - \xi^2})\,t\right]}{-2\sqrt{1 - \xi^2}} \qquad 0 < \xi < 1; \xi > 1 \qquad (e)$$

$$\text{Res} \,(\beta = r_1 = r_2) = i\omega t \exp\,(-\omega t) \qquad \xi = 1$$

According to Cauchy's residue theorem, the integral in Eq. (b) equals $+2\pi i \sum \text{Res}$ and $-2\pi i \sum \text{Res}$ when integration is clockwise and counterclockwise, respectively, around a closed path and the integral is analytic along the entire path, as in the case treated here. Thus one obtains the results

$$h(t) = \begin{cases} \dfrac{-2\pi i}{2\pi m \omega} \\[2mm] \quad \times \left\{\dfrac{\exp\left[i\omega(i\xi + \sqrt{1 - \xi^2})\,t\right]}{2\sqrt{1 - \xi^2}}\right. \\[3mm] \qquad \left. + \dfrac{\exp\left[i\omega(i\xi - \sqrt{1 - \xi^2})\,t\right]}{-2\sqrt{1 - \xi^2}}\right\} \quad t > 0 \\[2mm] 0 \hspace{5cm} t < 0 \end{cases} \; \left.\begin{matrix}\\[1cm]\end{matrix}\right\} 0 < \xi < 1$$

$$\dfrac{-2\pi i}{2\pi m \omega}$$

$$\quad \times \left(\dfrac{\exp\left\{i\omega[i(\xi + \sqrt{\xi^2 - 1})]\,t\right\}}{2i\sqrt{\xi^2 - 1}}\right.$$

$$\qquad \left. + \dfrac{\exp\left\{i\omega[i(\xi - \sqrt{\xi^2 - 1})]\,t\right\}}{-2i\sqrt{\xi^2 - 1}}\right) \quad t > 0 \left.\begin{matrix}\\\end{matrix}\right\} \; \xi > 1$$

$$0 \hspace{5cm} t < 0$$

$$\dfrac{-2\pi i}{2\pi m \omega}\,[i\omega t \exp\,(-\omega t)] \qquad t > 0 \left.\begin{matrix}\\\end{matrix}\right\} \; \xi = 1$$

$$0 \hspace{5cm} t < 0$$

$$(f)$$

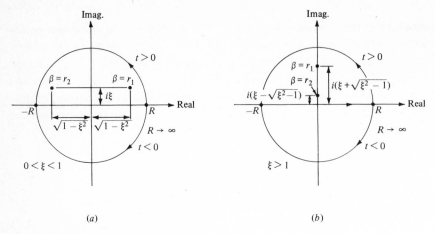

FIGURE E24-1
Poles for the integrand function of Eq. (*b*).

It is easily shown that Eqs. (*f*) reduce to

$$
h(t) = \begin{cases}
\dfrac{1}{\omega_D m} \exp(-\omega\xi t) \sin \omega_D t & t > 0 \\[2mm]
0 & t < 0
\end{cases} \; \Biggr\} \; 0 < \xi < 1 \\[6mm]
\left.
\begin{array}{l}
\dfrac{1}{2\omega m \sqrt{\xi^2 - 1}} \exp(-\omega\xi t) \\[2mm]
\quad \times \left[\exp(\omega\sqrt{\xi^2 - 1}\, t)\right. \\[2mm]
\qquad \left. - \exp(-\omega\sqrt{\xi^2 - 1}\, t)\right] \quad t > 0 \\[2mm]
0 \hspace{5.5cm} t < 0
\end{array}
\right\} \; \xi > 1 \\[6mm]
\left.
\begin{array}{l}
\dfrac{t}{m} \exp(-\omega t) \hspace{3cm} t > 0 \\[2mm]
0 \hspace{4.2cm} t < 0
\end{array}
\right\} \; \xi = 1
$$

$$(g)$$

where $\omega_D = \omega\sqrt{1 - \xi^2}$. Note that all of Eqs. (*g*) are contained in the single integral expression Eq. (*b*).

Since Eqs. (24-9) for $\xi > 0$ are indeed the inverse Fourier transform of Eq. (24-8), the direct transform relationship must also hold; that is, the first of Eqs. (24-16) must also be satisfied by Eqs. (24-8) and (24-9).   ////

24-3 RELATIONSHIP BETWEEN INPUT AND OUTPUT AUTOCORRELATION FUNCTIONS

The output or response function $v_r(t)$ shown in Fig. 24-1 is related to its corresponding input function $p_r(t)$ through the convolution integral relation

$$v_r(t) = \int_{-\infty}^{t} p_r(\tau)h(t - \tau) \, d\tau \qquad r = 1, 2, \ldots, \infty \qquad (24\text{-}17)$$

If the input process is assumed to have a zero mean value, that is,

$$E[p(t)] = 0 \qquad (24\text{-}18)$$

the mean value for the output process can be obtained by averaging Eq. (24-17) across the ensemble, which gives

$$E[v(t)] = E\left[\int_{-\infty}^{t} p(\tau)h(t - \tau) \, d\tau \right] = \int_{-\infty}^{t} E[p(\tau)]h(t - \tau) \, d\tau = 0 \qquad (24\text{-}19)$$

Thus it is shown that if the input ensemble has a zero mean value, the output ensemble will also have a zero mean value.

Consider now the ensemble average $E[v(t)v(t + \tau)]$, which can be evaluated by using Eq. (24-17) as shown in the relation

$$E[v(t)v(t + \tau)]$$
$$= E\left[\int_{-\infty}^{t} p(\theta_1)h(t - \theta_1) \, d\theta_1 \int_{-\infty}^{t+\tau} p(\theta_2)h(t + \tau - \theta_2) \, d\theta_2 \right] \qquad (24\text{-}20)$$

where θ_1 and θ_2 are dummy time variables. When a change of variables is introduced in accordance with the definitions

$$u_1 \equiv t - \theta_1 \qquad \theta_1 = t - u_1$$
$$u_2 \equiv t + \tau - \theta_2 \qquad \theta_2 = t + \tau - u_2 \qquad (24\text{-}21)$$

Eq. (24-20) becomes

$$E[v(t)v(t + \tau)]$$
$$= E\left[\int_{t+\infty}^{0} p(t - u_1)h(u_1) \, du_1 \int_{t+\tau+\infty}^{0} p(t + \tau - u_2)h(u_2) \, du_2 \right] \qquad (24\text{-}22)$$

When the limits of both integrals are inverted and use is made of the fact that $h(u_1)$ and $h(u_2)$ damp out for stable systems, Eq. (24-22) can be written in the form

$$E[v(t)v(t + \tau)] = E\left[\int\!\!\!\int_{0}^{\infty} p(t - u_1)p(t + \tau - u_2)h(u_1)h(u_2) \, du_1 \, du_2 \right] \qquad (24\text{-}23)$$

Since only the functions $p(t - u_1)$ and $p(t + \tau - u_2)$ change across the ensemble, Eq. (24-23) becomes

$$E[v(t)v(t + \tau)] = \int\int_0^\infty E[p(t - u_1)p(t + \tau - u_2)]h(u_1)h(u_2)\, du_1\, du_2 \qquad (24\text{-}24)$$

The ensemble average on the right-hand side of Eq. (24-24) is the autocorrelation function for stationary process $p(t)$ and is independent of time; therefore, the ensemble average on the left-hand side must also be independent of time. This shows that the output process is stationary and that its autocorrelation function is given by the relation

$$R_v(\tau) = \int\int_0^\infty R_p(\tau - u_2 + u_1)h(u_1)h(u_2)\, du_1\, du_2 \qquad (24\text{-}25)$$

If the input process $p(t)$ is gaussian, the output process $v(t)$ for a linear stable system will also be gaussian; therefore, in such cases the autocorrelation function given by Eq. (24-25) completely characterizes the process. To prove the first part of this statement consider Eq. (24-17) in the limiting form

$$v_r(t) = \lim_{\Delta\tau \to 0} \sum_{i=-\infty}^{t/\Delta\tau} p_r(\tau_i)h(t - \tau_i)\, \Delta\tau \qquad r = 1, 2, \ldots, \infty \qquad (24\text{-}26)$$

Since all terms $h(t - \tau_i)$ are known constants for fixed values of t, Eqs. (24-26) are identical in form with the linear transformations given by Eqs. (22-130), which have already been shown to retain the gaussian distribution.

To illustrate the application of Eq. (24-25), assume that the excitation $p(t)$ of the SDOF system represented by Eq. (24-1) is white noise; that is, its power spectral density function equals a constant S_0 which corresponds to the autocorrelation function

$$R_p(\tau) = 2\pi S_0 \delta(\tau) \qquad (24\text{-}27)$$

Further, assume that the system is undercritically damped, in which case the unit impulse function $h(t)$ is given by the first of Eqs. (24-9). Substituting this relation along with Eq. (24-27) into Eq. (24-25) gives

$$R_v(\tau) = \frac{2\pi S_0}{\omega_D^2 m^2} \int\int_0^\infty \delta(\tau - u_2 + u_1)$$

$$\times \exp\left[-\omega\xi(u_1 + u_2)\right] \sin \omega_D u_1 \sin \omega_D u_2\, du_1\, du_2 \qquad (24\text{-}28)$$

Completing the double integration and introducing the relation $k = m\omega^2$ leads to

$$R_v(\tau) = \frac{\pi\omega S_0}{2k^2\xi}\left(\cos \omega_D|\tau| + \frac{\xi}{\sqrt{1 - \xi^2}} \sin \omega_D|\tau|\right)$$

$$\times \exp\left(-\omega\xi|\tau|\right) \qquad -\infty < \tau < \infty \qquad (24\text{-}29)$$

If the system is overcritically damped, the unit-impulse function $h(t)$ is given by the second of Eqs. (24-9) and the above integration procedure leads to the relation

$$R_v(\tau) = \frac{\pi \omega S_0}{2k^2 \xi} [\phi \exp (\omega \sqrt{\xi^2 - 1} |\tau|) - \theta \exp (-\omega \sqrt{\xi^2 - 1} |\tau|)]$$

$$\times \exp (-\omega \xi |\tau|) \qquad -\infty < \tau < \infty \qquad (24\text{-}30)$$

where

$$\phi \equiv \frac{1}{2[\xi \sqrt{\xi^2 - 1} - (\xi^2 - 1)]} \qquad \theta \equiv \frac{1}{2[\xi \sqrt{\xi^2 - 1} + (\xi^2 - 1)]} \qquad (24\text{-}31)$$

If the white-noise input is gaussian, the output response $v(t)$ will also be gaussian, in which case Eqs. (24-29) and (24-30) completely characterize the processes they represent.

EXAMPLE E24-2   Consider the SDOF system

$$m\ddot{v} + c\dot{v} + kv = p(t) \qquad (a)$$

excited by a zero mean ergodic random process $p(t)$ having constant power spectral density equal to S_0 over the range $-\infty < \bar{\omega} < \infty$. Determine the average rate of energy dissipation in the system with time.

Since the damping force $c\dot{v}$ is the only nonconservative force in the system, the instantaneous rate of energy dissipation is given by $c\dot{v}^2$. Therefore, the average rate of energy dissipation Φ can be expressed as

$$\Phi = c \langle \dot{v}^2 \rangle = C R_{\dot{v}}(0) \qquad (b)$$

Using Eq. (23-43) gives

$$R_{\dot{v}}(\tau) = -R_v''(\tau) \qquad (c)$$

Substituting Eq. (24-29) into Eq. (c) gives

$$R_{\dot{v}}(\tau) = -R_v''(\tau) = \frac{\pi \omega S_0}{2k^2 \xi} \left(\omega^2 \cos \omega_D |\tau| - \frac{\omega^2 \xi}{\sqrt{1 - \xi^2}} \sin \omega_D |\tau| \right)$$

$$\times \exp (-\omega \xi |\tau|) \qquad (d)$$

from which

$$R_{\dot{v}}(0) = -R_v''(0) = \frac{\pi \omega^3 S_0}{2k^2 \xi} \qquad (e)$$

Substituting Eq. (e) into Eq. (b) and making use of the relations $k^2 = m^2 \omega^4$ and $c = 2m\omega\xi$ leads to

$$\Phi = \frac{\pi S_0}{m} \qquad (f)$$

Note that the average rate of energy dissipation is independent of the damping ratio ξ.   ////

24-4   RELATIONSHIP BETWEEN INPUT AND OUTPUT POWER SPECTRAL DENSITY FUNCTIONS

The power spectral density function for the output process $v(t)$ is related to its auto-correlation function through the Fourier transform relation

$$S_v(\bar{\omega}) = \frac{1}{2\pi} \int_{-\infty}^{\infty} R_v(\tau) \exp(-i\bar{\omega}\tau) \, d\tau \qquad (24\text{-}32)$$

Substituting Eq. (24-25) into Eq. (24-32) gives

$$S_v(\bar{\omega}) = \frac{1}{2\pi} \int_{-\infty}^{\infty} \left[\int\!\!\int_0^{\infty} R_p(\tau - u_2 + u_1)h(u_1)h(u_2) \, du_1 \, du_2 \right] \exp(-i\bar{\omega}\tau) \, d\tau \quad (24\text{-}33)$$

Interchanging the order of integration and introducing expanding limits of integration leads to

$$S_v(\bar{\omega}) = \frac{1}{2\pi} \lim_{s\to\infty} \left[\int_0^s h(u_1) \, du_1 \int_0^s h(u_2) \, du_2 \int_{-s}^s R_p(\tau - u_2 + u_1) \right.$$

$$\left. \times \exp(-i\bar{\omega}\tau) \, d\tau \right] \qquad (24\text{-}34)$$

When a change of variable $\theta \equiv \tau - u_2 + u_1$ is substituted, Eq. (24-34) changes to the form

$$S_v(\bar{\omega}) = \frac{1}{2\pi} \lim_{s\to\infty} \left[\int_0^s h(u_1) \exp(i\bar{\omega}u_1) \, du_1 \int_0^s h(u_2) \exp(-i\bar{\omega}u_2) \, du_2 \right.$$

$$\left. \times \int_{-s+u_1-u_2}^{s+u_1-u_2} R_p(\theta) \exp(-i\bar{\omega}\theta) \, d\theta \right] \qquad (24\text{-}35)$$

Since the unit-impulse-response functions $h(u_1)$ and $h(u_2)$ equal zero for $u_1 < 0$ and $u_2 < 0$, respectively, the lower limits of the first two integrals can be changed from zero to $-s$. Also since these functions must damp out with increasing values of u_1 and u_2 for the system to be stable, these terms can be dropped from the limits of the third integral. When use is made of the first of Eqs. (23-37) and (24-16), Eq. (24-35) reduces to the form

$$S_v(\bar{\omega}) = |H(i\bar{\omega})|^2 S_p(\bar{\omega}) \qquad (24\text{-}36)$$

where $H(i\bar{\omega})$ is the complex-frequency-response function for $v(t)$ based on a harmonic input function $p(t)$.

To consider once again the SDOF system represented by Eq. (24-1) when subjected to a white-noise excitation $p(t)$, that is, $S_p(\bar{\omega}) = S_0$, a substitution of Eq. (24-8) into (24-36) gives

$$S_v(\bar{\omega}) = \frac{S_0}{k^2[1 + (4\xi^2 - 2)(\bar{\omega}/\omega)^2 + (\bar{\omega}/\omega)^4]} \tag{24-37}$$

This power spectral density function is valid for both the under- and overcritically damped cases.

Plots of the output autocorrelation and power spectral density functions for a SDOF system, Eq. (24-1), subjected to white-noise excitation are shown in Fig. 24-3.

EXAMPLE E24-3   Derive Eq. (24-29) directly from Eq. (24-37) making use of the Fourier transform relation

$$R_v(\tau) = \int_{-\infty}^{\infty} S_v(\bar{\omega}) \exp(i\bar{\omega}\tau) \, d\bar{\omega} \tag{a}$$

Substituting Eq. (24-37) into Eq. (a) gives

$$R_v(\tau) = \frac{\omega S_0}{k^2} \int_{-\infty}^{\infty} \frac{\exp(i\omega\beta\tau)}{(\beta - r_1)(\beta - r_2)(\beta + r_1)(\beta + r_2)} \, d\beta \tag{b}$$

after introducing

$$\beta = \frac{\bar{\omega}}{\omega} \qquad r_1 = i\xi + \sqrt{1 - \xi^2} \qquad r_2 = i\xi - \sqrt{1 - \xi^2} \tag{c}$$

The integrand in Eq. (b) is an analytic function everywhere in the complex β plane except at points $\beta = r_1$, $\beta = r_2$, $\beta = -r_1$, and $\beta = -r_2$, where poles of order 1 exist. Points $\beta = r_1$ and $\beta = r_2$ are in the upper half plane, while points $\beta = -r_1$ and $\beta = -r_2$ are in the lower half plane. For positive values of τ, contour integration is carried out in the upper half plane; for negative values of τ, integration is carried out in the lower half plane. When Cauchy's residue theorem is used, the integral in Eq. (b) is easily carried out by procedures similar to those in Example E24-1, resulting in the relation

$$R_v(\tau) = \begin{cases} \dfrac{\pi\omega S_0}{2k^2\xi}\left(\cos\omega_D\tau + \dfrac{\xi}{\sqrt{1 - \xi^2}}\sin\omega_D\tau\right)\exp(-\omega\xi\tau) & \tau > 0 \\[4mm] \dfrac{\pi\omega S_0}{2k^2\xi}\left(\cos\omega_D\tau - \dfrac{\xi}{\sqrt{1 - \xi^2}}\sin\omega_D\tau\right)\exp(\omega\xi\tau) & \tau < 0 \end{cases} \tag{d}$$

Thus the validity of Eq. (24-29) is verified.   ////

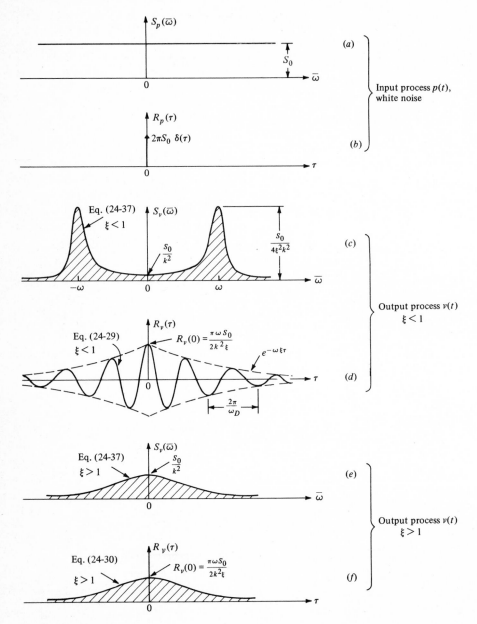

FIGURE 24-3
Output relations for SDOF system subjected to white-noise excitation.

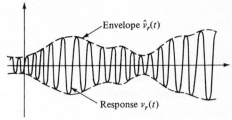

FIGURE 24-4
Sample function of a narrowband process $v(t)$.

24-5 RESPONSE CHARACTERISTICS FOR NARROWBAND SYSTEMS

Most structural systems have reasonably low damping ($\xi < 0.1$) and therefore are classified as narrowband systems. This classification results because the area under the response power spectral density function is highly concentrated near the natural frequency of the system, as shown in Fig. 24-3c. Such a concentration indicates that the predominant frequency components in a sample response function $v_r(t)$ will be contained in a relatively narrow band centered on the undamped natural frequency ω. Because of the beat phenomenon associated with two harmonics whose frequencies are close together, the response envelope for a narrowband system can be expected to show similar characteristics; however, since the predominant frequencies are spread over a narrow band, the beat behavior will be random in character, as shown in Fig. 24-4. Thus, it is correctly reasoned that the response will locally appear as a slightly distorted sine function with a frequency near the natural frequency of the system and with amplitudes that vary slowly in a random fashion. This same type of response can also be predicted from the autocorrelation function, Eq. (24-29), as plotted in Fig. 24-3d, since this function approaches the autocorrelation for the single harmonic process presented in Fig. 23-2 as damping approaches zero.

When it is noted that the sharply peaked output power spectral density function $S_v(\bar{\omega})$ shown in Fig. 24-3c is obtained by multiplying the similarly peaked transfer function $|H(i\bar{\omega})|^2$ by the constant power spectral density function S_0 of the white-noise input, it becomes clear that the response $v(t)$ is caused primarily by those frequency components in the input process $p(t)$ which are near the natural frequency of the system. Therefore in those cases when the input power spectral density function $S_p(\bar{\omega})$ is not a constant but is a slowly varying function of $\bar{\omega}$ in the vicinity of the natural frequency ω, a white-noise input process can be assumed with little loss in a predicting response provided the constant power spectral density S_0 is set equal to the intensity of $S_p(\bar{\omega})$ at $\bar{\omega} = \omega$; that is, let $S_p(\bar{\omega}) = S_p(\omega)$; thus, the output power spectral density function can be approximated by the relation

$$S_v(\bar{\omega}) = \frac{S_p(\omega)}{k^2[1 + (4\xi^2 - 2)(\bar{\omega}/\omega)^2 + (\bar{\omega}/\omega)^4]} \qquad \xi \ll 1 \qquad (24\text{-}38)$$

Note that as the damping ratio ζ approaches zero, the area under this function becomes more and more concentrated at the natural frequency ω and approaches infinity in the limit. This means that the mean square response of an undamped SDOF system is infinite when subjected to white-noise excitation of finite intensity. Such systems are, of course, classified as unstable systems.

To clarify further the response characteristics of narrowband linear systems subjected to a stationary gaussian excitation having zero mean values, consider the response conditional probability density function

$$p(v_2 \mid v_1) = \frac{p(v_1, v_2)}{p(v_1)} \qquad (24\text{-}39)$$

where $v_2 \equiv v(t + \tau)$ and $v_1 \equiv v(t)$. When use is made of the standard relation

$$p(v_1, v_2) = \frac{1}{2\pi\sigma_v{}^2\sqrt{1 - \rho_v{}^2}}$$
$$\times \exp\left[-\frac{1}{2\sigma_v{}^2(1 - \rho_v{}^2)}(v_1{}^2 - 2\rho_v v_1 v_2 + v_2{}^2)\right] \qquad (24\text{-}40)$$

$$p(v_1) = \frac{1}{\sqrt{2\pi}\,\sigma_v}\exp\left(-\frac{v_1{}^2}{2\sigma_v{}^2}\right) \qquad (24\text{-}41)$$

where
$$\sigma_v = R_v(0)^{1/2} \qquad \rho_v(\tau) = \frac{R_v(\tau)}{R_v(0)} \qquad (24\text{-}42)$$

Eq. (24-40) becomes

$$p[v_2 \mid v_1] = \frac{1}{\sqrt{2\pi}\,\sigma_v\sqrt{1 - \rho_v{}^2}}\exp\left[-\frac{(v_2 - \rho_v v_1)^2}{2(1 - \rho_v{}^2)\sigma_v{}^2}\right] \qquad (24\text{-}43)$$

This equation shows that when $v(t)$ is fixed, the expected value of $v(t + \tau)$ is $\rho_v(\tau)v(t)$ and its variance is $[1 - \rho_v(\tau)^2]\sigma_v{}^2$, thus lending support to the previously described response characteristics of narrowband systems.

Finally consider the joint probability density function

$$p[v(t), \dot{v}(t)] = \frac{1}{2\pi\sigma_v\sigma_{\dot{v}}}\exp\left[-\frac{1}{2}\left(\frac{v^2}{\sigma_v{}^2} + \frac{\dot{v}^2}{\sigma_{\dot{v}}{}^2}\right)\right] \qquad (24\text{-}44)$$

where
$$\sigma_{\dot{v}} = R_{\dot{v}}(0)^{1/2} = -R_v''(0)^{1/2} \qquad (24\text{-}45)$$

Since $R_v'(0) = 0$, the covariance of random variables v and $\dot{v}$ will also equal zero, which explains the uncoupled form of Eq. (24-44). This equation can now be used to find the probability that response $v(t)$ will cross a fixed level $\hat{v}$ with positive velocity within the time limits t and $t + dt$. To satisfy this condition $v(t)$ must conform to the relation

$$\Pr\left[v(t) < \hat{v} < v(t + dt)\right] = \Pr\left[0 < [\hat{v} - v(t)] < \dot{v}(t)\,dt\right] \qquad (24\text{-}46)$$

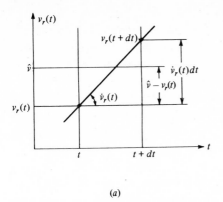

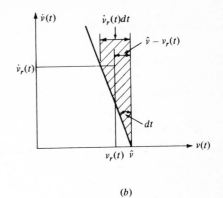

(a)   (b)

FIGURE 24-5
Velocity-displacement relations for positive slope crossings at level $\hat{v}$ within
time limits t and $t + dt$.

as illustrated graphically in Fig. 24-5a for one member of the ensemble. From the
$v \, \dot{v}$ plane shown in Fig. 24-5b, it is clear that those ensemble members which are
favorable to this condition must have values of $v(t)$ and $\dot{v}(t)$ which fall within the
shaded region. Therefore, if $Q(\hat{v}) \, dt$ represents the probability condition given by
Eq. (24-46), this term can be evaluated by simply integrating the joint probability
density function given by Eq. (24-44) over the shaded region, that is,

$$Q(\hat{v}) \, dt = \int_0^\infty \int_{\hat{v}-\dot{v} \, dt}^{\hat{v}} \frac{1}{2\pi\sigma_v \sigma_{\dot{v}}} \exp\left[-\frac{1}{2}\left(\frac{v^2}{\sigma_v{}^2} + \frac{\dot{v}^2}{\sigma_{\dot{v}}{}^2} \right) \right] d\dot{v} \, dv \qquad (24\text{-}47)$$

Substituting $dv = \dot{v} \, dt$ into this equation and completing the integration gives
Rice's relation[1]

$$Q(\hat{v}) = \frac{1}{2\pi} \frac{\sigma_{\dot{v}}}{\sigma_v} \exp\left(-\frac{1}{2} \frac{\hat{v}^2}{\sigma_v{}^2} \right) \qquad (24\text{-}48)$$

which is the probability density function for crossings of response $v(t)$ at level $\hat{v}$
with positive velocity per unit of time. With $\hat{v}$ set equal to zero, the probability
density function for positive crossings becomes

$$Q(0) = \frac{1}{2\pi} \frac{\sigma_{\dot{v}}}{\sigma_v} = \frac{1}{2\pi} \sqrt{ -\frac{R_v''(0)}{R_v(0)} } \qquad (24\text{-}49)$$

Substituting Eq. (24-29) into this relation gives the desired result

$$Q(0) = \frac{\omega}{2\pi} \qquad (24\text{-}50)$$

[1] S. O. Rice, Mathematical Analysis of Random Noise in N. Wax (ed.), "Selected
Papers on Noise and Stochastic Processes," Dover, New York, 1954.

thus showing that low-damped SDOF systems have the same average number of zero crossings when excited by white noise as when vibrating in a free undamped state. This fact is further evidence of the type of response which characterizes narrowband systems.

To establish the probability density function for the distribution of maxima in the response function $v(t)$ for narrowband systems, it can be assumed that on the average one maximum exists for each zero crossing; that is, for each time interval $T = 2\pi/\omega = 2\pi(\sigma_v/\sigma_{\dot{v}})$. The possibility of negative maxima exists, of course, but is highly improbable for this class of SDOF systems. As a direct result of this assumption, the probability density function for maxima occurring within the limits $\hat{v}$ and $\hat{v} + d\hat{v}$ per probable period must be given by the differential relation

$$p(\hat{v}) = -\frac{dQ(\hat{v})}{d\hat{v}}\left(2\pi\frac{\sigma_v}{\sigma_{\dot{v}}}\right) \qquad (24\text{-}51)$$

Substituting Eq. (24-48) into Eq. (24-51) gives the Rayleigh distribution

$$p(\hat{v}) = \frac{\hat{v}}{\sigma_v^2}\exp\left(-\frac{1}{2}\frac{\hat{v}^2}{\sigma_v^2}\right) \qquad (24\text{-}52)$$

which can be considered the probability density function for maxima or for the upper response envelope shown in Fig. 24-4. This distribution is a special case of that given previously by Eq. (23-106) for $\varepsilon = 0$.

24-6   NONSTATIONARY MEAN SQUARE RESPONSE RESULTING FROM ZERO INITIAL CONDITIONS

The response characteristics previously defined for output processes are based on steady-state conditions, that is, those conditions which result when input processes are assumed to start at time $t = -\infty$. In actual practice, however, input processes must be assumed to start at time $t = 0$. While such input processes may be assumed as stationary for $t > 0$, the resulting output processes will be nonstationary due to the usual zero initial conditions, which must be specified for $t = 0$. To illustrate the type of nonstationarity which results, consider the input process $p(t)$ to the SDOF system represented by Eq. (24-1) as stationary white noise of intensity S_0 starting at $t = 0$. This input process, as previously demonstrated, can be represented [see Eq. (23-86) and Fig. 23-8] by the relation

$$p_r(t) = \lim_{N\to\infty}\sum_{k=0}^{N-1} a_{kr}\eta(t - k\,\Delta t - \varepsilon_r) \qquad r = 1, 2, \ldots, \infty \qquad (24\text{-}53)$$

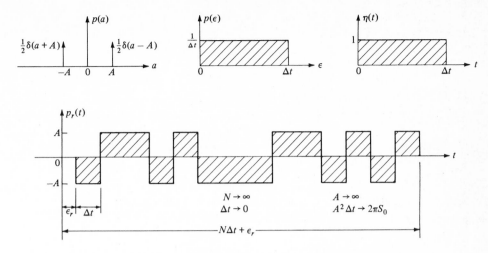

FIGURE 24-6
Stationary white-noise input process $p(t)$ for $t > 0$, Eq. (24-53).

which is plotted in Fig. 24-6, provided $A^2 \to \infty$ and $\Delta t \to 0$ in such a way that $A^2\,\Delta t \to C$ (a constant). As shown in Eq. (23-94), constant C equals $2\pi S_0$. Assuming an undercritically damped system ($\xi < 1$), response can be obtained by superposition through the time domain to give

$$v_r(j\,\Delta t) = \lim_{\Delta t \to 0} \sum_{k=0}^{j} \frac{a_{kr}\,\Delta t}{\omega_D m} \exp\left[-\omega\xi(j-k)\,\Delta t\right]\sin\omega_D(j-k)\,\Delta t \qquad (24\text{-}54)$$

and

$$v_r(j\,\Delta t)^2 = \lim_{\Delta t \to 0} \sum_{k=0}^{j}\sum_{g=0}^{j} \frac{a_{kr}a_{gr}\,\Delta t^2}{\omega_D{}^2 m^2} \exp\left[-\omega\xi(2j-k-g)\,\Delta t\right]$$

$$\times \sin\omega_D(j-k)\,\Delta t\,\sin\omega_D(j-g)\,\Delta t \qquad (24\text{-}55)$$

The ensemble average of the double summation term in this equation can be reduced immediately to the ensemble average of a single summation term since all covariances of random variables a_k and a_g ($k \neq g$) equal zero; therefore, the ensemble average is

$$E[v(j\,\Delta t)^2] = \lim_{\Delta t \to 0} \frac{\Delta t^2}{\omega_D{}^2 m^2} \sum_{k=0}^{j} \exp\left[-2\omega\xi(j-k)\,\Delta t\right]$$

$$\times \sin^2\omega_D(j-k)\,\Delta t\,\lim_{n\to\infty}\frac{1}{n}\sum_{r=1}^{n} a_{kr}{}^2 \qquad (24\text{-}56)$$

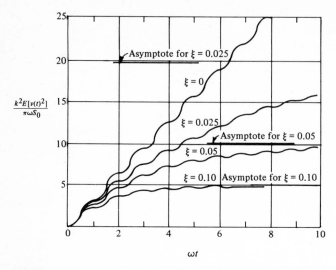

FIGURE 24-7
Nonstationary mean square response resulting from zero initial conditions, Eq. (24-58).

However since the second limiting term in this equation equals A^2 for all values of k, one obtains in the limit as $j\,\Delta t \to t$ and $k\,\Delta t \to \tau$ the relation

$$E[v(t)^2] = \frac{A^2\,\Delta t}{\omega_D{}^2 m^2}\exp\left(-2\omega\xi t\right)\int_0^t \exp\left(2\omega\xi\tau\right)\sin^2\omega_D(t-\tau)\,d\tau \qquad (24\text{-}57)$$

With the substitution of $2\pi S_0$ for $A^2\,\Delta t$ and k^2 for $\omega^4 m^2$, when the integration is completed, Eq. (24-57) becomes

$$E[v(t)^2] = \frac{\pi\omega S_0}{2\xi k^2}\left\{1 - \frac{\exp\left(-2\xi\omega t\right)}{\omega_D{}^2}\left[\omega_D{}^2 + \frac{(2\xi\omega)^2}{2}\sin^2 2\omega_D t\right.\right.$$

$$\left.\left. + \xi\omega\omega_D \sin 2\omega_D t\right]\right\} \qquad \xi < 1 \qquad (24\text{-}58)$$

This equation is plotted in Fig. 24-7 for various values of damping ratio ξ. Note the manner and relatively rapid rate at which the ensemble mean square value $E[v(t)^2]$ or variance $\sigma_v(t)^2$ approaches its steady-state value

$$R_v(0) = \frac{\pi S_0\omega}{2\xi k^2} \qquad (24\text{-}59)$$

If the power spectral density function for the stationary input process starting at time $t = 0$ is nonuniform but varies reasonably slowly in the vicinity of the un-damped frequency ω, the variance of the output process for low-damped systems

($\xi < 0.1$) can be approximated reasonably well using Eq. (24-58) provided the power spectral density intensity at $\bar{\omega} = \omega$ is substituted for S_0, that is,

$$E[v(t)^2] \doteq \frac{\pi \omega S_p(\omega)}{2\xi k^2} \left\{ 1 - \frac{\exp(-2\omega\xi t)}{\omega_D{}^2} \left[\omega_D{}^2 + \frac{(2\omega\xi)^2}{2} \sin^2 2\omega_D t \right. \right.$$
$$\left. \left. + \xi\omega\omega_D \sin 2\omega_D t \right] \right\} \quad \xi < 1 \quad (24\text{-}60)$$

The harmonic terms in this equation are relatively small and can be dropped with little loss of accuracy; thus, Eq. (24-60) reduces to the form

$$E[v(t)^2] \doteq \frac{\pi \omega S_p(\omega)}{2\xi k^2} [1 - \exp(-2\omega\xi t)] \quad \xi < 1 \quad (24\text{-}61)$$

As the damping ratio approaches zero, this equation becomes in the limit

$$E[v(t)^2] \doteq \frac{\pi \omega^2 S_p(\omega)}{k^2} t \quad \xi = 0 \quad (24\text{-}62)$$

Assuming that the stationary input process is gaussian, the nonstationary output process will also be gaussian, in which case the probability density function $p[v(t)]$ will be given by

$$p[v(t)] = \frac{1}{\sqrt{2\pi E[v(t)^2]}} \exp\left\{ -\frac{v(t)^2}{2E[v(t)^2]} \right\} \quad (24\text{-}63)$$

EXAMPLE E24-4   Consider the SDOF system

$$m\ddot{v} + c\dot{v} + kv = p(t) \quad (a)$$

excited by a zero-mean ergodic random process $p(t)$ having constant power spectral density equal to S_0 over the range $-\infty < \bar{\omega} < \infty$ and assume zero initial conditions are imposed on the system; that is, $v_r(0) = \dot{v}_r(0) = 0$ ($r = 1, 2, \ldots$). Calculate the ratio of the variance of $v(t)$ to the steady-state

Table 24-1   RATIO $\sigma^2_{v(t)}/\sigma^2_{(\infty)}$

			t	
ξ	2	5	10	15
0.02	0.395	0.714	0.919	0.987
0.05	0.715	0.957	0.998	0.99995

variance for $t = 2$, 5, 10, and 15 s, $\xi = 0.02$ and 0.05, and $\omega = 6.28$ rad/s ($f = 1$ Hz).

Using Eq. (24-61) gives

$$\frac{\sigma_{v(t)}^2}{\sigma_{v(\infty)}^2} = 1 - \exp{(-2\omega\xi t)} \qquad (b)$$

Substituting the above numerical values into Eq. (b) gives the results shown in Table 24-1. These results indicate the very rapid rate at which the response process $v(t)$ approaches its steady-state condition.   ////

24-7   FATIGUE PREDICTIONS FOR NARROWBAND SYSTEMS

The fatigue life of a narrowband SDOF system can easily be determined provided the material follows a prescribed SN relationship ($S =$ harmonic stress amplitude, $N =$ number of cycles to failure) and provided that Miner's linear-accumulative-damage criterion applies.[1]

Proceeding on this basis, assume that the SN relationship is known in the form

$$N = N(S) \qquad (24\text{-}64)$$

The accumulative damage (AD) can then be expressed in the discrete form

$$AD = \frac{n_1}{N(S_1)} + \frac{n_2}{N(S_2)} + \frac{n_3}{N(S_3)} + \cdots \qquad (24\text{-}65)$$

where $n_1, n_2, n_3, \ldots$ are the number of harmonic stress cycles applied to the material at amplitudes $S_1, S_2, S_3, \ldots$, respectively. Failure occurs when the accumulative damage reaches unity, that is, AD $= 1$.

If the system is responding as a narrowband system, the accumulative damage can be expressed in the continuous form

$$AD = \int_0^\infty \frac{n(S)}{N(S)}\, dS \qquad (24\text{-}66)$$

where $n(S)\, dS$ represents the number of harmonic stress cycles with amplitudes between S and $S + dS$. If a stationary response process of duration T is assumed, the total number of stress cycles will be equal to $\omega T/2\pi$, in which case

$$n(S)\, dS = \frac{\omega T}{2\pi}\, p(S)\, dS \qquad (24\text{-}67)$$

---

[1] M. A. Miner, Cumulative Damage in Fatigue, *J. Appl. Mech.*, ser. A, vol. 12, no. 1, pp. 159–164, 1945.

where $p(S)$ is the probability density function for stress amplitude S. Substituting Eq. (24-67) into Eq. (24-66) gives the accumulative damage in the form

$$AD = \frac{\omega T}{2\pi} \int_0^\infty \frac{p(S)}{N(S)} \, dS \qquad (24\text{-}68)$$

If $p(S)$ is of the Rayleigh form as represented by a narrowband process, that is,

$$p(S) = \frac{S}{\sigma_s^2} \exp\left(-\frac{S^2}{2\sigma_s^2}\right) \qquad (24\text{-}69)$$

where σ_s^2 is the variance of the critical stress $s(t)$, and if $N(S)$ takes on the familiar form

$$N(S) = \left(\frac{S_1}{S}\right)^b N_1 \qquad (24\text{-}70)$$

where S_1 and N_1 represent a convenient point on the SN curve and b is an even integer (usually $b > 10$), Eq. (24-68) becomes, after substituting Eqs. (24-69) and (24-70),

$$AD = \frac{\omega T}{2\pi N_1} \left(\frac{\sigma_s}{S_1}\right)^b 2^{b/2} \left(\frac{b}{2}\right)! \qquad (24\text{-}71)$$

Setting the accumulative damage equal to unity and solving for T gives the expected time to failure

$$T_{\text{failure}} = \frac{2\pi N_1}{\omega} \left(\frac{S_1}{\sigma_s}\right)^b \frac{2^{b/2}}{(b/2)!} \qquad (24\text{-}72)$$

For a linear system, the critical stress is related to displacement $v(t)$, Eq. (24-1), by the relation

$$s(t) = Cv(t) \qquad (24\text{-}73)$$

where C is a known constant. It follows therefore that

$$\sigma_s^2 = C^2 \sigma_v^2 \qquad (24\text{-}74)$$

Thus, if $p(t)$ is a white-noise process of intensity S_0, as shown in Fig. 24-3,

$$\sigma_s^2 = \frac{\pi \omega C^2 S_0}{2k^2 \xi} \qquad (24\text{-}75)$$

EXAMPLE 24-5   As represented by Eq. (2-21), a SDOF system excited by random support excitation $\ddot{v}_g(t)$ results in the equation of motion

$$m\ddot{v} + c\dot{v} + kv = -m\ddot{v}_g(t) = p_e(t) \qquad (a)$$

Stationary random process $\ddot{v}_g(t)$ has a uniform spectral density equal to 2 ft²/s³ over the frequency range $2 < \overline{\omega} < 100$ and $-100 < \overline{\omega} < -2$. The system has a natural frequency of 10 Hz and damping equal to 2 percent of critical, and the critical stress from a fatigue standpoint is given by

$$s(t) = 2 \times 10^5 v(t) \qquad (b)$$

where units are pounds and inches. If the material at the critical location satisfies the fatigue relation

$$N(S) = \left(\frac{60{,}000}{s}\right)^{12} \times 10^5 \qquad (c)$$

find the expected time to failure caused by excitation $\ddot{v}_g(t)$.

From Eq. (24-29) it is seen that the variance of $v(t)$ is

$$\sigma_{v(t)}{}^2 = R_v(0) = \frac{\pi \omega S_0}{2k^2 \xi} \qquad (d)$$

where S_0 is the power spectral density for a white-noise excitation $p_e(t)$. Since the natural frequency of the system ω falls within the bandwidth of support excitation, white-noise excitation can be assumed here of intensity

$$S_{p_e}(\overline{\omega}) = S_0 = m^2 S_{\ddot{v}_g(t)}(\overline{\omega}) \qquad (e)$$

Substituting Eq. (e) into Eq. (d) and making use of the notation $k^2 = \omega^4 m^2$ gives

$$\sigma_{v(t)}{}^2 = \frac{\pi S_{\ddot{v}_g(t)}(\omega)}{2\omega^3 \xi} \qquad (f)$$

Making use of Eq. (24-74) gives

$$\sigma_{s(t)}{}^2 = \frac{\pi C^2 S_{\ddot{v}_g(t)}(\omega)}{2\omega^3 \xi} \qquad (g)$$

where $C = 2 \times 10^5$ lb/in³, $S_{\ddot{v}(t)}(\omega) = 2$ ft²/s³, $\xi = 0.02$, and $\omega = 2\pi f = 62.8$ rad/s. Thus one obtains $\sigma_{s(t)} = 1.89 \times 10^4$ psi. When it is noted from Eq. (c) that $b = 12$, $N_1 = 10^5$, and $S_1 = 60{,}000$ psi, Eq. (24-72) yields the expected time to failure

$$T_{\text{failure}} = 0.27 \times 10^7 \text{ s} = 750 \text{ h} \qquad (h)$$

$////$

PROBLEMS

24-1 Consider the SDOF system represented by

$$m\ddot{v} + c\dot{v} + kv = p(t)$$

when excited by a gaussian zero-mean stationary process $p(t)$ having a constant power spectral density $S_0 = 2 \times 10^4$ lb²/s over two wide-frequency bands centered on $\pm \omega$, where ω is the natural circular frequency $\sqrt{k/m}$. The system is characterized by a mass m equal to 100 lb $\cdot$ s^2/ft, a natural frequency ω equal to 62.8 rad/s, and a damping ratio ξ equal to 2 percent of critical.

(a) Find the numerical value for the mean square displacement $E[v(t)^2]$.

(b) Find the numerical value for the mean square velocity $E[\dot{v}(t)^2]$.

(c) What is the joint probability density function for $v(t)$ and $\dot{v}(t)$? Find the numerical values for all constants in this function.

(d) What is the probability density function for the maxima of response process $v(t)$? Find the numerical values for all constants in this function. (Assume a Rayleigh distribution in this case since the parameter appearing in Fig. 23-12 is nearly equal to zero.)

(e) What is the numerical value of the mean extreme value of the process $v(t)$ as given by Eq. (23-117) when the duration T of the process is 30 s? [For this low-damped system, v as given by Eq. (23-116) is approximately equal to $\omega/2\pi$.]

(f) What is the numerical value of the standard deviation of the extreme values for process $v(t)$?

24-2 Approximately how much will the numerical values found in Prob. 24-1 change if the power spectral density function for the process $p(t)$ is changed from $S_p(\overline{\omega}) = S_0$ to

$$S_p(\overline{\omega}) = S_0 \exp(-0.0111 |\overline{\omega}|) \qquad -\infty < \overline{\omega} < \infty$$

24-3 The one-mass system shown in Fig. P24-1 is excited by support displacement $x(t)$. The spring and viscous dashpot are linear, having constant k and c, respectively. Let $\omega^2 = k/m$ and $\xi = c/2m\omega$.

(a) Obtain the unit-impulse-response function $h(t)$ for spring force $f_S(t) = k[y(t) - x(t)]$ when $x(t) = \delta(t)$.

(b) Obtain the complex-frequency-response function $H(i\overline{\omega})$ for force $f_S(t)$ which is the ratio of the complex amplitude of $f_S(t)$ to the complex amplitude of $x(t)$ when the system is performing simple harmonic motion at frequency $\overline{\omega}$.

(c) Verify that $h(t)$ and $H(i\overline{\omega})$ are Fourier transform pairs in accordance with Eqs. (24-16).

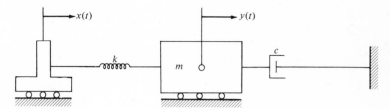

FIGURE P24-1
One-mass system of Prob. 3.

24-4 If the input $x(t)$ to the system defined in Prob. 24-2 is a stationary random process having a constant power spectral density S_0 over the entire frequency range $-\infty < \overline{\omega} < \infty$, derive the power spectral density and autocorrelation functions for response process $f(t)$ using Eqs. (24-36) and (24-25), respectively.

24-5   Show that the power spectral density and autocorrelation functions obtained in Prob. 24-3 are indeed Fourier transform pairs in accordance with Eqs. (23-37).

24-6   If $x(t)$ is the stationary random input to a linear system and $y(t)$ is the corresponding stationary random output, express the cross-correlation function $R_{xy}(\tau)$ in terms of $R_x(\tau)$ and $h(t)$.

24-7   If the input $x(t)$ to the linear system, shown in Fig. P24-2 produces the output $y(t)$

$$x(t) = \begin{cases} e^{-t} & t > 0 \\ 0 & t < 0 \end{cases} \qquad y(t) = \begin{cases} \dfrac{1}{a-b}(e^{-(b/a)t} - e^{-t}) & t \geq 0 \\ 0 & t \leq 0 \end{cases}$$

what is the complex-frequency-response function $H(i\bar{\omega})$ for the system?

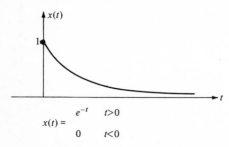

$$x(t) = \begin{array}{ll} e^{-t} & t > 0 \\ 0 & t < 0 \end{array}$$

$$y(t) = \begin{array}{ll} \frac{1}{(a-b)}[e^{-(b/a)t} - e^{-t}] & t > 0 \\ 0 & t < 0 \end{array}$$

FIGURE P24-2
Input and output functions of Prob. 7.

24-8   If a rectangular input $x(t)$ to a linear system produces a single sine-wave output $y(t)$, as shown in Fig. P24-3,

$$x(t) = \begin{cases} 1 & 0 < t < T \\ 0 & t < 0;\, t > T \end{cases} \qquad y(t) = \begin{cases} \sin \dfrac{2\pi t}{T} & 0 < t < T \\ 0 & t < 0;\, t > T \end{cases}$$

what would be the power spectral density function for the output process when the input is a stationary white-noise process of intensity S_0?

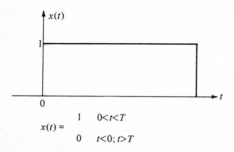
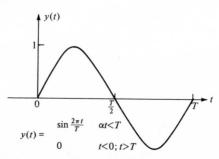

$$x(t) = \begin{array}{ll} 1 & 0 < t < T \\ 0 & t < 0;\, t > T \end{array}$$

$$y(t) = \begin{array}{ll} \sin \frac{2\pi t}{T} & \alpha t < T \\ 0 & t < 0;\, t > T \end{array}$$

FIGURE P24-3
Input and output functions of Prob. 8.

24-9 Consider two stationary processes $x(t)$ and $y(t)$ related through the differential equation

$$\ddot{x}_r(t) + A\dot{x}_r(t) + By_r(t) + C\dot{y}_r(t) = 0 \qquad r = 1, 2, \ldots$$

Express the power spectral density function for random process $x(t)$ in terms of the power spectral density function for process $y(t)$ and real constant A, B, and C.

25

STOCHASTIC RESPONSE OF LINEAR MDOF SYSTEMS

25-1 TIME-DOMAIN RESPONSE FOR LINEAR SYSTEMS

As shown in Chaps. 13 and 19, the dynamic response of linear MDOF systems (discrete or continuous) can be determined by solving the normal equations of motion

$$\ddot{Y}_n(t) + 2\omega_n\xi_n\dot{Y}_n(t) + \omega_n^2 Y_n(t) = \frac{P_n(t)}{M_n} \qquad n = 1, 2, \ldots \qquad (25\text{-}1)$$

where n is the mode number. Any response quantity $z(t)$ linearly related to the normal coordinates can be found using the relation

$$z(t) = \sum_n B_n Y_n(t) \qquad (25\text{-}2)$$

where coefficients B_n ($r = 1, 2, \ldots$) are obtained by standard methods of analysis. Usually, the rapid convergence rate of the series means that only a limited number of lower modes need be considered.

If random excitations on the system are assumed, each generalized forcing function $P_n(t)$ should be considered as a separate stochastic process. If the excitations are stationary, the response processes will also be stationary, in which case the interest is in obtaining the autocorrelation function for response $z(t)$, that is,

$$R_z(\tau) = E[z(t)z(t + \tau)] \qquad (25\text{-}3)$$

Substituting Eq. (25-2) into Eq. (25-3) gives

$$R_z(\tau) = E\left[\sum_m \sum_n B_m B_n Y_m(t) Y_n(t + \tau)\right] \qquad (25\text{-}4)$$

Solving for response through the time domain leads to

$$Y_n(t) = \int_{-\infty}^{t} P_n(\tau) h_n(t - \tau) \, d\tau \qquad (25\text{-}5)$$

where, for undercritically damped systems,

$$h_n(t) = \frac{1}{\omega_{D_n} M_n} \exp\left(-\xi_n \omega_n t\right) \sin \omega_{D_n} t \qquad \omega_{D_n} = \omega_n (1 - \xi_n^2)^{1/2} \qquad (25\text{-}6)$$

Substituting Eq. (25-5) into Eq. (25-4) gives

$$R_z(\tau) = E\left[\sum_m \sum_n \int_{-\infty}^{t} \int_{-\infty}^{t+\tau} B_m B_n P_m(\theta_1) P_n(\theta_1)\right.$$

$$\left. \times h_m(t - \theta_1) h_n(t + \tau - \theta_2) \, d\theta_1 \, d\theta_2\right] \qquad (25\text{-}7)$$

where θ_1, θ_2, and τ are dummy time variables. With the change of variables

$$u_1 \equiv t - \theta_1 \qquad u_2 \equiv t + \tau - \theta_2$$

$$du_1 = -d\theta_1 \qquad du_2 = -d\theta_2 \qquad (25\text{-}8)$$

and recognition that $h_n(t)$ damps out for stable systems, Eq. (25-7) can be written in the form

$$R_z(\tau) = \sum_m \sum_n R_{z_m z_n}(\tau) \qquad (25\text{-}9)$$

where

$$R_{z_m z_n}(\tau) \equiv \int\limits_{0}^{\infty}\int B_m B_n R_{P_m P_n}(\tau - u_2 + u_1) h_m(u_1) h_n(u_2) \, du_1 \, du_2 \qquad (25\text{-}10)$$

$R_{P_m P_n}(\tau)$ is the covariance function for random variables $P_m(t)$ and $P_n(t + \tau)$, and $R_{z_m z_n}(\tau)$ is the covariance function for modal responses $z_m(t)$ and $z_n(t)$. From this derivation it is clear that if the covariance function $R_{P_m P_n}(\tau)$ is known for all combinations of m and n, the integrations in Eq. (25-10) and the summations in Eq. (25-9) can be completed to obtain the desired autocorrelation function for response $z(t)$.

For systems that are lightly damped, the usual case in structural engineering, response process $z_m(t)$ produced by mode m is almost statistically independent of response $z_n(t)$ produced by mode n; that is, the cross terms in Eq. (25-9) are nearly equal to zero. Therefore, the autocorrelation function for total response can usually be approximated by the relation

$$R_z(\tau) \doteq \sum_m R_{z_m z_m}(\tau) \qquad (25\text{-}11)$$

where $R_{z_m z_m}(\tau)$ is the autocorrelation function for process $z_m(t)$. When τ is made equal to zero, Eq. (25-11) can be written in terms of standard deviations, that is,

$$\sigma_z = (\sigma_{z1}{}^2 + \sigma_{z2}{}^2 + \sigma_{z3}{}^2 + \cdots)^{1/2} \qquad (25\text{-}12)$$

Since the mean extreme-values of response for processes $z(t)$ and $z_m(t)$ $(m = 1, 2, \ldots)$ are proportional to their respective standard deviations σ_z and σ_{z_m}, Eq. (25-12) lends support to the common root-sum-square method of weighting the maximum normal-mode responses when estimating maximum total response.

25-2 FREQUENCY-DOMAIN RESPONSE FOR LINEAR SYSTEMS

The power-spectral-density function for response $z(t)$ is obtained by taking the Fourier transform of the autocorrelation function, that is,

$$S_z(\bar{\omega}) = \frac{1}{2\pi} \int_{-\infty}^{\infty} R_z(\tau) \exp\left(-i\bar{\omega}\tau\right) d\tau \qquad (25\text{-}13)$$

Substituting Eq. (25-10) into Eq. (25-9) and then Eq. (25-9) into Eq. (25-13) gives

$$S_z(\bar{\omega}) = \frac{1}{2\pi} \int_{-\infty}^{\infty} \left\{ \sum_m \sum_n \iint_0^{\infty} B_m B_n R_{P_m P_n}(\tau - u_2 + u_1) \right. $$
$$\left. \times\, h_m(u_1) h_n(u_2)\, du_1\, du_2 \right\} \exp\left(-i\bar{\omega}\tau\right) d\tau \qquad (25\text{-}14)$$

or

$$S_z(\bar{\omega}) = \frac{1}{2\pi} \sum_m \sum_n B_m B_n \left[\lim_{T \to \infty} \int_0^T h_m(u_1)\, du_1 \int_0^T h_n(u_2)\, du_2 \right. $$
$$\left. \times \int_{-T}^{T} R_{P_m P_n}(\tau - u_2 + u_1) \exp\left(-i\bar{\omega}\tau\right) d\tau \right] \qquad (25\text{-}15)$$

Since $h(u_1)$ and $h(u_2)$ equal zero for u_1 and u_2 less than zero, the lower limits of the first two integrals in Eq. (25-15) can be changed from zero to $-T$. After substituting the change of variable

$$\gamma \equiv \tau - u_2 + u_1 \qquad (25\text{-}16)$$

Eq. (25-15) becomes

$$S_z(\bar{\omega}) = \frac{1}{2\pi} \sum_m \sum_n B_m B_n \left[\lim_{T \to \infty} \int_{-T}^{T} h_m(u_1) \exp\left(i\bar{\omega}u_1\right) du_1 \right. $$
$$\times \int_{-T}^{T} h_n(u_2) \exp\left(-i\bar{\omega}u_2\right) du_2 $$
$$\left. \times \int_{-T-u_2+u_1}^{T-u_2+u_1} R_{P_m P_n}(\gamma) \exp\left(-i\bar{\omega}\gamma\right) d\gamma \right] \qquad (25\text{-}17)$$

Since $R_{P_rP_s}(\tau)$ damps out with increasing values of $|\tau|$, the limits of the last integral in Eq. (25-17) can be changed to $\int_{-T}^{T}$. With use of Eq. (23-56) and the first of Eqs. (24-16), Eq. (25-17) becomes

$$S_z(\bar{\omega}) = \sum_m \sum_n S_{z_m z_n}(\bar{\omega}) \qquad (25\text{-}18)$$

where

$$S_{z_m z_n}(\bar{\omega}) \equiv B_m B_n H_m(-i\bar{\omega}) H_n(i\bar{\omega}) S_{P_m P_n}(\bar{\omega}) \qquad (25\text{-}19)$$

is the cross-spectral density function for modal responses $z_m(t)$ and $z_n(t)$, $S_{P_m P_n}(\bar{\omega})$ is the cross-spectral density function for processes $P_m(t)$ and $P_n(t)$, and

$$H_m(-i\bar{\omega}) = \frac{1}{K_m[1 - 2i\xi_m(\bar{\omega}/\omega_m) - (\bar{\omega}/\omega_m)^2]}$$

$$H_n(i\bar{\omega}) = \frac{1}{K_n[1 + 2i\xi_n(\bar{\omega}/\omega_n) - (\bar{\omega}/\omega_n)^2]} \qquad (25\text{-}20)$$

For lightly damped systems, the cross terms in Eq. (25-18) are nearly equal to zero, in which case $S_z(\bar{\omega})$ simplifies to the approximate form

$$S_z(\bar{\omega}) \doteq \sum_m S_{z_m z_m}(\bar{\omega}) \qquad (25\text{-}21)$$

where

$$S_{z_m z_m}(\bar{\omega}) = \sum_m B_m{}^2 |H_m(i\bar{\omega})|^2 S_{P_m P_m}(\bar{\omega}) \qquad (25\text{-}22)$$

$$|H_m(i\bar{\omega})|^2 = \frac{1}{K_m{}^2[1 + (4\xi_m{}^2 - 2)(\bar{\omega}/\omega_m)^2 + (\bar{\omega}/\omega_m)^4]} \qquad (25\text{-}23)$$

and $S_{P_m P_m}(\bar{\omega})$ is the power spectral density function for process $P_m(t)$.

If all input processes are gaussian, response $z(t)$ will also be gaussian, in which case $S_z(\bar{\omega})$ completely characterizes the process.

25-3 RESPONSE TO DISCRETE LOADINGS

It was shown in Chap. 13 that if a linear structure is subjected to discrete applied loadings $p_i(t)$ $(i = 1, 2, \ldots)$, the generalized forcing function for the nth mode becomes

$$P_n(t) = \sum_i \phi_{in} p_i(t) = \boldsymbol{\phi}_n{}^T \mathbf{p}(t) \qquad (25\text{-}24)$$

where constants ϕ_{in} are the components of nth-modal displacements at points i in the directions of corresponding forces $p_i(t)$. If each discrete force $p_i(t)$ is a stationary gaussian process as defined by $S_{p_i}(\bar{\omega})$ or $R_{p_i}(\tau)$, the cross-spectral density and covariance functions for $P_m(t)$ and $P_n(t)$ become

$$S_{P_m P_n}(\bar{\omega}) = \sum_i \sum_k \phi_{im} \phi_{kn} S_{p_i p_k}(\bar{\omega}) = \boldsymbol{\phi}_m{}^T \mathbf{S}_p \boldsymbol{\phi}_n \qquad (25\text{-}25)$$

$$R_{P_m P_n}(\tau) = \sum_i \sum_k \phi_{im} \phi_{kn} R_{p_i p_k}(\tau) = \boldsymbol{\phi}_m{}^T \mathbf{R}_p \boldsymbol{\phi}_n \qquad (25\text{-}26)$$

The power spectral density and autocorrelation functions for response $z(t)$ are now obtained by substituting Eqs. (25-25) and (25-26) into Eqs. (25-19) and (25-10), respectively.

EXAMPLE E25-1   A uniform inverted L-shaped member of mass $\bar{m}$ per unit length and flexural stiffness EI in its plane is discretized as shown in Fig. E25-1. This model is subjected to simultaneous stationary random base accelerations $a_x(t)$ and $a_y(t)$ having power spectral and cross-spectral densities given by

$$S_{a_x a_x}(\bar{\omega}) = S_0 \qquad S_{a_y a_y}(\bar{\omega}) = {}^1\!/_2 S_0 \qquad S_{a_x a_y}(\bar{\omega}) = S_{a_y a_x}(\bar{\omega}) = C S_0 \qquad (a)$$

where C is a real constant. Assuming modal damping of the uncoupled form, where the damping ratio in each normal mode equals ξ, determine the variance of base moment $\mathfrak{M}(t)$ expressed in terms of $\bar{m}$, L, EI, ξ, S_0, and C. What is the range of possible numerical values for C?

Using the flexibility and mass matrices shown in Fig. E25-1 gives the following mode shapes and frequencies:

$$\phi_1^T = [-0.807 \quad 1.000 \quad 0.307] \qquad \omega_1^2 = 0.197 \frac{EI}{\bar{m}L^4}$$

$$\phi_2^T = [-1.000 \quad -0.213 \quad -0.280] \qquad \omega_2^2 = 2.566 \frac{EI}{\bar{m}L^4} \qquad (b)$$

$$\phi_3^T = [-0.306 \quad -0.304 \quad 1.000] \qquad \omega_3^2 = 21.858 \frac{EI}{\bar{m}L^4}$$

From the first of Eqs. (13-15), the normal-coordinate generalized masses are

$$M_1 = 3.84 \frac{\bar{m}L}{2} \qquad M_2 = 1.29 \frac{\bar{m}L}{2} \qquad M_3 = 2.41 \frac{\bar{m}L}{2} \qquad (c)$$

and from the fourth of Eqs. (13-15) the corresponding generalized forces are

$$P_n(t) = -\phi_n^T \mathbf{m} \begin{bmatrix} a_y \\ a_x \\ a_x \end{bmatrix} \qquad n = 1, 2, 3$$

From this equation it follows that

$$P_m(t) P_n(t) = \phi_m^T \mathbf{m} \begin{bmatrix} a_y \\ a_x \\ a_x \end{bmatrix} [a_y \quad a_x \quad a_x] \mathbf{m} \phi_n$$

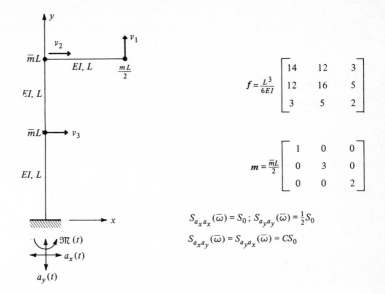

FIGURE E25-1
Discrete model of uniform inverted L-shaped member.

from which the power and cross-spectral density functions become

$$S_{P_m P_n}(\bar{\omega}) = \phi_m{}^T \mathbf{m} \begin{bmatrix} S_{a_y a_y}(\bar{\omega}) & S_{a_y a_x}(\bar{\omega}) & S_{a_y a_x}(\bar{\omega}) \\ S_{a_x a_y}(\bar{\omega}) & S_{a_x a_x}(\bar{\omega}) & S_{a_x a_x}(\bar{\omega}) \\ S_{a_x a_y}(\bar{\omega}) & S_{a_x a_x}(\bar{\omega}) & S_{a_x a_x}(\bar{\omega}) \end{bmatrix} \mathbf{m} \phi_n$$

or $\quad S_{P_m P_n}(\bar{\omega}) = \phi_m{}^T \mathbf{m} \begin{bmatrix} {}^1\!/_2 S_0 & CS_0 & CS_0 \\ CS_0 & S_0 & S_0 \\ CS_0 & S_0 & S_0 \end{bmatrix} \mathbf{m} \phi_n \qquad m, n = 1, 2, 3 \qquad (d)$

To find the moment $\mathfrak{M}_n(t)$ contributed by the nth normal mode, apply the force vector

$$\begin{bmatrix} f_{1n} \\ f_{2n} \\ f_{3n} \end{bmatrix} = \omega_n{}^2 \mathbf{m} \phi_n Y_n(t)$$

to the member as shown in Fig. E25-2. Summing the moments about the base gives

$$\mathfrak{M}_n(t) = \omega_n{}^2 [-L \quad 2L \quad L] \mathbf{m} \phi_n Y_n(t) \qquad n = 1, 2, 3$$

Thus for this case Eq. (25-2) becomes

$$\mathfrak{M}(t) = \sum_{n=1}^{3} B_n Y_n(t) \qquad (e)$$

where

$$B_n = \omega_n{}^2 [-L \quad 2L \quad L] \mathbf{m} \phi_n \qquad n = 1, 2, 3 \qquad (f)$$

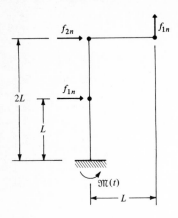

FIGURE E25-2

When Eqs. (25-18) and (25-19) are used, the spectral density for base moment becomes

$$S_{\mathfrak{M}}(\bar{\omega}) = \sum_{m=1}^{3} \sum_{n=1}^{3} B_m B_n H_m(-i\bar{\omega}) H_n(i\bar{\omega}) S_{P_m P_n}(\bar{\omega}) \qquad (g)$$

where $H_m(-i\bar{\omega})$ and $H_n(i\bar{\omega})$ are given by Eq. (25-20). Integrating this expression with respect to $\bar{\omega}$ from $-\infty$ to $+\infty$ gives the variance for base moment as

$$\sigma_{\mathfrak{M}(t)}^2 = \sum_{m=1}^{3} \sum_{n=1}^{3} \left[B_m B_n S_{P_m P_n} \int_{-\infty}^{\infty} H_m(-i\bar{\omega}) H_n(i\bar{\omega}) \, d\bar{\omega} \right] \qquad (h)$$

where $S_{P_m P_n}$ has the frequency-invariant form given by Eq. (d). By using Eqs. (25-20), the integral in Eq. (h) can be completed to give

$$\sigma_{\mathfrak{M}(t)}^2 = \sum_{m=1}^{3} \sum_{n=1}^{3} \frac{4\pi\xi B_m B_n S_{P_m P_n}}{M_n M_m (\omega_n + \omega_m)[(\omega_n - \omega_m)^2 + 4\xi^2 \omega_n \omega_m]} \qquad (i)$$

Assuming a low-damped system, that is, $\xi < 0.1$, the cross terms in this equation will be relatively small and can be neglected, which leads to the approximate expression

$$\sigma_{\mathfrak{M}(t)}^2 \doteq \frac{\pi}{2\xi} \sum_{n=1}^{3} \frac{B_n^2 S_{P_n P_n}}{M_n^2 \omega_n^3} \qquad (j)$$

Upon substitution from Eqs. (b) to (d) and (f), Eq. (j) becomes

$$\sigma_{\mathfrak{M}(t)}^2 \doteq \frac{\bar{m}^2 L^2}{\xi} \sqrt{\frac{EI}{m}} (9.3 + 3.2C) S_0 \qquad (k)$$

If excitations $a_x(t)$ and $a_y(t)$ are statistically independent, $S_{a_x a_y} = C S_0 = 0$. Therefore, $C = 0$ in this case. However, if $a_x(t)$ and $a_y(t)$ are fully correlated statistically, then $a_y(t) = \alpha a_x(t)$, where α is a real constant. From the definitions of power and cross-spectral densities given by Eqs. (23-27) and (23-60), respectively, it follows that

$$S_{a_y a_y} = \alpha^2 S_{a_x a_x} \qquad S_{a_x a_y} = \alpha S_{a_x a_x}$$

From these relations and Eqs. (a) $C = k = \pm 1/\sqrt{2}$. Thus the range of possible numerical values for C is

$$-\frac{1}{\sqrt{2}} \leq C \leq +\frac{1}{\sqrt{2}} \qquad (l)$$

Substituting the maximum and minimum values of C into Eq. (k) gives

$$\sigma_{\mathfrak{M}(t)}^2 = \begin{cases} 7.0 \, \dfrac{\overline{m}^2 L^2}{\xi} \sqrt{\dfrac{EI}{\overline{m}}} \, S_0 & C = -\dfrac{1}{\sqrt{2}} \\[4mm] 11.6 \, \dfrac{\overline{m}^2 L^2}{\xi} \sqrt{\dfrac{EI}{\overline{m}}} \, S_0 & C = +\dfrac{1}{\sqrt{2}} \end{cases} \qquad (m)$$

////

25-4   RESPONSE TO DISTRIBUTED LOADINGS

If the distributed applied loading $p(x,t)$ on a linear structure is random with respect to both x and t, the generalized forcing function for the nth mode is of the form

$$P_n(t) = \int \phi_n(x) p(x,t) \, dx \qquad (25\text{-}27)$$

where $\phi_n(x)$ is simply the continuous form of ϕ_{in} defined in Sec. 25-3. If $p(x,t)$ is a stationary gaussian process defined by $S_p(x,\alpha,\overline{\omega})$ or $R_p(x,\alpha,\tau)$, the cross-spectral density and covariance functions for $P_m(t)$ and $P_n(t)$ become

$$S_{P_m P_n}(\overline{\omega}) = \iint \phi_m(x) \phi_n(\alpha) S_p(x,\alpha,\overline{\omega}) \, dx \, d\alpha \qquad (25\text{-}28)$$

$$R_{P_m P_n}(\tau) = \iint \phi_m(x) \phi_n(\alpha) R_p(x,\alpha,\tau) \, dx \, d\alpha \qquad (25\text{-}29)$$

where α is a dummy space variable. The power spectral density and autocorrelation functions for response $z(t)$ are now obtained by substituting Eqs. (25-28) and (25-29) into Eqs. (25-19) and (25-10), respectively.

It should now be quite apparent to the reader how one determines the stochastic response of linear structures subjected to stationary gaussian excitations which are random with respect to all three space coordinates as well as time.

PROBLEMS

25-1 Consider the linear system with two degrees of freedom shown in Fig. P25-1, where $p_1(t)$ and $p_2(t)$ are two different zero-mean stationary random processes. The system has discrete masses and springs as indicated and may be assumed to be undercritically damped, with linear viscous damping of the uncoupled form yielding modal damping ratios $\xi_1 = \xi_2 = \xi$. If the power and cross-spectral density functions for processes $p_1(t)$ and and $p_2(t)$ are

$$S_{p_1 p_1}(\bar\omega) = S_0 \qquad S_{p_2 p_2}(\bar\omega) = A S_0 \qquad S_{p_1 p_2}(\bar\omega) = (B + iC)S_0$$

over the entire frequency range $-\infty < \bar\omega < \infty$, where A, B, and C are real constants (A must be positive, but B and C may be positive or negative), express the power spectral density function for spring force $f_s(t)$ in terms of constants k, m, ξ, S_0, A, B, and C. Write an expression, involving constants A, B, and C, which gives the ranges of possible values for constants B and C. *Note:* The range of possible values for B cannot be expressed independently of the range of possible values for A.

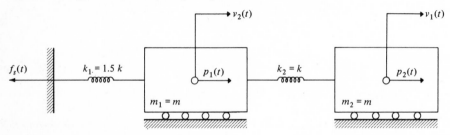

FIGURE P25-1
Two-mass system of Prob. 1.

25-2 A uniform simple beam of length L, stiffness EI, and mass $\bar m$ per unit length is subjected to zero-mean stationary random vertical-support motions at each end. Let $v(x,t)$ represent the total vertical displacement of the member; the vertical displacements are given by $v(0,t)$ and $v(L,t)$. Assume viscous damping of the uncoupled form with all normal modes having the same damping ratio ξ ($0 < \xi < 1$). If the power and cross-spectral density functions for vertical-support accelerations are given by

$$S_{a_1 a_1}(\bar\omega) = S_0 \qquad S_{a_2 a_2}(\bar\omega) = 0.5 S_0 \qquad S_{a_1 a_2}(\bar\omega) = (0.4 + 0.2i)S_0$$

over the entire frequency range $-\infty < \bar\omega < \infty$, where subscripts a_1 and a_2 are used to represent $\ddot v(0,t)$ and $\ddot v(L,t)$, respectively, find the power spectral density functions for (a) displacement $v(x,t)$, (b) moment $\mathfrak{M}(x,t)$, and (c) shear $\mathfrak{V}(x,t)$. Give your answers in series form expressed in terms of L, EI, $\bar m$, ξ, S_0, and $\bar\omega$. Discuss the relative rates of convergence of these series.

25-3 The tapered vertical cantilever member shown in Fig. P25-2 is subjected to a distributed zero-mean gaussian stationary random loading $p(x,t)$ having the power spectral density function

$$S_p(x,\alpha,\bar\omega) = S(\bar\omega)e^{-(A/L)|x-\alpha|}$$

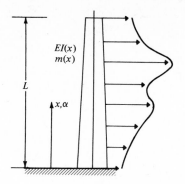

FIGURE P25-2
Cantilever member of Prob. 3.

where $S(\bar{\omega})$ is a known function of $\bar{\omega}$ with units of $lb^2 \cdot s/ft^2$ and A is a known positive real constant. The functions $m(x)$ and $EI(x)$ are known, and viscous damping can be assumed of the uncoupled form yielding the same damping ratio ξ in each normal mode. *Outline* how you would make an analysis of the stochastic response of this structure. Explain in sufficient detail to show that you could actually obtain correct numerical results if requested to do so.

PART FIVE

Analysis of Structural Response to Earthquakes

SEISMOLOGICAL BACKGROUND

26-1 INTRODUCTORY NOTE

Although dynamic loadings acting on structural systems may result from any of several different source mechanisms, including wind or wave action and vehicular motions, the type of dynamic input which is of greatest importance to the structural engineer undoubtedly is that produced by earthquakes. The significance of the earthquake problem stems partly, of course, from the terrible consequences of a major earthquake in a heavily populated area. Since the design of economic and attractive structures which can successfully withstand the forces generated by a severe ground motion is a challenge demanding the best in structural engineering art and science, on this basis alone it would be reasonable to use the field of earthquake engineering as the framework on which to demonstrate the application of the theories and techniques presented in Parts One to Four.

However, the significance of the earthquake problem has now extended beyond the immediate need for earthquake-resistant construction in seismic regions. In planning for the development of the nuclear power-generating industry, stringent seismic criteria have been adopted which must be considered in the design of nuclear stations to be built in any part of the United States, and these requirements have led to greatly increased interest in, as well as practical uses of, structural-dynamics theory. Finally, it may be pertinent to paraphrase the comment made by Newmark and Rosenblueth:[1] Earthquakes systematically bring out the mistakes made in design and construction—even the most minute mistakes; it is this aspect of earthquake engineering that makes it challenging and fascinating, and gives it an educational value far beyond its immediate objectives.

The essential background for study or practice in the field of earthquake engineering is, of course, the earthquake itself. The detailed study of earthquakes and earthquake mechanisms lies in the province of seismology, but in his studies the earthquake engineer must take a different point of view than the seismologist. Seismologists have focused their attention primarily on the global or long-range effects of earthquakes and therefore are concerned with very small-amplitude ground motions which induce no significant structural responses. Engineers, on the other hand, are concerned mainly with the local effects of large earthquakes, where the ground motions are intense enough to cause structural damage. These so-called *strong-motion earthquakes* are too violent to be recorded by the typical seismographs used by seismologists and have necessitated the development of special types of strong-motion seismographs. Nevertheless, even though the objectives of an earthquake engineer differ from those of a seismologist, there are many topics in seismology which are of immediate interest to him. A brief summary of the more important ones is presented in this chapter.

26-2 SEISMICITY

The basic mechanisms within the earth which give rise to earthquakes are not yet fully understood, and the various theories proposed concerning these mechanisms tend to be in conflict. For present purposes it will be sufficient to point out that the underlying causes of earthquakes are closely related to the global tectonic processes, which are continually producing mountain ranges and ocean trenches at the earth's surface. The crustal plates whose movements characterize these processes tend to be outlined, at least in part, by earthquake-occurrence maps, such as Fig. 26-1, which shows the location of earthquakes occurring in 1966. The seismicity, or rate of earthquake occurrence, which is demonstrated by this map, is generally representative, as well, of the earth's seismicity during all of the present century, that is, during the entire history of significant seismographic records.

The two principal world zones, or belts, of earthquake activity are clearly shown on this map, the Circum-Pacific belt around the Pacific Ocean, which includes

---

[1] N. M. Newmark and E. Rosenblueth, *Fundamentals of Earthquake Engineering*, Prentice-Hall, Inc., Englewood Cliffs, N.J., 1971.

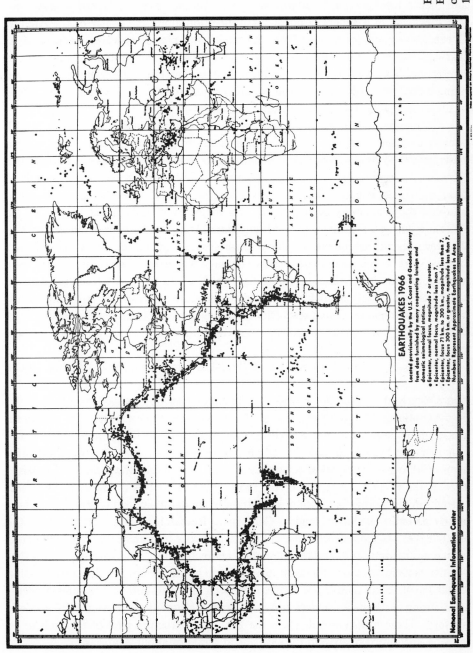

FIGURE 26-1
Earthquakes
occurring during
1966.

the great majority of all earthquakes (both destructive and small), and the Alpide belt which extends from the Himalaya mountain range through Iran and Turkey and thence through the Mediterranean Sea. Another sharply defined seismic belt, which extends in a generally north-south direction along the center of the Atlantic Ocean, clearly locates a boundary between crustal plates; however, because of its marine location this belt obviously has little interest for the structural engineer. An interesting fact about the Circum-Pacific belt is that fault displacements observed along it tend generally to indicate a common counterclockwise rotation of the Pacific basin crustal plates relative to the land masses on all sides.

It is important to keep in mind that the earthquakes indicated in Fig. 26-1 all occurred during a 1-year interval, which is much too short a time to provide a basis for establishing seismic design requirements in the different regions of the earth. Obviously if a structure is to have a design life of 50 years, it would be desirable to have earthquake-occurrence statistics over a period several times that long in order to establish the probability of occurrence of a major earthquake in the vicinity of the structure during its planned life. An additional factor to be considered is that Fig. 26-1 includes all earthquakes of global seismological significance, but a vast majority of these were far too small to have caused damage even to structures in their immediate vicinity. Data on earthquake intensities which could produce structural damage are much less readily available, and it is not possible at present to construct world-wide or even regional design seismicity maps which would indicate the probability of occurrence of earthquake motions of a specified intensity within a specified return period.

However, although significant earthquakes might be expected within, say, a 200-year interval in many of the regions indicated to be nonseismic in Fig. 26-1, this figure clearly shows the areas where earthquakes may be expected most frequently. In the continental United States, the most active seismic region is along the California coast and is largely associated with the San Andreas fault. This fault and the sub-sidiary faults branching from it, shown in Fig. 26-2, has been the source of most major California earthquakes during historic times, including the great San Francisco earthquake of 1906. It is one of the most active as well as the most studied fault systems in the world; its location is apparent in topographic features (Fig. 26-3) over nearly its entire extent on land in California. Research on this fault zone and the earthquakes associated with it has contributed greatly to present knowledge of earth-quake mechanisms and earthquake-motion characteristics. An interesting fact is that relative movements along this fault, corresponding with the counterclockwise rotation of the Pacific basin mentioned above, have been observed both in fault breaks occur-ring during earthquakes and in continual creep deformations measured by geodetic surveys. These measurements show the geological structure west of the fault to be moving northward relative to the east side at a rate of about 2 in/year.

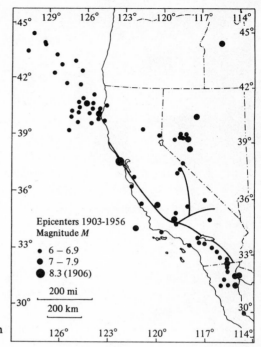

FIGURE 26-2
Principal earthquakes and faults in California.

26-3 ELASTIC-REBOUND THEORY OF EARTHQUAKES

It was from a study of the ruptures which occurred along the San Andreas fault during the San Francisco earthquake of 1906 that H. F. Reid first put into clear focus the elastic-rebound theory of earthquake generation. Many seismologists had already surmised that earthquakes result somehow from fractures, or faulting, of the earth's crust. However, Reid's investigation of the large-amplitude shearing displacements which resulted from this earthquake for dozens of miles along the fault (a typical displacement is shown in Fig. 26-4) led him to conclude that the specific source of the earthquake vibration energy is the release of accumulated strain in the earth's crust, the release itself resulting from the sudden shear-type rupture.

The essential concept of this elastic-rebound mechanism, which still provides the most satisfactory explanation for the types of earthquakes causing intense, potentially damaging surface motions, is portrayed in Fig. 26-5. The active fault zone is shown in the center, and, as with the San Andreas system, the geological structure to the left is assumed to be moving northward at a constant rate. If a series of fences were built perpendicularly across the fault (Fig. 26-5a), this continual north-ward drift would gradually distort the fence lines as shown in Fig. 26-5b. Also shown

FIGURE 26-3
Aerial view of part of San Andreas fault in California.

FIGURE 26-4
Fence offset by San Andreas fault slip,
San Francisco earthquake, April 18, 1906.

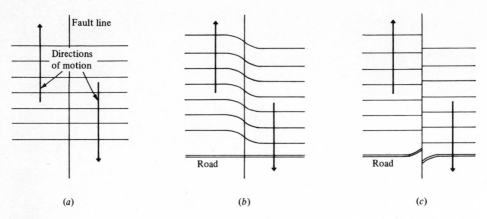

FIGURE 26-5
Elastic-rebound theory of earthquake generation: (*a*) before straining; (*b*) strained
(before earthquake); (*c*) after earthquake.

in the sketch is a road which is assumed to have been built after the fence-line deformations developed. Eventually, the continuing deformation of the crustal structure
will lead to stresses and strains which exceed the material strength. A rupture will then
initiate at some critical point in the fault zone and will propagate rapidly throughout
the length of the highly stressed material. The resulting release of strain and the
corresponding displacements lead to the conditions depicted in Fig. 26-5*c*, with large
offsets visible across the fault on both the road and the fence lines. With the release
of strain, the fence lines would become straight, but the road (which was built over
strained basement rock) would be locally curved.

26-4   EARTHQUAKE WAVES

The elastic-rebound theory postulates that the source of an earthquake is the sudden
displacement of the ground on both sides of the fault which results from a rupture
of the crustal rock. The earthquake itself, however, is the vibratory wave system which
emanates from this disturbance. Classical seismology consists largely of the study of
these elastic waves as they are propagated through the earth. Two types of body
waves are generated by the rupture: dilatation waves (called P for primary) and
shear waves (called S for secondary). The P waves travel faster than the S waves; so
if their velocities (which depend on the material through which they propagate) are
known, the distance from an observation point to the source can be calculated from the
difference in arrival times of the two types of waves. By study of the arrival times of

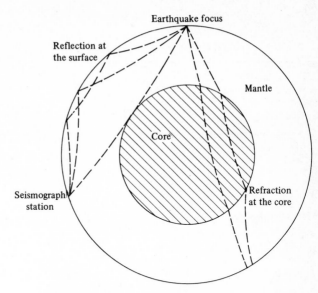

FIGURE 26-6
Paths of some P-type earthquake waves from the focus.

waves from a common source recorded at many stations widely distributed around the earth, the position of the source and the pattern of wave reflections and refractions from internal material discontinuities can be inferred, as depicted in Fig. 26-6. Nearly all that is presently known about the interior structure of the earth has been deduced from such seismological observations.

Teleseismic records of earthquake waves like these have little significance in structural engineering design, as explained earlier, because of the infinitesimal amplitudes of the ground motions involved. It is only near the rupture point (*focus* or *hypocenter*) that the motions are large enough to cause structural damage. At this range the wave generation mechanism may be described conveniently by reference to a penny-shaped crack located on the fault surface, as shown in Fig. 26-7, following a general line of reasoning due to G. W. Housner. Suppose that the state of stress in the area of this crack zone has reached the rupture point. When the rupture occurs, the release of strain adjacent to the crack surface will be accompanied by a sudden relative displacement of the two sides. This displacement initiates a displacement wave which propagates radially from the source; a record of this simple displacement pulse as it passes a recording station located at a moderate distance from the focus would be like that shown in Fig. 26-8. The velocity and acceleration records corresponding to this displacement pulse are also shown. Comparison of these idealized ground-motion records with the actual seismograph acceleration record

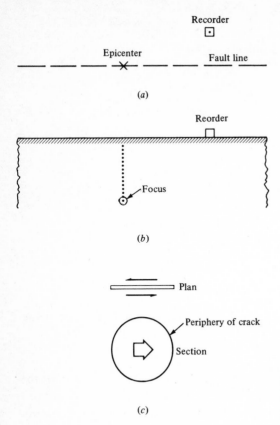

FIGURE 26-7
Idealized point-source earthquake rupture: (a) plan; (b) section along fault line; (c) penny-shaped crack at focus.

obtained from the Port Hueneme earthquake of 1957 (shown in Fig. 26-9, together with the velocity and displacement diagrams obtained by integration of the accelerogram) demonstrates that the Port Hueneme earthquake also was essentially a single-displacement pulse; thus it may be inferred that the mechanism which generated this earthquake was similar to the rupture of the simple penny-shaped crack.

Records of typical earthquakes, such as the El Centro earthquake of 1940 (one component of which is shown in Fig. 26-10), are much more complicated than the Port Hueneme record, and it is probable that the generating mechanism is correspondingly more complex. A hypothesis which provides a satisfactory explanation of the typical record is that the earthquake involves a sequence of ruptures along the fault surface. Each successive rupture is the source of a simple earthquake wave of the Port Hueneme type, but because they occur at different locations and times, the motions observed at a nearby station will be a random combination of simple records which could look much like Fig. 26-10.

In this context, it is important to note that seismologists define the focus of the

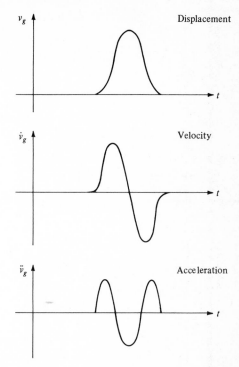

FIGURE 26-8
Idealized ground motion: point source.

earthquake as the point within the earth at which the *first* rupture of the fault surface takes place and the *epicenter* as the point at the surface directly above the focus. If the earthquake results from a sequence of ruptures along the fault line, it is evident that the focus may not coincide with the center of energy release. In a major earthquake, which may be associated with a fault break hundreds of miles in length, the distance of a building from the epicenter may be of little importance; the significant factor is the distance to the nearest point along the rupture surface.

26-5 MEASURES OF GROUND MOTION CHARACTERISTICS

To an earthquake engineer, the most important aspect of an earthquake's ground motions is the effect they will have on structures, that is, the stresses and deformations or the amount of damage they would produce. This damage potential is, of course, at least partly dependent on the "size" of the earthquake, and a number of measures of size are used for different purposes. The most important measure of size from a seismological point of view is the amount of strain energy released at the source, and this is indicated quantitatively as the *magnitude*. By definition, magnitude is the

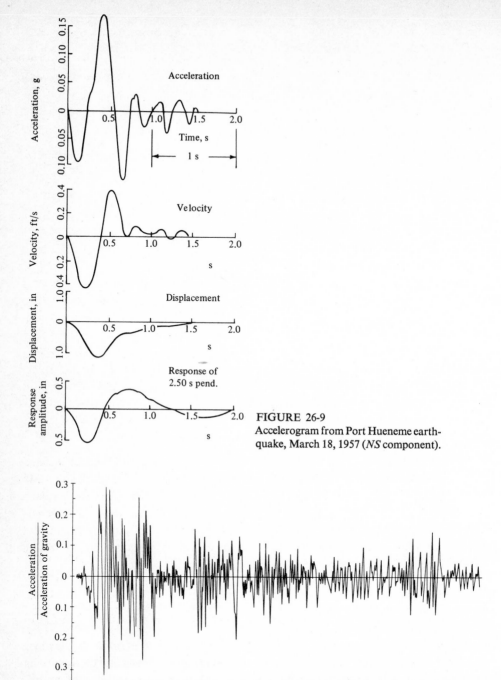

FIGURE 26-9
Accelerogram from Port Hueneme earthquake, March 18, 1957 (*NS* component).

FIGURE 26-10
Accelerogram from El Centro earthquake, May 18, 1940 (*NS* component).

(base 10) logarithm of the maximum amplitude, measured in micrometers (10^{-6} m) of the earthquake record obtained by a Wood-Anderson seismograph, corrected to a distance of 100 km. This magnitude rating has been related empirically to the amount of earthquake energy released E by the formula

$$\log E = 11.8 + 1.5M \qquad (26\text{-}1)$$

in which M is the magnitude. By this formula, the energy increases by a factor of 32 for each unit increase of magnitude. More important to engineers, however, is the empirical observation that earthquakes of magnitude less than 5 are not expected to cause structural damage, whereas for magnitudes greater than 5, potentially damaging ground motions will be produced.

The magnitude of an earthquake by itself is not sufficient to indicate whether structural damage can be expected. This is a measure of the size of the earthquake at its source, but the distance of the structure from the source has an equally important effect on the amplitude of its response. The severity of the ground motions observed at any point is called the *earthquake intensity*; it diminishes generally with distance from the source, although anomalies due to local geological conditions are not uncommon. The oldest measures of intensity are based on observations of the effects of the ground motions on natural and man-made objects. In the United States, the standard measure of intensity for many years has been the Modified Mercalli (MM) scale. This is a 12-point scale ranging from I (not felt by anyone) to XII (total destruction). Results of earthquake-intensity observations are typically compiled in the form of isoseismal maps like that shown in Fig. 26-11. Although such subjective intensity ratings are very valuable in the absence of any instrumented records of an earthquake, their deficiencies in providing criteria for the design of earthquake-resistant structures are obvious.

Basic information on the characteristics of earthquake motions which could be used for earthquake engineering purposes did not become available until the first strong-motion-recording accelerographs were developed and a network of such instruments was installed by the United States Coast and Geodetic Survey. The accelerogram of Fig. 26-10 (together with the perpendicular horizontal component and the vertical component) was recorded by one of these early instruments. The rate at which such instrumental information was gathered was very slow for many years because the number and distribution of the instruments were very limited. Gradually more extensive networks have been installed in Japan, Mexico, the most active seismic regions of the United States, and various other parts of the world, and much new and significant information is now being obtained. Unfortunately, however, the distribution of instruments is still quite limited, and destructive earthquakes in most parts of the world provide no strong-motion records. Consequently,

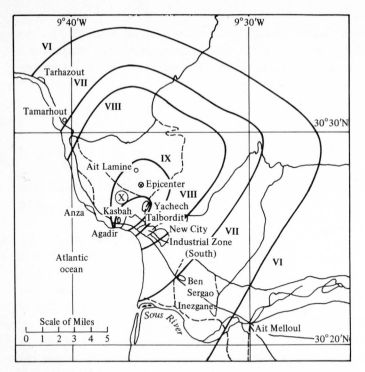

FIGURE 26-11
Isoseismal map of Agadir earthquake, 1960 (Modified Mercalli intensity scale).

basic data concerning the influence of such factors as magnitude, distance, and local soil conditions on the characteristics of earthquake motions are still very scarce.

The three components of ground motion recorded by a strong-motion accelerograph provide a complete description of the earthquake which would act upon any structure at that site. However, the most important features of the record obtained in each component (such as Fig. 26-10), from the standpoint of its effectiveness in producing structural response, are the amplitude, the frequency content, and the duration. The amplitude generally is characterized by the peak value of acceleration or sometimes by the number of acceleration peaks exceeding a specified level. (It is worth noting that the ground velocity may be a more significant measure of intensity than the acceleration, but it generally is not available without supplementary calculations.) The frequency content can be represented roughly by the number of zero crossings per second in the accelerogram, and the duration by the length of time between the first and the last peaks exceeding a given threshold level. It is evident, however, that all these quantitative measures taken together provide only a very limited description of the ground motion and certainly do not quantify its damage-producing potential adequately.

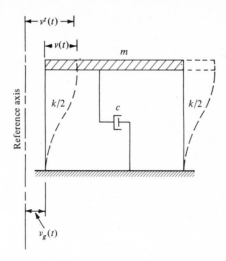

FIGURE 26-12
Basic SDOF dynamic system.

A more precise measure of the intensity of a given ground motion can be obtained by evaluating the response it would produce in a simple oscillator, such as the SDOF frame shown in Fig. 26-12. The response of this frame to a specified ground acceleration $\ddot{v}_g(t)$ may be expressed by means of the Duhamel integral [Eq. (7-14)] if it is noted that the effective loading in this case is given by $p_{\text{eff}}(t) = -m\ddot{v}_g(t)$ [Eq. (2-21)]. Thus

$$v(t) = \frac{1}{m\omega_D} \int_0^t -m\ddot{v}_g(\tau) \exp\left[-\xi\omega(t - \tau)\right] \sin \omega_D(t - \tau) \, d\tau \qquad (26\text{-}2)$$

When the difference between the damped and the undamped frequencies is neglected, as is permissible for the small damping ratios found in practical structures ($\xi < 20$ percent), and when it is noted that the negative sign has no real significance in an earthquake excitation, this can be reduced to

$$v(t) = \frac{1}{\omega} \int_0^t \ddot{v}_g(\tau) \exp\left[-\xi\omega(t - \tau)\right] \sin \omega(t - \tau) \, d\tau \qquad (26\text{-}3)$$

Finally, when the maximum value of the response relative to the ground is taken as the measure of the earthquake intensity, this maximum may be expressed as

$$v_{\max} = \frac{1}{\omega} S_v \qquad (26\text{-}4)$$

in which S_v is given by

$$S_v(\xi, \omega) = \left[\int_0^t \ddot{v}_g(\tau) \exp\left[-\xi\omega(t - \tau)\right] \sin \omega(t - \tau) \, d\tau\right]_{\max} \qquad (26\text{-}5)$$

and is called the spectral pseudo-velocity response of the ground motion $\ddot{v}_g(t)$.

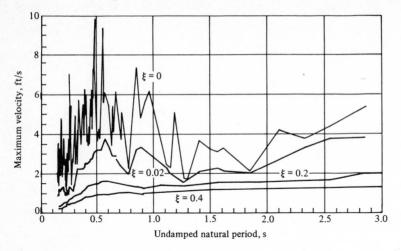

FIGURE 26-13
Pseudovelocity response spectrum, El Centro earthquake, May 18, 1940 (*NS* component).

As indicated by Eq. (26-5), S_v depends not only on the ground-motion history but also on the frequency of vibration and the damping of the oscillator. Thus for any given earthquake record, by assuming a specific value of damping in the structure it is possible to calculate values of S_v for a full range of vibration frequencies. A graph showing these spectral velocity-response values plotted as a function of frequency (or of the reciprocal quantity, period of vibration) is called a pseudo-velocity response spectrum of the earthquake motion. Such spectra generally are computed for several different damping ratios and all plotted on a single graph, as shown in Fig. 26-13 for the acceleration record of Fig. 26-10.

A closely related measure of the oscillator response to the ground motion $\ddot{v}_g(t)$ is the spectral displacement, S_d, which is the maximum displacement relative to the ground, and according to Eq. (26-4) is given by

$$S_d = \frac{S_v}{\omega} \qquad (26\text{-}6)$$

In addition, the spectral acceleration S_a, defined by

$$S_a = \omega S_v = \omega^2 S_d \qquad (26\text{-}7)$$

is a measure of the maximum spring force developed in the oscillator

$$f_{S,\text{max}} = k S_d = \omega^2 m S_d = m S_a$$

Plots of the spectral displacement and acceleration obviously could be constructed in a form similar to the pseudo-velocity spectrum of Fig. 26-13, but the simple relationships existing between these three quantities make it more convenient to

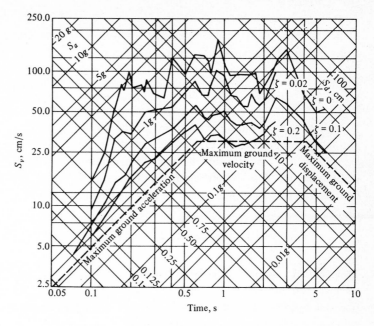

FIGURE 26-14
Response spectra for El Centro earthquake, 1940.

present them all in a single plot on four-way log paper, as shown in Fig. 26-14 (again for the accelerogram of Fig. 26-10). On this graph, the abscissas represent the logarithm of the period of vibration, the ordinates show the logarithm of the pseudo-velocity response, and log S_a and log S_d are represented by distances measured at 45° to the base.

It is evident that the response spectra provide a much more meaningful measure of the intensity of an earthquake motion than any single quantity, e.g., the peak acceleration, does. In fact these response spectra show directly to what extent any given real SDOF structure (with specified period of vibration and damping) would respond to this ground motion. The only limitation in its application is that the response is assumed to be linear elastic because such behavior is inherent in the Duhamel integral [Eq. (26-2)]. Thus the response spectra cannot define the extent of damage to be expected from a given earthquake, inasmuch as damage involves inelastic deformations. Nevertheless, the amount of elastic deformation to be expected is a very significant indication of a ground-motion intensity. Moreover, the response spectrum indicates the maximum deformation for structures of all periods of vibration; hence the integral of the response spectrum taken over an appropriate period range is probably the best overall measure of the ground-motion intensity. Housner has called this quantity the *response-spectrum intensity* and has defined it as the integral of the pseudo-velocity response spectrum taken over the range of structural vibration periods from 0.1 to 2.5 s, that is,

$$\text{SI}(\xi) = \int_{0.1}^{2.5} S_v(\xi, T) \, dT \qquad (26\text{-}8)$$

As indicated, it can be evaluated for any desired damping ratio in the structure.

The response parameter S_v is called the spectral pseudo velocity because it does not actually represent the maximum velocity of the oscillator; as may be seen from Eq. (26-6), it provides a direct means of evaluating the true maximum relative displacement, but it is only an approximation of the relative velocity. An expression for the true relative velocity can be obtained by differentiating Eq. (26-3), with the result

$$\dot{v}(t) = \int_0^t \ddot{v}_g(\tau) \exp \left[-\xi\omega(t - \tau) \right] \cos \omega(t - \tau) \, d\tau - \xi\omega v(t) \qquad (26\text{-}9)$$

This can then be used to construct a true-velocity response spectrum by evaluating its maximum for a range of frequencies and for various specified damping ratios. In practice, however, the spectral pseudo velocity is more useful because it leads to the response displacements, which are of greatest interest in structural design.

The total energy of the structural system can be obtained by combining expressions for the strain energy and the kinetic energy, that is,

$$E(t) = T(t) + V(t) = \frac{1}{2}m[\dot{v}(t)]^2 + \frac{1}{2}k[v(t)]^2 \qquad (26\text{-}10)$$

where the displacements and velocity are given, respectively, by Eqs. (26-3) and (26-9). Now a maximum-energy response spectrum can be constructed by evaluating the maximum energy given by Eq. (26-10) for a range of frequencies and damping values. In practice it is convenient to express the energy in terms of the square root of twice its value per unit mass, that is,

$$\sqrt{\frac{2E(t)}{m}} = \{[\omega v(t)]^2 + [\dot{v}(t)]^2\}^{1/2} \qquad (26\text{-}11)$$

The maximum value of this quantity clearly provides an upper bound to both the pseudo-velocity response spectrum and the true-velocity response spectrum, each of which is represented by one of the terms under the square-root sign. Note that for the undamped case, represented by setting $\xi = 0$ into Eqs. (26-3) and (26-9), Eq. (26-11) can be reduced to

$$\sqrt{\frac{2E(t)}{m}} = \left\{ \left[\int_0^t \ddot{v}_g(\tau) \cos \omega\tau \, d\tau \right]^2 + \left[\int_0^t \ddot{v}_g(\tau) \sin \omega\tau \, d\tau \right]^2 \right\}^{1/2} \qquad (26\text{-}12)$$

Another approach to characterizing the ground-motion history is provided by its Fourier spectrum, defined as

$$F(\omega) = \int_{\tau = -\infty}^{\infty} \ddot{v}_g(\tau) \exp \left(-i\omega\tau \right) \, d\tau \qquad (26\text{-}13)$$

If the duration of the ground motion is from $\tau = 0$ to $\tau = t_1$, this can be written

$$F(\omega) = \int_0^{t_1} \ddot{v}_g(\tau) \cos \omega\tau \, d\tau - i \int_0^{t_1} \ddot{v}_g(\tau) \sin \omega\tau \, d\tau$$

Finally, the amplitude of the Fourier spectrum, that is, the Fourier amplitude spectrum $|F(\omega)|$, is given by

$$|F(\omega)| = \left\{ \left[\int_0^{t_1} \ddot{v}_g(\tau) \cos \omega\tau \, d\tau \right]^2 + \left[\int_0^{t_1} \ddot{v}_g(\tau) \sin \omega\tau \, d\tau \right]^2 \right\}^{1/2} \qquad (26\text{-}14)$$

Comparison of Eqs. (26-14) and (26-12) shows that the Fourier amplitude spectrum is a measure of the total energy of an undamped system evaluated at the end of the earthquake $t = t_1$. Of course, the maximum energy generally is achieved at some time before the end of the earthquake; thus Eq. (26-12) usually is greater than the Fourier amplitude spectrum. The Fourier spectrum is a complete measure of the earthquake input because its transform [the inverse transform of Eq. (26-13)]:

$$\ddot{v}_g(t) = \frac{1}{2\pi} \int_{\omega = -\infty}^{\infty} F(\omega) \exp (i\omega t) \, d\omega \qquad (26\text{-}15)$$

recovers that motion. However, for engineering design purposes the response spectra, which represent the maximum response developed during the earthquake, are of greater significance.

26-6 SELECTION OF DESIGN EARTHQUAKES

To the structural designer, the only purpose in studying seismology is to enable him to predict the characteristics of the earthquake input for which his structures should be designed. The earthquake loading is unique among the types of loads that he must consider because a great earthquake would generally cause greater stresses and deflections in various critical components of his structure than all the other loadings combined, yet the probability of such an earthquake's occurring within the expected life of his structure is very low. In order to deal effectively with this combination of extreme loading and low probability, a strategy based on dual design criteria usually is adopted:

1 A moderate earthquake which reasonably may be expected at the building site during the life of the structure is taken as the basis of design. The building should be proportioned to resist this intensity of ground motion without significant damage to the basic structure.

2 The most severe earthquake which possibly could occur at the site is applied as a test of the structural safety. Because this earthquake is very unlikely to occur within the life of the structure, the designer is economically justified in permitting it to cause significant structural damage; however, collapse and loss of life must be avoided.

In order to establish the ground-motion characteristics of the design earthquake and of the maximum probable earthquake for any given building site, it is necessary first to study the earthquake history of the region for as long a period as any type of seismic information is available. Only from such established data can an estimate be made of the magnitudes of the earthquakes which may be expected to affect the site and of their probable frequency of occurrence (or *return period* between events). Because earthquakes occur very infrequently, the statistical data which can be compiled give at best only a crude estimate of the seismicity of the site. Other supporting evidence should be obtained from field geological studies, which serve to locate potentially active faults and to identify the tectonic characteristics of the local geological structure.

The end result of the seismological study usually is the definition of the design earthquake and of the maximum possible earthquake in terms of their magnitudes and distances from the building site. For example, it might be predicted that one earthquake of magnitude 7 should occur on a given fault at a distance of, say, 25 km during the projected 50-year life of a building. At the same site, the maximum shaking that possibly could occur might result from a magnitude 8.5 earthquake at a distance of perhaps only 10 km. The expected return period on this latter earthquake might be several hundred years, and so the probability that it would occur during the life of the building is very low, but the design should ensure the safety of the structure even in this extreme event.

Of course, the magnitude and distance from the construction site of the design earthquake do not directly indicate the ground motions for which a structure should be designed. In order to establish the design requirements, it is necessary to relate these basic seismological parameters to the intensity and other characteristics of the accelerations expected at the site. Many procedures have been suggested for doing this, and a full discussion of them would go beyond the needs of this chapter; it will be sufficient here merely to describe briefly some of the popular techniques.

One of the simplest ways to define the expected ground motions is to make use of the accelerogram of a past earthquake which had the proper magnitude and was recorded at an appropriate distance. For example, the strong-motion record from the El Centro earthquake, May 1940 *NS* (Fig. 26-10), has been used in many cases to represent the motions of a magnitude-7.0 design earthquake at a distance of about 7 km. However, experience has shown that there may be drastic differences between the records of earthquakes having similar magnitudes and distances, and the structural responses produced by such records may vary even more widely. Thus, the use of a

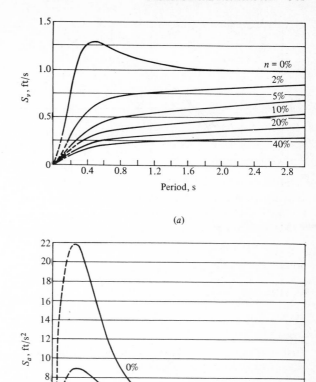

FIGURE 26-15
Smoothed average-earthquake response spectra: (*a*) velocity S_v; (*b*) acceleration response S_a.

single record such as Fig. 26-10 to define a design earthquake leaves considerable uncertainty as to the significance of the response it produces.

Obviously an "average" earthquake would be a more meaningful design input, and the most effective way of describing an average earthquake is by means of its response spectra. For example, Housner[1] developed the design spectrum shown in

---

[1] G. W. Housner, Behavior of Structures during Earthquakes, *Proc. ASCE*, vol. 85, no. EM-4, October, 1959.

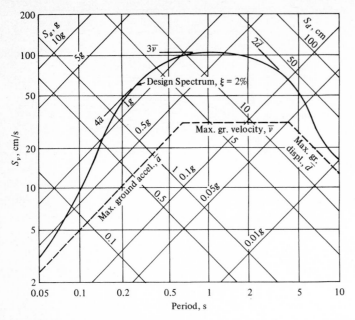

FIGURE 26-16
Design response spectrum constructed from maximum ground-motion character-istics. (*After Newmark and Rosenblueth.*)

Fig. 26-15 by computing the response spectra for two components each of four different earthquake records and then normalizing, averaging, and smoothing the resulting curves. This spectrum has been normalized so that the long-period value of the undamped spectral velocity has a value of 1.0 ft/s, which is about 1/2.7 of the intensity of the accelerogram in Fig. 26-10. Design response spectra also have been constructed from estimates of peak values of the maximum ground-motion accelera-tion, velocity, and displacement. When these peak ground-motion values are designated by $\bar{a}$, $\bar{v}$, and $\bar{d}$, respectively, it may be assumed that the corresponding maximum response values for a lightly damped oscillator are about $4\bar{a}$, $3\bar{v}$, and $2\bar{d}$.[1] On the basis of these assumptions, together with the knowledge that the acceleration response tends toward $\bar{a}$ for very stiff (short-period) structures while the displacement response tends toward $\bar{d}$ for very flexible (long-period) structures, it is possible to sketch a reasonable smoothed design response spectrum like that shown in Fig. 26-16. It should be noted, however, this approach required the local ground-motion values ($\bar{a}$, $\bar{v}$, and $\bar{d}$) to be estimated from the basic earthquake magnitude and distance before the response spectrum could be drawn.

Although a design response spectrum provides an adequate basis for propor-

[1] Newmark and Rosenblueth, *op. cit.*, p. 228.

tioning a structure during preliminary design phases, it generally is necessary to have an explicit description of the ground motion before completing the design of a major structure. An actual time-history record is especially important for structures in which nonlinear response must be considered (where the response spectrum is not applicable) or involving various types of structural interactions. One approach to obtaining an appropriate ground-motion record has been to modify and distort an actual earthquake record so that it represents an event of different magnitude and distance. For example, the intensity of motion may be adjusted by means of a simple amplitude-scaling factor, the frequency content may be modified by a change of time scale, and the duration of the earthquake may be changed by truncating or duplicating portions of the record. Such techniques have proved quite effective in providing one or two accelerograms appropriate to a given earthquake magnitude and distance. However, because all significant earthquake motions are generated by processes that are essentially random, the characteristics of the ground motions associated with a particular magnitude and distance can vary widely from one event to the next. Consequently it is most desirable to have a statistically significant sample of earthquake motions appropriate to the given magnitude and distance, rather than one or two, so that the structural performance can be evaluated on a statistical basis.

In response to this need, it now is becoming common practice to derive artificial earthquake records which represent the desired design earthquake. These ground-motion records are generated by random processes which may be repeated indefinitely to provide the desired sample of input motions. One technique for producing artificial ground-motion records has been to simulate with a very simple mathematical model the succession of ruptures along an assumed fault line and the propagation of the vibratory waves from each successive source to the observation point. Such simulation requires rather complex and expensive computational processes and seems more refined than is justified by the assumptions which must be made concerning the expected earthquake.

A simpler approach, which gives results that are equally reliable, is based on the hypothesis that the source of the ground motion is a random sequence of impulses generated at some distance and propagated to the point of observation through the basement rock structure. The generation of artificial accelerograms by this approach is described in some detail in Chap. 28.

27

DETERMINISTIC ANALYSIS OF EARTHQUAKE RESPONSE

27-1 EARTHQUAKE INPUT MECHANISMS

The definition of an appropriate ground-motion history, as described in Chap. 26, is the most difficult and uncertain phase of the problem of predicting structural response to earthquakes. Once a suitable support-excitation history has been established, the calculation of the stresses and deflections it would produce in any given structure is a standard problem of structural dynamics, which can be carried out by the techniques described in Parts One to Four. The deterministic analysis of the response to a specified earthquake motion will be discussed in this chapter because a deterministic understanding of the earthquake response leads to valuable insight into the structural behavior. However, seismic design requirements can often be described more effectively in probabilistic terms than by a single accelerogram, and for this reason nondeterministic earthquake-response analysis procedures are presented in Chap. 28. The only special feature of the earthquake problem, compared with any other form of dynamic loading, is that the excitation is applied in the form of support motions rather than by external loads; thus the essential subject of the present discussion is the methods of defining the effective external-load history resulting from a given form of support motion.

Earthquake ground motions usually are expressed in terms of three components of translational accelerations. The response of any linear system to these three components of input can be computed, of course, by superposing the responses calculated separately for each component. Thus the standard analytical problem is reduced to the evaluation of the structural response to a single component of support translation. In a more general case, the support point will be subjected to rotations in addition to the translational motions, as the earthquake waves propagate through the foundation soils. Thus, a complete description of the earthquake input should, in principle, include three components of support rotations as well as translations. Unfortunately, no measurements have been made of the magnitude and character of the ground-rotation components, and consequently this effect (in the few instances where it has been considered) has been accounted for only by tentative order-of-magnitude analyses in which rotational motions were hypothesized from the translational components.

Another assumption inherent in the usual treatment of earthquake excitations is that the same motion acts simultaneously at all parts of the structure's foundation. If rotational motions are neglected, this assumption is equivalent to considering the foundation soil or rock to be rigid. Such a hypothesis is not consistent with the concept of earthquake waves propagating through the earth's crust from the point of fault rupture; however, if the base dimensions of the structure are small relative to the vibration wavelengths in the basement rock, the hypothesis is acceptable. For example, if the velocity of wave propagation is 6,000 ft/s, a wave of 3 Hz frequency will have a length of 2,000 ft and a building with a base dimension of 100 ft will be subjected to essentially the same motions over its entire length. On the other hand, a suspension bridge or a dam having a length of, say, 1,500 ft obviously would be subjected to drastically differing motions along its length. No direct measurements have been made during an earthquake of such relative movements at different points on the earth's surface; however, it is evident that they must take place, and preliminary analyses have demonstrated that they could contribute significantly to the dynamic response stresses. Therefore, it is important to develop analysis procedures capable of dealing with multiple support excitation, that is, with different earthquake inputs applied at separate points of support.

One final factor that should be considered in defining the effective forces developed in a structure by an earthquake is that the ground motions at the base of the structure may be influenced by the motions of the structure itself. In other words, the motion introduced at the base of the structure may be different from the free-field motions that would have been observed without the structure. This soil-structure interaction effect will be of slight importance if the foundation rock is firm and the building is relatively flexible; in this case, the structure can transmit little energy into the soil, and the free-field motion is an adequate measure of the foundation displacements. On the other hand, if a heavy, stiff structure (such as a nuclear-reactor power station) is supported on a deep, soft soil layer, considerable energy will be transferred from the structure to the soil and the base motions may differ drastically from the free-field conditions. This soil-structure interaction effect is independent of, and in addition to, the effect mentioned in Chap. 26 that the soil layer might have on the characteristics of the free-field motions; in general, both the ground-motion modification and the interaction effects of a soft surficial soil layer can be important and must be accounted for in an earthquake-response analysis.

It is the purpose of this chapter to discuss the deterministic earthquake-response analysis of various types of structural systems, considering successively each of these different input conditions: (1) simple single-component translation of the base, (2) rigid-base rotations, (3) relative movements of different support points, and (4) the case of soil-structure interaction where the motion of the base of the structure does not directly follow the specified free-field motion. In addition, the most significant aspects of the nonlinear earthquake response of buildings will be discussed briefly.

27-2 EXCITATION BY RIGID-BASE TRANSLATION

Lumped SDOF Systems

The simplest form of earthquake response problem involves a SDOF lumped-mass system subjected to identical translations of all support points. An example of such a system, shown in Fig. 27-1, was used to define the concept of an earthquake-response spectrum in Chap. 26. The equation of motion of this structure is

$$m\ddot{v}^t + c\dot{v} + kv = 0 \qquad (27\text{-}1)$$

where the superscript t denotes total displacement. The effective earthquake force which produced the dynamic response of this system results from the fact that the inertia-force term in Eq. (27-1) depends on the total motion, while the damping and elastic forces depend only on relative motion. When it is noted that $v^t = v_g + v$, Eq. (27-1) can be written in terms of the relative displacements as

$$m\ddot{v} + c\dot{v} + kv = p_{\text{eff}} \qquad (27\text{-}2)$$

where the effective earthquake force is given by

$$p_{\text{eff}} = -m\ddot{v}_g \qquad (27\text{-}3)$$

On the other hand, if the damping and elastic forces are expressed in terms of the difference between the total motion and the ground motion, the equation can be written in terms of total displacements:

$$m\ddot{v}^t + c\dot{v}^t + kv^t = c\dot{v}_g + kv_g \qquad (27\text{-}4)$$

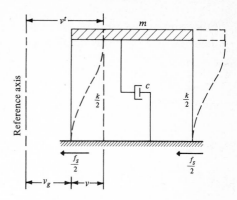

FIGURE 27-1
Lumped SDOF system subjected to base
translation.

Here the right-hand side is the effective force expressed in terms of the damping and stiffness coefficients. In principle, the earthquake-response problem could be solved by either of Eqs. (27-1) or (27-4); in practice, however, Eq. (27-4) is seldom used because its effective-force expression is more complicated and because the earthquake motions generally are presented in terms of the ground accelerations.

For the purposes of this discussion, it is convenient to express the earthquake-displacement response in terms of the Duhamel integral solution of Eq. (27-2) [as was described for Eq. (26-3)], although it should be noted that the response may actually be calculated in practice by step-by-step integration or by a frequency-domain analysis. With the Duhamel integral approach, the relative displacements given by Eq. (27-2) can be written

$$v(t) = \frac{1}{\omega} V(t) \qquad (27\text{-}5)$$

in which $V(t)$ is the earthquake-response integral, defined as

$$V(t) = \int_0^t \ddot{v}_g(\tau) \exp\left[-\xi\omega(t-\tau)\right] \sin\omega(t-\tau)\, d\tau \qquad (27\text{-}6)$$

It should be noted that the undamped frequency has been used in this expression in place of the damped value; the approximation introduced thereby is trivial in comparison with the uncertainties in the assumed earthquake motion $\ddot{v}_g(t)$. Also it will be noted that the negative sign of the effective-load term in Eq. (27-3) has been ignored here; generally the sense of the response has little significance in an earthquake analysis.

Evaluation of the *relative* motion as given by Eq. (27-5) is one of the principal objectives of an earthquake-response analysis; of course the total motion could be obtained by adding the ground displacements, but this is seldom of interest. The other primary measure of earthquake response is the force associated with the elastic deformations. For the SDOF lumped-mass system, this may be written directly as

$$f_S(t) = kv(t) \qquad (27\text{-}7a)$$

where $v(t)$ is given by Eq. (27-5). However, in the analysis of more complicated structures it is often convenient to use an alternate formulation of these forces. This can be derived by noting that in free undamped vibrations, the equation of dynamic equilibrium is of the form

$$f_I + f_S = 0 \qquad (27\text{-}8)$$

in which, for harmonic motions,

$$f_I = m\ddot{v}(t) = -m\omega^2 v(t) \qquad (27\text{-}9)$$

Hence, substituting from Eqs. (27-5) and (27-9) into Eq. (27-8) gives the elastic force

$$f_S(t) = m\omega^2 v(t) = m\omega V(t) \qquad (27\text{-}7b)$$

The equivalence of Eqs. (27-7a) and (27-7b) is apparent from the fact that $\omega^2 = k/m$.

Equation (27-7b) does not imply that $\omega^2 v(t)$ is the total acceleration of the mass, because in general a damping force is acting in the system in addition to the inertia force. An expression for the total acceleration can be derived by solving Eq. (27-1) and noting that $c/m = 2\xi\omega$; thus

$$\ddot{v}^t(t) = -2\xi\omega\dot{v}(t) - \omega^2 v(t)$$

Now if it is assumed that the damping term may be neglected on the basis that the damping force contributes little to the equilibrium relationship, the total acceleration is given approximately by

$$\ddot{v}^t(t) \doteq -\omega^2 v(t) \doteq \omega V(t) \qquad (27\text{-}10)$$

The time-varying earthquake response of a lumped SDOF system is expressed by Eqs. (27-5), (27-7b), and (27-10), all of which contain the response integral $V(t)$. The numerical evaluation of this integral for any given earthquake input to obtain the complete response history of a specific structure is a major computational task. On the other hand, if the response spectrum of the ground motion is available, the *maximum* response of the system can be obtained from it very easily. When it is noted that the spectral velocity, is, by definition, the maximum value of the response integral, that is,

$$S_v(\xi, T) \equiv V_{\max}(\xi, T) \qquad (27\text{-}11)$$

the maximum structural responses can be obtained directly from the response spectrum values corresponding to the period and damping ratio of the structure. Thus

$$v_{\max} = \frac{1}{\omega} S_v(\xi, T) = S_d(\xi, T)$$

$$f_{s,\max} = m\omega S_v(\xi, T) = m S_a(\xi, T)$$

$$\ddot{v}^t_{\max} \doteq S_a(\xi, T)$$

Expressed in words, the displacement-response spectrum gives the maximum displacement, the product of the mass and the spectral acceleration gives the maximum elastic force, and the spectral acceleration also is an approximation of the maximum acceleration.

EXAMPLE E27-1 Assuming that the structure of Fig. 27-1 has the following properties:

$$m = 2 \text{ kips·s}^2/\text{in} \qquad k = 60 \text{ kips/in} \qquad c = 0.438 \text{ kips·s/in}$$

compute the maximum displacement and maximum base shear force produced in this structure by the earthquake having the velocity response spectrum shown in Fig. E27-1. The first step in the analysis is to determine the vibration period and damping ratio of the structure:

$$\omega = \sqrt{\frac{k}{m}} = \sqrt{\frac{60}{2}} = 5.48 \text{ rad/s}$$

$$T = \frac{2\pi}{\omega} = \frac{2\pi}{5.48} = 1.147 \text{ s}$$

$$\xi = \frac{c}{2m\omega} = \frac{0.438}{2(2)(5.48)} = 0.02$$

From Fig. E27-1, the spectral velocity for this period and damping ratio is $S_v = 0.75$ ft/s. Hence the maximum displacement produced by this earthquake is

$$v_{max} = \frac{S_v}{\omega} = \frac{0.75}{5.48} = 0.137 \text{ ft}$$

The maximum base shear force may be computed in either of two ways:

$$\mathcal{V}_{0,max} = kv_{max} = 60(0.137)(12) = 98.6 \text{ kips}$$
$$\mathcal{V}_{0,max} = mS_a = m\omega S_v = 2(5.48)(0.75)(12) = 98.6 \qquad ////$$

Generalized SDOF Systems

Any structure of arbitrary form can be treated as a SDOF system if it is assumed that its displacements are restricted to a single shape, as explained in Part One. This generalized-coordinate approach can be used effectively in earthquake engineering; the only special problem to be considered in this case is the evaluation of the generalized force resulting from the support excitation.

The formulation of the generalized-coordinate equation of motion will be

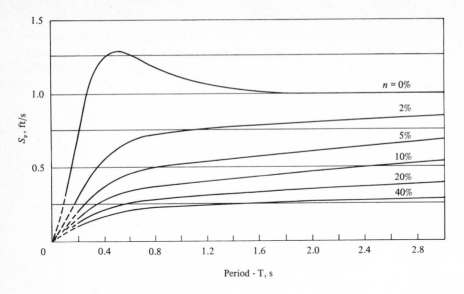

FIGURE E27-1
Average velocity response spectrum.

explained with reference to the tower structure of Fig. 27-2. The equilibrium of this system involves inertia, damping, and elastic forces, which are distributed along the axis, and may be expressed as

$$f_I(x,t) + f_D(x,t) + f_S(x,t) = 0 \qquad (27\text{-}12)$$

The basic assumption of the SDOF approximation is that the displacements are given by the product of a single shape function $\psi(x)$ and a generalized coordinate amplitude $Y(t)$, that is,

$$v(x,t) = \psi(x)Y(t) \qquad (27\text{-}13)$$

When a virtual displacement of this form $\delta v = \psi(x)\delta Y$ is applied, the principle of virtual work leads to the SDOF equilibrium relationship

$$f_I^* \, \delta Z + f_D^* \, \delta Z + f_S^* \, \delta Z = 0 \qquad (27\text{-}14)$$

in which

$$f_I^* = \int_0^L f_I(x,t)\psi(x) \, dx$$

$$f_D^* = \int_0^L f_D(x,t)\psi(x) \, dx \qquad (27\text{-}15)$$

$$f_S^* = \int_0^L f_S(x,t)\psi(x) \, dx$$

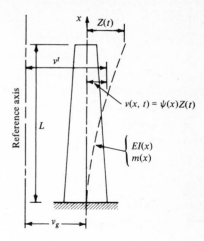

FIGURE 27-2
Generalized SDOF system with rigid-
base translation.

Because the distributed damping and elastic forces are assumed to depend
only on the relative motions, the corresponding generalized forces here are the same
as for the situation discussed in Part One, where the dynamic load was applied
externally, that is,

$$f_D^* = c^*\dot{Z} \qquad f_S = k^*Z$$

where c^* and k^* are given by expressions such as Eqs. (2-38) and (2-39). However,
the local inertia forces depend on the total acceleration, that is,

$$f_I(x,t) = m(x)\ddot{v}^t(x,t)$$

Thus since

$$v^t(x,t) = v(x,t) + v_g(t) = \psi(x)Z(t) + v_g(t)$$

the generalized inertia force is found to be

$$f_I^* = \ddot{Z}(t) \int_0^L m(x)[\psi(x)]^2 \, dx + \ddot{v}_g \int_0^L m(x)\psi(x) \, dx$$

Substituting all these generalized-force expressions into Eq. (27-14) then leads to the
final equation of motion

$$m^*\ddot{Z}(t) + c^*\dot{Z}(t) + k^*Z(t) = -\mathscr{L}\ddot{v}_g(t) \qquad (27\text{-}16)$$

in which

$$m^* = \int_0^L m(x)[\psi(x)]^2 \, dx \qquad (27\text{-}17)$$

$$\mathscr{L} = \int_0^L m(x)\psi(x) \, dx \qquad (27\text{-}18)$$

Equation (27-17) is equivalent to the generalized-mass expression of Eq. (2-47), while the quantity $\mathscr{L}$ given by Eq. (27-18) is the earthquake-excitation factor representing the extent to which the earthquake motion tends to excite response in the assumed shape $\psi(x)$.

Ignoring the sign of the effective earthquake force in Eq. (27-16) and dividing by the generalized mass leads to

$$\ddot{Z} + 2\xi\omega\dot{Z} + \omega^2 Z = \frac{\mathscr{L}}{m^*}\ddot{v}_g(t) \qquad (27\text{-}19)$$

By analogy with the foregoing analysis of the lumped SDOF system, the solution of Eq. (27-19) may now be written

$$Z(t) = \frac{\mathscr{L}}{m^*\omega}V(t) \qquad (27\text{-}20)$$

and hence the local displacements [from Eq. (27-13)] are

$$v(x,t) = \frac{\psi(x)\mathscr{L}}{m^*\omega}V(t) \qquad (27\text{-}21)$$

It will be noted by comparison of Eqs. (27-5) and (27-20) that the factor $\mathscr{L}/m^*$ characterizes the difference between the lumped and the generalized SDOF response; this factor depends on the mass distribution of the structure as well as its assumed shape function and generally is significantly different from 1.

In principle, the elastic forces produced by the earthquake motions can be evaluated from the structural displacements of Eq. (27-21) acting on the structural stiffness properties. However, when expressed in this way, the forces in this generalized-coordinate analysis depend on derivatives of the displacements or, in other words, on derivatives of the assumed shape functions $\psi(x)$. Thus the local forces obtained from such an analysis usually are less accurate than the displacements because the derivatives of the assumed shapes are poorer approximations than the shapes themselves. A more dependable formulation of the elastic forces can be obtained by expressing them in terms of the inertia forces of free vibration, following the general approach described above for the lumped-mass case. The equilibrium condition in undamped free vibration is obtained by omitting the damping term from Eq. (27-12); thus since the inertia force in free harmonic motion is

$$f_I(x,t) = m(x)\ddot{v}(x,t) = -\omega^2 m(x)v(x,t)$$

the resulting equation may be written

$$-\omega^2 m(x)v(x,t) + f_S(x,t) = 0 \qquad (27\text{-}22)$$

Now if the displacements are assumed to be of the form given by Eq. (27-13), the force balance implied by Eq. (27-22) will not be satisfied, in general; that is, the

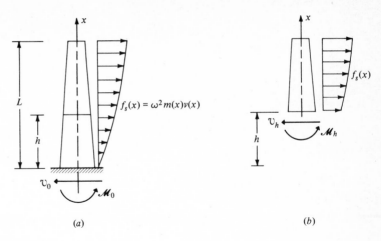

FIGURE 27-3
Elastic-force response of generalized SDOF system: (a) base forces; (b) section forces.

assumed shape will not satisfy equilibrium locally. However, if a virtual displacement of this form is introduced, the virtual-work principle can be used to obtain the global-equilibrium relationship

$$\delta Z \int_0^L [-\omega^2 m(x)v(x,t) + f_S(x,t)]\psi(x)\, dx = 0$$

On the other hand, even though it is valid only in an integrated or weighted-average sense, Eq. (27-22) provides the best available estimate of the elastic forces developed during the earthquake response, that is,

$$f_S(x,t) = \omega^2 m(x)v(x,t) = m(x)\psi(x)\frac{\mathscr{L}}{m^*}\omega V(t) \qquad (27\text{-}23)$$

From these distributed elastic forces, which are depicted in Fig. 27-3, any desired force resultant can be obtained by standard methods of statics. For example, the base shear $\mathcal{V}_0$ is given by

$$\mathcal{V}_0(t) = \int_0^L f_S(x,t)\, dx = \frac{\mathscr{L}}{m^*}\omega V(t) \int_0^L m(x)\psi(x)\, dx \qquad (27\text{-}24)$$

that is,

$$\mathcal{V}_0(t) = \frac{\mathscr{L}^2}{m^*}\omega V(t)$$

Similarly the base moment is given by $\mathcal{M}_0(t) = \int_0^L f_S(x,t)x\, dx$, that is,

$$\mathcal{M}_0(t) = \frac{\mathscr{L}}{m^*}\omega V(t) \int_0^L m(x)\psi(x)x\, dx \qquad (27\text{-}25)$$

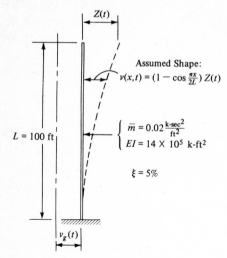

$Z(t)$

Assumed Shape:
$v(x,t) = (1 - \cos\frac{\pi x}{2L}) Z(t)$

$L = 100$ ft

$\begin{cases} \bar{m} = 0.02 \frac{\text{k-sec}^2}{\text{ft}^2} \\ EI = 14 \times 10^5 \text{ k-ft}^2 \end{cases}$

$\xi = 5\%$

$v_g(t)$

FIGURE E27-2
SDOF idealization of uniform cantilever
column.

Expressions for moment and shear at any arbitrary section h can be written similarly:

$$\mathcal{V}_h(t) = \frac{\mathcal{L}}{m^*} \omega V(t) \int_h^L m(x)\psi(x) \, dx$$

$$\mathcal{M}_h(t) = \frac{\mathcal{L}}{m^*} \omega V(t) \int_h^L m(x)\psi(x)x \, dx$$

Of course, the evaluation of time-varying-response expressions such as Eqs. (27-21) and (27-23) requires the numerical integration of the earthquake-response integral $V(t)$. However, maximum response values can be determined readily from the earthquake-response spectra, as explained in the discussion of the lumped SDOF systems, by merely selecting spectral values appropriate to the period of vibration and damping of the structure. For example, the maximum local displacements and local elastic forces are given, respectively, by

$$v_{\max}(x) = \psi(x) \frac{\mathcal{L}}{m^*} S_d(\xi, T)$$

$$f_{S,\max}(x) = m(x)\psi(x) \frac{\mathcal{L}}{m^*} S_a(\xi, T)$$

EXAMPLE E27-2 The earthquake-response analysis of a generalized SDOF structure will be demonstrated by subjecting the uniform cantilever column of Fig. E27-2 to a base motion $v_g(t)$ for which the velocity-response spectrum is shown in Fig. E27-1. It will be assumed that the displaced shape of the column is given by $\psi(x) = 1 - \cos(\pi x/2L)$; hence the generalized properties of this structure are as given in Example E2-3, and for the numerical values shown in Fig. E27-2 they are

$$m^* = 0.228\overline{m}L = 0.456 \text{ kips·s}^2/\text{ft}$$

$$k^* = \frac{\pi^4}{32}\frac{EI}{L^3} = 4.26 \text{ kips/ft}$$

$$\mathscr{L} = 0.364\overline{m}L = 0.728 \text{ kips·s}^2/\text{ft}$$

From these values, the circular frequency of the column is

$$\omega = \sqrt{\frac{k^*}{m^*}} = 3.056 \text{ rad/s}$$

Hence the period is

$$T = \frac{2\pi}{\omega} = 2.056 \text{ s}$$

With this period and the given damping ratio ($\xi = 5$ percent), the spectral velocity shown by Fig. E27-1 is $S_v = 0.62$ ft/s. Hence the maximum generalized-coordinate displacement [by analogy with Eq. (27-20)] is

$$Z_{max} = \frac{\mathscr{L}}{m^*\omega} S_v = 0.324 \text{ ft}$$

and so the maximum displacements of the column are

$$v_{max} = 0.324 \left(1 - \cos\frac{\pi x}{2L}\right) \quad \text{ft}$$

Similarly the maximum base shear [by analogy with Eq. (27-24)] is

$$\mathcal{V}_{0,\,max} = \frac{\mathscr{L}^2}{m^*} \omega S_v = 2.202 \text{ kips}$$

and the maximum distributed earthquake forces acting on the column are given by

$$f_{S,\,max} = \frac{\overline{m}\psi(x)}{\mathscr{L}} \mathcal{V}_{0,\,max} = 0.060 \left(1 - \cos\frac{\pi x}{2L}\right) \quad \text{kips/ft} \qquad ////$$

Lumped MDOF Systems

The formulation of the earthquake-response analysis of a lumped MDOF system can be carried out in matrix notation in a manner entirely analogous to the foregoing development of the lumped SDOF equations. Thus the equations of motion of the

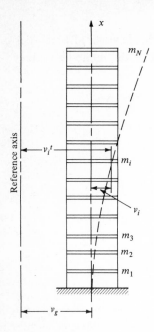

FIGURE 27-4
Lumped MDOF system with rigid-base translation.

multistory building shown in Fig. 27-4 can be written by analogy with Eq. (27-1) as

$$\mathbf{m}\ddot{\mathbf{v}}^t + \mathbf{c}\dot{\mathbf{v}} + \mathbf{k}\mathbf{v} = \mathbf{0} \qquad (27\text{-}26)$$

and again the effective earthquake force can be derived by expressing the total displacements as the sum of the relative motions plus the displacements resulting directly from the support motions. For the system of Fig. 27-4 this relationship may be written

$$\mathbf{v}^t = \mathbf{v} + \{\mathbf{1}\}v_g \qquad (27\text{-}27)$$

in which $\{\mathbf{1}\}$ represents a column of ones. This vector expresses the fact that a unit static translation of the base of this structure produces directly a unit displacement of all degrees of freedom. Of course, this simple relationship is a consequence of the type of support displacement which has been applied as well as of the structural configuration; for other forms of structures or support motions this static-displacement vector would be different. Thus, the structure shown in Fig. 27-4 should be considered as a special case, even though the vast majority of practical analyses are assumed to be of this type.

Substituting Eq. (27-27) into (27-26) leads to the relative-response equations of motion

$$\mathbf{m}\ddot{\mathbf{v}} + \mathbf{c}\dot{\mathbf{v}} + \mathbf{k}\mathbf{v} = \mathbf{p}_{\text{eff}}(t) \qquad (27\text{-}28)$$

in which

$$\mathbf{p}_{\text{eff}}(t) = -\mathbf{m}\{\mathbf{1}\}\ddot{v}_g(t) \qquad (27\text{-}29)$$

Equation (27-28) could be solved directly by numerical integration of the coupled equations; however, in analyzing the earthquake response of linear structures, it generally is much more efficient to transform to a system of normal (modal) coordinates because the support motions tend to excite strongly only the lowest modes of vibration. Thus good approximations of the earthquake response of systems having dozens or even hundreds of degrees of freedom can often be obtained by carrying out the analysis for only a few normal coordinates.

The transformation to normal coordinates has been described in adequate detail in Chap. 13. If it is assumed that the damping matrix is of a form which satisfies the same orthogonality conditions as the mass and stiffness matrices,[1] the result is a set of N uncoupled modal equations of the form

$$M_n \ddot{Y}_n + C_n \dot{Y}_n + K_n Y_n = P_n(t) \qquad (27\text{-}30)$$

in which M_n, C_n, and K_n are the generalized properties associated with mode n [see Eqs. (13-15)], Y_n is the amplitude of this modal response, and the generalized force resulting from the earthquake excitation [neglecting the negative sign in Eq. (27-29)] is given by

$$P_n = \phi_n{}^T \mathbf{p}_{\text{eff}}(t) = \mathscr{L}_n \ddot{v}_g(t) \qquad (27\text{-}31)$$

in which, for the structure of Fig. 27-4, the modal earthquake-excitation factor is given by

$$\mathscr{L}_n \equiv \phi_n{}^T \mathbf{m}\{\mathbf{1}\} \qquad (27\text{-}32)$$

It will be recognized that this is the matrix equivalent of Eq. (27-18), which was derived for the generalized SDOF system; of course the modal excitation factor is different for each mode because it contains the mode shape ϕ_n.

By analogy with the derivation of the generalized SDOF response, it may be seen that the response of each mode of the MDOF system is given by

$$Y_n(t) = \frac{\mathscr{L}_n}{M_n \omega_n} V_n(t) \qquad (27\text{-}33)$$

where the modal earthquake-response integral is of the form of Eq. (27-6) and is dependent on the damping ξ_n and frequency ω_n of the nth mode of vibration. The relative displacement vector produced in this mode then is given by

$$\mathbf{v}_n(t) = \phi_n \frac{\mathscr{L}_n}{M_n \omega_n} V_n(t) \qquad (27\text{-}34)$$

Finally, the relative-displacement vector due to all modal responses is obtained by superposition, that is,

$$\mathbf{v}(x,t) = \mathbf{\Phi} \mathbf{Y}(t) = \mathbf{\Phi} \left\{ \frac{\mathscr{L}_n}{M_n \omega_n} V_n(t) \right\} \qquad (27\text{-}35)$$

[1] The analysis of systems where this is not true will be discussed briefly in Sec. 27-5.

in which $\boldsymbol{\Phi}$ is made up of all mode shapes for which the modal response is excited significantly by the earthquake, and the term in braces represents a vector of such terms defined for each mode considered in the analysis.

The elastic forces associated with the relative displacements can be obtained directly by premultiplying by the stiffness matrix

$$\mathbf{f}_S(t) = \mathbf{k}\mathbf{v}(t) = \mathbf{k}\boldsymbol{\Phi}\mathbf{Y}(t) \qquad (27\text{-}36)$$

However, as mentioned in the discussion of the SDOF systems, it frequently is more convenient to express these forces in terms of the equivalent inertia forces developed in undamped free vibrations. The equivalence of the elastic and inertia forces is expressed by the eigenproblem relationship, which may be written

$$\mathbf{k}\boldsymbol{\Phi} = \mathbf{m}\boldsymbol{\Phi}\boldsymbol{\Omega}^2 \qquad (27\text{-}37)$$

in which $\boldsymbol{\Omega}^2$ is a diagonal matrix of the squared modal frequencies ω_n^2. Substituting Eq. (27-37) into Eq. (27-36) results in the alternate expression for the elastic forces

$$\mathbf{f}_S(t) = \mathbf{m}\boldsymbol{\Phi}\boldsymbol{\Omega}^2\mathbf{Y}(t) = \mathbf{m}\boldsymbol{\Phi}\left\{\frac{\mathscr{L}_n}{M_n}\omega_n V_n(t)\right\} \qquad (27\text{-}38)$$

It will be noted that the elastic-force vector associated with each mode in this equation, that is,

$$\mathbf{f}_{Sn}(t) = \mathbf{m}\boldsymbol{\phi}_n \frac{\mathscr{L}_n}{M_n}\omega_n V_n(t) \qquad (27\text{-}39)$$

is given by the matrix equivalent of the generalized SDOF expression of Eq. (27-23). It must be emphasized that Eq. (27-38) is a completely general expression for the elastic forces developed in a damped structure subjected to arbitrarily varying ground motions; the fact that it was derived from an expression for undamped free vibrations does not limit its applicability.

When the distribution of these effective elastic forces at any time t during the earthquake has been determined as illustrated, for example, in Fig. 27-5, the value of any desired force resultant at that same time can be computed by standard statics procedures. For example, the base shear force $\mathcal{V}_0(t)$ of the system in Fig. 27-5 is given by the sum of all the story forces, that is,

$$\mathcal{V}_0(t) = \sum_{i=1}^{N} f_{Si}(t) = [\mathbf{1}]\,\mathbf{f}_S(t)$$

where $[\mathbf{1}]$ represents a row vector of ones. Substituting Eq. (27-38) into this expression leads to

$$\mathcal{V}_0(t) = \sum_{n=1}^{N} \frac{\mathscr{L}_n^2}{M_n}\omega_n V_n(t) \qquad (27\text{-}40)$$

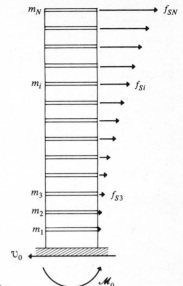

FIGURE 27-5
Elastic forces in lumped MDOF system.

in which it has been noted that $[\mathbf{1}]\mathbf{M}\boldsymbol{\Phi} = [\mathscr{L}_1 \quad \mathscr{L}_2 \quad \cdots \quad \mathscr{L}_N]$. Similarly, the resultant overturning moment at the base of the building is

$$\mathscr{M}_0(t) = \sum_{i=1}^{N} x_i f_{Si}(t) = [\mathbf{x}]\mathbf{f}_S(t)$$

in which x_i is the height of mass i above the base and $[\mathbf{x}]$ is a row vector of these heights. Substituting Eq. (27-38) into this yields the expression for the base moment

$$\mathscr{M}_0(t) = [\mathbf{x}]\mathbf{M}\boldsymbol{\Phi}\boldsymbol{\Omega}^2\mathbf{Y}(t) = [\mathbf{x}]\mathbf{M}\boldsymbol{\Phi} \left\{ \frac{\mathscr{L}_n}{M_n} \omega_n V_n(t) \right\} \qquad (27\text{-}41)$$

The quantity $\mathscr{L}_n{}^2/M_n$ in Eq. (27-40) has the dimensions of mass and is sometimes called the *effective modal mass* of the structure because it can be interpreted as the part of the total mass responding to the earthquake in each mode. This interpretation of the expression is valid only for structures of the type shown in Fig. 27-4, having masses lumped along a vertical axis; for such structures, the total mass M_T is given by

$$M_T = [\mathbf{1}]\mathbf{m}\{\mathbf{1}\} \qquad (27\text{-}42)$$

Now it can be proved that the sum of all modal effective masses is equal to the total mass by expressing the vector of ones $\{\mathbf{1}\}$ in modal coordinates as

$$\{\mathbf{1}\} = \boldsymbol{\Phi}\mathbf{Y}$$

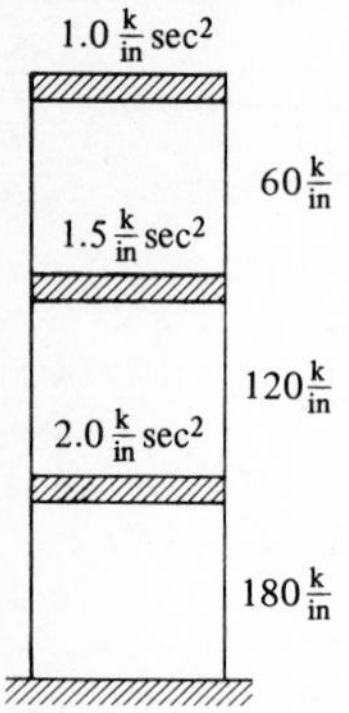

$$\mathbf{\Phi} = \begin{bmatrix} 1.000 & 1.000 & 1.00 \\ 0.644 & -0.601 & -2.57 \\ 0.300 & -0.676 & 2.47 \end{bmatrix}; \qquad \Omega^2 = \begin{bmatrix} 21.0 & 0 & 0 \\ 0 & 96.6 & 0 \\ 0 & 0 & 212.4 \end{bmatrix}$$

$$\omega_n = \begin{Bmatrix} 4.58 \\ 9.82 \\ 14.59 \end{Bmatrix}; \quad T_n = \begin{Bmatrix} 1.37 \\ 0.640 \\ 0.431 \end{Bmatrix} \text{sec}; \quad M_n = \begin{Bmatrix} 1.801 \\ 2.455 \\ 23.10 \end{Bmatrix}; \quad \mathcal{L}_n = \begin{Bmatrix} 2.566 \\ -1.254 \\ 2.08 \end{Bmatrix}$$

FIGURE E27-3
Building frame and its vibration properties.

where each modal amplitude Y_n can be evaluated by multiplying both sides by $\boldsymbol{\phi}_n^T\mathbf{m}$ and applying the mass orthogonality relationship, that is,

$$\boldsymbol{\phi}_n^T\mathbf{m}\{\mathbf{1}\} = \boldsymbol{\phi}_n^T\mathbf{m}\mathbf{\Phi}\mathbf{Y} = M_n Y_n$$

Since the left-hand triple matrix product is $\mathcal{L}_n$, each modal amplitude can be expressed as $Y_n = \mathcal{L}_n/M_n$ and the ones vector is given by

$$\{\mathbf{1}\} = \mathbf{\Phi}\begin{Bmatrix} \dfrac{\mathcal{L}_n}{M_n} \end{Bmatrix} \tag{27-43}$$

Substituting this into Eq. (27-42) gives

$$M_T = [\mathbf{1}]\mathbf{m}\mathbf{\Phi}\begin{Bmatrix} \dfrac{\mathcal{L}_n}{M_n} \end{Bmatrix} = [\mathcal{L}_1 \;\; \mathcal{L}_2 \;\; \cdots \;\; \mathcal{L}_N]\begin{Bmatrix} \dfrac{\mathcal{L}_n}{M_n} \end{Bmatrix} = \sum_{n=1}^{N} \frac{\mathcal{L}_n^{\,2}}{M_n} \qquad \text{Q.E.D.} \tag{27-44}$$

Hence each modal contribution $\mathcal{V}_{0n}(t)$ to the base shear of Eq. (27-40) may be looked upon as the reaction of the effective modal mass to the effective modal acceleration of the ground $\omega_n V_n(t)$.

EXAMPLE E27-3 To demonstrate the earthquake response analysis of a MDOF structure, the three-story building shown in Fig. E27-3 will be considered. This is the structure of Example E12-1 with its stiffness reduced by a factor of 10 to provide frequencies typical of a taller building in which the higher modes would contribute more to the response. The vibration properties, generalized masses, and modal excitation factors are also shown in the figure; in addition it is assumed that the damping is 5 percent critical in each mode.

From the given frequency and damping values, the first-mode response integral $V_1(t)$ was calculated for the entire history of a certain earthquake motion $v_g(t)$; the maximum value of this integral occurred at $t_1 = 3.08$ s. Then the second- and third-mode response integrals were evaluated at the same time; the values of the three modal response integrals at this time were

$$\mathbf{V}(t_1) = -\begin{bmatrix} 1.74 \\ 1.22 \\ 0.77 \end{bmatrix} \text{ ft/s}$$

When these values and the other modal properties are introduced into Eq. (27-33), the normal coordinates at this time are

$$\mathbf{Y}(t_1) = \left\{ \frac{\mathscr{L}_n}{M_n \omega_n} V_n(t_1) \right\} = \begin{bmatrix} 0.541 \\ 0.0635 \\ 0.00475 \end{bmatrix} \text{ ft}$$

and the resulting displacements are

$$\mathbf{v}(t_1) = \mathbf{\Phi Y}(t_1) = \begin{bmatrix} 0.541 + 0.064 + 0.005 \\ 0.348 - 0.038 - 0.012 \\ 0.162 - 0.043 + 0.012 \end{bmatrix} = \begin{bmatrix} 0.610 \\ 0.298 \\ 0.131 \end{bmatrix} \text{ ft}$$

where the individual modal displacements have been shown as a matter of interest. Similarly, the elastic-force vector at this time is

$$\mathbf{f}_S(t_1) = \mathbf{m\Phi} \left\{ \frac{\mathscr{L}_n}{M_n} \omega_n V_n(t_1) \right\} = \begin{bmatrix} 11.35 + 6.13 + 1.01 \\ 10.95 - 5.53 - 3.90 \\ 6.80 - 8.29 + 5.00 \end{bmatrix} = \begin{bmatrix} 18.49 \\ 1.52 \\ 3.51 \end{bmatrix} \text{ kips}$$

and the base shear force is given by the sum of the story forces:

$$\mathcal{V}_0(t_1) = 23.52 \text{ kips} \qquad ////$$

To evaluate the earthquake response of a lumped MDOF system at any time t using Eq. (27-35) or (27-38) involves the evaluation of the earthquake-response integral at that time for each significant response mode. Hence, the evaluation of the maximum response requires that each modal response be computed in this way for each time during the earthquake history, in order that the maximum value can be identified. This obviously constitutes a major computational task and makes an approximate analysis based on the ground-motion response spectra an attractive alternative.

For each individual mode of the structure, the maximum response can be obtained directly from the response spectrum as described for the SDOF systems. For example, from Eq. (27-34) the maximum displacement in mode n is given by

$$\mathbf{v}_{n,\,\text{max}} = \boldsymbol{\phi}_n \frac{\mathscr{L}_n}{M_n} S_d(\xi_n, T_n) \qquad (27\text{-}45)$$

where $S_d(\xi_n, T_n)$ is the spectral displacement corresponding to the damping and period of the nth mode of vibration. Similarly, from Eq. (27-39) the maximum elastic-force vector in mode n is given by

$$\mathbf{f}_{Sn,\,\text{max}} = \mathbf{m}\boldsymbol{\phi}_n \frac{\mathscr{L}_n}{M_n} S_a(\xi_n, T_n) \qquad (27\text{-}46)$$

where $S_a(\xi_n, T_n)$ is the spectral acceleration for the nth mode. However, the maximum total response cannot be obtained, in general, by merely adding the modal maxima because these maxima usually do not occur at the same time. In most cases, when one mode achieves its maximum response, the other modal responses are less than their individual maxima. Therefore, although the superposition of the modal spectral values obviously provides an upper limit to the total response, it generally over-estimates this maximum by a significant amount.

A number of different formulas have been proposed to obtain a more reasonable estimate of the maximum response from the spectral values. The simplest and most popular of these is the square root of the sum of the squares of the modal responses. Thus if the maximum modal displacements are given by Eq. (27-45), the maximum total displacement is approximated by

$$\mathbf{v}_{\text{max}} \doteq \sqrt{(\mathbf{v}_1)_{\text{max}}^2 + (\mathbf{v}_2)_{\text{max}}^2 + \cdots} \qquad (27\text{-}47)$$

where the terms under the radical sign represent vectors of the modal displacements squared. Similarly the maximum story forces could be approximated from the modal maxima of Eq. (27-46) as follows:

$$\mathbf{f}_{S,\,\text{max}} \doteq \sqrt{(\mathbf{f}_{S1})_{\text{max}}^2 + (\mathbf{f}_{S2})_{\text{max}}^2 + \cdots} \qquad (27\text{-}48)$$

EXAMPLE E27-4   The response spectrum analysis of a MDOF structure will be illustrated by evaluating the response of the building of Example E27-3 to an earthquake having a response spectrum 3 times that shown in Fig. E27-1. (With this increase, the intensity of the response spectrum motion is about equivalent to that of the earthquake considered in Example E27-3.)

When 5 percent critical damping is assumed, and when the periods given in Example E27-3 are used, the modal spectral velocities of Fig. E27-1 multiplied by 3 are

$$\mathbf{S}_v = \begin{bmatrix} 1.74 \\ 1.41 \\ 1.20 \end{bmatrix} \text{ft/s}$$

Hence the modal maximum displacements, which are given by

$$\mathbf{v}_{n, \text{max}} = \boldsymbol{\phi}_n \frac{\mathscr{L}_n}{M_n} \frac{S_{vn}}{\omega_n}$$

are

$$\mathbf{v}_{1, \text{max}} = \begin{bmatrix} 0.541 \\ 0.348 \\ 0.162 \end{bmatrix} \qquad \mathbf{v}_{2, \text{max}} = \begin{bmatrix} 0.074 \\ 0.044 \\ 0.050 \end{bmatrix} \qquad \mathbf{v}_{3, \text{max}} = \begin{bmatrix} 0.008 \\ 0.019 \\ 0.018 \end{bmatrix}$$

Superposing the modal maxima by the root-sum-square procedure gives for the approximate total maximum displacement

$$\mathbf{v}_{\text{max}} \doteq \begin{bmatrix} 0.546 \\ 0.351 \\ 0.170 \end{bmatrix} \text{ft}$$

in which it is evident that the higher modes contribute very little.

Similarly, the modal maximum forces are given by

$$\mathbf{f}_{Sn, \text{max}} = \mathbf{m}\boldsymbol{\phi}_n \frac{\mathscr{L}_n}{M_n} \omega_n S_{vn}$$

and are

$$\mathbf{f}_{S1, \text{max}} = \begin{bmatrix} 11.35 \\ 10.95 \\ 6.80 \end{bmatrix} \qquad \mathbf{f}_{S2, \text{max}} = \begin{bmatrix} 7.08 \\ 6.39 \\ 9.58 \end{bmatrix} \qquad \mathbf{f}_{S3, \text{max}} = \begin{bmatrix} 1.57 \\ 6.08 \\ 7.79 \end{bmatrix}$$

Superposing these by the root-sum-square method gives for the approximate total earthquake story forces

$$\mathbf{f}_{S, \text{max}} = \begin{bmatrix} 13.47 \\ 14.06 \\ 14.10 \end{bmatrix} \text{kips} \qquad (a)$$

Finally, the modal maximum base shears, which are given by

$$\mathcal{V}_{0n} = \frac{\mathscr{L}_n^2}{M_n} \omega_n S_{vn}$$

are $\mathcal{V}_{01, \text{max}} = 29.13$; $\mathcal{V}_{02, \text{max}} = 8.87$; $\mathcal{V}_{03, \text{max}} = 3.28$ kips; taking the root-sum-square of these gives the approximate maximum base shear

$$\mathcal{V}_{0, \text{max}} = 30.6 \text{ kips} \qquad (b)$$

This example demonstrates clearly that the maximum base shear (b) cannot be obtained by simply summing the maximum story forces (a); these forces are *not* concurrent.

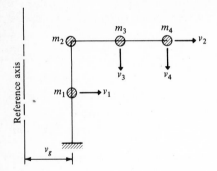

FIGURE 27-6
General lumped MDOF system with
rigid-base translation.

It is interesting to note in this example that the effective modal masses for this structure are

$$\left\{ \frac{\mathscr{L}_n^2}{M_n} \right\} = \begin{bmatrix} 3.656 \\ 0.641 \\ 0.187 \end{bmatrix}$$

and their sum is 4.48 which except for slight slide rule error is the same as the sum of the story masses. This equality applies to all building-type structures, as was noted earlier.   ////

In both Eqs. (27-47) and (27-48) only the significant modal contributions need be included, and because each term is squared, very few modes need be considered in most cases. It is important to note that this approximate superposition procedure must be applied directly to the response quantity in question. As was shown in Example E27-4, to estimate the maximum base shear it was necessary to compute the modal base shears and superpose them:

$$\mathcal{V}_{0,\,\text{max}} \doteq \sqrt{(\mathcal{V}_{01})_{\text{max}}^2 + \mathcal{V}_{02})_{\text{max}}^2 + \cdots}$$

The base shear could not be found by summing the maximum forces $\mathbf{f}_{S,\,\text{max}}$ over the height of the building because the signs of the local force quantities are lost in the squaring process.

It was pointed out at the beginning of this discussion of lumped MDOF systems that the type of system shown in Fig. 27-4, having a vertical axis and subjected to horizontal excitation, represents a special class of earthquake problem for which the relationship between the total and relative motions takes the simple form of Eq. (27-27). In a more general case, where the relative displacements are not all measured parallel to the ground motion, an example of which is shown in Fig. 27-6, the total displacement may be expressed as the sum of the relative displacement and the pseudostatic displacements $\mathbf{v}^s$ that would result from a static-support displacement, that is,

$$\mathbf{v}^t = \mathbf{v} + \mathbf{v}^s \qquad (27\text{-}49)$$

The pseudostatic displacements can be expressed conveniently by an influence coefficient vector $\mathbf{r}$ which represents the displacements resulting from a unit support displacement; thus $\mathbf{v}^s = \mathbf{r}v_g$ and

$$\mathbf{v}^t = \mathbf{v} + \mathbf{r}v_g \qquad (27\text{-}50)$$

From comparison of Eqs. (27-27) and (27-50) it is evident that $\mathbf{r}$ is a vector of ones for the structure of Fig. 27-4; for the system of Fig. 27-6 it would be given by $\mathbf{r}^T = \begin{bmatrix} 1 & 1 & 0 & 0 \end{bmatrix}$.

This generalization affects only the effective-force vector generated by the earthquake motion; that is, in place of Eq. (27-29), which was derived for the special static-displacement influence vector, the generalized expression is

$$\mathbf{p}_{\text{eff}}(t) = -\mathbf{mr}\ddot{v}_g(t) \qquad (27\text{-}51)$$

Similarly, the generalized form of the modal earthquake-excitation factor, replacing Eq. (27-32), would be

$$\mathscr{L}_n = \boldsymbol{\phi}_n{}^T \mathbf{mr} \qquad (27\text{-}52)$$

With this generalized definition of $\mathscr{L}_n$, the response equations [Eqs. (27-33) to (27-39)] are now fully applicable to generalized forms of lumped-mass structures. Of course it must be noted that the elastic forces act in the directions of the corresponding displacements $\mathbf{v}$; hence new expressions for the force resultants (such as base shear or moment) would have to be derived, appropriate to the given structural configuration.

EXAMPLE E27-5 The earthquake response of a structure for which a unit static support movement does not cause a unit displacement of each degree of freedom will be demonstrated by analysis of the structure shown in Fig. E27-4. The mass and stiffness matrices defined for the two specified degrees of freedom are shown, as are the eigenvectors and eigenvalues describing its free vibration.

From these data, the modal earthquake-response parameters are

$$\begin{bmatrix} M_1 \\ M_2 \end{bmatrix} = \begin{bmatrix} 2.557 \\ 3.834 \end{bmatrix} m \qquad \begin{bmatrix} \mathscr{L}_1 \\ \mathscr{L}_2 \end{bmatrix} = \begin{bmatrix} 1.293 \\ 3.000 \end{bmatrix} m$$

$$\begin{bmatrix} \omega_1 \\ \omega_2 \end{bmatrix} = \begin{bmatrix} 5.49 \\ 16.86 \end{bmatrix} \text{rad/s} \qquad \begin{bmatrix} T_1 \\ T_2 \end{bmatrix} = \begin{bmatrix} 1.144 \\ 0.373 \end{bmatrix} \text{s}$$

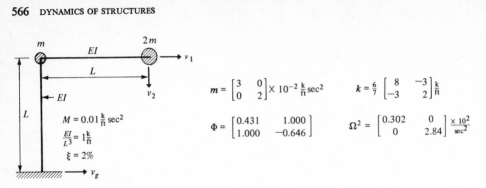

$$m = \begin{bmatrix} 3 & 0 \\ 0 & 2 \end{bmatrix} \times 10^{-2} \frac{k}{ft} \sec^2 \qquad k = \frac{6}{7} \begin{bmatrix} 8 & -3 \\ -3 & 2 \end{bmatrix} \frac{k}{ft}$$

$$\Phi = \begin{bmatrix} 0.431 & 1.000 \\ 1.000 & -0.646 \end{bmatrix} \qquad \Omega^2 = \begin{bmatrix} 0.302 & 0 \\ 0 & 2.84 \end{bmatrix} \frac{\times 10^2}{\sec^2}$$

FIGURE E27-4
Two DOF frame and its vibration properties.

When it is assumed that this structure is subjected to an earthquake motion corresponding to the velocity-response spectrum of Fig. E27-1 (note the 2 percent critical modal damping ratios), the following modal spectral velocities are obtained:

$$\begin{bmatrix} S_{v1} \\ S_{v2} \end{bmatrix} = \begin{bmatrix} 0.75 \\ 0.55 \end{bmatrix} \text{ft/s}$$

Accordingly, from Eq. (27-46) the maximum modal response forces are

$$\mathbf{f}_{S1,\,max} = \{m\phi_1\} \frac{\mathcal{L}_1}{M_1} \omega S_{v1} = \begin{bmatrix} 1.293 \\ 2.000 \end{bmatrix} \frac{1.293}{2.557} (5.49)(0.75 \times 10^{-2})$$

$$= \begin{bmatrix} 2.69 \\ 4.16 \end{bmatrix} \times 10^{-2} \text{ kip}$$

$$\mathbf{f}_{S2,\,max} = \begin{bmatrix} 3.000 \\ -1.292 \end{bmatrix} \frac{3.000}{3.834} (16.86)(0.55 \times 10^{-2})$$

$$= \begin{bmatrix} 21.77 \\ 9.37 \end{bmatrix} \times 10^{-2} \text{ kip}$$

Applying the root-sum-square method to these modal results gives the approximate maximum response forces

$$\mathbf{f}_{S,\,max} = \begin{bmatrix} 21.94 \\ 10.25 \end{bmatrix} \times 10^{-2} \text{ kip} \qquad ////$$

Similar comments apply to the system of Fig. 27-7, which consists of a rigid rectangular slab supported by identical columns at three corners. When the degrees of freedom of this system are defined as the x and y translations of the center of mass, together with the rotation about that center, that is, $\mathbf{v}^T = [v_1 \ v_2 \ v_3]$, and when

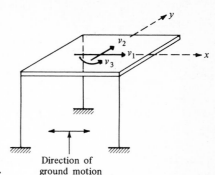

FIGURE 27-7
Rigid slab subjected to base translation.

Direction of
ground motion

it is assumed that the earthquake motions act in the direction of the x axis, the influence vector for this case is given by $\mathbf{r}^T = \begin{bmatrix} 1 & 0 & 0 \end{bmatrix}$. The modal earthquake-excitation factors for this structure are then obtained by substituting this vector into Eq. (27-52), and the response is given finally by Eqs. (27-33) to (27-39).

EXAMPLE E27-6   Because the earthquake-response analysis of a rigid slab structure of this type involves several features of special interest, the example structure of Fig. E27-5 will be discussed in some detail. It is assumed that the three columns supporting the slab are rigidly attached to the foundation and to the slab, so that the resistance at the top of each column to lateral displacement in any direction is $12EI/L^3 = 5$ kips/ft. The torsional stiffness of the columns is negligible.

For the purpose of this example, the three degrees of freedom of the

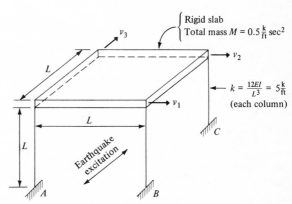

FIGURE E27-5
Slab supported by three columns.

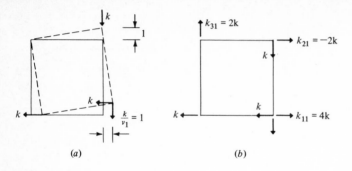

FIGURE E27-6
Evaluation of stiffness coefficients for $v_1 = 1$: (*a*) displacement $v_1 = 1$ and resisting column forces; (*b*) column forces and equilibrating stiffness coefficients.

slab are represented by the displacement components of the corners as shown. The total mass of the slab is $m = 0.5$ kip·s²/ft and is distributed uniformly over the area. The structure is subjected to an earthquake having the response spectrum of Fig. E27-1 and acting in the direction parallel with coordinate v_3. It is desired to determine the maximum displacements of the slab due to this earthquake.

The mass and stiffness matrices of this system can be evaluated by direct application of the definitions of the influence coefficients. Considering first the stiffness matrix, a unit displacement $v_1 = 1$ is applied while the other coordinates are constrained, as shown in Fig. E27-6a. The forces exerted by the columns in resisting this displacement are shown in this sketch, and the equilibrating forces corresponding to the degrees of freedom are shown in Fig. 27-6b. By applying unit displacements of the other two coordinates, the remaining stiffness coefficients can be determined similarly.

The mass matrix is evaluated by applying a unit acceleration separately to each degree of freedom and determining the resulting inertia forces in the slab. For example, Fig. E27-7a shows the unit acceleration $\ddot{v}_2 = 1$ and the slab inertia forces resisting this acceleration, while Fig. E27-7b shows the mass influence coefficients which equilibrate these inertial forces. The other mass coefficients can be found by unit accelerations of the other two coordinates. The complete stiffness and mass matrices for the system are

$$\mathbf{k} = \frac{12EI}{L^3} \begin{bmatrix} 4 & -2 & 2 \\ -2 & 3 & -2 \\ 2 & -3 & 3 \end{bmatrix} \qquad \mathbf{m} = \frac{m}{6} \begin{bmatrix} 4 & -1 & 3 \\ -1 & 4 & -3 \\ 3 & -3 & 6 \end{bmatrix}$$

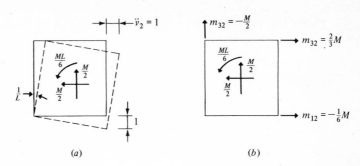

FIGURE E27-7
Evaluation of mass coefficients for $\ddot{v}_2 = 1$. (a) Acceleration $\ddot{v}_2 = 1$ and resisting inertia forces; (b) slab inertia forces and equilibrating mass coefficients.

When the eigenproblem $(\mathbf{k} - \omega^2\mathbf{m})\hat{\mathbf{v}} = \mathbf{0}$ is solved, the mode shapes and frequencies of the system are found to be

$$\mathbf{\Phi} = \begin{bmatrix} 0.366 & 1.000 & -1.366 \\ 1.000 & 1.000 & 1.000 \\ 1.000 & -1.000 & 1.000 \end{bmatrix} \qquad \omega^2 = \begin{bmatrix} 25.36 \\ 30.00 \\ 94.64 \end{bmatrix} (\text{rad/s})^2$$

Study of these mode shapes reveals that the first and third represent rotations about points on the symmetry diagonal while the second is simple translation along this diagonal. Obviously these motions could have been identified more easily by a more appropriate coordinate system; translation of the center of mass in the direction of the two diagonals plus rotation about the center of mass would have been a better choice of coordinates.

The frequencies, periods of vibration, and the spectral velocities given by Fig. E27-1 (assuming 5 percent damping) for the three modes of this structure are

$$\omega = \begin{bmatrix} 5.036 \\ 5.477 \\ 9.464 \end{bmatrix} \text{rad/s} \qquad \mathbf{T} = \begin{bmatrix} 1.25 \\ 1.15 \\ 0.65 \end{bmatrix} \text{s} \qquad \mathbf{S}_v = \begin{bmatrix} 0.55 \\ 0.54 \\ 0.48 \end{bmatrix} \text{ft/s}$$

Also the generalized masses M_n and modal earthquake-excitation factors $\mathcal{L}_n = \boldsymbol{\phi}_n{}^T\mathbf{mr}$ where $\mathbf{r}^T = \begin{bmatrix} 0 & 0 & 1 \end{bmatrix}$ are

$$\mathbf{M} = \begin{bmatrix} 0.5 \\ 1.0 \\ 0.5 \end{bmatrix} \text{kips}\cdot\text{s}^2/\text{ft} \qquad \mathcal{L} = \begin{bmatrix} 0.3415 \\ -0.5000 \\ -0.0915 \end{bmatrix} \text{kips}\cdot\text{s}^2/\text{ft}$$

Hence the maximum modal displacements are found, from

$$v_{n, \text{max}} = \phi_n \frac{\mathcal{L}_n}{M_n} \frac{S_{vn}}{n}$$

to be

$$v_{1, \text{max}} = \begin{bmatrix} 0.0272 \\ 0.0745 \\ 0.0745 \end{bmatrix} \text{ft} \qquad v_{2, \text{max}} = \begin{bmatrix} 0.0493 \\ 0.0493 \\ -0.0493 \end{bmatrix} \text{ft} \qquad v_{3, \text{max}} = \begin{bmatrix} -0.0124 \\ 0.0091 \\ 0.0091 \end{bmatrix} \text{ft}$$

An approximation of the maximum displacement in each coordinate could be determined from these results by the root-sum-square method. ////

Comparison with Uniform Building Code Requirements

It is of interest to compare the foregoing formulation of expressions for the forces developed in a building due to seismic excitation with the seismic design requirements of a typical building code. For example, in the Uniform Building Code (UBC)[1] the principal seismic provision defines the effective intensity of the design earthquake in terms of the maximum shear force which it produces at the base of the building. The expression for this code base shear force $\mathcal{V}_0$ is of the form

$$\mathcal{V}_{0, \text{max}} = k\hat{C}W \qquad (a)$$

where W is the weight of the building, $\hat{C}$ is the base-shear coefficient, and k is a factor which depends on the type of structural framing system. This factor is intended to account for the relative energy-absorbing capacity of the framing type and varies from $2/3$ for a rigid jointed frame which resists lateral forces by flexure of the columns and girders to $4/3$ for a box-type structure assembled from shear panels in the horizontal and vertical planes. The base-shear coefficient is expressed as a function of the fundamental period of vibration T of the structure as

$$\hat{C} = \frac{0.05}{\sqrt[3]{T}} \qquad (b)$$

Also contained in the UBC provisions is a zone factor which reduces the design forces for zones of less seismicity; the zone factor of unity implied in Eq. (a) is intended for the regions of highest seismicity.

An analytical expression corresponding to the code formula of Eq. (a) above can easily be derived from Eq. (27-40) by considering only the fundamental mode and

---

[1] Published by the International Conference of Building Officials, Pasadena, California.

expressing the maximum response value in terms of the spectral acceleration for this mode $S_{a1} = \omega_1 V_{1,\,max}$; thus

$$\mathcal{V}_{0,\,max} \doteq \mathcal{V}_{01,\,max} = \frac{\mathscr{L}_1^2}{M_1} S_{a1} \equiv \frac{\mathscr{L}_1^2}{M_1} g \frac{S_{a1}}{g} \qquad (c)$$

where g is the acceleration of gravity. Comparison of Eqs. (a) and (c) reveals the equivalence of the following terms:

Code: $\qquad\qquad\qquad\qquad\qquad\qquad\qquad\qquad\quad \hat{C} \qquad\quad W$

Analytical: $\qquad\qquad\qquad\qquad\qquad\qquad\qquad\quad \dfrac{S_{a1}}{g} \quad \dfrac{\mathscr{L}_1^2}{M_1} g$

Thus, the base-shear coefficient is equivalent to the spectral acceleration expressed as a fraction of gravity, while the total weight is considered to be equivalent to the effective weight in the first mode. Actually, the first-mode effective weight must be less than the total, and usually is about two-thirds to three-fourths of the total, so that this weight assumption is rather conservative. However, the earthquake forces defined by the code expression depend mainly on the formula for base-shear coefficient [Eq. (b)]; thus the conservatism in W is not really significant.

The other major provision of the UBC seismic code defines how the total base-shear force is distributed over the height of the building, as follows:

$$f_{Si,\,max} = \frac{w_i x_i}{\Sigma w_i x_i} \, \mathcal{V}_{0,\,max} \qquad (d)$$

where f_{Si} = lateral force at floor level i

$\qquad w_i$ = weight at level i

$\qquad x_i$ = height of level i above the building base

A corresponding analytical expression can be derived from Eq. (27-39) by writing it for the first mode and stating its maximum value in terms of the spectral acceleration; thus

$$\mathbf{f}_{S,\,max} = \mathbf{m}\boldsymbol{\phi}_1 \frac{\mathscr{L}_1}{M_1} S_{a1} \equiv \frac{\mathbf{m}\boldsymbol{\phi}_1}{\mathscr{L}_1} \frac{\mathscr{L}_1^2}{M_1} S_{a1} \qquad (e)$$

If it is assumed that only the first-mode response is significant in the total lateral force, with substitution from Eq. (c) the lateral force at level i given by Eq. (e) is

$$f_{si} = \frac{m_i \phi_{1i}}{\Sigma m_i \phi_{1i}} \, \mathcal{V}_{0,\,max} \qquad (f)$$

in which the earthquake-excitation factor for a lumped-mass system has been written in the form $\mathscr{L}_1 = \Sigma m_i \phi_{1i}$.

Comparison of Eqs. (f) and (d) shows that the code expression represents the response of a lumped-mass system which is constrained to deflect with a straight-line shape, that is, $\phi_{1i} = x_i/L$. This assumed shape has been incorporated into the code because observations of the vibrations of a large number of buildings demonstrate that the first-mode shape generally is quite close to a straight line. In summary, then, it may be seen that the principal UBC seismic force requirements are equivalent to the results of a first-mode response spectrum analysis if the first-mode shape is a straight line and if the base-shear coefficient is taken as the first-mode spectral acceleration (multiplied by the ratio of total weight to effective modal weight if a refined comparison is to be made). The UBC also includes a provision for applying a larger part of the lateral force to the top of a tall building, to account in part for the influence of higher-mode responses, but this refinement is beyond the scope of the present comparison.

Distributed-Parameter Systems

The formulation of the earthquake-response equations for systems having continuously distributed properties can be carried out by procedures which are completely analogous to those described above. The decoupled normal-coordinate equations of motion take the same form as for the lumped-mass system and may be expressed as

$$\ddot{Y}_n + 2\xi_n\omega_n\dot{Y}_n + \omega_n^2 Y_n = \frac{P_n(t)}{M_n} = \frac{\mathscr{L}_n}{M_n}\ddot{v}_g(t) \qquad (27\text{-}53)$$

However, the generalized mass associated with the distributed mass $m(x)$ now is given by

$$M_n = \int_0^L \phi_n^2(x)m(x)\,dx \qquad (19\text{-}5)$$

and the modal earthquake-excitation factor takes the integral form equivalent to the previous triple matrix product

$$\mathscr{L}_n = \int_0^L \phi_n(x)m(x)r(x)\,dx \qquad (27\text{-}54)$$

In this equation, $r(x)$ is the static-displacement influence function representing the displacements resulting from a unit displacement of the ground $v_g = 1$; thus

$$v^s(x) = r(x)v_g \qquad (27\text{-}55)$$

By using these expressions for M_n and $\mathscr{L}_n$ in Eq. (27-33) the amplitude of each modal response can be evaluated. Then the total displacement response can be obtained by superposition, using the continuous equivalent of Eq. (27-35) as follows:

$$v(x,t) = \sum_{n=1}^{\infty} \phi_n(x)Y_n(t) = \sum_{n=1}^{\infty} \phi_n(x)\frac{\mathscr{L}_n}{M_n\omega_n}V_n(t) \qquad (27\text{-}56)$$

Of course in practice only the significant modal responses are included in the super-position even though in principle an infinite number of modes might be considered. The elastic-force distribution is given similarly by an expression analogous to Eq. (27-38):

$$f_s(x,t) = \sum_{n=1}^{\infty} m(x)\phi_n(x)\omega_n^2 Y_n(t) = \sum_{n=1}^{\infty} m(x)\phi_n(x)\frac{\mathscr{L}_n}{M_n}\omega_n V_n(t) \qquad (27\text{-}57)$$

Equations (27-56) and (27-57) express the time history of earthquake response for arbitrary distributed-parameter systems. The procedure for approximating the maximum earthquake response of this type of structure by response-spectrum super-position is entirely equivalent to that described earlier for the lumped-mass systems and need not be discussed further.

Although the analysis procedure for structures with distributed properties outlined above is completely general in principle, its use in practice is limited by the fact that the vibration mode shapes and frequencies can be obtained only for very simple systems. For this reason, the more complicated distributed-parameter systems usually are discretized by the finite-element method so that their analysis can be car-ried out in matrix form. The matrix equations for the earthquake-response analysis of finite-element systems are identical in form to the lumped-mass equations described above, except that where the consistent-mass formulation is used, the mass matrix no longer is diagonal. If the off-diagonal coefficients in the mass matrix, which introduce coupling between the support displacements and the response degrees of freedom, are denoted $\mathbf{m}_g$, the equations of motion become

$$\mathbf{m}\ddot{\mathbf{v}}^t + \mathbf{m}_g\ddot{v}_g + \mathbf{c}\dot{\mathbf{v}} + \mathbf{k}\mathbf{v} = 0 \qquad (27\text{-}58)$$

With the total acceleration expressed in terms of the relative and the pseudostatic components by Eq. (27-50), this equation can be written in the form of Eq. (27-28) but with the effective force now given by

$$\mathbf{p}_{\text{eff}}(t) = -(\mathbf{mr} + \mathbf{m}_g)\ddot{v}_g(t) \qquad (27\text{-}59)$$

The corresponding modal earthquake-excitation factor then becomes

$$\mathscr{L}_n = \phi_n^T\mathbf{mr} + \phi_n^T\mathbf{m}_g \qquad (27\text{-}60)$$

Once this factor has been evaluated, the rest of the analysis is carried out exactly as for a lumped-mass system. In most cases there are few nonzero terms in the mass-coupling matrix $\mathbf{m}_g$, and the second term in Eq. (27-60) usually contributes little to the earthquake-excitation factor; however, it should be included in the formulation for completeness.

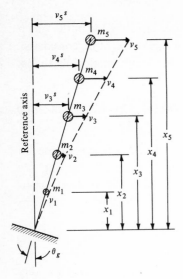

FIGURE 27-8
Tower with lumped point masses subjected to base rotation.

27-3 EXCITATION BY RIGID-BASE ROTATION

The ground-motion input mechanism is accounted for in the earthquake-response equations of motion, such as Eq. (27-58), by the effective-force term on the right-hand side. This term arises from expressing the total-displacement vector as the sum of a pseudostatic displacement plus the dynamic motions, and consequently the specific nature of the earthquake input is contained in the pseudostatic displacements or, more precisely, in the pseudostatic influence coefficients $\mathbf{r}$. Thus to deal with any form of earthquake input other than the rigid-base translation discussed in the preceding section, it is necessary only to define the matrix $\mathbf{r}$ appropriate to the new support-motion condition.

Consider now the case of a vertical-axis structure subjected to small-amplitude base rotation, as shown in Fig. 27-8; the pseudostatic displacements in this case clearly are given by a rigid rotation about the base, that is,

$$\mathbf{v}^s = \begin{bmatrix} x_1 \\ x_2 \\ \vdots \\ x_5 \end{bmatrix} \theta_g \qquad (27\text{-}61)$$

Consequently the influence-coefficient matrix is merely the listing of the height of the masses above the base, as follows:

$$\mathbf{r}^T = \begin{bmatrix} x_1 & x_2 & x_3 & x_4 & x_5 \end{bmatrix} \qquad (27\text{-}62)$$

When this vector $\mathbf{r}$ is used in expressing the effective earthquake forces [by Eq. (27-51)] and the modal earthquake-excitation factor [by Eq. (27-52)], the response

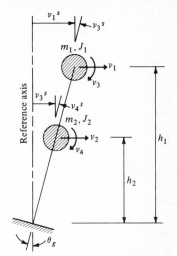

FIGURE 27-9
Tower with lumped masses having rotational inertia subjected to base rotation.

resulting from a base-rotation input can be calculated in exactly the same way as described above for the systems with rigid-base translations.

The only other factor that need be pointed out about the rotational support excitation is that effective earthquake moments are induced in proportion to the rotational inertia of the lumped masses as well as the effective translational forces which result from the translational inertia. In the example shown in Fig. 27-9, the two masses m have rotational inertias J; to account for these rotational effects in the equilibrium equations, it is necessary to include the rotational as well as translational degrees of freedom in the displacement vector. The pseudostatic influence-coefficient vector for all of these degrees of freedom is given by

$$\mathbf{r}^T = \begin{bmatrix} h_1 & h_2 & 1 & 1 \end{bmatrix}$$

where the h terms are the distances of the masses from the base-rotation point. There are many other types of systems in which rotational-inertia effects may be important to the free-vibration behavior (Fig. 27-7, for example); in any such case the contribution of rotational inertia to the effective forces resulting from the support rotation should be considered in the analysis.

27-4 MULTIPLE-SUPPORT EXCITATION

When a structure is supported at more than one point and has different ground motions applied at each, the total response of the structure can be obtained by superposition of the response due to each independent support input. The formulation of the equations expressing the response to each input component is somewhat different, however, from that described above for systems having only a single input component, because in the present case only one support point at a time can move while all the independent support points are considered fixed against displacement. Consequently, the resulting motion of the supports relative to each other induces pseudostatic stresses in the structure which must be considered in addition to the dynamic-response stresses.

When the motion of the single support which is permitted to move at one time is denoted by v_g, the equation of motion of a general discrete-parameter system can be expressed in a form equivalent to Eq. (27-58) as

$$\mathbf{m}\ddot{\mathbf{v}}^t + \mathbf{m}_g\ddot{v}_g + \mathbf{c}\dot{\mathbf{v}}^t + \mathbf{c}_g v_g + \mathbf{k}\mathbf{v}^t + \mathbf{k}_g v_g = 0 \qquad (27\text{-}63)$$

in which $\mathbf{c}_g$ and $\mathbf{k}_g$ are damping- and elastic-coupling vectors expressing the forces developed in the active degrees of freedom by the motions of the support. In this equation, the damping and elastic forces have been expressed in terms of the total motions of the structure (including the support motion) in order to permit a more general characterization of the system's damping properties than was assumed in Eq. (27-58). When terms are rearranged to put the effective earthquake forces on the right-hand side, this equation of motion in terms of the total displacements becomes

$$\mathbf{m}\ddot{\mathbf{v}}^t + \mathbf{c}\dot{\mathbf{v}}^t + \mathbf{k}\mathbf{v}^t = -\mathbf{m}_g\ddot{v}_g - \mathbf{c}_g\dot{v}_g - \mathbf{k}_g v_g \qquad (27\text{-}64)$$

As noted earlier, however, a more convenient form of the effective-force vector can be derived if the response is expressed in terms of the dynamic component only, that is, if the pseudostatic motions are removed from the total displacements. The relationship between these various displacement components still is expressed by Eqs. (27-49) or (27-50); however, in the present case, the pseudostatic displacements cannot be evaluated by rigid-body kinematics. Instead, they must be computed from the static-equilibrium condition obtained by omitting the dynamic (time-dependent) forces from Eq. (27-64). The result is

$$\mathbf{k}\mathbf{v}^s = -\mathbf{k}_g v_g \qquad (27\text{-}65)$$

in which the displacement $\mathbf{v}^s$ by definition is the pseudostatic vector. Solving for these displacements leads to

$$\mathbf{v}^s = -\mathbf{k}^{-1}\mathbf{k}_g v_g$$

from which it is apparent that the pseudostatic influence vector is given by

$$\mathbf{r} = -\mathbf{k}^{-1}\mathbf{k}_g \qquad (27\text{-}66)$$

Finally substituting Eqs. (27-66) and (27-50) into Eq. (27-64) and simplifying leads to the equations of motion expressed in terms of the dynamic response

$$\mathbf{m}\ddot{\mathbf{v}} + \mathbf{c}\dot{\mathbf{v}} + \mathbf{k}\mathbf{v} = -(\mathbf{m}\mathbf{r} + \mathbf{m}_g)\ddot{v}_g - (\mathbf{c}\mathbf{r} + \mathbf{c}_g)\dot{v}_g \qquad (27\text{-}67)$$

No effective-force contribution from the structural-stiffness properties appears in this equation because

$$kr + k_g = 0$$

as can be verified by substituting from Eq. (27-66) for $\mathbf{r}$.

If the damping matrix were proportional to the stiffness matrix, the velocity-dependent term on the right-hand side of Eq. (27-67) would also disappear and the effective-force expression would reduce to the form of Eq. (27-59). Hence the assumption that the damping forces depend only on relative motions [on which Eq. (27-59) was based] is valid for a system with stiffness-proportional damping; however, it clearly does not apply to the case of mass-proportional damping. On the other hand, the damping contribution to the effective earthquake forces can be expected to be small, and therefore it frequently is neglected regardless of the type of damping involved. In this case, the effective earthquake force acting in a system with multiple support excitation is given by Eq. (27-59) with the pseudostatic influence coefficients given by the statically indeterminate analysis expressed by Eq. (27-66).

Although it is most convenient in practice to evaluate the response to multiple support excitation by separately calculating the response to each input component and superposing the results, it should be evident that Eq. (27-67) can easily be extended to represent the combined response to several support motions. It is necessary merely to express all the ground motions in the form of a vector $\mathbf{v}_g(t)$ and to include all the corresponding columns in the pseudostatic influence-coefficient matrix $\mathbf{r}$ and in the coupling matrices $\mathbf{m}_g$ and $\mathbf{c}_g$.

When the effective earthquake force is assumed to be given by Eq. (27-59), the analysis of the displacement response to each support-point-motion history can be carried out exactly as described for the system subjected to rigid-base translation. The analysis requires the evaluation of an earthquake-excitation factor $\mathscr{L}_n$ [Eq. (27-60)] for each response mode and for each input component considered. In this evaluation, the vibration mode-shape vector ϕ_n describes the free vibrations of the structure with all support-point displacements constrained, while each pseudostatic influence-coefficient vector $\mathbf{r}$ [given by Eq. (27-66)] represents the displacements resulting from a unit movement of one support with all other supports fixed. The dynamic (or relative) displacement response resulting from a single support-excitation history $\ddot{v}_g(t)$ is given by Eq. (27-35), in which the modal response integral $V_n(t)$ has been evaluated for the given ground-motion input. The total response to this one excitation then can be obtained by adding the pseudostatic displacements to these dynamic displacements, that is,

$$\mathbf{v}^t(t) = \mathbf{v}(t) + \mathbf{v}^s(t) = \mathbf{v}^t + \mathbf{r}v_g(t)$$

This displacement analysis for one component of the multiple support excitation is entirely analogous to that for a rigid-base input. The analysis of the elastic forces resulting from the response can also be shown to be equivalent to the rigid-base case, as follows. The total elastic forces in the system are given by the product of its stiffness matrices and the total displacements

$$\mathbf{f}_S{}^t = \mathbf{k}\mathbf{v}^t + \mathbf{k}_g v_g$$

However, if the total displacements are expressed in terms of the pseudostatic and dynamic components, this becomes

$$\mathbf{f}_S{}^t = \mathbf{k}\mathbf{v} + (\mathbf{k}\mathbf{r} + \mathbf{k}_g)v_g$$

Here the pseudostatic forces, represented by the second term on the right, vanish because of the definition of $\mathbf{r}$ [Eq. (27-66)]; thus the elastic modal forces depend only on the dynamic displacements and are given by Eq. (27-38).

The force analysis of this system differs from that of the rigid-base case, however, in that the support forces $\mathbf{f}_g$ cannot be obtained directly from the elastic modal forces $\mathbf{f}_S$ or from the dynamic displacements $\mathbf{v}$. They depend, in addition, on the relative support displacements and must be expressed in terms of the structural-stiffness submatrices pertaining to the supports, that is,

$$\mathbf{f}_g{}^t = \mathbf{k}_g{}^T \mathbf{v}^t + \mathbf{k}_{gg} v_g$$

in which $\mathbf{k}_g{}^T$ expresses all support forces due to unit displacements of the active degrees of freedom and $\mathbf{k}_{gg}$ gives all support forces due to a unit displacement of the support point. With $\mathbf{v}^t = \mathbf{v} + \mathbf{r}v_g$ this can be expressed as

$$\mathbf{f}_g{}^t(t) = \mathbf{k}_g{}^T \mathbf{v}(t) + (\mathbf{k}_{gg} - \mathbf{k}_g{}^T \mathbf{k}^{-1} \mathbf{k}_g)v_g(t) \qquad (27\text{-}68)$$

The first term in this result is the dynamic support force vector due to the dynamic response, while the second is the pseudostatic forces; it is this second term which results from the statically indeterminate nature of the multiple-support-excitation problem and which must not be overlooked in the stress analysis of the system. After the elastic nodal forces have been computed from Eq. (27-38) and the support forces from Eq. (27-68), the resultant forces at any sections of interest can be obtained by standard methods of statics.

27-5 INFLUENCE OF FOUNDATION MEDIUM IN EARTHQUAKE RESPONSE

Modeling of the Foundation Medium

In all the cases of earthquake excitation discussed above, it has been assumed that the earthquake motions were introduced as specified quantities at the structural support points. In effect, these input displacements were assumed to depend only on the earthquake-generation and wave-propagation mechanism and were not influenced by the response of the structure. In actual fact, however, the structure and the soil on which it is founded form a combined dynamic-response mechanism, and there may be significant feedback from the structure into the soil layer. In this case, the earthquake input cannot be expressed independently of the properties of the structure, as has been assumed in the foregoing paragraphs.

As mentioned earlier, the extent to which the structural response may alter the characteristics of the earthquake motions observed at the foundation level depends on the relative mass and stiffness properties of the soil and the structure. Thus the physical properties of the foundation medium may be an important factor in the intensity of the earthquake responses of structures supported on it. In general, it is convenient to consider that the soil layer has two separate influences on the structural response to earthquakes: (1) The soil is considered without the structure, and its effect on the characteristics of the vibratory waves propagated upward from the basement rock to the surface is evaluated; the resulting soil-surface accelerations without the structure are termed the *free-field motions*. (2) The structural response to the free-field motions is computed taking account of the interaction of soil and structure if the interaction is considered to be significant. (If the interaction is not important, the free-field motion may be applied at the structural supports and the response analyzed directly, as described in Secs. 27-2 to 27-4.)

Whether the foundation medium is to be considered merely as a modifying influence affecting the characteristics of the free-field motions observed at its surface or is to be used in a full soil-structure interaction analysis, it must be represented in the analysis by an appropriate mathematical model. Various types of soil models may be used for such analyses, depending on the geometry of the boundaries and interfaces between different material properties, as well as on the form of the foundation system supporting the structure. For example, if the soil is horizontally layered and extends uniformly over a broad area, it may be modeled as a simple one-dimensional system, as shown in Fig. 27-10a. This simple soil model also could be used in a soil-structure interaction analysis if the structural foundation were rigid and extended over a large surface area.

More generally, however, the soil does not extend uniformly in all directions; e.g., it may be confined in a long, narrow valley, as shown in Fig. 27-10b. In this case, a two-dimensional finite-element model could be used to represent the soil, as shown in the sketch. The same soil model could be used to evaluate soil-structure interaction if the structure extended uniformly for a long distance in the direction of the axis of the valley.

Another type of two-dimensional soil model which has found considerable use in practical analyses is the axisymmetric system shown in Fig. 27-10c. In this case, it is assumed that the soil boundaries and material interfaces are rotationally symmetric about the vertical axis, so that the radial and vertical coordinates are sufficient

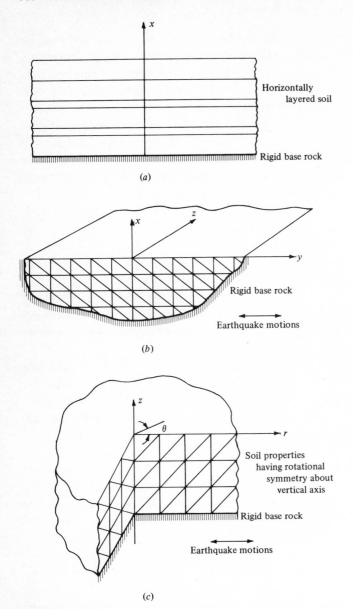

FIGURE 27-10
Mathematical modeling of foundation media: (*a*) one-dimensional model of foundation system; (*b*) two-dimensional finite-element model of foundation system; (*c*) axisymmetric finite-element model of foundation system.

to define the geometry of the soil system as well as its finite-element idealization. Evidently soil-structure interaction can also be modeled with this type of system if the structure is rotationally symmetric about the same axis. It should be noted, however, that the horizontal components of earthquake excitation are not axisymmetric; to deal with this type of loading it is necessary to express both the effective earthquake forces and the response displacements as the first term in a harmonic expansion relative to the tangential coordinate.

In the most general case involving arbitrary geometric form of the soil system and of the structure it supports, both soil and structure must be described in a general three-dimensional coordinate system. The general soil system may be idealized as an assemblage of three-dimensional finite elements and the structure as a combination of finite elements of one, two, and three dimensions, in accordance with the actual configuration.

Soil Modification of Earthquake Motions

If design-earthquake records which are representative of the free-field motions to be expected at a proposed construction site can be generated artificially or selected from available accelerograms, it is obvious that no special evaluation need be made of any modifying influence caused by the surficial soil layer. Presumably this influence is contained in the records, and the analysis of the response to these given free-field motions can be dealt with directly. Frequently, however, it is not possible to obtain or generate ground-motion records which are representative of the specific surficial soil conditions at the site, and in this case it is desirable to take account of the foundation medium by direct analysis.

If the earthquake motions in the basement rock underlying the soil layer are known, the free-field motion can be determined by treating the soil layer exactly like any other structural system for which the support motions are known. By using a discretized model suited to the geometric form of the soil deposit, as discussed above, and assuming that the earthquake input is a rigid-base translation, the response at the soil surface can be calculated with the equations of Sec. 27-2. Of course, the purpose of the analysis is to establish the time history of the surface motions; therefore a time-history analysis must be used rather than the response-spectrum approach. It should be noted, however, that there is little reason for obtaining the free-field motions independently of the structural-response analysis when the base-rock input is known. In this case, it usually is more efficient to include both soil and structure in a single idealization and to take account of the soil-structure interaction directly.

On the other hand, if the surface motions are known for a type of soil system which differs significantly from the soil at the proposed building site, the separate evaluation of the soil-modification effects can be very useful. In this case, the basement-rock motions are first determined by an inverse (downward) analysis from the given ground-motion record, and then the free-field motions at the surface of the design-site soil deposit are computed by a direct (upward) analysis. It is worth noting that both the inverse and the direct soil-modification analyses could be performed by wave-propagation procedures if the soil systems were horizontally layered so that they could be idealized as one-dimensional models. However, for more general soil-deposit geometries, the analyses must be performed by matrix procedures with discretized finite-element models, and only this general case will be described in detail here.

Although the upward propagation of earthquake motions from basement rock to soil surface can be evaluated by the time-domain procedures presented in Sec. 27-2, it is much more convenient and effective to solve the inverse problem of determining the base motions from the surface record by a frequency-domain analysis. Hence it is convenient to use the frequency-domain procedure for both purposes, as will be explained here.

It has been shown in Sec. 7-4 that the response $y(t)$ of any linear vibratory system to an arbitrary excitation $x(t)$ can be obtained from the complex frequency response of the system $H(\bar{\omega})$ by means of the inverse Fourier transform, as follows:

$$y(t) = \frac{1}{2\pi} \int_{-\infty}^{\infty} H(\bar{\omega}) X(\bar{\omega}) \exp(i\bar{\omega}t) \, d\bar{\omega} \qquad (27\text{-}69a)$$

in which $X(\bar{\omega})$ is the Fourier transform of the forcing function $x(t)$, that is,

$$X(\bar{\omega}) = \int_{-\infty}^{\infty} x(t) \exp(-i\bar{\omega}t) \, dt \qquad (27\text{-}69b)$$

In the present case, the forcing function is bedrock acceleration history, which will be designated $\ddot{v}_b(t)$; its Fourier transform will be denoted

$$B(\bar{\omega}) = \int_{-\infty}^{\infty} \ddot{v}_b(t) \exp(-i\bar{\omega}t) \, dt \qquad (27\text{-}70)$$

The desired response is the acceleration at the surface of the soil layer $\ddot{v}_g(t)$ and the complex-frequency response is the transfer function expressing the amplitude of harmonic motion at the soil surface due to a harmonic-acceleration input from the basement rock. This transfer function will be designated $Q(\bar{\omega})$ and, assuming that the properties of the soil layer are such that there is no damping coupling, it can be conveniently derived by the mode-superposition procedure.

Thus, if the discrete-coordinate degrees of freedom $\mathbf{v}_c$ of the soil medium are expressed in terms of its free-vibration mode shapes ϕ as

$$\mathbf{v}_c = \mathbf{\Phi Y} \qquad (27\text{-}71)$$

the equations of motion of the idealized soil can be transformed to a set of independent equations, each having the form

$$\ddot{Y}_n + 2\xi_n\omega_n\dot{Y}_n + \omega_n{}^2 Y_n = -\mathcal{L}_n\ddot{v}_b \quad (27\text{-}72)$$

in which Y_n is the amplitude of the response of mode n, where each mode shape $\boldsymbol{\phi}_n$ has been normalized to provide a unit generalized mass $M_n = \boldsymbol{\phi}_n{}^T\mathbf{m}\boldsymbol{\phi}_n = 1$, and where the modal earthquake-excitation factor is given by

$$\mathcal{L}_n = \boldsymbol{\phi}_n{}^T\mathbf{mr} \quad (27\text{-}52)$$

The complex displacement-response function for this modal coordinate is $-\mathcal{L}_n H_n(\overline{\omega})$, where, from Eq. (5-11),

$$H_n(\overline{\omega}) = \frac{1}{\omega_n{}^2} \frac{1}{1 - \beta_n{}^2 + 2i\xi_n\beta_n} \quad (27\text{-}73)$$

Hence the complex acceleration-response function is $\mathcal{L}_n H_n(\overline{\omega})\overline{\omega}^2$. Therefore the modal transfer function expressing the acceleration in mode n at the surface due to a unit base acceleration is

$$Q_n(\overline{\omega}) = \phi_{gn}\mathcal{L}_n H_n(\overline{\omega})\overline{\omega}^2 \quad (27\text{-}74)$$

in which ϕ_{gn} is the surface displacement in mode n. Then the total transfer function is obtained by superposing the modal transfer functions, that is,

$$Q(\overline{\omega}) = \sum_{n=1}^{N} Q_n(\overline{\omega}) \quad (27\text{-}75)$$

and by analogy with Eq. (27-69), the *relative* surface acceleration $\ddot{v}_g$ is given by

$$\ddot{v}_g(t) = \frac{1}{2\pi} \int_{-\infty}^{\infty} Q(\overline{\omega})B(\overline{\omega}) \exp(i\overline{\omega}t)\, d\overline{\omega} \quad (27\text{-}76)$$

However, the desired free-field motion is the *total* surface acceleration, which includes the relative acceleration plus the pseudostatic contribution

$$\ddot{\tilde{v}}_g(t) = \ddot{v}_g(t) + r_g\ddot{v}_b(t) \quad (27\text{-}77)$$

In this expression r_g is the displacement at the surface due to a unit static displacement of the bedrock. The final expression for total surface acceleration therefore becomes

$$\ddot{\tilde{v}}_g(t) = \frac{1}{2\pi} \int_{-\infty}^{\infty} [Q(\overline{\omega}) + r_g]B(\overline{\omega}) \exp(i\overline{\omega}t)\, d\overline{\omega} \quad (27\text{-}78)$$

Equation (27-78) expresses the analysis of free-field motions at the soil surface due to a specified basement-rock motion $\ddot{v}_b(t)$. From the properties of Fourier transform pairs, the frequency functions can be written in terms of a time integral

$$[Q(\overline{\omega}) + r_g]B(\overline{\omega}) = \int_{-\infty}^{\infty} \ddot{\tilde{v}}_g(t) \exp(-i\overline{\omega}t)\, dt \equiv G(\overline{\omega}) \quad (27\text{-}79)$$

in which the Fourier transform of the surface-acceleration history has been denoted $G(\bar{\omega})$. After dividing through by the left-hand factor, Eq. (27-79) becomes

$$B(\bar{\omega}) = [Q(\bar{\omega}) + r_g]^{-1} G(\bar{\omega}) \qquad (27\text{-}80)$$

Finally taking the inverse transform of $B(\bar{\omega})$ [noting its definition in Eq. (27-70)] leads to

$$\ddot{v}_b(t) = \frac{1}{2\pi} \int_{-\infty}^{\infty} [Q(\bar{\omega}) + r_g]^{-1} G(\bar{\omega}) \exp(i\bar{\omega}t) \, d\bar{\omega} \qquad (27\text{-}81)$$

which expresses the basement-rock motions in terms of the Fourier transform of the free-field motions.

Equations (27-81) and (27-78) completely define the influence of any given soil system in modifying an arbitrary earthquake motion transmitted through it, expressing the surface accelerogram in terms of the base-acceleration history or vice versa. Although this discussion has assumed a single component of input and output, it should be evident that the analysis can easily be expanded to treat multiple input and output. In this case, $[Q(\bar{\omega}) + r_g]$ becomes a square matrix of dimension corresponding to the number of input and output components. It also should be noted that the evaluation of the transfer function $Q(\bar{\omega})$ can be performed very easily by wave-propagation methods if the soil layer can be idealized as a one-dimensional system. In this case there is no need to use mode superposition; the harmonic input-output relationship for a soil having any number of layers of different properties can be written directly. Finally it must be emphasized again that the numerical evaluation of the integral expressions in Eqs. (27-78) and (27-81) should be carried out in discrete form, using the fast Fourier transform technique.

Soil-Structure Interaction: Equations of Motion

Discretized soil model It was noted earlier that, in addition to the effect a soil layer may have on the characteristics of the free-field motions observed at the surface, it also may interact with a building or other structure founded on it so as to cause important changes in the structural response. The equations of motion for analysis of the soil-structure interaction problem will be derived with reference to Fig. 27-11. It is assumed that both the soil and the structure are represented by discretized models, the degrees of freedom within the soil mass being designated by v_a, the contact degrees of freedom associated with both building and structure by v_g, those in the building alone by v^t, and the basement-rock displacements by v_b, as shown in the figure. In the present discussion, no limitations will be placed on the number of degrees of freedom associated with each node of the foundation or the added structure; these may include translation and rotation in three components as well as higher-order deformation degrees of freedom.

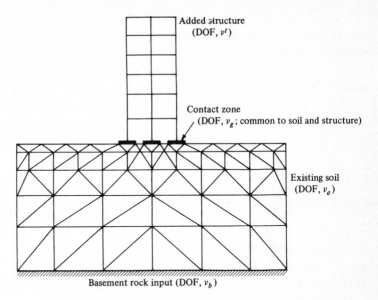

Added structure
(DOF, v^t)

Contact zone
(DOF, v_g; common to soil and structure)

Existing soil
(DOF, v_a)

Basement rock input (DOF, v_b)

FIGURE 27-11
Definition of degrees of freedom of typical soil-structure system.

From the configuration of the system shown in Fig. 27-11 it is evident that the equations of motion for the combined system could be written directly in the form of Eq. (27-64). In fact, this would be the most efficient approach if the basement-rock motions were known so that the effective-load term on the right-hand side could be evaluated directly. However, in most cases only the free-field motions are known and it is preferable to express the effective earthquake forces in terms of them. For this purpose, the (free-field) equations of motion of the soil layer are expressed by an equation like Eq. (27-64), which is written

$$\tilde{\mathbf{m}}_c \ddot{\tilde{\mathbf{v}}}_c + \tilde{\mathbf{c}}_c \dot{\tilde{\mathbf{v}}}_c + \tilde{\mathbf{k}}_c \tilde{\mathbf{v}}_c = -\tilde{\mathbf{m}}_b \ddot{\mathbf{v}}_b - \tilde{\mathbf{c}}_b \dot{\mathbf{v}}_b - \tilde{\mathbf{k}}_b \mathbf{v}_b \qquad (27\text{-}82)$$

in which $\tilde{\mathbf{m}}_c$, $\tilde{\mathbf{c}}_c$, and $\tilde{\mathbf{k}}_c$ represent the properties of the existing foundation material, $\tilde{\mathbf{v}}_c$ is the free-field motion of this system, $\mathbf{v}_b$ is the input motion of the basement rock, and $\tilde{\mathbf{m}}_b$, $\tilde{\mathbf{c}}_b$, and $\tilde{\mathbf{k}}_b$ are the coupling terms expressing forces in the foundation material due to motions of the basement rock.

When the building is superposed on the foundation, the response is changed even though the effective forces due to the basement-rock motion remain the same. The equations of motion of the combined system are obtained by merely adding appropriate terms to the left side of Eq. (27-82), with the result

$$[\tilde{\mathbf{m}}_c + \mathbf{m}_c]\{\ddot{\tilde{\mathbf{v}}}_c + \ddot{\mathbf{v}}_c^t\} + [\tilde{\mathbf{c}}_c + \mathbf{c}_c]\{\dot{\tilde{\mathbf{v}}}_c + \dot{\mathbf{v}}_c^t\} + [\tilde{\mathbf{k}}_c + \mathbf{k}_c]\{\tilde{\mathbf{v}}_c + \mathbf{v}_c^t\}$$
$$= -\tilde{\mathbf{m}}_b \ddot{\mathbf{v}}_b - \tilde{\mathbf{c}}_b \dot{\mathbf{v}}_b - \tilde{\mathbf{k}}_b \mathbf{v}_b$$

in which $\mathbf{m}_c$, $\mathbf{c}_c$, and $\mathbf{k}_c$ are the properties of the *added* building and $\mathbf{v}_c{}^t$ is the *added* response resulting from putting the building on this site.[1] Although it completely describes the soil-structure interaction, this form of the response equation is not often used because the effective-force input is expressed in terms of the basement-rock motions. Instead the equivalent free-field forces on the left side of Eq. (27-82) are substituted for the forces expressed in terms of the rock motions, as follows:

$$[\tilde{\mathbf{m}}_c + \mathbf{m}_c]\{\ddot{\tilde{\mathbf{v}}}_c + \ddot{\mathbf{v}}_c{}^t\} + [\tilde{\mathbf{c}}_c + \mathbf{c}_c]\{\dot{\tilde{\mathbf{v}}}_c + \dot{\mathbf{v}}_c{}^t\} + [\tilde{\mathbf{k}}_c + \mathbf{k}_c]\{\tilde{\mathbf{v}}_c + \mathbf{v}_c{}^t\}$$
$$= +\tilde{\mathbf{m}}_c\ddot{\tilde{\mathbf{v}}}_c + \tilde{\mathbf{c}}_c\dot{\tilde{\mathbf{v}}}_c + \tilde{\mathbf{k}}_c\tilde{\mathbf{v}}_c \qquad (27\text{-}83)$$

This equation expresses the total motion of the combined soil-structure system, using an effective-force vector given by the foundation material acted upon by the free-field motions. It could be used as one approach to the dynamic-response analysis, but in many cases it is more convenient to write the equation of motion in terms of the *added* response $\mathbf{v}_c{}^t$ only. This modification involves transferring those terms on the left side of Eq. (27-83) which are associated with the free-field motions to the right-hand side, leading to

$$[\tilde{\mathbf{m}}_c + \mathbf{m}_c]\ddot{\mathbf{v}}_c{}^t + [\tilde{\mathbf{c}}_c + \mathbf{c}_c]\dot{\mathbf{v}}_c{}^t + [\tilde{\mathbf{k}}_c + \mathbf{k}_c]\mathbf{v}_c{}^t$$
$$= -\mathbf{m}_c\ddot{\tilde{\mathbf{v}}}_c - \mathbf{c}_c\dot{\tilde{\mathbf{v}}}_c - \mathbf{k}_c\tilde{\mathbf{v}}_c \qquad (27\text{-}84)$$

Here the *added* motion of the *combined* soil-structure system is given, due to an effective force resulting from the added structure acted upon by the free-field motion.

Now this effective-force expression can be simplified by taking advantage of the fact that it involves only the added structure acted upon by the free-field motions. For this purpose, the added and the free-field displacement vectors will be partitioned as follows:

$$\mathbf{v}_c{}^t = \begin{bmatrix} \mathbf{v}^t \\ \mathbf{v}_g{}^t \\ \mathbf{v}_a{}^t \end{bmatrix} \qquad \tilde{\mathbf{v}}_c = \begin{bmatrix} \mathbf{0} \\ \tilde{\mathbf{v}}_g \\ \tilde{\mathbf{v}}_a \end{bmatrix} \qquad (27\text{-}85)$$

in which the three partitions refer respectively to the degrees of freedom in the super-structure, those common to the building and foundation material, and the noncontact degrees of freedom in the foundation material, as shown in Fig. 27-11. Of course, no free-field motions are defined for the building superstructure. All the physical-property matrices are partitioned correspondingly, the building and foundation mass matrix partitions being

$$\mathbf{m}_c = \begin{bmatrix} \mathbf{m} & \mathbf{m}_g & \mathbf{0} \\ \mathbf{m}_g{}^T & \mathbf{m}_{gg} & \mathbf{0} \\ \mathbf{0} & \mathbf{0} & \mathbf{0} \end{bmatrix} \qquad \tilde{\mathbf{m}}_c = \begin{bmatrix} \mathbf{0} & \mathbf{0} & \mathbf{0} \\ \mathbf{0} & \tilde{\mathbf{m}}_{gg} & \tilde{\mathbf{m}}_{ga} \\ \mathbf{0} & \tilde{\mathbf{m}}_{ag} & \tilde{\mathbf{m}}_{aa} \end{bmatrix} \qquad (27\text{-}86)$$

---

[1] Note that braces are used to identify the response vectors in this equation, whereas brackets enclose the physical property matrices.

Similar expressions can be written for the damping and stiffness matrices. Hence it is clear that the effective forces of Eq. (27-84) are given by the central column of the partitioned physical-property matrices of the added system, acted upon by the free-field motions of the *contact* degrees of freedom only. With this simplification, Eq. (27-84) can be written

$$[\tilde{m}_c + m_c]\ddot{v}_c{}^t + [\tilde{c}_c + c_c]\dot{v}_c{}^t + [\tilde{k}_c + k_c]v_c{}^t$$

$$= -\begin{bmatrix} m_g \\ m_{gg} \\ 0 \end{bmatrix}\ddot{\tilde{v}}_g - \begin{bmatrix} c_g \\ c_{gg} \\ 0 \end{bmatrix}\dot{\tilde{v}}_g - \begin{bmatrix} k_g \\ k_{gg} \\ 0 \end{bmatrix}\tilde{v}_g \qquad (27\text{-}87)$$

Now these equations of motion can be expressed in a still more convenient form by writing the response $v_c{}^t$ as the sum of a dynamic component v_c and a pseudo-static component $v_c{}^s$, following the general procedure presented earlier for the system without soil interaction. As before, the pseudostatic displacements can be derived from the static-equilibrium relationship which remains when the dynamic terms are omitted from the equation of motion. Thus the statics equation derived from Eq. (27-87) is

$$[\tilde{k}_c + k_c]v_c{}^s = -\begin{bmatrix} k_g \\ k_{gg} \\ 0 \end{bmatrix}\tilde{v}_g$$

and by solving this, the pseudostatic displacements are found to be

$$v_c{}^s = r_c\tilde{v}_g \qquad (27\text{-}88)$$

in which

$$r_c = -[\tilde{k}_c + k_c]^{-1}\begin{bmatrix} k_g \\ k_{gg} \\ 0 \end{bmatrix} \qquad (27\text{-}89)$$

This is the influence-coefficient matrix which expresses the pseudostatic displacements resulting from the application of unit static free-field displacements at the contact degrees of freedom. The physical significance of this matrix can be explained by visualizing each vector of r_c as the displacements developed in the combined soil-structure system due to the action of a static effective-force vector acting on the contact nodes. Each static effective-force vector is chosen so that if it acted on the foundation system alone, it would produce a unit displacement of the specified contact node while constraining the other contact nodes against displacement. Force vectors which would produce the desired displacement patterns could be derived easily by application of the principles of static structural analysis; however, they are not actually needed in the dynamic-response analysis: the necessary influence matrix is completely defined by Eq. (27-89).

When the response degrees of freedom are expressed in the form

$$\mathbf{v}_c{}^t = \mathbf{v}_c + \mathbf{r}_c\tilde{\mathbf{v}}_g \qquad (27\text{-}90)$$

Eq. (27-87) can be simplified to

$$[\tilde{\mathbf{m}}_c + \mathbf{m}_c]\ddot{\mathbf{v}}_c + [\tilde{\mathbf{c}}_c + \mathbf{c}_c]\dot{\mathbf{v}}_c + [\tilde{\mathbf{k}}_c + \mathbf{k}_c]\mathbf{v}_c$$

$$= -\left\{[\tilde{\mathbf{m}}_c + \mathbf{m}_c]\mathbf{r}_c + \begin{bmatrix} \mathbf{m}_g \\ \mathbf{m}_{gg} \\ \mathbf{0} \end{bmatrix}\right\}\ddot{\tilde{\mathbf{v}}}_g \qquad (27\text{-}91)$$

which may be considered as the ultimate soil-structure interaction equation of motion. The stiffness-dependent term has dropped out from the effective-force vector on the right-hand side of this equation because the pseudostatic displacements were defined so that

$$[\tilde{\mathbf{k}}_c + \mathbf{k}_c]\mathbf{r}_c + \begin{bmatrix} \mathbf{k}_g \\ \mathbf{k}_{gg} \\ \mathbf{0} \end{bmatrix} = 0 \qquad (27\text{-}92)$$

as can be verified by substitution of Eq. (27-89). The damping-dependent term has also been omitted in Eq. (27-91), these forces being assumed negligible either because the damping matrix is proportional to the stiffness matrix, which would impose a damping condition equivalent to the stiffness condition of Eq. (27-92), or because the damping coefficients themselves are negligible.

It is of interest to note that the interaction equations (27-91) can be reduced directly to the equations of motion derived previously for systems subjected to specified support motions. It is necessary only that the physical properties associated with the contact points of the added structure be small compared with the corresponding properties of the foundation system; that is,

$$\mathbf{m}_{gg} \ll \tilde{\mathbf{m}}_{gg} \qquad \mathbf{c}_{gg} \ll \tilde{\mathbf{c}}_{gg} \qquad \mathbf{k}_{gg} \ll \tilde{\mathbf{k}}_{gg}$$

In this case, the equations associated with the foundation degrees of freedom in Eq. (27-87) demonstrate that the *added* motions of these degrees of freedom will be zero (that is, there is no interaction), whereas the equations associated with the superstructure reduce directly to Eq. (27-64).

Elastic-half-space model   An important limitation of the discretized model of the foundation medium considered in the foregoing discussion is that such models necessarily have finite boundaries. Thus a very important question to be considered in the modeling process is the extent of the foundation zone which must be included. Where natural physical boundaries exist, such as hard rock underlying a soft surficial soil deposit, the extent of the model will be obvious and its boundary behavior can be expected to simulate the prototype system adequately. On the other hand, if the

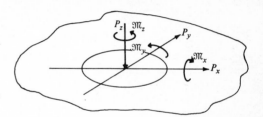

FIGURE 27-12
Forces applied to elastic half-space
foundation medium.

structure is founded on a broad, deep, and uniform soil mass, the boundaries of the mathematical model must be far removed from the structure if reflections from the fictitious model boundaries are not to induce spurious responses in the structure. In effect, the model boundaries tend to retain the vibration energy within the system, and unless they are at a great distance, they will inhibit the radiation energy loss from the structure.[1] Under these circumstances, a very large number of degrees of freedom may be required to simulate the behavior of the soil system adequately. Frequently the number of degrees of freedom in the foundation far exceeds the number representing the structure which is the real subject of the investigation; thus the overall efficiency of this type of model may be very poor.

In cases where the foundation medium is rather uniform over a broad, deep zone and where the contact surface between the supported structure and the soil deposit may be considered as a rigid plate, this limitation of the discretized model can be avoided by treating the soil as an elastic half space. The function of the foundation medium is to resist the forces applied to it by the base of the building. During an earthquake, a rigid base slab may be subjected to displacements in six degrees of freedom, and the resistance of the soil may be expressed by the six corresponding resultant-force components, shown in Fig. 27-12. Hence the structural behavior of the elastic half space is represented completely by a set of force-displacement relationships defined for these degrees of freedom.

To simulate the static behavior of the soil-structure system, it is evident that the foundation medium could be modeled by six linear springs acting in the rigid-base degrees of freedom. Appropriate static spring constants can be evaluated for the elastic half space by the methods of continuum mechanics. However, the dynamic resistance of the soil mobilized during an earthquake includes inertial and damping effects in addition to the static spring stiffness. The appropriate dynamic-force–displacement relationships of an elastic half space subjected to a harmonic excitation of the rigid building foundation can also be evaluated by methods of continuum

---

[1] In some simple systems, so-called *quiet boundaries* are provided to absorb the vibration energy which reaches them and thus to model better the radiation energy loss. Such devices are beyond the scope of the present discussion.

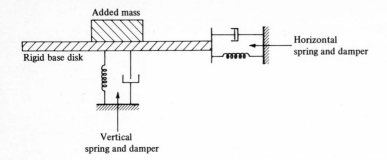

FIGURE 27-13
Lumped-parameter model for elastic half-space foundation (translational motion in vertical plane).

mechanics, and the results show that both the relative response amplitude and the phase of the resisting forces are functions of the frequency of applied displacements. Consequently, the dynamic behavior of the half space should be modeled in each degree of freedom by a spring-dashpot device having frequency-dependent properties.

In principle, such a frequency-dependent foundation model could be used directly in a frequency-domain analysis of the soil-structure interaction problem. However, the analysis of a complicated soil-structure idealization involving a large number of degrees of freedom can be greatly simplified if the foundation component of the model is assumed to be frequency-independent; moreover, if the analysis is to be extended into the nonlinear range, the frequency-domain approach is not applicable. For these reasons, it is advantageous to represent the elastic half-space foundation medium by frequency-independent components, with their properties selected to reproduce the true frequency-dependent behavior as well as possible.

In cases where the foundation slab is to support a mechanical system operating at relatively high frequencies, the inertial resistance of the soil is small and the half space can be modeled by a simple spring-dashpot device in each degree of freedom. The spring constant should be selected to give the correct static displacements, and then the dashpot coefficient is chosen to provide the best possible agreement with the theoretical resultant base-force amplitude and phase. In modeling the foundation system for an earthquake-response analysis, however, where the highest significant response frequencies are no more than a few times the fundamental soil-structure frequency, better results can be obtained by introducing the virtual mass of the soil as an additional model parameter. The model for each degree of freedom then consists of the spring constant defined by the static-load displacement plus an appropriate virtual mass together with a dashpot coefficient reduced somewhat from that required in the massless model. (The lumped-parameter model for the vertical motion is shown in Fig. 27-13.) With the proper selection of these dynamic properties, the response of the lumped-parameter foundation model can be controlled to within a few percent of the theoretical half-space performance over the frequency range of interest.

The continuum-mechanics analysis of the frequency-dependent forces acting between a rigid slab and the surface of an elastic half-space foundation medium has been carried out for circular and rectangular slab shapes and is based on the assumption that the dynamic-stress distribution acting on the slab is the same as that developed in a static loading. Values of the frequency-independent constants recommended to approximate these theoretical results in an earthquake-response analysis are listed in Table 27-1. These values are to be used in formulating equations of motion of the soil-structure system in the form of Eq. (27-91). The foundation medium thus is represented by diagonal matrices $\tilde{\mathbf{m}}_c$, $\tilde{\mathbf{c}}_c$, and $\tilde{\mathbf{k}}_c$ defined for the surface (contact) degrees of freedom $\tilde{\mathbf{v}}_g$, the diagonal elements being the lumped-parameter coefficients from Table 27-1. Of course the virtual-mass coefficients of the soil model (contained in $\tilde{\mathbf{m}}_c$) contribute to the effective earthquake forces on the right-hand side of Eq. (27-91) as well as to the inertial-force resistance on the left. It should be noted that all degrees of freedom of this type of soil model are at the contact surface; there are no internal soil degrees of freedom $\tilde{\mathbf{v}}_a$.

Soil-Structure Interaction: Response Analysis

The earthquake response of any soil-structure system can be evaluated by numerical integration of the general equations of motion, Eq. (27-91). However, the mathematical models of such systems usually include a large number of degrees of freedom, and the analysis may be both tedious and expensive if it is done directly in these

Table 27-1 **EQUIVALENT DISCRETE PROPERTIES FOR ELASTIC HALF-SPACE**
Supporting a rigid circular plate at the surface

Degree of freedom	Spring constant†	Viscous damper	Added mass
Vertical	$K \equiv \dfrac{4Gr}{1-v}$	$1.79\sqrt{K\rho r^3}$	$1.50\,\rho r^3$
Horizontal	$18.2Gr\dfrac{(1-v^2)}{(2-v)^2}$	$1.08\sqrt{K\rho r^3}$	$0.28\,\rho r^3$
Rocking*	$2.7Gr^3[v=0]$	$0.47\sqrt{K\rho r^5}$	$0.49\,\rho r^5$
Torsion*	$5.3Gr^3$	$1.11\sqrt{K\rho r^5}$	$0.70\,\rho r^5$

SOURCE: Adapted from N. M. Newmark and E. Rosenblueth, *Fundamentals of Earthquake Engineering*, Prentice-Hall, Inc., Englewood Cliffs, N.J., 1971, p. 98.

* Note that these are rotational properties for these degrees of freedom.
† r = radius of plate; G = modulus of rigidity; v = Poisson's ratio; ρ = mass density.

coordinates. As mentioned earlier, it generally is desirable to transform the equations of motion to the normal (free-vibration mode) coordinates before performing an earthquake-response analysis because the earthquake tends to excite only relatively few modes. It should be emphasized that no approximation is inherent in the use of normal coordinates; if all N coordinates are included in the analysis, the computed response will be the same whether it is evaluated in normal coordinates or in the original nodal coordinates of the discretized system. However, the normal coordinates are much more efficient in describing the displaced shape of the structure, and generally a very good approximation of the response is given by the first few modal terms. In this case, the results are approximate only to the extent that significant motions are contained in the modal components which have been truncated.

The relationship between the discretized system coordinates $\mathbf{v}_c$ and the modal coordinates $\mathbf{Y}$ is given by

$$\mathbf{v}_c = \mathbf{\Phi}\mathbf{Y} \qquad (27\text{-}71)$$

where $\mathbf{\Phi}$ now represents the significant undamped free-vibration mode shapes of the combined soil-structure system. In general only a very limited number p of the modes will contribute to the earthquake response problem, that is, $p \ll N$, and so $\mathbf{\Phi}$ is a rectangular matrix with many more rows than columns. Now if Eq. (27-71) is substituted into Eq. (27-91) and both sides are premultiplied by $\mathbf{\Phi}^T$, the result is the set of p normal-coordinate equations of motion which may be written

$$\ddot{\mathbf{Y}} + \mathbf{C}\dot{\mathbf{Y}} + \mathbf{\Omega}^2\mathbf{Y} = \mathscr{L}\ddot{\tilde{\mathbf{v}}}_g \qquad (27\text{-}93)$$

where $\mathscr{L}$ is a matrix of earthquake-excitation factors defined for each mode of vibration and for each component of the free-field excitation $\ddot{\tilde{\mathbf{v}}}_g$, as follows:

$$\mathscr{L} = -\mathbf{\Phi}^T\left\{[\tilde{\mathbf{m}}_c + \mathbf{m}_c]\mathbf{r}_c + \begin{bmatrix} \mathbf{m}_g \\ \mathbf{m}_{gg} \\ \mathbf{0} \end{bmatrix}\right\} \qquad (27\text{-}94)$$

In Eq. (27-93) it has been assumed for convenience that the mode shapes are normalized so that the generalized mass for each mode is unity, that is,

$$\boldsymbol{\phi}_n^{\ T}[\tilde{\mathbf{m}}_c + \mathbf{m}_c]\boldsymbol{\phi}_n = 1$$

Correspondingly, the generalized stiffness for each mode becomes the modal frequency squared, and the diagonal array of these frequencies squared is denoted here by

$$\mathbf{\Omega}^2 = \begin{bmatrix} \omega_1^{\ 2} & & & \\ & \omega_2^{\ 2} & & \\ & & \ddots & \\ & & & \omega_p^{\ 2} \end{bmatrix} \qquad (27\text{-}95)$$

The generalized damping matrix in Eq. (27-93) is defined by

$$\mathbf{C} = \mathbf{\Phi}^T[\tilde{\mathbf{c}}_c + \mathbf{c}_c]\mathbf{\Phi} \qquad (27\text{-}96)$$

and if the damping matrix of the combined soil-structure system is such that the mode shapes are orthogonal with respect to it, Eq. (27-96) has the form

$$\mathbf{C} = \begin{bmatrix} 2\xi_1\omega_1 & & & \\ & 2\xi_2\omega_2 & & \\ & & \ldots & \\ & & & 2\xi_p\omega_p \end{bmatrix} \qquad (27\text{-}97)$$

In this case, Eq. (27-93) becomes a set of independent modal response equations which can be solved separately, that is, the response can be determined by mode superposition. The time history of displacements or forces developed in this special class of structure can then be evaluated by the same procedures described earlier for structures without soil interaction [Eqs. (27-34) and (27-38)]. Also, the maximum response of this soil-structure system without modal coupling can be approximated by the response-spectrum technique [Eqs. (27-47) and (27-48)].

However, in most soil-structure interaction problems it is not reasonable to assume that the damping matrix will satisfy the modal orthogonality condition. Generally the damping of the foundation medium will be relatively high compared with that of the structure, and the damping forces will tend to cause coupling between the undamped free-vibration modes. Damping coupling is especially significant in systems where the foundation medium is represented by an elastic half-space model because of the very large viscous damper required to simulate the radiated-energy loss. In such systems, the generalized damping matrix of Eq. (27-96) is not diagonal; instead it has off-diagonal terms which cause coupling between the modal coordinates in Eq. (27-93) so that mode superposition is not applicable. Nevertheless, the transformation to normal coordinates still leads to important savings in the response analysis because the essential dynamic behavior of a system with damping coupling still can be expressed in terms of a limited number of undamped mode shapes. As noted earlier, the only approximation involved in the use of normal coordinates results from the truncation of the higher modes, and these higher modes are no more significant in a system with damping coupling than in an uncoupled case.

It is evident in Eq. (27-96) that explicit damping matrices must be defined for both the foundation medium $\tilde{\mathbf{c}}_c$ and the structure $\mathbf{c}_c$ before the coupled generalized damping matrix of Eq. (27-93) can be evaluated. In most cases these can be conveniently derived by assuming modal damping ratios separately for the building and for the foundation medium. Then explicit damping matrices can be derived for each of these subsystems, using the methods presented in Chap. 13. Frequently a simple mass- or stiffness-proportional damping matrix will be assumed for the building, while the foundation medium is assumed to be an elastic half space for which the explicit damping coefficients are taken from Table 27-1. When the coupled generalized damping matrix has been established by means of Eq. (27-96), the normal-coordinate response $\mathbf{Y}(t)$ to any given earthquake input is computed by the direct step-by-step integration of the coupled equations of motion (27-93). Then the displacements and forces expressed in the original coordinates are given by equations equivalent to Eqs. (27-35) and (27-38), that is

$$\mathbf{v}(t) = \mathbf{\Phi}\mathbf{Y}(t) \qquad (27\text{-}98)$$

$$\mathbf{f}_S(t) = \mathbf{m}\mathbf{\Phi}\mathbf{\Omega}^2\mathbf{Y}(t) \qquad (27\text{-}99)$$

These expressions represent the essential response of the coupled soil-structure system. The error resulting from using only a limited number of undamped modal coordinates in the analysis can be examined in a practical way by comparing results of two analyses obtained using different numbers of such coordinates. It should be apparent that the response-spectrum technique cannot be applied to a system with damping coupling because the response spectrum is derived for independent modal responses and cannot account for the interaction between modal coordinates.

27-6 NONLINEAR RESPONSE TO EARTHQUAKES

Need for Nonlinear Analysis

In the preceding description of the deterministic analysis of earthquake response, it has been assumed that the structure is a linear system, that is, its properties do not change during the earthquake. However, it was mentioned in Chap. 26 in the discussion of basic seismic design criteria that an extreme earthquake should be included among the design-load conditions and that such severe ground motions should be expected to cause significant structural damage. Clearly this criterion envisages nonlinear earthquake response because the structural stiffness must undergo changes as the result of significant damage.

Furthermore, it can easily be demonstrated that an earthquake of only moderate intensity will cause significant overstress in a building designed according to typical seismic code requirements. For example, it was pointed out in Sec. 27-2, in the comparison of the dynamic-response analysis procedure with the Uniform Building Code requirements, that the base-shear coefficient specified by the code is equivalent to an earthquake-acceleration response spectrum. In Fig. 27-14 the UBC base-shear coefficient may be compared with response spectra which have been plotted for three earthquakes; the response spectra of two of them have been divided by 4 to facilitate the comparison. It is evident from this figure that even a moderate earthquake will

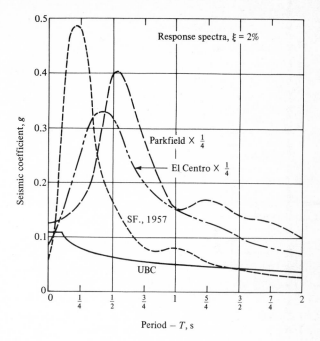

FIGURE 27-14
Comparison of UBC seismic coefficient with response spectra of actual earthquakes.

produce structural forces which are several times greater than the code design requirements; therefore damage may be expected from moderate earthquakes in a building designed only for the code forces.

A more detailed comparison of code-force effects with the response to a moderate earthquake has been made for the "standard" 20-story building shown in Fig. 27-15. The stresses and deflections of this structure, due to the lateral seismic forces specified by the UBC, were calculated with a static-frame-analysis computer program. Then the mode-superposition procedure was used to calculate the response to the El Centro (May 1940) earthquake accelerogram shown in Fig. 27-16 with a dynamic-frame-analysis program. The story displacements and the moments in specified girders and columns obtained by the two analyses are shown in Fig. 27-17. The dynamic-response results are *envelope* values, that is, the maximum value achieved at any time during the dynamic response, and thus they are not concurrent in general. This figure again shows that a moderate earthquake causes significantly greater stresses and deflections than the code forces do, by a factor of about 4 in this case. Of course, this comparison is not completely valid for a typical structure because it has been assumed that the dynamic structural response is linear elastic. In actuality, a code-designed

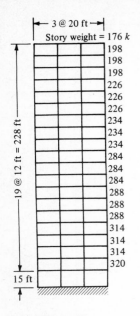

Story weight = 176 k
3 @ 20 ft
198
198
198
226
226
226
234
234
234
284
284
284
288
288
288
314
314
314
320

19 @ 12 ft = 228 ft

15 ft

Relative stiffness of columns and girders

Ratio $(EI) : (EI)_0$		
Columns		Girders
exterior	interior	
1.0	2.0	4.0
1.5	3.0	6.0
3.0	6.0	
4.5	9.0	8.0
6.0	12.0	
10.0	20.0	10.0
12.0	24.0	

FIGURE 27-15
Dimensions and properties of standard building.

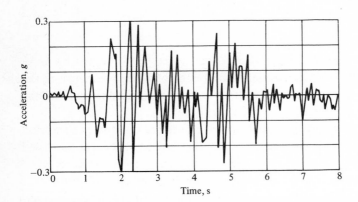

FIGURE 27-16
El Centro earthquake accelerogram, May 18, 1940 (*NS* component).

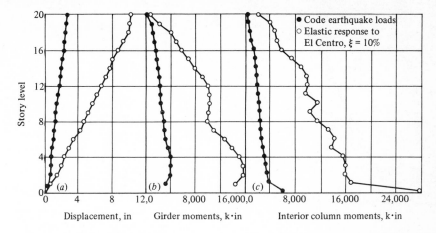

FIGURE 27-17
Comparison of elastic dynamic earthquake response with code effects.

structure would have been overstressed by this ground motion, and the true response could be determined only by a nonlinear analysis.

Method of Nonlinear Analysis

Detailed descriptions of numerical techniques for the analysis of nonlinear MDOF structures have been presented in Chap. 13. In general, such analyses involve the step-by-step integration of the equations of motion, dividing the response history into short time increments and assuming the properties of the structure to remain constant during each increment but changing in accordance with the deformation state existing at the end of the increment. Thus the nonlinear analysis actually is treated as a sequence of linear analyses of a successively changing structure.

The most difficult aspect of this analysis process is the evaluation of the stiffness properties to be applied during a time increment on the basis of the deformation state developed at the end of the preceding increment. In the nonlinear analysis of the building of Fig. 27-15, it was assumed that plastic hinges would develop at the ends of the columns and girders when the end moment reached the yield value specified for that member. During deformations beyond yield, the rotation stiffness was reduced to 5 percent of the original value, but when the deformation reversed, the original stiffness was restored (thus the moment-curvature relationship was of the typical bilinear elastoplastic form). The stiffness matrix of the complete structure was obtained for each time increment by assembling the stiffnesses of the columns and girders, each having been evaluated for the bilinear property corresponding to the

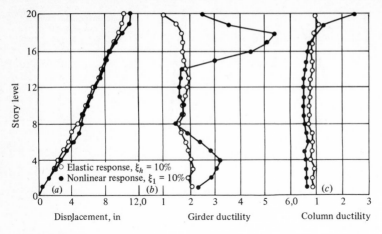

FIGURE 27-18
Comparison of elastic and nonlinear dynamic earthquake response.

current state of deformation. The incremental response was then evaluated by the linear-acceleration method described in Chap. 13, and the new deformations added to those existing at the end of the preceding time increment.

Typical Inelastic-Response Behavior

A sequence of analyses was carried out for the building of Fig. 27-15 subjected to the El Centro earthquake, assuming different values and distributions of the strengths of the structural members. The first case considered was a "standard" design, in which the column yield moments were specified to be 6 times the corresponding design moments produced by the static code seismic forces, while the girder yield moments were set at twice their code-force design values. Damping was assumed to be 10 percent of critical.

Results of the nonlinear analysis of this structure are compared with the elastic (infinite-strength) results (discussed earlier) in Fig. 27-18. The maximum displacements of the nonlinear system are essentially the same as those obtained in the linear analysis—just a few percent larger. However, the deformations of the columns and girders are significantly changed by the inelastic yielding mechanism. These results are expressed in terms of the member ductility ratio (DR), which is defined as the ratio of the maximum end rotation to the end rotation at the incipient yield condition, that is,

$$DR = \frac{\theta_{max}}{\theta_y} = \frac{\gamma_{max} + \theta_y}{\theta_y} \qquad (27\text{-}100)$$

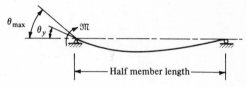

FIGURE 27-19
Definition of ductility ratio.

where the symbols are defined in Fig. 27-19. It will be noted that a ductility ratio less than unity means that the member has not yielded. The elastic-response results in Fig. 27-18 also are presented as the ratio of the maximum end rotation to the member yield rotation; however, for the linear structure there is no yielding associated with ratios greater than 1, of course.

The results in Fig. 27-18 show that significant yielding develops in the girders of the nonlinear structure, primarily in the lower and upper stories, while its columns remain elastic except in the top stories. This is considered to be typical behavior of a building in which the strength has been proportioned according to normal lateral-force design requirements. The yielding in the lower stories results directly from the high-intensity base-motion input, whereas the relatively high yielding indicated in the upper stories is due to a whiplash or wave-reflection effect. Of course, this analysis defines the ductility "requirements" imposed by the earthquake; it is up to the designer to detail the members and joints so that their ductility *capacity* is at least equal to the earthquake requirement.

Influence of Strength Variations

The fact that the building considered in Fig. 27-18 yielded almost entirely in the girders while the columns remained essentially elastic is due to the relative strength distribution produced by the normal design procedures. In effect, the energy absorbed during the yielding of the girders has tended to protect the columns from overstress. Changes in the relative strengths of columns and girders can be expected to cause the yielding to shift from one type of element to the other.

The results of analyses in which various levels of girder strength were considered are presented in Fig. 27-20.[1] In all cases, the column strengths were the same as in Fig. 27-18, but girder strengths 1.5 and 6 times the design moment levels were considered in addition to the standard strength factor of 2. As might be expected, the girder ductility ratios can be seen to vary inversely with the girder strengths. However, the lateral displacements in the upper stories increase with increasing girder

---

[1] It should be mentioned that damping was neglected in these analyses, hence the "standard" building results are slightly different from those shown in Fig. 27-18.

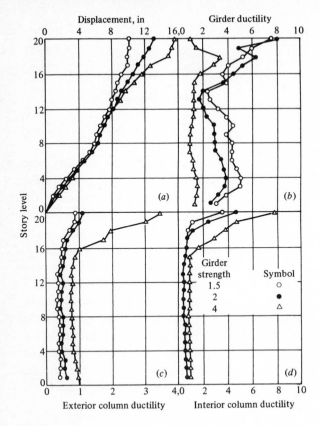

FIGURE 27-20
Effect of girder strength on nonlinear response.

strengths. This somewhat surprising result is explained by the column yielding behavior; it is evident that an increase of girder strength forces more yielding into the upper-story columns, and these increased column deformations are reflected directly in the story displacements.

The balance between girder and column yielding can also be influenced by changes in the column strengths, as shown in Fig. 27-21. In these cases, the girder strengths were maintained as in Fig. 27-18, but column strengths 2 and 10 times the design moment levels were considered in addition to the standard strength factor of 6. It is apparent in Fig. 27-21 that the increase of column strength had little effect on the building behavior, which is not surprising inasmuch as the standard columns were strong enough to undergo only very slight yielding. On the other hand, in the case with reduced column strengths the yield requirements in the girders were greatly reduced as a result of the greatly increased column yielding. The column ductility

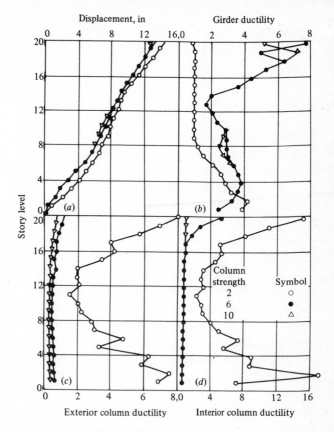

FIGURE 27-21
Effect of column strength on nonlinear response.

ratios computed here, which were as high as 16 in certain locations, would impose a very difficult requirement upon the structural designer.

Although it is not the purpose of this presentation to discuss the design of earthquake-resistant structures, a few conclusions can be drawn from the results of these nonlinear analyses. First, it is evident that a reasonable balance should be maintained between the strengths of columns and girders, but that a weak-girder–strong-column philosophy is to be recommended. This is because the girder ductility requirements tend to be more moderate, and more important, since local yielding of the girders does not seriously affect the vertical-load-carrying capacity of the structure while local column yielding could easily lead to collapse. Second, it may be inferred that local weak zones should be avoided in the structural frame because damage energy will be attracted there and the rest of the structure consequently will work at less than capacity. The most efficient design will have balanced strength so that yielding will be distributed evenly and no excessive damage will result at any point.

Ductility-Factor Method

Although the nonlinear analysis used in these examples is not difficult in concept, it requires a large expenditure of computing effort to evaluate the behavior of even rather simple structures because the stiffness matrix must be reformed and decomposed for each step of the incremental process. Moreover, the selection of appropriate member stiffnesses and strengths is an iterative procedure, so that a succession of different designs must be analyzed before the final results can be accepted. Therefore a complete nonlinear analysis is seldom used except as a final check on the adequacy of completed design.

In order to obtain a reasonable measure of the nonlinear earthquake behavior of a structure without carrying out a true nonlinear analysis, the ductility-factor method has been developed. The basic assumption of this method is that the deflections produced by a given earthquake input are essentially the same, whether the structure responds elastically or yields significantly. This behavior is demonstrated in Fig. 27-18, where the inelastic-response displacements are nearly identical with the elastic displacements even though ductility ratios of over 5 were developed in some of the girders. If the nonlinear member deformations can be assumed to be identical with the elastic-response deformations, the inelastic behavior can be interpreted directly from a linear elastic-response analysis. Consider, for example, the member force-displacement relation shown in Fig. 27-22. If the maximum deflection δ_{max} developed in this member is the same regardless of its strength property, the ratio of the maximum deformation to the elastic-limit deformation is equal to the ratio of the force developed in purely elastic response to the member yield forces, that is,

$$\frac{\delta_{max}}{\delta_y} = \frac{f_{max}}{f_y}$$

The ratio δ_{max}/δ_y is called the *ductility factor* μ of the member, that is,

$$\mu = \frac{\delta_{max}}{\delta_y}$$

Hence it is clear that the design strength required in a member can be expressed in terms of the elastic earthquake-response force and a ductility factor specified for the member,

$$f_y = \frac{1}{\mu} f_{max}$$

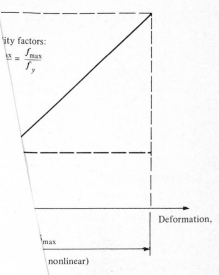

lity factors:

$$\frac{}{} = \frac{f_{max}}{f_y}$$

Deformation,

$_{max}$

nonlinear)

FIGURE 27-22
Definition of ductility factor μ.

be seen that the ap-
t the concentrations
btained. In general,
ility-factor approach.
distributed uniformly
o local strain concen-
chieving a reasonably
e with systems having
liberately designed to

response, therefore, by per-
providing strengths in all the
orces reduced by a ductility
tion considered acceptable
en design can be evaluated
the ratio of the maximum

$$\mu = \frac{f_{max}}{f_y}$$

perties:

.24 kips · s/in

e caused by an earthquake

acture is increased to $k =$
 change.) Comment on the
g earthquake resistance.
7-2 has the properties $\overline{m} =$
eflected shape is $\psi(x) = 1 -$

can be demonstrated
esented in Fig. 27-17.
ecified columns and
nents by the member
design decision, but
times the indicated
y factors shown in

ese same member

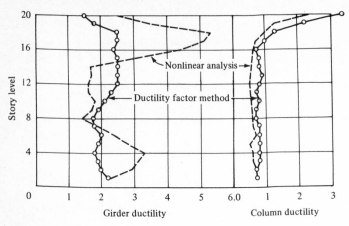

FIGURE 27-23
Approximate member ductility compared with exact.

strengths, have been replotted here for comparison, and it may
proximate results are of the correct order of magnitude but tha
of girder yielding in the lower and upper stories have not been
this is about the best accuracy that can be expected from the duc
In effect, the method assumes that ductile deformations will be
over the entire structure; hence any mechanism which leads t
trations will not be accounted for. It provides a basis for a
balanced design in regular structures, but clearly it cannot cop
pronounced strength discontinuities, such as a building de
yield only at the first-story level.

PROBLEMS

27-1 Assume that the structure of Fig. 27-1 has the following pr

$$m = 3.2 \text{ kips} \cdot \text{s}^2/\text{in} \qquad k = 48 \text{ kips/in} \qquad c = $$

Determine the maximum displacement and base shear forc
having the response spectrum of Fig. E27-1.

27-2 Repeat Prob. 27-1 assuming that the stiffness of the str
300 kips/in. (Note that both T and ξ are affected by thi
effectiveness of increasing stiffness as a means of increasin

27-3 Assume that the uniform cantilever column of Fig. E2
$0.016 \text{ kip} \cdot \text{s}^2/\text{ft}^2$ and $EI = 10^6 \text{ kips} \cdot \text{ft}^2$, and that its d

$\cos(\pi x/2L)$. If this structure is subjected to the ground motion having the response spectrum of Fig. E27-1 (with $\xi = 2\%$),

(a) Determine the maximum tip displacement, base moment, and base shear.

(b) Determine the maximum displacement, moment, and shear at midheight.

27-4 Repeat Prob. 27-3 assuming the same response shape but considering the following nonuniform mass and stiffness properties:

$$m(x) = 0.01(2 - x/L) \text{ kip} \cdot \text{s}^2/\text{ft}^2 \qquad EI(x) = 5 \times 10^5 (1 - x/L)^2 \text{ kip} \cdot \text{ft}^2$$

Use Simpson's rule with $\Delta x = L/2$ to evaluate the generalized property integrals.

27-5 A building similar to that shown in Fig. E27-3 has the following mass matrix and vibration properties:

$$\mathbf{m} = 2 \begin{bmatrix} 1 & 0 & 0 \\ 0 & 1 & 0 \\ 0 & 0 & 1 \end{bmatrix} \text{kip} \cdot \text{s}^2/\text{ft} \qquad \mathbf{\Phi} = \begin{bmatrix} 1.000 & 1.000 & 1.00 \\ 0.548 & -1.522 & -6.26 \\ 0.198 & -0.872 & 12.10 \end{bmatrix} \qquad \omega = \begin{bmatrix} 3.88 \\ 9.15 \\ 15.31 \end{bmatrix} \text{rad/s}$$

Determine the displacements, overturning moments, and shear force at each story at a time t_1 during an earthquake when the response integrals for the three modes are

$$\mathbf{V}(t_1) = \begin{bmatrix} 1.38 \\ -0.50 \\ 0.75 \end{bmatrix} \text{ft/s}$$

The height of each story is 12 ft.

27-6 For the structure and earthquake of Prob. 27-5, the acceleration response spectrum values for the three modes are

$$\mathbf{S}_a = \begin{bmatrix} 9.66 \\ 5.15 \\ 12.88 \end{bmatrix} \text{ft/s}^2$$

(a) For each mode of vibration, calculate the maximum displacement, overturning moment, and shear force at each story level.

(b) By the root-sum-square method, determine approximate total maximums for each of the response quantities of part a.

27-7 For preliminary design purposes, the tall building shown in Fig. P27-1 will be assumed to behave as a uniform shear beam. Thus its mode shapes and frequencies are given by

$$\phi_n(x) = \sin \frac{2n - 1}{2} \left(\frac{\pi x}{L} \right)$$

$$\omega_n = \frac{2n - 1}{2} \pi \left(\frac{12 \sum EI}{m_i h L^2} \right)$$

where the values of the properties are shown in the figure.

(*a*) Determine the effective modal mass $\mathscr{L}_n{}^2/M_n$ for each of the first five modes. What fraction of the total mass is associated with each mode?

(*b*) Compute the approximate maximum top displacement, base shear, and base overturning moment by the root-sum-square method, assuming that the velocity response spectrum value for each mode is 1.6 ft/s.

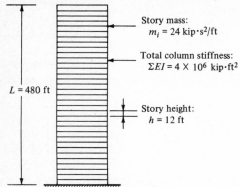

Story mass:
$m_i = 24$ kip·s²/ft

Total column stiffness:
$\Sigma EI = 4 \times 10^6$ kip·ft²

$L = 480$ ft

Story height:
$h = 12$ ft

FIGURE P27-1
Uniform shear building.

27-8 A structure is idealized as the two-degree-of-freedom system shown in Fig. P27-2; also shown are its vibration mode shapes and frequencies. Assuming $\xi = 2$ percent in each mode and using the response spectrum of Fig. E27-1, compute the approximate (root-sum-square) maximum moment at the column base assuming the direction of the earthquake motions is

(*a*) Horizontal.

(*b*) Vertical.

(*c*) Along the inclined axis *ZZ*.

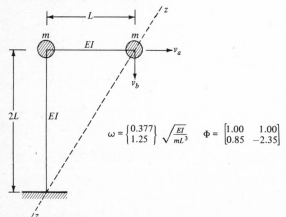

$$\omega = \begin{Bmatrix} 0.377 \\ 1.25 \end{Bmatrix} \sqrt{\frac{EI}{mL^3}} \qquad \Phi = \begin{bmatrix} 1.00 & 1.00 \\ 0.85 & -2.35 \end{bmatrix}$$

FIGURE P27-2
SDOF plane frame.

27-9 A 6-in concrete slab is supported by four W8 × 40 columns which are located and oriented as shown in Fig. P27-3. Also shown are the structure's mass matrix and vibration properties, based on the assumption that the slab is rigid, that the columns are weightless, and that the clear height of the columns is 12 ft. The mass matrix and mode shapes are expressed in terms of the slab centroid coordinates that are shown.

Assuming that an earthquake having the response spectrum of Fig. E27-1 acts in the direction of coordinate v_1, determine the maximum dynamic displacement at the top of each column in the *first* mode of vibration.

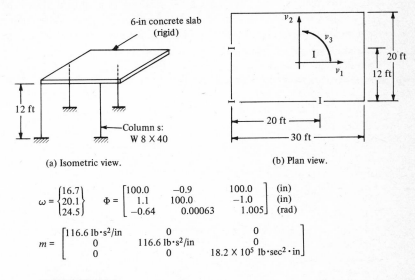

(a) Isometric view. (b) Plan view.

$$\omega = \begin{Bmatrix} 16.7 \\ 20.1 \\ 24.5 \end{Bmatrix} \qquad \Phi = \begin{bmatrix} 100.0 & -0.9 & 100.0 \\ 1.1 & 100.0 & -1.0 \\ -0.64 & 0.00063 & 1.005 \end{bmatrix} \begin{matrix} \text{(in)} \\ \text{(in)} \\ \text{(rad)} \end{matrix}$$

$$m = \begin{bmatrix} 116.6 \text{ lb·s}^2/\text{in} & 0 & 0 \\ 0 & 116.6 \text{ lb·s}^2/\text{in} & 0 \\ 0 & 0 & 18.2 \times 10^5 \text{ lb·sec}^2 \cdot \text{in} \end{bmatrix}$$

FIGURE P27-3
Rigid deck frame.

27-10 A uniform bridge deck is simply supported with an 80-ft span, as shown below. Also shown are the mass and stiffness properties as well as an idealized earthquake-velocity-response spectrum. Assuming that this same earthquake acts simultaneously on *both* end supports in the vertical direction,

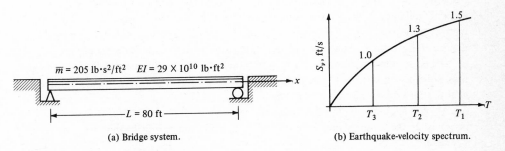

$\overline{m} = 205$ lb·s^2/ft^2 $EI = 29 \times 10^{10}$ lb·ft^2

(a) Bridge system. (b) Earthquake-velocity spectrum.

FIGURE P27-4
Bridge subjected to vertical earthquake motions.

(a) Compute the maximum moment at midspan for each of the first three modes of vibration.

(b) Compute the approximate (root-sum-square) maximum midspan moment due to these three modes.

27-11   Repeat Prob. 27-10, assuming that only the right-hand support is subjected to this vertical motion. Note that $r(x) = x/L$ in this case.

27-12   The service platform for a space rocket is idealized as a lumped mass tower, as shown below. Also shown are the shapes and frequencies of its first two modes of vibration. Determine the maximum moment developed at the base of this tower due to a harmonic horizontal ground acceleration $\ddot{v}_g = A \sin \bar{\omega}t$ where $A = 5$ ft/s^2 and $\bar{\omega} = 8$ rad/s. Consider only the steady-state response of the first two modes, and neglect damping.

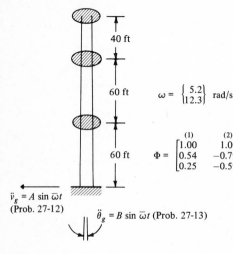

$$\omega = \begin{Bmatrix} 5.2 \\ 12.3 \end{Bmatrix} \text{ rad/s}$$

$$\Phi = \begin{bmatrix} 1.00 & 1.00 \\ 0.54 & -0.79 \\ 0.25 & -0.59 \end{bmatrix}$$

$\ddot{v}_g = A \sin \bar{\omega}t$
(Prob. 27-12)

$\ddot{\theta}_g = B \sin \bar{\omega}t$ (Prob. 27-13)

FIGURE P27-5
Lumped mass tower subjected to earthquake.

27-13   Repeat Prob. 27-12 assuming that the harmonic ground motion applied at the base is a rotation θ_g rather than horizontal translation. In this case $\ddot{\theta}_g = B \sin \bar{\omega}t$ where $B = 0.06$ rad/s^2 and $\bar{\omega} = 8$ rad/s.

27-14   A rigid bar of length L and total uniformly distributed mass m has an additional lumped mass $m/2$ at each end. This bar is rigidly attached to the top of a weightless column of length L and has a lateral spring support at midheight, as shown in Fig. P27-6. The mass matrix for the rigid bar and the stiffness matrix for the entire system including the support degrees of freedom are shown in the figure, together with the vibration properties.

This system is subjected to a ground motion for which the spectral velocity at the first mode period is 2.7 ft/s. Determine the first mode maximum response of coordinate v_2 if the earthquake motion is applied

(a) At both support points simultaneously.

(b) Only at the column base (coordinate v_{gb}), while the spring support (v_{ga}) is fixed against motion.

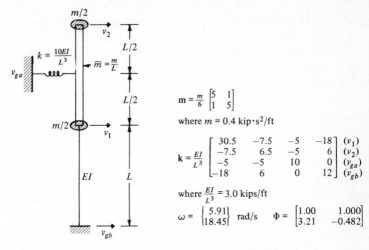

$$m = \frac{m}{6} \begin{bmatrix} 5 & 1 \\ 1 & 5 \end{bmatrix}$$

where $m = 0.4$ kip·s^2/ft

$$k = \frac{EI}{L^3} \begin{bmatrix} 30.5 & -7.5 & -5 & -18 \\ -7.5 & 6.5 & -5 & 6 \\ -5 & -5 & 10 & 0 \\ -18 & 6 & 0 & 12 \end{bmatrix} \begin{matrix} (v_1) \\ (v_2) \\ (v_{ga}) \\ (v_{gb}) \end{matrix}$$

where $\frac{EI}{L^3} = 3.0$ kips/ft

$$\omega = \begin{Bmatrix} 5.91 \\ 18.45 \end{Bmatrix} \text{ rad/s} \qquad \Phi = \begin{bmatrix} 1.00 & 1.000 \\ 3.21 & -0.482 \end{bmatrix}$$

FIGURE P27-6
System with multiple supports.

27-15 A single-story building is to be built on a soil base which is essentially an elastic half-space (Fig. P27-7a). For the purpose of earthquake-response analysis of the building-foundation system, the soil will be modeled as a spring, mass, damper system, as shown in Fig. P27-7b. (Note for simplicity that only horizontal motions are included in this interaction problem.)

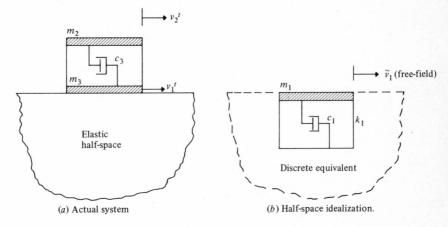

(a) Actual system   (b) Half-space idealization.

FIGURE P27-7
Modeling horizontal component of soil-structure interaction.

Write the 2DOF matrix equations of motion for this system, defining the effective earthquake force in terms of the free field acceleration $\ddot{\tilde{v}}_1$, and expressing the total response in terms of the dynamic and pseudostatic components $\mathbf{v}_c^{\ t} = \mathbf{v}_c + \mathbf{v}_c^{\ s}$. Note that the displacement influence matrix $\mathbf{r}_c$ is to be evaluated in terms of the system properties so that the final expressions may be reduced to their simplest form.

Do not try to solve the equations of motion.

NONDETERMINISTIC ANALYSIS OF EARTHQUAKE RESPONSE

28-1 Stochastic Modeling of Strong Ground Motions

Since seismic waves are initiated by irregular slippage along faults followed by numerous random reflections, refractions, and attentuations within the complex ground formations through which they pass, stochastic modeling of strong ground motions seems appropriate. If unlimited ground-motion data were available, representative stochastic models could be established directly by statistical analyses. Unfortunately, strong-motion data in the form of accelerograms are limited. Therefore, one is forced to hypothesize forms of models and to use the available strong-motion data primarily in checking the appropriateness of these forms.

Stationary White Noise

In view of the irregular manner in which slippage undoubtedly occurs along a fault, strong ground motions at some distance from the fault might be considered as the superposition of short-duration random pulses arriving randomly in time. Therefore, since accelerograms usually have a phase of nearly constant intensity during the period of most severe oscillation, one might consider modeling this phase with a white-noise process of limited duration. Housner, Rosenblueth, Bycroft, Thomson,[1] and others considered this possibility in their earlier investigations.

For analytical purposes one may wish to generate sample functions which approach white noise. This procedure can be carried out digitally by first sampling a sequence of pairs of statistically independent random numbers $x_1, x_2, x_3, x_4, \ldots, x_{n-1}, x_n$, all of which have a uniform probability distribution over the range $0 < x < 1$. A new sequence of pairs of statistically independent random numbers $y_1, y_2, y_3, y_4, \ldots, y_{n-1}, y_n$ are then generated using the relations

$$y_i = (-2 \ln x_i)^{1/2} \cos 2\pi x_{i+1}$$
$$y_{i+1} = (-2 \ln x_i)^{1/2} \sin 2\pi x_{i+1} \tag{28-1}$$

which as shown in Example E22-7 possess a gaussian distribution with a mean of zero and a variance of unity.

A sample function $a_r(t)$ can now be established by assigning the values $y_1, y_2, \ldots, y_n$ to n successive ordinates spaced at equal intervals $\Delta\varepsilon$ along a time abscissa and by assuming a linear variation of ordinates over each interval. Usually, the initial ordinate y_0 is assumed equal to zero and is located at $t = t_{0r}$, where t_0 is a random variable having a uniform probability density function of intensity $1/\Delta\varepsilon$ over the interval $0 < t_0 < \Delta\varepsilon$.

A complete ensemble of m such sample functions $a_r(t)$ $(r = 1, 2, \ldots, m)$ can be obtained by repeating this procedure m times, thereby creating a stationary process characterized by the autocorrelation function

$$R_a(\tau) = \begin{cases} \dfrac{2}{3} - \left(\dfrac{\tau}{\Delta\varepsilon}\right)^2 + \dfrac{1}{2}\left(\dfrac{|\tau|}{\Delta\varepsilon}\right)^3 & -\Delta\varepsilon \leq \tau \leq \Delta\varepsilon \\[2mm] \dfrac{4}{3} - 2\dfrac{|\tau|}{\Delta\varepsilon} + \left(\dfrac{\tau}{\Delta\varepsilon}\right)^2 - \dfrac{1}{6}\left(\dfrac{|\tau|}{\Delta\varepsilon}\right)^3 & \begin{aligned} -2\Delta\varepsilon &\leq \tau \leq -\Delta\varepsilon \\ \Delta\varepsilon &\leq \tau \leq 2\Delta\varepsilon \end{aligned} \\[2mm] 0 & \tau \leq -2\Delta\varepsilon; \tau \geq 2\Delta\varepsilon \end{cases} \tag{28-2}$$

[1] G. W. Housner, Properties of Strong Ground Motion Earthquakes, *Bull. Seismol. Soc. Am.*, vol. 45, no. 3, pp. 197–218, July 1955; E. Rosenblueth and J. Bustamante, Distribution of Structural Response to Earthquakes, *Proc. Pap.* 3173, *J. Eng. Mech. Div. ASCE*, vol. 88, no. EM3, pp. 75–106, June 1962; G. N. Bycroft, White Noise Representation of Earthquakes, *Proc. Pap.* 2434, *J. Eng. Mech. Div. ASCE*, vol. 86, no. EM2, pp. 1–16, April 1960; W. T. Thomson, Spectral Aspect of Earthquakes, *Bull. Seismol. Soc. Am.*, vol. 49, pp. 91–98, 1959.

as given in Example E23-2. If the intensity of this process is now changed by multiplying each ordinate y_i by the normalization factor $(2\pi S_0/\Delta\varepsilon)^{1/2}$, where S_0 is a constant, the autocorrelation function for the new process becomes

$$
R_a(\tau) = \begin{cases}
\dfrac{2\pi S_0}{\Delta\varepsilon}\left[\dfrac{2}{3} - \left(\dfrac{\tau}{\Delta\varepsilon}\right)^2 + \dfrac{1}{2}\left(\dfrac{|\tau|}{\Delta\varepsilon}\right)^3\right] & -\Delta\varepsilon \le \tau \le \Delta\varepsilon \\[3mm]
\dfrac{2\pi S_0}{\Delta\varepsilon}\left[\dfrac{4}{3} - 2\dfrac{|\tau|}{\Delta\varepsilon} + \left(\dfrac{\tau}{\Delta\varepsilon}\right)^2 - \dfrac{1}{6}\left(\dfrac{|\tau|}{\Delta\varepsilon}\right)^3\right] & \begin{aligned}-2\Delta\varepsilon &\le \tau \le -\Delta\varepsilon \\ \Delta\varepsilon &\le \tau \le 2\Delta\varepsilon\end{aligned} \\[3mm]
0 & \tau \le -2\Delta\varepsilon;\ \tau \ge 2\Delta\varepsilon
\end{cases}
\tag{28-3}
$$

Taking the Fourier transform of Eq. (28-3) in accordance with Eq. (23-37) gives the power spectral density function

$$
S_a(\bar\omega) = S_0\ \frac{6 - 8\cos\bar\omega\,\Delta\varepsilon + 2\cos 2\bar\omega\,\Delta\varepsilon}{(\bar\omega\,\Delta\varepsilon)^4} \qquad -\infty < \bar\omega < \infty \tag{28-4}
$$

As reported by Ruiz and Penzien,[1] this function is flat to within 5 percent error for $\bar\omega\,\Delta\varepsilon < 0.57$ and to within 10 percent error for $\bar\omega\,\Delta\varepsilon < 0.76$. The function drops to 50 percent its maximum value S_0 at $\bar\omega\,\Delta\varepsilon = 2$.

It is significant that as $\Delta\varepsilon$ approaches zero, Eq. (28-3) approaches

$$
R_a(\tau) = 2\pi S_0\delta(\tau) \tag{28-5}
$$

Therefore in the limit as $\Delta\varepsilon \to 0$, this process becomes gaussian white noise of intensity S_0 over the infinite frequence range $-\infty < \bar\omega < \infty$.

Stationary Filtered White Noise

Fourier analyses of existing strong-motion accelerograms reveal that the Fourier amplitude spectra are not constant with frequency even over a limited band. They are somewhat oscillatory in character, may peak at one or several frequencies, and damp out with increasing frequency—all of which suggest that a stationary filtered white noise of limited duration could be more representative of actual strong ground motions provided the filter transfer characteristics are properly selected. Kanai and Tajimi[2] have suggested the filter transfer function

$$
|H_1(i\bar\omega)|^2 = \frac{1 + 4\xi_g{}^2(\bar\omega/\omega_g)^2}{[1 - (\bar\omega/\omega_g)^2]^2 + 4\xi_g{}^2(\bar\omega/\omega_g)^2} \tag{28-6}
$$

[1] P. Ruiz and J. Penzien, Probabilistic Study of Behavior of Structures during Earthquakes, *Univ. Calif. Berkeley Earthquake Eng. Res. Cent.* 69-3, 1969.
[2] K. Kanai, Semi-empirical Formula for the Seismic Characteristics of the Ground, *Univ. Tokyo Bull. Earthquake Res. Inst.*, vol. 35, pp. 309–325, 1957; H. Tajimi, A Statistical Method of Determining the Maximum Response of a Building Structure during an Earthquake, *Proc. 2d World Conf. Earthquake Eng. Tokyo and Kyoto*, vol. II, pp. 781–798, July 1960.

With this transfer function, the power spectral density function for the filtered process $a_1(t)$ would be

$$S_{a_1}(\bar{\omega}) = |H_1(i\bar{\omega})|^2 S_a(\bar{\omega}) \qquad -\infty < \bar{\omega} < \infty \qquad (28\text{-}7)$$

where $S_a(\bar{\omega})$ is the power spectral density function for process $a(t)$. Parameters ω_g and ξ_g appearing in the transfer function may be thought of as some characteristic ground frequency and characteristic damping ratio, respectively. Kanai has suggested 15.6 rad/s for ω_g and 0.6 for ξ_g as being representative of firm soil conditions. Other numerical values should be selected, as appropriate, when significantly different soil conditions are present.

It should be recognized that the above filter attenuates the higher-frequency components and amplifies those frequency components in the neighborhood of $\bar{\omega} = \bar{\omega}_g$. Since it does not change the amplitudes as $\bar{\omega} \to 0$, some difficulty may arise with the very low-frequency components. The cause of this difficulty can easily be recognized by noting that the power spectral density functions for ground velocity and ground displacement are obtained by dividing Eq. (28-7) by $\bar{\omega}^2$ and $\bar{\omega}^4$, respectively. Thus, strong singularities are present at $\bar{\omega} = 0$ which cause the stationary variances of ground velocity and ground displacement to be unbounded. These undesirable singularities can be removed by passing process $a_1(t)$ through another filter which greatly attenuates the very low-frequency components. An appropriate filter for this process is one having the transfer function

$$|H_2(i\bar{\omega})|^2 = \frac{(\bar{\omega}/\omega_1)^4}{[1 - (\bar{\omega}/\omega_1)^2]^2 + 4\xi_1{}^2(\bar{\omega}/\omega_1)^2} \qquad (28\text{-}8)$$

where the frequency parameter ω_1 and the damping parameter ξ_1 are selected to give the desired filter characteristics. The output process $a_2(t)$ from this filter has a power spectral density function of the form

$$S_{a_2}(\bar{\omega}) = |H(i\bar{\omega})|^2 S_a(\bar{\omega}) \qquad -\infty < \bar{\omega} < \infty \qquad (28\text{-}9)$$

where
$$|H(i\bar{\omega})|^2 \equiv |H_1(i\bar{\omega})|^2 |H_2(i\bar{\omega})|^2 \qquad (28\text{-}10)$$

Equation (28-9) has the form shown in Fig. 28-1.

The first filtering of process $a(t)$ described above can be accomplished by solving the differential equations

$$\ddot{y}_r + 2\omega_g\xi_g\dot{y}_r + \omega_g{}^2 y_r = -a_r(t) \qquad r = 1, 2, \ldots \qquad (28\text{-}11)$$

for y_r and $\dot{y}_r$ using a digital computer and standard numerical-integration techniques and then obtaining a_{1_r} using the relation

$$a_{1_r} = -2\omega_g\xi_g\dot{y}_r - \omega_g{}^2 y \qquad (28\text{-}12)$$

Likewise, the second filtering can be accomplished by solving the differential equations

$$\ddot{z}_r + 2\bar{\omega}_1\xi_1\dot{z}_r + \bar{\omega}_1{}^2 z_r = -a_{1_r} \qquad r = 1, 2, \ldots \qquad (28\text{-}13)$$

for z_r and then letting

$$a_{2_r} = z_r \qquad (28\text{-}14)$$

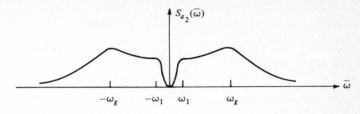

FIGURE 28-1
Power spectral density function for filtered stationary white noise.

All members of the desired stationary filtered white-noise process can be obtained by repeating this procedure m times.

A more direct method of obtaining the desired stationary filtered process $a_2(t)$ is to lump the area under the power spectral density function $S_{a_2}(\bar{\omega})$ at equal frequency intervals $\Delta\bar{\omega}$ and to let these areas equal one-half the squared amplitudes of a set of discrete harmonics, that is, let $[S_{a_2}(-i\,\Delta\bar{\omega}) + S_{a_2}(i\,\Delta\bar{\omega})]\,\Delta\bar{\omega} = A_{ir}{}^2/2$.

In this case

$$a_{2_r}(t) = \sum_i [4S_{a_2}(i\,\Delta\bar{\omega})\,\Delta\bar{\omega}]^{1/2} \sin(i\,\Delta\bar{\omega}\,t + \phi_{ir}) \qquad \begin{matrix} r = 1, 2, \ldots \\ i = 1, 2, \ldots \end{matrix} \qquad (28\text{-}15)$$

where ϕ is a random phase angle having a uniform probability density function over the range $0 < \phi < 2\pi$. With this method, $S_{a_2}(\bar{\omega})$ can be expressed in the form of Eq. (28-7), in which case the summation in Eq. (28-15) should be started with $i = h$, where $h\,\Delta\bar{\omega} = \omega_1$. It can also be expressed in the form of Eq. (28-9), in which case the summation in Eq. (28-15) could be started with $i = 1$. This direct method can obviously permit any arbitrary form for $S_{a_2}(\bar{\omega})$ without posing difficulty.

Nonstationary Filtered White Noise

To obtain an even more representative process for strong ground motions, the nonstationary character of actual accelerograms can be considered. Real accelerograms often show a short phase of intensity buildup to some maximum level. The intensity then remains fairly constant for some time, after which it decays in an exponential fashion. This appearance suggests using a nonstationary process of the type described in Sec. 23-12, namely, a process $a(t)$ given by

$$a(t) = \zeta(t)a_2(t) \qquad (28\text{-}16)$$

where $a_2(t)$ is the stationary filtered process previously described and $\zeta(t)$ is an intensity function having an appropriate form based on statistical analyses of real accelerograms. One form which has been suggested[1] is that given in Fig. 28-2.

[1] P. C. Jennings, G. W. Housner, and N. C. Tsai, Simulated Earthquake Motions, Rept. *Earthquake Eng. Res. Lab., California Institute of Technology*, April 1968; A. K. Chopra, V. V. Bertero, and S. A. Mahin, Response of the Olive View Medical Center, Main Building, during the San Fernando Earthquake, *5th World Conf. Earthquake Eng., Rome, June 1973, Prepr. 4.*

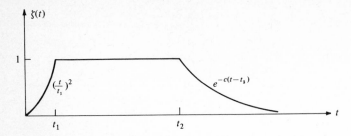

FIGURE 28-2
Intensity function $\zeta(t)$ for nonstationary process $a(t)$.

Constants t_1, t_2, and c should be assigned only after considering such factors as earthquake magnitude, epicentral distance, etc.

28-2 ANALYSIS OF LINEAR SYSTEMS

SDOF Systems

If a stationary white-noise process of intensity S_0 is assumed for ground acceleration $\ddot{v}_g(t)$, the response of a linear SDOF system to this support acceleration is governed by the equation

$$\ddot{v} + 2\xi\omega\dot{v} + \omega^2 v = -\ddot{v}_g(t) \qquad (28\text{-}17)$$

where v is the mass displacement relative to the moving support. The principles set forth in Chap. 24 give for an undercritically damped system

$$R_v(\tau) = \frac{\pi S_0}{2\omega^3 \xi}\left(\cos \omega_D|\tau| + \frac{\xi}{\sqrt{1-\xi^2}} \sin \omega_D|\tau|\right) \exp\left(-\omega\xi|\tau|\right) \qquad (28\text{-}18)$$

$$R_{\dot{v}}(\tau) = \frac{\pi S_0}{2\omega \xi}\left(\cos \omega_D|\tau| - \frac{\xi}{\sqrt{1-\xi^2}} \sin \omega_D|\tau|\right) \exp\left(-\omega\xi|\tau|\right) \qquad (28\text{-}19)$$

$$S_v(\bar{\omega}) = \frac{S_0\omega^{-4}}{[1 - (\bar{\omega}/\omega)^2]^2 + 4\xi^2(\bar{\omega}/\omega)^2} \qquad (28\text{-}20)$$

$$S_{\dot{v}}(\bar{\omega}) = \frac{S_0(\bar{\omega}/\omega)^2\omega^{-2}}{[1 - (\bar{\omega}/\omega)^2]^2 + 4\xi^2(\bar{\omega}/\omega)^2} \qquad (28\text{-}21)$$

$$\sigma_v^2 = \frac{\pi S_0}{2\omega^3 \xi} \qquad (28\text{-}22)$$

$$\sigma_{\dot{v}}^2 = \frac{\pi S_0}{2\omega \xi} \qquad (28\text{-}23)$$

If a stationary filtered white-noise process having the power spectral density $S_{a_2}(\overline{\omega})$ is assumed for ground acceleration $\ddot{v}_g(t)$, Eqs. (28-20) and (28-21) are still valid provided $S_{a_2}(\overline{\omega})$ is substituted for S_0. Means and standard deviations of extreme values can be estimated using Eqs. (23-117) and (23-118).

MDOF Systems

The linear response of discrete MDOF systems subjected to the same stationary acceleration $\ddot{v}_g(t)$ at all support points can be determined using normal-mode super-position as described in Chap. 25. The generalized forcing function $P_n(t)$ shown in Eq. (25-1) and defined by Eq. (25-24) becomes

$$P_n(t) = -\ddot{v}_g(t) \sum_i m_i \phi_{in} = \phi_n{}^T \mathbf{m}\{\mathbf{1}\}\ddot{v}_g(t) \quad n = 1, 2, \ldots \quad (28\text{-}24)$$

and Eqs. (25-25) and (25-26) can be expressed in the form

$$S_{P_m P_n}(\overline{\omega}) = S_{\ddot{v}_g}(\overline{\omega}) \sum_i \sum_k m_i m_k \phi_{im} \phi_{kn} = S_{\ddot{v}_g}(\overline{\omega}) \phi_m{}^T \{\mathbf{1}\}\{\mathbf{1}\}^T \mathbf{m}\phi_n \quad (28\text{-}25)$$

$$R_{P_m P_n}(\tau) = R_{\ddot{v}_g}(\tau) \sum_i \sum_k m_i m_k \phi_{im} \phi_{kn} = R_{\ddot{v}_g}(\tau) \phi_m{}^T \mathbf{m}\{\mathbf{1}\}\{\mathbf{1}\}^T \mathbf{m}\phi_n \quad (28\text{-}26)$$

For a distributed-mass system with $m = m(x)$, Eqs. (28-24) to (28-26) become

$$P_n(t) = -\ddot{v}_g(t) \int m(x)\phi_n(x)\, dx \quad (28\text{-}27)$$

$$S_{P_m P_n}(\overline{\omega}) = S_{\ddot{v}_g}(\overline{\omega}) \iint m(x)m(\alpha)\phi_m(x)\phi_n(\alpha)\, dx\, d\alpha \quad (28\text{-}28)$$

$$R_{P_m P_n}(\tau) = R_{\ddot{v}_g}(\tau) \iint m(x)m(\alpha)\phi_m(x)\phi_n(\alpha)\, dx\, d\alpha \quad (28\text{-}29)$$

where α is a dummy space coordinate.

The power spectral density and autocorrelation functions for response $z(t)$ are now obtained by substituting Eqs. (28-25) and (28-26) or (28-28) and (28-29) into Eqs. (25-19) and (25-10), respectively.

28-3 ANALYSIS OF NONLINEAR SYSTEMS

The stochastic response of nonlinear SDOF or MDOF systems cannot be obtained by the methods previously presented, which employ the principle of superposition. For these complex systems, which are often history-dependent due to hysteresis effects, one is usually forced to generate an ensemble of ground-motion accelerograms,

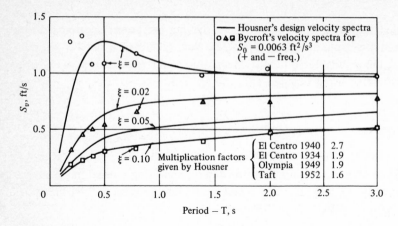

FIGURE 28-3
Mean extreme-values of pseudo relative velocity for linear SDOF systems
(stationary white-noise excitation).

by the techniques previously described, to determine deterministically the time-history response of the nonlinear system to each input accelerogram and then to examine the output response process using Monte Carlo methods. Usually, one is interested primarily in the mean and the standard deviation of the extreme-values of response.[1]

This general method of stochastic analysis can also be used for linear systems; however, in this case, the direct method previously described is usually preferable.

28-4   EXTREME-VALUE RESPONSE OF SDOF SYSTEMS

Stationary White Noise Excitation

Consider the SDOF linear system represented by Eq. (28-17) subjected to earthquake ground motion $\ddot{v}_g(t)$. Figure 28-3 shows Housner's pseudo-velocity design-spectrum curves for this system for different damping ratios, that is, $\xi = 0, 0.02, 0.05, 0.10$. Since these curves were obtained by normalizing eight components of recorded ground accelerations (two components each of El Centro 1940, El Centro 1934, Olympia 1949, and Taft 1952) to a common intensity level and by averaging the eight pseudo-velocity response spectra derived therefrom, one can consider the ordinates in Fig. 28-3 as representing mean extreme-values of relative pseudo velocity.

---

[1] J. Penzien and S. C. Liu, Nondeterministic Analysis of Nonlinear Structures Subjected to Earthquake Excitations, *Proc. 4th World Conf. Earthquake Eng., Santiago, Chile*, vol. I, sec. A-1, pp. 114–129, January 1969.

The multiplication factors given in this figure increase the ordinates to intensity levels corresponding to the earthquakes indicated.

Using an analog computer, Bycroft studied the possibility of using a white-noise process to represent earthquake ground motions at a given intensity level.[1] In these studies, Bycroft noted the extreme-values of response for a SDOF system using 20 separate bursts of stationary white-noise input of 25 s duration each. It was necessary in these studies to limit the input bandwidth having constant power spectral density to the range 0 to 35 Hz. To compare his mean extreme-values with Housner's earlier published velocity spectra, Bycroft normalized his results to that power spectral density of input S_0 which would give full agreement with Housner's results for $T_n = 3$ s and $\xi = 0.20$. This normalization criterion resulted in a value of S_0 equal to 0.75 ft²/Hz over the frequency range $0 < f < \infty$. A further normalization of these same results so that they can be compared with Housner's design velocity spectra requires that $S_0 = 0.0063$ ft²/rad·s³ over the frequency range $-\infty < p < \infty$. Bycroft's mean extreme-values normalized to this intensity level are shown in Fig. 28-3. These results would seem to indicate that white noise is a reasonable simulation of earthquake ground accelerations.

Stationary Filtered White Noise Excitation

Many investigators have used stationary filtered white noise to simulate earthquake ground accelerations. In one of these investigations, Liu and Penzien[2] used a single filter having the transfer function given by Eq. (28-6) with $\omega_g = 15.6$ rad/s and $\xi_g = 0.6$. Fifty sample functions of band-limited white noise were generated by the digital-computer methods of Sec. 28-1 with $S_0 = 0.00614$ ft²/s³ and $\Delta\varepsilon = 0.025$ s. These sample functions, each having 30 s duration, were then filtered by digital-computer techniques to provide an ensemble of 50 artificial accelerograms.

Complete time histories of response for the linear SDOF system when subjected separately to each of the 50 input accelerations were established by deterministic methods. The extreme-values of relative displacement were noted in each case and were averaged to obtain mean values. These mean values of displacement were then converted to mean extreme-values of pseudo velocity by multiplying by ω. These values are plotted in Fig. 28-4a, where they may be compared with Housner's design-spectrum curves in Fig. 28-4b. The close agreement of these two sets of curves lends support to using filtered stationary white noise in the simulation of strong-earthquake ground motions.

Complete time histories of relative displacement response $v(t)$ for the ordinary

---

[1] Bycroft, White Noise Representation of Earthquakes, *loc. cit.*
[2] Penzien and Liu, Nondeterminate Analysis of Nonlinear Structures, *loc. cit.*

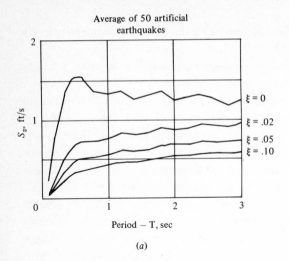

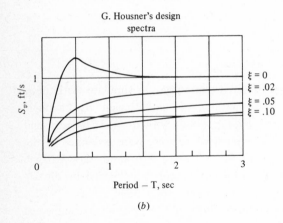

FIGURE 28-4
Mean extreme-values of pseudo relative velocity for linear SDOF systems (filtered stationary white-noise excitation).

elastoplastic and stiffness-degrading models were established by standard numerical-integration procedures when subjected separately to support accelerations $\ddot{v}_g(t)$ corresponding to the filtered process described above but after normalizing by a factor of $(2.90)^2$ so that the process intensity S_0 would represent the intensity of the NS component of the 1940 El Centro earthquake $[S_0 = (2.90)^2(0.00614) = 0.0516$ ft^2/s$^3]$.

The basic parameters of these nonlinear models, which are comparable to those used for the linear models, are shown in Fig. 28-5. In all cases T and ξ represent the period of vibration and viscous-damping ratio, respectively, in the initial elastic range.

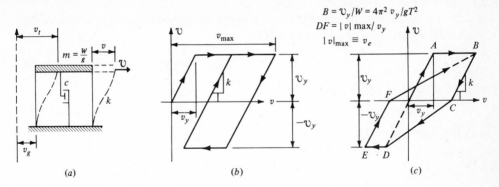

FIGURE 28-5
Nonlinear models of one DOF system.

The static-force–deflection relations for the elastoplastic and stiffness-degrading models are shown in Fig. 28-5b and 28-5c, respectively. The strength ratio B and ductility factor DF are defined for these models in accordance with the relations $B \equiv v_y/W$ and $DF \equiv |v(t)|_{max}/v_y$. It is significant that in addition to loss of stiffness following any yielding, the stiffness-degrading model permits hysteresis loops to be formed even at very low amplitudes of oscillation. Therefore, this model dissipates more energy in the lower-amplitude ranges of response than the equivalent elastoplastic model does.

The response of the elastoplastic and stiffness-degrading models are considered for two different periods, $T = 0.3$ and 2.7 s, and for two different damping ratios, $\xi = 0.02$ and 0.10; thus, the response of eight different nonlinear models as presented in Table 28-1 are discussed. Strength ratios B are based on the assumption that the yield resistance v_y equals twice the design load as specified in the 1973 edition of the Uniform Building Code[1] for moment-resisting frames, that is, $B = 2KC = (2 \times 967)(0.05)T^{-1/3}$. The mean and standard deviation of the 50 extreme-values in each case are shown in Table 28-1.

Probability distribution functions $P(|v|_{max})$ based on 50 extreme-values for each of the eight nonlinear models are shown in Fig. 28-6 in the form of Gumbel plots. For comparison, probability distribution functions are also presented for the four corresponding linear elastic models, that is, models having the same corresponding initial stiffnesses and viscous-damping ratios. These models are identified by the arabic numerals 1 to 12 in Fig. 28-6 and have the properties listed in Table 28-1.

Two probability distribution functions are shown in Fig. 28-6 for each of the 12 structural models, namely, a wavy-line function, which is a plot of the actual

[1] International Conference of Building Officials, *op. cit.*

extreme-values determined for process $v(t)$, and a straight-line function, which is the theoretical distribution (Type I) of the form

$$P(|v|_{max}) = \exp\left[-\exp\left(-\hat{v}\right)\right] \quad (28\text{-}30)$$

where $\hat{v}$ is the reduced extreme-value defined by

$$\hat{v} \equiv \alpha(|v|_{max} - u) \quad (28\text{-}31)$$

As explained in Sec. 23-11, constants α and u can be determined using the relations

$$\frac{1}{\alpha} = \frac{\sigma_{|v|max}}{\sigma_{\hat{v}}} \qquad u = |v|_{max} - \frac{\bar{\hat{v}}}{\alpha} \quad (28\text{-}32)$$

where $\sigma_{|v|max}$ and $|v|_{max}$ represent the standard deviation and mean value, respectively, for the 50 extreme-values of $v(t)$ and $\sigma_{\hat{v}}$ and $\bar{\hat{v}}$ represent the standard deviation and mean value, respectively, for the reduced extreme-values $\hat{v}$. The numerical values of $\sigma_{\hat{v}}$ and $\bar{\hat{v}}$ depend upon the number of observed extreme-values. When this number equals 50, as considered here, $\sigma_{\hat{v}} = 1.1611$ and $\bar{\hat{v}} = 0.548$.[1] The numerical values for $\sigma_{|v|max}$ and $|\bar{v}|_{max}$ and the corresponding values for $1/\alpha$ and u are given in Table 28-1.

The probability distribution scale on Gumbel extreme-value charts as shown in Fig. 28-6 varies in such a manner that Eq. (28-30) plots as a straight line with its ordinate $|v|_{max}$ at the origin $(\hat{v} = 0)$ representing the most probable extreme-value and with its slope proportional to the standard deviation of the extreme-values.

Table 28-1

Case No.	Structural type*	Period T, s	Damping ratio, ξ	Strength ratio, B	Yield displ. v_y, in	$\sigma_{|v|max}$, in	$|\bar{v}|_{max}$, in	u, in	$\frac{1}{\alpha}$
1	E	0.3	0.02	—	—	0.115	0.768	0.722	0.085
2	EP	0.3	0.02	0.10	0.088	1.613	3.214	2.450	1.390
3	SD	0.3	0.02	0.10	0.088	0.711	2.480	2.144	0.613
4	E	0.3	0.10	—	—	0.050	0.354	0.330	0.043
5	EP	0.3	0.10	0.10	0.088	0.910	1.947	1.517	0.784
6	SD	0.3	0.10	0.10	0.088	0.360	1.327	1.157	0.310
7	E	2.7	0.02	—	—	3.07	14.15	12.73	2.59
8	EP	2.7	0.02	0.048	3.42	5.51	16.35	13.75	4.75
9	SD	2.7	0.02	0.048	3.42	5.83	14.32	11.56	5.02
10	E	2.7	0.10	—	—	1.31	8.77	8.24	0.97
11	EP	2.7	0.10	0.048	3.42	4.56	11.57	9.41	3.94
12	SD	2.7	0.10	0.048	3.42	3.26	9.98	8.45	2.80

* E — Elastic
EP — Elasto-plastic
SD — Stiffness degrading

[1] E. J. Gumbel and P. G. Carlson, Extreme Values in Aeronautics, op. cit.; E. J. Gumbel, Probability Tables for the Analysis of Extreme-Value Data, op. cit.

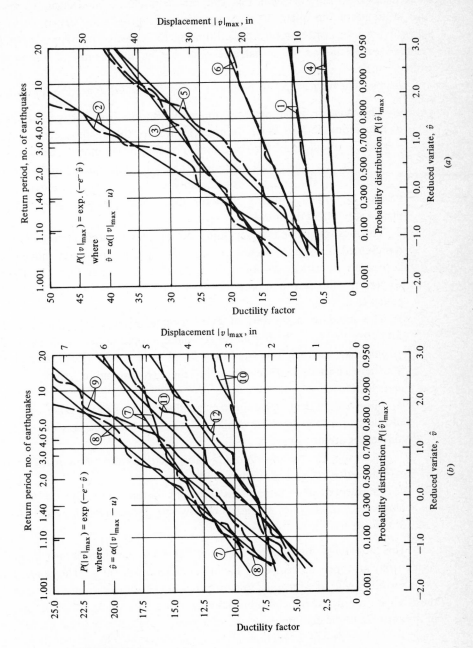

FIGURE 28-6
Probability distribution for extreme-values of relative displacement.

Note that the extreme-values in Fig. 28-6 for the nonlinear models can be measured also in terms of the ductility factor and that the probability distribution can be measured in terms of the return period, that is, the expected number of earthquakes required to produce a single extreme-value having the magnitude shown by the ordinate scale.

The significant features to be noted in Fig. 28-6a and b are the following:

1 The most probable extreme-values of response for short-period structures as represented in Fig. 28-6a are much greater for the elastoplastic and stiffness-degrading models than for their corresponding linear models, are appreciably greater for the elastoplastic models than for their corresponding stiffness-degrading models, and are considerably greater for those models having 2 percent of critical damping than for their corresponding models having 10 percent of critical damping.

2 The most probable extreme-values of response for long-period structures as represented in Fig. 28-6b are considerably greater for those models having 2 percent of critical damping than for their corresponding models having 10 percent of critical damping; however, these values differ very little from one model to another.

3 The standard deviations of extreme-value response for the short-period structures are considerably larger for the elastoplastic and stiffness-degrading models than for their corresponding linear models and are appreciably larger for the elastoplastic models than for their corresponding stiffness-degrading models.

4 The standard deviations of extreme-value response for long-period structures correlate in a manner quite similar to short-period structures except that the differences are not so great.

5 Increasing the viscous-damping ratio increases the standard deviations of extreme-value response for each model type.

6 The theoretical extreme value functions as represented by Eq. (28-30) and plotted as straight lines in Fig. 28-6 show very good correlations with the actual distributions.

The probability distribution functions for extreme-values shown in Fig. 28-6 result from an input process $\ddot{v}_g(t)$ having a duration of 30 s. The corresponding extreme-values will, of course, be less for processes of shorter duration. To illustrate these effects, a ratio of the ensemble average of extreme-values for an input process of duration T_0 to the ensemble average of extreme-values for an input process of 30 s is plotted in Fig. 28-7 as a function of the duration ratio $T_0/30$.

It is quite evident, from curve 2 in Fig. 28-7a, that the mean peak response of typically damped, linear, short-period structures ($T = 0.3$ s) increases very slowly with duration beyond approximately 6 s. Long-period structures are, of course, more

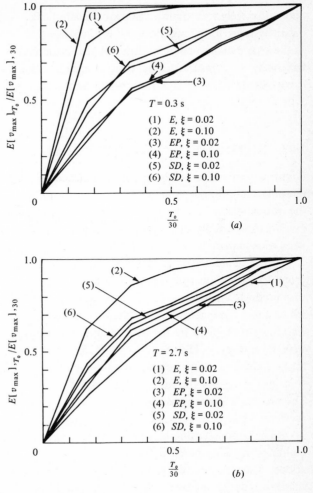

FIGURE 28-7
Duration effect of stationary process on mean peak response of linear and nonlinear structures.

sensitive to duration, as shown by curve 2 in Fig. 28-7b. This curve indicates that the magnitude of mean peak response for a 15-s duration process is approximately 95 percent of the magnitude observed for a 30-s duration process. As shown in Fig. 28-7a and 28-7b, elastoplastic and stiffness-degrading structures are much more sensitive to duration than elastic structures are; thus, it is apparent that realistic durations must be used for stationary inputs when investigating the response of nonlinear structures.

As demonstrated above, stationary processes of short duration can be used quite effectively to establish the probabilistic peak response of both linear and non-linear systems to strong-motion earthquakes of a given intensity level. However, as the true dynamic characteristics of real structures become better known, damage will likely be measured using various accumulative-damage criteria, in which case it may be desirable to use appropriate nonstationary processes for the excitation.

28-5 EXTREME-VALUE RESPONSE OF MDOF SYSTEMS

Numerous investigators have suggested the use of nonstationary processes to represent strong ground motions. One such process was established by Ruiz[1] to study the probabilistic response of multistory shear buildings. Selected results of his investigation are presented here to provide an example of the stochastic response of MDOF systems.

Ruiz generated a ground-acceleration process $\ddot{v}_g(t)$ to simulate the expected ground motions on firm soil at a distance of about 45 mi from the epicenter of a magnitude 8.3 earthquake. Twenty sample functions of band-limited stationary white noise were generated by digital-computer methods. These sample functions were then multiplied by the deterministic intensity function $\zeta(t)$ shown in Fig. 28-2 with $t_1 = 0$, $t_2 = 11.5$ s, and $c = 0.155$ s^{-1}. The resulting nonstationary waveforms were then filtered once using the methods described in Sec. 28-1 with the filter transfer function of the form given by Eq. (28-6) with $\omega_g = 15.7$ rad/s and $\xi_g = 0.6$. The process was normalized to an intensity level corresponding to an expected peak acceleration of 0.3g.

The complete time history of the elastoplastic response of an eight-story shear building was determined deterministically for each of the 20 input ground accelerations. The eight lumped masses of this building were of equal magnitude and equally spaced, and the relative story elastic spring constants were adjusted so that the fundamental mode shape of the building was linear. The lateral drift of each story was related to its shear force through a bilinear hysteretic-force–deflection relation independent of axial forces acting in the columns. Yielding in all stories was assumed to start simultaneously as the static lateral loading, distributed in accordance with the Uniform Building Code, increased monotonically. Yielding in each story was assumed to start when this loading reached a level twice as great as the design loading. The yielding stiffness in each story was set at 10 percent of its initial elastic stiffness. Viscous damping introduced into the normal modes in the uncoupled form was controlled by specifying the same damping ratio in each mode.

Probability distribution functions based on the 20 extreme-values of drift in

---

[1] Ruiz and Penzien, *op. cit.*

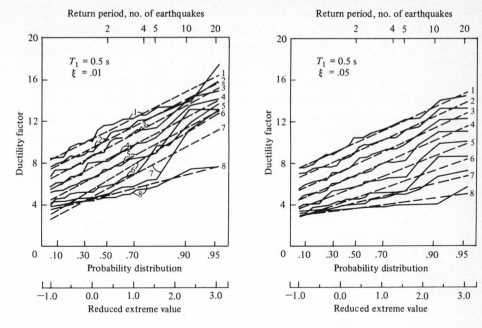

FIGURE 28-8
Probability distributions for story ductility factors.

each story are presented in the form of Gumbel plots (Type I) for two different shear buildings in Fig. 28-8. The two buildings identified in this figure have fundamental periods of 0.5 and 2.0 s; thus, they represent a stiff building and a flexible building. Both buildings are assigned damping ratios of 5 percent in all modes. These probability distribution functions are similar to those shown in Fig. 28-6 and described in Sec. 28-4 for the SDOF system; therefore, no additional description of the meaning of these plots is necessary.

From the results of Fig. 28-8a, the following observations are made with regard to stiff shear buildings:

1 The most probable ductility factors decrease monotonically toward the top of the structure.
2 The standard deviations of the ductility factors are almost the same for all stories except the top story, where a large reduction is observed.
3 The estimated probability distribution functions show very good agreement with the theoretical distributions represented by the straight lines.

Likewise from the results in Fig. 28-8b, the following observations are made with regard to flexible shear buildings:

1 The most probable ductility factor decreases toward the middle stories and then increases toward the top stories, reaching a value in the top story of magnitude comparable with that in the first story.

2 The standard deviations of the ductility factors decrease toward the upper stories but with a slight increase in the top story.

3 The agreement between the estimated probability distribution and the theoretical straight-line distribution is not as good as for the stiff shear building.

From a comparison of Fig. 28-8*a* and *b*, it is clear that the most probable ductility factors and their standard deviations are higher for the stiffer structures.

INDEX

Accelerometer, 62
Argand diagram, 33, 34
Artificial earthquake records, 543, 611, 613
Autocorrelation function, 444
Axial deformations, equations of motion, 305
Axial force, influence on stiffness, 182, 296, 360
Axial wave propagation:
  boundary conditions, 369, 370
  discontinuity, 371, 375
  equations of motion, 365, 366

Bertero, V. V., 615
Betti's law, 154
Binomial probability distribution, 397
Buckling analysis, 27, 183
  harmonic excitation, 184
  matrix iteration, 222
Bustamente, J., 612
Bycroft, G. N., 612, 619

Carlson, P. G., 474, 622
Cartwright, D. E., 472
Cauchy's residue theorem, 487
Central-limit theorem, 401
Chain structure, 226
Chopra, A. K., 615
Complex frequency response, 84, 484
Conditional probability, 406
Consistent mass:
  definition, 163
  matrix, 254
Contour integration, 487
Convolution integral, 101
Correlation coefficient, 413, 443
Covariance, 413, 416, 443
Cross-correlation function, 456
Cross-spectral density, 456

D'Alembert's principle:
  application of, 30

D'Alembert's principle
   definition of, 9
Damped mode shapes, 198
Damping:
   continuous system, 301
   equivalent, 74
   evaluation of, 69
      energy loss per cycle, 73
      free vibration decay, 70
      half-power method, 71
      resonant amplification, 70
   hysteretic, 76
   MDOF properties, 165
Damping coupling, 198
Damping influence coefficient, 148
Damping matrix, 148
Damping orthogonality, 195
Damping ratio, 45
Degrees of freedom:
   definition, 5
   selection of, 145
Design earthquake, 539
Design response spectrum, 541, 542
Dirac delta function, 391
Direct stiffness method, 158
Discrete Fourier transform, 114, 584
Discrete loadings, stochastic, 511
Discretization, 5
   finite element, 7
   generalized displacements, 6
   lumped mass, 5
Displacement meter, 62
Distributed loadings, stochastic, 515
Distributed parameter systems, 291
Ductility factor, 602
Duhamel integral:
   definition, 101
   evaluation of, 102
Dynamic direct-stiffness method, 345
Dynamic loading, general, 100
Dynamic magnification factor, 54, 57
Dynamic matrix:
   general, 210
   symmetric form, 252
Dynamic stiffness:
   axial deformation, 357
   discrete system, 184
   flexural, 346, 349
      and axial deformation, 358
   including rigid axial displacement, 353, 355

Earthquake-excitation factor, 552, 557, 565
Earthquake response:
   deterministic analysis, 544
   general, 519
   generalized SDOF systems, 549
   lumped MDOF systems, 555
   rigid base rotation, 574
   stochastic: extreme values, 619, 626
      MDOF systems, 617
      nonlinear systems, 617
      SDOF systems, 616
Earthquake:
   focus, 529
   hypocenter, 529
   intensity, 533
   magnitude, 531
   response integral, 547
   spectrum, 536
   waves (P & S waves), 528
Effective modal mass, 559
Eigenproblem:
   eigenvalues, 182
   iterative solution, 245
Eigenvalue-economizer procedures, 235
Eigenvectors:
   left hand, 243
   right hand, 243
Elastic-rebound theory, 525
Energy:
   kinetic, 130
   potential, 130
Epicenter, 531
Equations of constraint, 284
Equations of motion:
   axial deformations: formulation, 305
      uncoupling, 338
   definition, 9
   discrete systems: damped, 194
      undamped, 146
   distributed parameter systems:
      formulation, 293
      uncoupling, 331
   variational formulation, 271
Equilibrium, dynamic, 146
Extreme values, 619, 626

Fast Fourier Transform, 114, 584
Fatigue predictions, 502

Finite element:
  discretization, 7
  interpolation functions, 8
  nodal points, 8
  stiffness, 155
Flexibility:
  influence coefficients, 151
  matrix, 152
Focus, earthquake, 529
Foundation modeling, 578
Foundation-structure interaction, 578
Fourier:
  amplitude spectrum, 539
  integral, 111
  inverse transform, 111
  series: exponential form, 83
    trigonometric form, 80
  spectrum, 539
  transform, 111
    discrete, 114, 584
Free field motion, 579, 584
Free vibrations:
  amplitude, 43
  damped, 44
  SDOF system, 41
  undamped, 42
    critically, 44
    overdamped, 48
    underdamped, 45
Frequencies:
  analysis of, 176, 182
  circular undamped, 43
  damped, 45
Frequency equation, 177
Frequency ratio, 53
Frequency vector, 177

Gaussian distribution:
  M-random variables, 430
  one-random variable, 392
  two-random variables, 404, 413
Generalized coordinates, 7, 271, 272
Generalized properties:
  continuous systems: load, 332
    mass, 332
  SDOF systems: combined stiffness, 31, 36
    damping, 26, 31, 34
    geometric stiffness, 31, 36

Generalized properties: SDOF systems:
    load, 26, 31, 36
    mass, 26, 31, 34
    stiffness, 26, 31, 35
Generalized SDOF systems, 23
  distributed flexibility, 29
  rigid body assemblage, 23
Geometric stiffness, 31
  coefficients, 149
    consistent, 169
    linear approximation, 167
  matrix, 150, 169, 171
Gravitational force, influence of, 20
Gumbel, E. J., 474, 622

Hamilton's principle, 11
  application of, 19, 34, 273
  definition of, 11
Harmonic loading, 52
Holzer method, vibration analysis, 226
Holzer-Myklestad method, vibration analysis, 232
Housner, G. W., 529, 541, 612, 615
Hypocenter, 529

Impulsive loads, 87
  approximate analysis, 96
  definition, 87
  rectangular pulse, 91
  sine wave, 88
  triangular impulse, 92
Incident wave, 373
Incremental equilibrium equations, 119, 262
Inertia force, 5, 10
Input-output relations:
  autocorrelation function, 489
  power spectral density function, 492
Intensity:
  earthquake (Modified Mercalli), 533
  response-spectrum (Housner), 537
International Conference of Building Officials, 621
Interpolation functions, 8
Inverse iteration, 246
Iteration, matrix: direct, 243
    inverse, 246
    (*See also* Matrix iteration)

Jacobian transformation, 405, 414, 422, 431
Jennings, P. C., 615

Kanai, K., 613
Kinetic energy, 11
Kolousek, V., 349

Lagrange multipliers, 284
Lagrange's equations of motion, 273
Leckie, F., 235
Linear transformation of random variables, 432
Liu, S. C., 618
Loading:
    consistent nodal loads, 166
    evaluation, 165
    static resultants, 166
Loading types:
    impulse, 3
    periodic, 2
    prescribed, 2
    random, 2
    simple harmonic, 2
Logarithmic decrement of damping, 47
Longuet-Higgins, M. S., 472

Magnification factor, 54, 57
Magnitude (Richter), 531
Mahin, S. A., 615
Mass influence coefficient, 149
Mass matrix:
    consistent, 161, 163
    general, 149
    lumped, 160
Matrix iteration:
    general, 243
    subspace, 250
    with shifts, 247
Maxwell's law of reciprocal deflections, 155
Mean value, 395, 412, 416, 443
Miner's fatigue criterion, 502
Mode shapes:
    evaluation, 179
    matrix, 180
Mode superposition, 194, 199
Modified Mercalli (MM) intensity, 533
Mohr's circle, 425

Narrowband systems, 495
Newmark, N. M., 522, 542
Newton's second law, 9
Nodal points, 8
Nonlinear response analysis:
    earthquake, 594, 599
    MDOF, 260
    SDOF, 118
Nonstationary random process, zero initial
    conditions, 498
Normal coordinate:
    definition, 192
    equations of motion:
        continuous system, 328
        discrete system, 193
Normal distribution:
    M-random variables, 430
    one-random variable, 392
    two-random variables, 404, 413
Normalizing mode shapes, 187
Nowacki, W., 349
Numerical analysis, frequency domain, 113, 582
Numerical integration:
    simple summation, 103
    Simpson's rule, 103
    trapezoidal rule, 103

Orthogonality conditions:
    axial vibration modes, 325
    continuous systems, 321
    discrete systems, 185
Orthonormal, 188

Penzien, J., 613, 618
Pestel, E., 235
Phase angle, 57
Pile driving (stress analysis), 377
Positive definite, property of, 153
Potential energy, 11
    minimum, 11
Power spectral density, 448
Principal axes, 423, 426
Probability:
    density, 390
    joint, 403
    marginal, 405
Probability distribution, 393, 406
    extreme values, 472

Probability distribution:
   maxima, 468
Probability theory, 389
Pseudostatic displacement, 303, 576
Pseudo-velocity spectrum, 535
Purified displacement vector, 216, 218

Random processes:
   definition, 436
   derivatives, 454
   ergodic, 438
   multiple independent variables, 476
   nonstationary, 463, 476
   stationary, 438
   stationary gaussian, 457
   superposition, 455
Random variable, 389, 403
   averages, 394, 412
   single, 389
   transformation, 393, 422
Random walk:
   one-dimensional, 396
   two-dimensional, 419
Rayleigh damping:
   continuous systems, 337
   discrete systems, 195
Rayleigh method:
   approximate frequency analysis, 131
   critical buckling load, 32
   discrete coordinate systems, 237
   improved, 137, 138
   principle of, 129
   selection of shape, 133
Rayleigh probability distribution, 428
Rayleigh-Ritz method, 239
Reflected wave, 373
Refracted wave, 373
Reid, H. F., 525
Residue theorem (Cauchy), 487
Resonance, 60, 73
Response ratio, 54
Response spectrum:
   definitions, 94
   design, 541, 542
   intensity, 537
Rice's equation, 473
Root-sum-square method, 562–564
Rosenblueth, E., 522, 542, 612
Rotatory inertia, influence of, 298
Ruiz, P., 613

Scatter diagram, 416
SDOF system:
   damped response: frequency domain, 108
      time domain, 107
   definition, 15
   stochastic response, 482
   undamped: complementary solution, 52
      general solution, 53
      particular solution, 53
Seismicity, 552
Shear building:
   description, 226
   wave propagation, 382
Shear, influence on beam vibrations, 298
Shock spectra, 94
Simpson's rule, 103
Single degree of freedom (see SDOF
      system)
Soil modification, earthquake motion, 581
Soil-structure interaction:
   damping coupling, 592
   definition, 584
   elastic half space model, 588, 591
   response analysis, 591
Spectral acceleration, 536
Spectral displacement, 536
Stability matrix, 223, 224
State vector, 231, 233
Static condensation, 172
Steady state response, 54
Step-by-step integration:
   conditional stability, 124, 265
   linear acceleration method, 121, 263
Stiffness influence coefficient, 147, 152
Stiffness matrix, 148
   dynamic: axial deformation, 357
      flexural, 346
      flexural and axial, 358
      with rigid axial motion, 353
Stochastic modeling, strong ground motion,
      611
Stochastic response of MDOF systems:
   frequency domain, 510
   time-domain, 508
Stodola method (matrix iteration), 209
   convergence, 213
   higher modes, 215, 218
   highest mode, 220
Strain energy, 153
Structural-property matrices, evaluation, 151

Support excitation:
  continuous systems, 302
  influence of, 21
  multiple (discrete), 575
Strong motion earthquakes:
  actual, 522
  artificial, 543
  stochastic modeling: filtered white noise,
    613
    white noise, 611
Sweeping matrix, 216, 219, 220, 257

Tajimi, H., 613
Thompson, W. T., 612
Transfer function, 482
Transfer matrix:
  concept, 230, 234
  field, 230, 233
  point, 231, 233
Transient response, 54
Transmissibility, 65
Trapezoidal rule, 103
Tsai, N. C., 615

Unconditionally stable linear acceleration
    method (see Wilson θ method)
Unconstrained structures, 255
Undamped free vibrations, MDOF systems,
    176
Uniform building code seismic
    requirements, 570
Unit impulse response function, 483

Variance, 395, 412, 416, 443
Variational formulation (equations of
    motion), 271
Vianello method, 223
Vibration analysis:
  continuous systems: axial deformations,
    323
    axial force effects, 317
    basic case, 308
    rotatory inertia effects, 318
    shear effects, 318
  determinental solution, 177
  practical methods, 208
  Rayleigh method, 129
Vibration isolation, 64
Vibration mode shapes:
  damped, 199
  undamped, 179
Virtual displacements, principle of, 10
Virtual work analysis, 19

Wave propagation:
  analysis, 364
  pile driving, 377
  reflected, 373
  refracted, 373
  shear building, 382
White noise, 464
Wilson θ method:
  definition of, 266
  performance of, 268